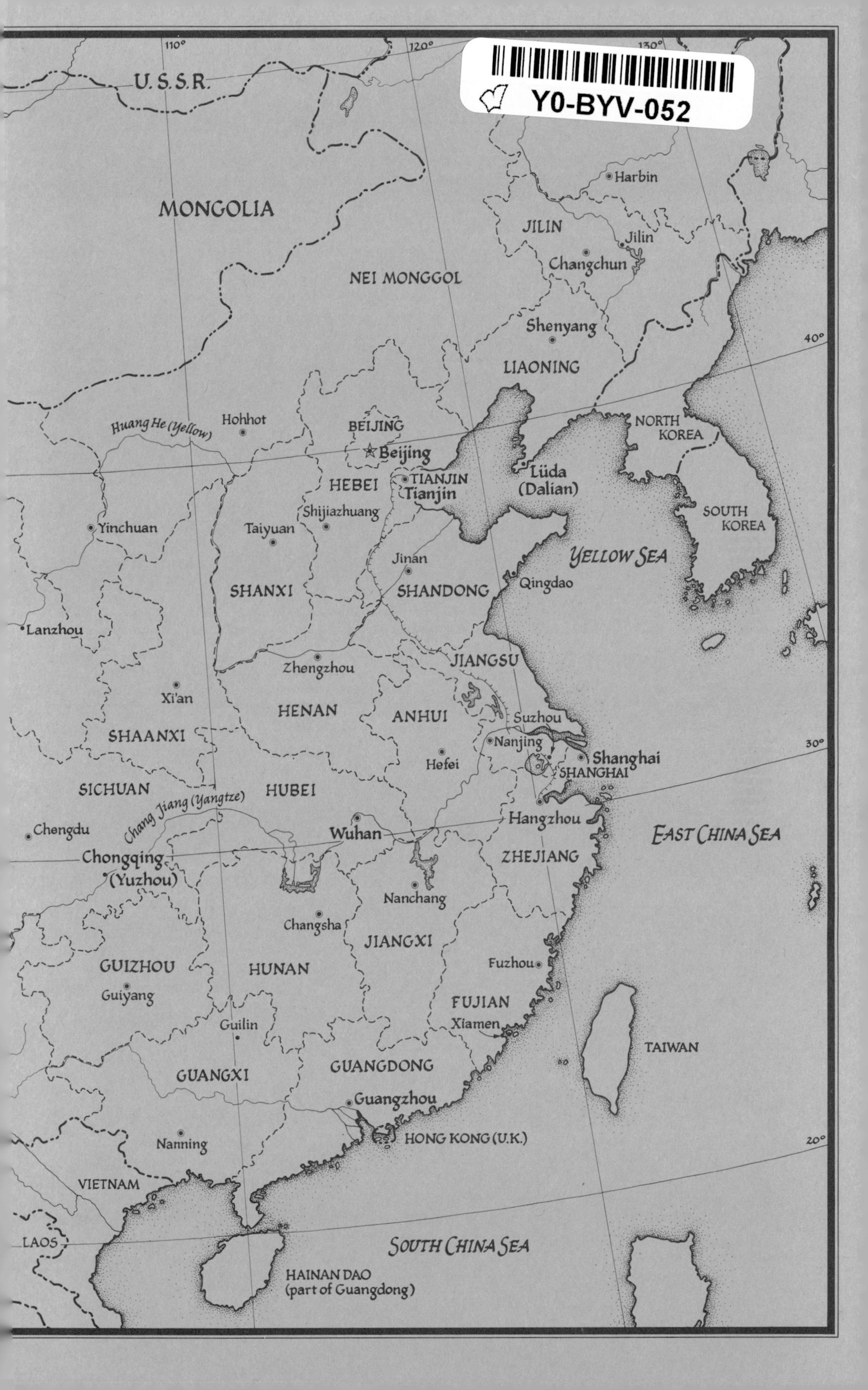

110°
120°
130°
40°
30°
20°
U.S.S.R.
MONGOLIA
NEI MONGGOL
Harbin
JILIN
Jilin
Changchun
Shenyang
LIAONING
Hohhot
Huang He (Yellow)
BEIJING
Beijing
TIANJIN
Tianjin
Lüda
(Dalian)
NORTH
KOREA
HEBEI
Shijiazhuang
Yinchuan
Taiyuan
SOUTH
KOREA
YELLOW SEA
Jinan
SHANXI
SHANDONG
Qingdao
Lanzhou
JIANGSU
Zhengzhou
Xi'an
HENAN
ANHUI
Suzhou
SHAANXI
Nanjing
Shanghai
Hefei
SHANGHAI
SICHUAN
HUBEI
Chang Jiang (Yangtze)
Hangzhou
Chengdu
Wuhan
EAST CHINA SEA
Chongqing
(Yuzhou)
ZHEJIANG
Nanchang
Changsha
JIANGXI
Fuzhou
GUIZHOU
HUNAN
Guiyang
FUJIAN
Xiamen
Guilin
TAIWAN
GUANGXI
GUANGDONG
Guangzhou
Nanning
HONG KONG (U.K.)
VIETNAM
LAOS
SOUTH CHINA SEA
HAINAN DAO
(part of Guangdong)

*SCIENCE IN CONTEMPORARY CHINA*

# SCIENCE
# IN CONTEMPORARY
# CHINA

*Edited by Leo A. Orleans*

*With the Assistance of Caroline Davidson*

STANFORD UNIVERSITY PRESS
*Stanford, California* 1980

Stanford University Press
Stanford, California



Printed in the United States of America
ISBN 0-8047-1078-3
LC 79-65178

# *FOREWORD*

To describe the state of science in an industrialized country is becoming an increasingly complex task. In the United States, to follow just the most important scientific developments in understanding nature, man, and society requires the perusal of innumerable documents and reports of the National Science Foundation and other government agencies that sponsor scientific research and training; publications of foundations, high technology corporations, private research institutes, and universities and colleges; and, of course, a large number of journals and books that discuss in depth some of the specific facets of scientific research.

A similar information explosion is complicating the study of science and scientific institutions in most other industrialized countries. In addition to the vast scientific literature, disaggregated knowledge is distributed among individuals, "opposite number" laboratories, professional organizations, academies, scientific attachés, and others. Under these circumstances, personal contacts and exchanges become indispensable if we want to understand the more intricate and subtle aspects of any significant scientific establishment. It is therefore not surprising that there is no single scientific "Baedecker" to help us grasp the state of science in any nation.

There exists no scientific Baedecker for China either, but for quite different reasons. There has probably never been adequate information about all aspects of Chinese science and technology. More important, however, the years of isolation of the United States and the People's Republic of China have proved rather costly to our understanding of each other's scientific efforts and institutions. Over the last decade, however, as contacts were gradually renewed following the Shanghai Communiqué of 1972, and as new knowledge began to accumulate, the need for a "map of Chinese science" became increasingly apparent—a map that would take account of the immensity of the country, of the historical and institutional context, and of the major developments that have taken place since the beginning and, in particular, since the end of the Cultural Revolution. All the more so, because with the exception of the economically oriented study *Science and Technology in the People's Republic of China* (published in 1977 by the Organization for Economic Cooperation and Development), almost two decades had elapsed since the appearance in 1961 of *Sciences in Communist China,* a publication of the American Association for the Advancement of Science.

Since the start of the dramatic changes and improvements in Chinese-American relations, the Committee on Scholarly Communication with the People's Republic of China (CSCPRC) has played a special role in the scholarly domain, sponsoring some 80 delegations (involving over 1,000 scholars) between the two countries. Taking advantage of the tripartite sponsorship by the National Academy of Sciences, the Social Science Research Council, and the American Council of Learned Societies, the Committee has been able to act as a unique broker in acquainting (and sometimes reacquainting) individual scholars and scholarly institutions in both countries. It was therefore only natural for the CSCPRC both to appreciate the need for an up-to-date review of Chinese science and to provide the supportive environment for such an ambitious effort.

A Steering Committee was formed that had little to "steer," since it was fortunate in having on hand a knowledgeable Study Director, Leo A. Orleans, an agency willing to fund the study, and authors who were highly qualified, whether they had been members of a CSCPRC delegation or not.

Given the enormity of the task of gaining a comprehensive and realistic understanding of China's scholarly enterprises and considering the individual disciplinary and institutional perspectives of the authors, no one should be surprised that a rather uneven coverage emerged. But while none of the chapters in this volume would claim the perspicacity and wisdom of a Tocqueville, they contain extremely useful information and evaluations for anyone interested in a general "map" of Chinese science at the end of the 1970's.

We take pleasure in acknowledging that without the cordial hospitality of our colleagues in China and the cooperation of their institutions—particularly their Scientific and Technical Association—we would have been unable to gain the insights these chapters represent. We hope that Chinese and American scientists will continue a friendly dialogue that will lead to a better understanding of each other's institutions and the ways in which progress in science and technology contributes to our respective societies. We would also like to acknowledge the role of scientists and engineers of Chinese origin—our colleagues in American institutions. They provided both a useful perspective and a friendly ambiance for renewed communication.

In spite of its universality, natural science has given rise to rather diverse subcultures in different countries. Their study requires understanding not only science but also the historical, social, and institutional context in which science evolved and is being carried on. The present volume belongs thus to a new genre, studies of science as an important human and societal activity, in countries at different stages of development.

Walter A. Rosenblith

# PREFACE

Several matters of form deserve brief comment. First, although the debate over the use of the Wade-Giles versus the pinyin transliteration system is not fully resolved within the academic community, we have decided, looking to the future, to relinquish the more familiar Wade-Giles system in favor of the Beijing-sanctioned pinyin. For the reader's convenience, the Index of Personal Names (p. 585) includes the Wade-Giles transliteration in parentheses. Second, the many names of scientific and technical journals mentioned have been rendered as follows: after the Chinese name of the periodical, the English version follows in parentheses, printed in italic if it is part of the official title of the publication and rendered in roman type if, instead, the English is a supplied translation rather than a part of the title. Last, after some consideration, it was decided that a standard subject index would be superfluous for this volume. Instead, the subheadings of each chapter have been included in the table of contents to serve as a guide to the topics discussed herein, and are supplemented by the Index of Personal Names and the Index of Institutions.

## *ACKNOWLEDGMENTS*

We doubt that when Walter A. Rosenblith, Provost of the Massachusetts Institute of Technology, agreed to be Chairman of the Steering Committee for development of this volume, he could anticipate the problems and predicaments that awaited him. We thank him for "being there" whenever his help and his advice were needed. We would also like to recognize the assistance rendered by the other members of the Steering Committee: Philip M. Arnold, Roy M. Hofheinz, Carl Kisslinger, Alexander Leaf, Robert Metcalf, and Chi Wang.

This publication would not exist were it not for Mary B. Bullock, Staff Director of the Committee on Scholarly Communication with the People's Republic of China, and for her predecessor, Anne G. Keatley. Both realized the need for this study and pursued their convictions through innumerable bureaucratic obstacles to obtain the approval and the funding for this project; we thank them for their vision and their persistence. Indeed, we are grateful not only to Mary for her continued support but also to her whole staff, which was always there to listen, to sympathize, and to assist. Special thanks go to Patricia Tsuchitani, who helped get the project off the ground; to Richard Glover, who waded through tons of paper in the process of supplying the authors with relevant materials; to Lois Edwards, who, between her other duties, managed to take care of innumerable details associated with this work; and to Marybeth Jones, for her careful and intelligent typing of parts of the manuscript.

We are grateful to the National Science Foundation for funding, and to Aaron Segal, of the Foundation's Division of International Programs, for his personal interest and support; to J. G. Bell, of Stanford University Press, for his attentive guidance through the lengthy and complex production process; and to John Ziemer, our copy editor and indexer, for his eagle eye, zealous attention to detail, and indefatigable energy, all of which were invaluable in preparing this volume for the printer. Finally, we are most grateful to the 90-odd reviewers of the chapters, who unfortunately must remain anonymous. We hope they can obtain some solace from Mao Zedong's admonition that "we must all learn the spirit of absolute selflessness."

One does not usually thank one's authors, but in this instance we feel we should. Most were understanding and patient, and some were even good-humored, as their chapters moved through what must have seemed an unusually lengthy review and editorial process. We believe we are safe in saying that without them this study would not have been possible.

L.O.
C.D.

# CONTENTS

# *CONTRIBUTORS*

*John D. Baldeschwieler,* a graduate of Cornell University and the University of California at Berkeley, is Professor of Chemistry at the California Institute of Technology. He is currently involved in developing instrumentation for the study of the extended x-ray absorption fine structure (EXAFS) and in the perturbed angular correlation method to study the targeting of phospholipid vesicles in vivo. He is a member of the National Academy of Sciences, the American Academy of Arts and Sciences, the American Philosophical Society, and several national science committees and boards. A member of the Committee on Scholarly Communication with the People's Republic of China's (CSCPRC) 1978 Delegation in Pure and Applied Chemistry, he is the editor of the delegation report: *Chemistry and Chemical Engineering in the People's Republic of China* (1979).

*Nicolaas Bloembergen* is Gordon McKay Professor of Applied Physics and Rumford Professor of Physics at Harvard University. He is a fellow of the American Academy of Arts and Sciences and the American Physical Society and a member of the National Academy of Sciences and several natural science committees. Dr. Bloembergen received the National Medal of Science in 1974. In 1975 he traveled to China with the CSCPRC Solid State Physics Delegation.

*Baruch Boxer* is Professor of Geography and Coordinator of International Environmental Studies at Rutgers University. His research focuses on comparative scientific and policy dimensions of international environmental problems. Recent publications include *Environmental Protection in the People's Republic of China* (coedited) and "Mediterranean Action Plan: An Interim Evaluation," *Science,* 202: 585–90. He held a Rockefeller Foundation Fellowship in Environmental Affairs in 1978–79. In 1980 he visited China at the invitation of the State Council's General Office of Environmental Protection.

*Kwang-chih Chang* was born in Beijing, did his undergraduate work at National Taiwan University, and received his Ph.D. in anthropology from Harvard. He taught at Yale from 1961 to 1977 and since then has been Professor of Anthropology at Harvard. Dr. Chang is a member of the National Academy of Sciences and the author of *The Archaeology of Ancient China* (1963, 1968, 1977 editions), *Shang Civilization* (1980), and other books and articles.

*Edward C. T. Chao* has been a geologist with the U.S. Geological Survey since 1949. Born in China, he received his degree in geochemistry and petrology from the University of Chicago. Dr. Chao is a member of numerous professional societies and the recipient of several scientific awards. He was principal investigator of lunar samples (1969–79) and since 1973 has been the USGS representative for relations with the People's Republic of China.

*Jerome Alan Cohen* is Professor of Law, Director of East Asian Legal Studies, and Associate Dean at Harvard Law School, where he has been since 1964. He is the author of several books and numerous articles on China, as well as editor of *Studies in East Asian Law, Harvard University.* He has visited China many times and spent the 1979–80 academic year serving as Honorary Visiting Professor at the Law Faculty of Hong Kong University and as a consultant on Chinese law relating to trade and investment.

*Caroline Davidson* is a social historian and free-lance writer and editor. She has an M.A. in History from King's College, Cambridge, and a Diploma in Journalism Studies from University College, Cardiff. Before moving to Washington, D.C., in 1978, Ms. Davidson worked for Reuter's News Agency in London.

*Albert Feuerwerker,* Professor of History and Director of the Center for Chinese Studies at the University of Michigan, is a specialist on modern Chinese history. In addition to works on economic and political history, he has written extensively on the historiography of the People's Republic of China.

*Leo Goldberg* has served as Director of three leading observatories in the United States—Michigan, Harvard, and Kitt Peak National Observatory, where he is now Distinguished Research Scientist. His research has dealt with a broad range of problems in astrophysics in such fields as atomic physics, the sun, and the interstellar medium. He is currently studying mass loss from cool stars.

*Harry Harding* is an Associate Professor of Political Science at Stanford University. He is the author of *Organizing China: The Problem of Bureaucracy, 1949–1976* and numerous other essays and monographs on contemporary Chinese politics, Chinese military policy, Chinese foreign policy, and Sino-American relations. In 1979–80 he was Coordinator of the East Asia Program at the Wilson Center in Washington, D.C.

*Jack R. Harlan* received his Ph.D. in genetics from the University of California at Berkeley in 1942. He did research in grass breeding and forage crops at USDA/Oklahoma State University from 1942 to 1966 and since then has been Professor of Genetics at the Crop Evolution Laboratory, University of Illinois. He is a member of the National Academy of Sciences and has done research for the USDA in Africa, Asia, Latin America, and the Middle East.

*Arthur Kelman* has been Chairman of the Department of Plant Pathology at the University of Wisconsin-Madison since 1965, following 16 years of teaching and research at the North Carolina State University. He is a Fellow of the American Phytopathological Society, a member of the National Academy of Sciences and the American Academy of Arts and Sciences, and President of the American Phytopathological Society and the International Society for Plant Pathology.

*Saunders Mac Lane* taught at Harvard and Cornell universities and since 1947 has been at the University of Chicago, where he is currently the Max Mason Distinguished Service Professor. He is a former member of the National Science Board (1974–80) and Vice-President of the National Academy of Sciences, where he is also Chairman of the Report Review Committee. In 1976 Professor Mac Lane headed the CSCPRC Delegation of Pure and Applied Mathematics to the People's Republic of China.

*Robert L. Metcalf* is Professor of Biology, Entomology, Environmental Studies, and Veterinary Pharmacology and former head of the Department of Zoology at the University of Illinois, where he has taught since 1968. Previously, he was Professor of Entomology and Entomologist at the University of California at Riverside, where for three years he was Vice-Chancellor. Dr. Metcalf is a member of numerous national and international organizations and committees, the recipient of many awards, and the author of several hundred scientific publications. He is a member of the National Academy of Sciences and the American Academy of Arts and Sciences.

*Jerry Norman* studied Chinese and linguistics at the University of California at Berkeley, where he received his Ph.D. in 1969. He has done research on Chinese dialects and is also interested in China's minority languages, especially those of the Altaic group. Professor Norman participated in both the 1974 and the 1977 CSCPRC linguistics delegations to the People's Republic of China. At present he teaches Chinese linguistics at the University of Washington, Seattle.

*Leo A. Orleans* is China Specialist at the Library of Congress. He is the author of *Every Fifth Child: The Population of China* and numerous monographs and articles dealing with various aspects of China's society and economic development. Raised in China, he revisited the country as a member of the CSCPRC's first medical delegation.

*Clifton W. Pannell,* Professor of Geography at the University of Georgia, received his Ph.D. from the University of Chicago in 1971. He is the author of a monograph on urban structure in Taichung, Taiwan, and recently has been engaged in studies of Chinese urban and regional land use based on remotely sensed data.

*Dwight H. Perkins* is Professor of Modern China Studies and of Economics and Chairman, Department of Economics, Harvard University. He is the author of

*Agricultural Development in China, 1368–1968,* and many other books and articles. He led the CSCPRC-sponsored delegation on rural small-scale industry (1975) and was deputy leader of the CSCPRC's first economics delegation (1979).

*Richard J. Reed* received his B.S. from the California Institute of Technology in 1945 and Sc.D. from the Massachusetts Institute of Technology in 1949. Since 1954 he has been Professor of Atmospheric Sciences at the University of Washington. He is a past president of the American Meteorological Society and a member of the National Academy of Sciences.

*Walter A. Rosenblith* has put his education in communications engineering and physics to use in the study of the biophysics of sensory communication and of the electrical activity of the brain. At the Massachusetts Institute of Technology he has been Provost since 1971. He is a member of the Academies of Sciences and Engineering and of the Institute of Medicine.

*Nathan Sivin* is Professor of Chinese Culture and of the History of Science at the University of Pennsylvania, a fellow of the American Academy of Arts and Sciences, and a member of the Académie Internationale d'Histoire des Sciences. He has published studies on every aspect and period of science in traditional China and is coauthor with Joseph Needham of the forthcoming volume *Science and Civilisation in China.* He is general editor of the Science, Medicine, and Technology in East Asia monograph series and editor and publisher of the periodical *Chinese Science.* He visited China with the U.S. Astronomy Delegation in October 1977 and as a guest of the Chinese Academy of Sciences in September 1979.

*Vaclav Smil* teaches interdisciplinary courses on energy and China at the University of Manitoba, Winnipeg, Canada. His research interests focus on China's energy, economy, food, and environment, on energy in the Third World, and on renewable resources. He is the author of *China's Energy* and an editor of *Energy in the Developing World.*

*Richard P. Suttmeier* is Associate Professor and Chairman, Department of Government, Hamilton College, Clinton, New York. He is the author of *Science, Technology, and China's Drive for Modernization* (1980) and numerous other works on science policy and organization as these relate to Chinese modernization. He was a member of the CSCPRC Pure and Applied Chemistry Delegation, which visited China in May and June 1978.

*Bohdan O. Szuprowicz* is founder of 21st Century Research, a business intelligence firm specializing in the high technology, electronics, and strategic materials trade of the Soviet bloc, China, and the Third World. He is a graduate of the Imperial College of Science and Technology of the University of London, the coauthor of *Doing Business with the People's Republic of China* (1978), and the author of numerous other publications.

*H. M. Temin,* Professor of Viral Oncology and Cell Biology, McArdle Laboratory, University of Wisconsin-Madison, was elected to the National Academy of Sciences in 1974 and received the Nobel Prize in Medicine in 1975. In the spring of

1977 he traveled to the People's Republic of China with a delegation from the state of Wisconsin.

*Chang-Lin Tien* is currently Professor and Chairman, Department of Mechanical Engineering, University of California at Berkeley. Born in Wuhan, China, he spent his childhood in Shanghai and moved with his family to Taiwan in 1949. After graduation from National Taiwan University, he came to the United States and received his M.A. and Ph.D. from Princeton University. Professor Tien is a member of the National Academy of Engineering and the recipient of numerous awards and honors. He has also written over 150 referred research papers in the area of heat transfer and infrared radiation.

*Myron E. Wegman,* Dean Emeritus, School of Public Health, University of Michigan, was head of Pediatrics at the Louisiana State University Medical School and Charity Hospital and Secretary General, Pan-American Health Organization. He visited China as a member of a medical delegation in 1973 and in 1979 as a member of the Committee on Advanced Studies in China—both sponsored by the CSCPRC. He is the editor of *Public Health in the People's Republic of China.*

*Thomas B. Wiens,* Senior Economist at Mathtech, Inc., Bethesda, Maryland, is a specialist in agricultural economics. He visited China with the Vegetable Farming Systems Delegation in 1977 and spent the first seven months of 1980 doing research in China under the National Education Exchange Program, sponsored by the CSCPRC. Dr. Wiens has recently published papers on Chinese agricultural technology and the economics of China's vegetable-farming system.

*L. C. L. Yuan* was born in China, where he completed his undergraduate education. He received his Ph.D. in physics from the California Institute of Technology in 1940 and did research at the RCA Laboratories and Princeton University before joining the Brookhaven National Laboratory in 1949, were he is now a Senior Research Physicist. Dr. Yuan is the recipient of numerous awards and honors for his research achievements in several areas of high energy physics.

*Leo A. Orleans*

# *INTRODUCTION*

As a nation that has over 20 percent of the world's population and a land area second only to that of the Soviet Union, China can be seen as a major world power. But as a nation with a per capita gross national product of about U.S.$ 400 and not even 1 percent of the adult population with completed higher education, it can also be viewed as a poor developing country. This is the incongruity of China. After 30 years, the People's Republic of China (PRC) continues to be a nation of contrasts, where some of the world's most competent scientists may be working within a few hundred miles of peasants whose life-style has changed little in generations. But although China's place on the developmental scale may exist only in the eye of the beholder, its current status and potential are highly significant in the realms of international politics and economics, for it is upon such external perceptions of China that other nations make and implement their policies.

Two important criteria in judging a country's development are its level of modern science and the extent to which its economy is based on modern technology. In today's competitive international climate, however, concern with image often receives undue priority. As a result, developing countries often strive to acquire a modern production facility or to set up a research institute staffed by Western-trained scholars, while paying less attention to what is most appropriate in terms of their resources and needs. Unfortunately, backward rural societies cannot leapfrog into the late twentieth century, and the successful diffusion of modern science and technology is a slow, continuous process that must go hand in hand with a country's overall social and economic progress.

The PRC is by no means a typical developing country. China's long and distinguished history, its ancient civilization, its complex social and political systems that go back to premodern times, its people's long-standing respect for knowledge and scholarship, have all contributed to a real sense of national pride and identity—attributes that often have to be created artificially in other developing countries. Joseph Needham's *Science and Civilisation in China* documents the high level of cultural and technological development in traditional China. In our own century, thousands of Chinese intellectuals have selected careers in science as a means of advancing both themselves and their country. With the establishment of the People's Republic in 1949, China not only had a certain tradition in the sciences but already had a substantial number of Western-trained, urban-based scientists and engineers.

Many of Mao Zedong's notions have been compatible with contemporary development theories. He emphasized the need to expose China's whole population to basic science and technology; to adapt programs to local conditions, resources, and needs; to place more stress on middle-level manpower; and, most of all, to "walk on two legs"—to use a variety of approaches to solve basic problems. Mao wanted to see greater emphasis placed on "science for the masses," so as to alter some of the peasants' traditional resistance to change and make them more amenable to accepting "scientific methods." To smooth out the differences between the city and the countryside and between mental and physical labor, Mao initiated some disruptive movements, notably the Great Leap Forward of 1958–59 and the much more damaging Great Proletarian Cultural Revolution of 1966–69. We cannot know where China's science and technology might be today had the scientific establishment and higher education not suffered so severely from these interruptions.

But all that is prologue. Mao's death in September 1976 was quickly followed by the appointment of Hua Guofeng as the new Chairman, the purge of the Gang of Four, the second ascendency of Deng Xiaoping in July 1977, and the all-out effort to make up for the years when economic growth was subordinated to politics and ideology. At the National Science Conference in March 1978, China defined its developmental goals in the "four modernizations" policy, in which science and technology, one of the "modernizations," was seen as playing a key role in propelling the other three: agriculture, industry, and national defense.

By 1979 many developments that would have seemed almost inconceivable a few years earlier were being reported by Chinese news sources and foreign visitors. The Chinese were even publishing economic statistics (unavailable for some two decades) and reporting demonstrations for human rights and improved living standards. New scientific institutes and publications were flourishing, and Chinese scientists were touring the West and frankly discussing China's scientific and human problems. Scholarly, governmental, and commercial contacts with China were formalized, and there was a concomitant demand for more and better information about China's science and technology.

*

*Science in Contemporary China* is being published at a most propitious time. It coincides not only with China's ambitious new domestic policies, but also with significant changes in world politics that are causing China to play a more prominent international role—a role that will depend largely on its ability to develop and absorb advanced scientific knowledge and technological expertise. To appreciate the obstacles facing China as it enters the 1980's and to judge its future potential, it is imperative to understand its recent scientific past. The present volume provides material for such an understanding, with a special emphasis on the decade of the 1970's. A group of distinguished American scientists present their personal evaluations of their own scientific disciplines in China, thus providing a perspective that should be useful to other scientists, China scholars, policymakers, and businessmen, as well as other informed readers.

In a sense, the volume is the fruit of newly improved relations between China and the United States. Since 1972 it has been possible for Americans to make first-hand observations of China's science and technology. Prior to that, only a handful of Western scientists managed to visit the PRC, so that by default it fell to a small number of social scientists, working from a distance and relying on Chinese sources, occasionally to report on science policies and scientific developments in China. By the early 1970's, however, China began increasing contacts with major non-Communist industrial powers. For the United States the harbinger of change was the visit to China by the American table-tennis team in 1971. Early the following year President Richard Nixon's historic visit to Beijing culminated in the February 1972 signing of the Shanghai Communiqué, which called for the facilitation of "people-to-people contacts and exchanges" in the fields of science, technology, culture, sports, and journalism. The most important contacts evolved in science and technology, and starting with a trickle, the number of U.S. scientists and engineers (many of Chinese extraction) visiting the PRC increased to a steady flow. In addition to observing and reporting on Chinese science, they undoubtedly contributed to moving China and the United States toward normalization of relations on January 1, 1979.

The American organization that initially served as a focal point for resuming scientific exchanges with China was the Committee on Scholarly Communication with the People's Republic of China (CSCPRC) at the National Academy of Sciences. It was formed jointly in 1966 by the Academy, the Social Science Research Council, and the American Council of Learned Societies to explore and encourage communication between Chinese and American scholars in the natural, medical, and social sciences, as well as the humanities. Following the Shanghai Communiqué, the CSCPRC received endorsement and support from both governments, and from 1972 through 1979 sponsored 35 U.S. delegations to China and hosted 42 Chinese delegations in the United States.

Initially, scientific exchanges were important mostly for their symbolic value and were accurately described as "scientific tourism" by Philip Handler, President of the National Academy of Sciences. But as delegation reports and articles by individual scientists began to accumulate, the first-hand impressions and evaluations added considerably to Western knowledge of Chinese science. Since China has only a small number of highly trained scientists and since most important research—especially basic research—is concentrated in a few institutes of the Chinese Academy of Sciences, even three- to four-week visits by competent observers could provide valuable information. By 1978 it appeared that the time had come to examine the scattered reports and impressions that had become available, analyze them, and prepare a publication summarizing the current state of Chinese science and technology as the PRC embarked on its fourth decade. Such a state-of-the-art evaluation of Chinese science has not been attempted since the publication of *Sciences in Communist China* (1961), a collection of papers from a 1960 conference of the American Association for the Advancement of Science.

The present study was initiated by the CSCPRC and funded by the Division of International Programs of the National Science Foundation. Since it was important for the authors to go beyond their experiences in China and, in some instances, to update their observations, an intensive effort was made to provide them with a variety of supplemental sources covering both the periods before and after their visits. The materials included articles, chapters of books, and published and unpublished reports, as well as a large volume of translations and abstracts from Chinese scientific literature, mostly produced by the U.S. Joint Publications Research Service. Literally thousands of pages of such materials were reproduced and sent out to the collaborating authors. It is only fair to point out, however, that since first-hand impressions make the most lasting imprints, the different conclusions reached by the authors of the papers often reflect the fluctuating political climate that was in evidence during the specific year of their visit.

The reader may observe gaps or unevenness in coverage, but for the most part the chapters reflect the emphasis the Chinese themselves place on the different fields. This explains the heavy stress on agriculture and applied plant sciences, for example, and the omission of biology as a separate discipline—the Chinese cover it under the umbrella of agricultural sciences and biomedicine. To a lesser extent the coverage was dictated by the quantity and quality of available data on specific fields and the need to keep the volume down to a manageable size.

Developments in U.S.-China relations have progressed at an unprecedented rate since normalization in January 1979 and are continuing to do so. Cooperation in science and technology has become increasingly complex, involving important governmental agreements stemming from cabinet-level missions to the PRC, extensive participation by the U.S. business community, and independent agreements between academic and research institutions. The CSCPRC, now only one of many organizations pursuing scientific and technological exchanges, is finally sending individual scholars to travel, study, and do research in China, conducting bilateral scientific symposia, and sponsoring reciprocal lectures by American and Chinese scientists. The United States has reached a new plateau in its relations with China, and Americans have reached a new stage in their study of Chinese science. The "viewing of science from horseback," to paraphrase Mao, is over and an era of on-the-spot, collaborative work with Chinese scientists has begun.

Once again, then, *Science in Contemporary China* can be viewed as a kind of benchmark, marking the base from which Chinese science and technology is taking off into the last two decades of the twentieth century.

Ideally, a tome of this size should contain a handy summary to satisfy readers who are short of time or have only a modest curiosity about the subjects covered. It is difficult, however, to integrate into a cohesive summary the thoughts of 28 authors, who are not only from different disciplines but have very different personalities and points of view. Nevertheless, some common themes and conclusions do emerge from their work.

It is clear that the adverse effects of the Cultural Revolution and the subsequent rule of the Gang of Four (1966–76) on the educational system will continue to hamper China's progress in science and technology throughout the 1980's. Since 1949, China's older generation of Western-trained scholars and to a lesser extent those trained in the Soviet Union, have occupied key positions in the Chinese Academy of Sciences (CAS) and its research institutes, and have been responsible for most of the scientific and technological progress. The best of these scientists are as good as can be found anywhere in the world, but they are aging and their numbers are rapidly dwindling. Who will take their place? The current effort to raise the quality of higher education and to upgrade the competence of the middle-level professional cohort (aged 30–45), and the program of sending promising students to the United States and other industrial nations for advanced training, are the first steps in a process that will take many years.

There is less agreement among our authors about the familiar conflict between basic and applied research—an argument well known to most developing nations. Mao believed that scientists should focus all their efforts on meeting China's practical needs. This resulted in periods of considerable friction between administrators and scientists but nevertheless research was, for the most part, dictated by national priorities. Some of our authors deplore this virtual absence of basic research in their discipline in China; others believe that during China's "catch-up" phase basic research should be undertaken only when the scientific lag in a particular field is less than five years. Whichever side is right, Beijing seems to be gradually relaxing restrictions on basic research. No doubt the government's emphasis will continue to be related to immediate economic plans and social goals, but Chinese scientists are now confident that basic and long-range research will not be neglected.

Because China's first priority after 1949 was the improvement of people's living conditions, science had to play an important role in improving public health and increasing food production. The achievements in health have received international notice, and progress in agriculture has been much better than many observers anticipated. In both fields the research and innovations by China's scientists have elicited considerable interest on the part of their Western colleagues.

As social conditions improved, China was able to concentrate an increasing proportion of its scientific and technological resources on its three other priorities, which have essentially remained unchanged over the years. One is the rapid development of the Chinese economy, and many of the most capable Chinese scientists have spent years endeavoring to make China's industry, mining, transport, and other economic sectors more efficient and productive. A second is the emphasis on national defense, a priority that protected scientists working in military-related nuclear and space research even during the height of the Cultural Revolution. The third is international recognition and prestige, a goal that permits some scientists to pursue work that may not be practical but can attract worldwide attention. One example is the excellent work of Chinese surgeons in the reimplantation of severed limbs—seemingly a low-priority activity in a nation with only one Western-trained physician for every 10,000 people. Another, criticized by some Western physicists, is the effort going into high energy physics. Yet another is the orbiting of a small satellite whose sole apparent purpose beyond demonstrating China's capabilities in this area was to advertise China's ideology to the world below by playing the song "The East Is Red."

Practical considerations, however, have led to great imbalances within specific disciplines. In chemistry, for example, there has been a high-quality effort in such areas as pharmaceutical research, polymers, materials, and gas chromatography—all important in China's economic production plans—but little or no work in quantum and theoretical chemistry, molecular spectroscopy, organometallic chemistry, and other areas of chemistry that have no immediate economic benefits.

There seems to be unanimous agreement about the obsolescence of equipment in Chinese laboratories. But although some of our authors believe that future progress in their field will depend on how much instrumentation China can import from abroad, others are impressed by how much the Chinese are able to accomplish despite this handicap—for example, by how completely Chinese scientists were able to identify and analyze most mineral and rock samples without modern petrographic microscopes, modern x-ray equipment, and electron microprobes. All agree, however, that China's inadequate and mostly antiquated computer technology is a major handicap—a view widely held by the Chinese themselves. Computers represent a priority area in China's eight-year plan, and the country is increasing both domestic production and imports of larger and more sophisticated equipment.

Perhaps most indicative of the recent changes that have taken place in

China are the developments in the social sciences. Struggling to survive even during the more moderate periods after 1949, social sciences, and especially sociology and political science, were severely criticized during the Great Leap and essentially discarded during the Cultural Revolution. The real renaissance in the social sciences came about only after the fall of the Gang of Four and the establishment of the independent Academy of Social Sciences in May 1977. It will take many years for the Chinese to overcome the serious human, material, and intellectual constraints that have accumulated over 30 years of neglect, but there is no doubt that social sciences are no longer simply an extension of Beijing's propaganda arm, and are now seen by the government as making important contributions to China's overall modernization.

Although there is understandably some degree of disagreement among our authors about what the United States can gain in pursuing scientific exchanges with a nation that is from five to 30 years behind the West in most disciplines, the majority anticipate real scientific benefits from continuing and expanding cooperative research and maintaining a variety of contacts with Chinese scientists.

Astronomy, meteorology, geophysics, and oceanography all benefit from global observation and cooperation; the work of Chinese scientists should fill many information gaps. Zoologists, botanists, geophysicists, and earth scientists are excited at the prospect of being able to make firsthand observations and studies in China. In the medical field there is great interest in medical burn therapy and in the reimplantation of severed limbs, where the Chinese have done some significant, original work; in herbal pharmacology and acupuncture, which have piqued Western curiosity; and in a new contraceptive technology on which Chinese scientists have published some interesting reports. In addition, some medical researchers are eager to carry out epidemiological studies in China. Agricultural scientists want to increase their existing cooperation in such fields as the exchange of germ plasm and biological pest control. Similarly China's much publicized successful earthquake predictions have stimulated interest among Western seismologists. Quite naturally, social scientists specializing in Chinese studies are especially enthusiastic about their prospects for research in the country and about having access to resources (albeit selective) previously inaccessible to them. Furthermore, the value of China's science and technology must be viewed in the perspective of its impact stemming from thousands of years of history and culture. There is much that can be learned from China's experience about the workability of human ideas and institutions. The list is in no sense exhaustive, but it does adequately demonstrate that the benefits of scientific exchanges between China and the United States are not as one-sided as one might first imagine.

Finally, since most of the authors set down their observations and judgments in the early stages of de-Maoization, they have all expressed understandable concern about the drastic political and institutional changes that have occurred since then, and especially about the significant developments that have taken place in some fields of scientific knowledge. The only consolation that can be offered to both author and reader is that this problem is endemic to contemporary China studies: China never stands still.

As a postscript, it must be stressed that during the 1980's the vital question for China (and for the world) is not whether it will be able to achieve its stated purpose of catching up with the advanced countries by the year 2000. That goal is unrealistic and almost incidental. The rhetoric is essentially for domestic consumption and is characteristic of slogans designed to stimulate enthusiasm among the people by introducing a concrete reason for hard work and sacrifice. In connection with national modernization, China has many more immediate and serious problems than catching up and overtaking the rest of the world. Two of them stand out above all others.

China's first concern is nothing less than a matter of survival, for all its publicized plans presume that it will be able to maintain a reasonable balance between food and population. Despite rationing, China has admitted that between 1957 and 1977 there was a decline in the per capita production of grain; and despite earlier birth control campaigns, only in the mid-1970's did officials publicly acknowledge a direct relationship between population growth and China's ability to achieve the stated economic goals. Thanks to the priority being given to the allocation of resources to agriculture, and to the apparent initial success of intense propaganda and even coercion in reducing China's birthrate to a level unheard of in other developing countries, there are grounds for cautious optimism. But the balance will continue to be precarious; and a bad crop year or any relaxation of pressures in family planning could create serious setbacks.

China's second concern for the 1980's and 1990's is the impact that modernization is bound to have on Chinese institutions, life-styles, and values. Is it possible to introduce sophisticated science and technology into the economy without radically altering the material expectations of China's workers and peasants? Can the PRC be opened up to sustained Western contact without far-reaching effects on its political system? Do China's leaders agree on a particular balance between modernization and revolution? Can China achieve its goal of "four modernizations" without also introducing the "fifth modernization"—a liberalization of the system? The answers that the Chinese leaders eventually find for these questions are likely to be crucial not only to China's developmental course, but to its stability.

China must overcome innumerable domestic and international difficulties. Yet this is no reason for pessimism; one should never underestimate the talents and resources of the Chinese people.

*SCIENCE IN CONTEMPORARY CHINA*

*Nathan Sivin*

# *SCIENCE IN CHINA'S PAST*

One of the most interesting developments in historical scholarship over the last generation has been the collapse of the notion that the ancestry of modern science is exclusively European and that before modern times no other civilization was able to do science except under European influence. We have gradually come to understand that scientific traditions differing from the European tradition in fundamental respects—from techniques, to institutional settings, to views of nature and man's relation to it—existed in the Islamic world, India, and China, and in smaller civilizations as well. It has become clear that these traditions and the tradition of the Occident, far from being separate streams, have interacted more or less continuously from their beginnings to the point where they were replaced by local versions of the modern science that they have all helped to form.

Central to this clearing of the air has been the study of China. There the record of technical endeavor has been fullest, most continuous, most accessible, and most exactly dated.[1] That is only what one would expect of a state that was administered by a full-fledged bureaucracy 2,000 years ago and a country where books were being routinely printed 400 years before the Gutenberg Bible.

[1] For further reading, see Nathan Sivin, "An Introductory Bibliography of Traditional Chinese Science: Books and Articles in Western Languages," in Shigeru Nakayama and Nathan Sivin, *Chinese Science: Explorations of an Ancient Tradition,* MIT East Asian Science Series, 2 (Cambridge, Mass.: MIT Press, 1973), pp. 279–310. A broad but thorough exploratory survey of the field with full bibliographies can be found in Joseph Needham et al., *Science and Civilisation in China,* 7 vols. in multiple fascicles projected (Cambridge, Eng.: Cambridge University Press, 1954– ). For an abridged version, see Colin A. Ronan, *The Shorter Science and Civilisation in China* (Cambridge, Eng.: Cambridge University Press, 1978– ); the length of this version has not been announced, but the first volume corresponds to the first two of Needham. For a complete bibliography of Needham's writings outside biology to about 1973, see Mikuláš Teich and Robert Young, *Changing Perspectives in the History of Science: Essays in Honour of Joseph Needham* (London: Heinemann, 1973), pp. 472–78.

Understanding the technical traditions of other cultures is not just a matter for exotic tastes. To the contrary, our sense of Europe's development has been transformed. Consider, for example, Francis Bacon's influential attempt, shortly after 1600, to explain that great efflorescence of human knowledge and activity that we now call the Renaissance:

> Again, it is well to observe the force and virtue and consequences of discoveries, and these are to be seen nowhere more conspicuously than in those three which were unknown to the ancients, and of which the origin, though recent, is obscure and inglorious; namely, printing, gunpowder, and the magnet. For these three have changed the whole face and state of things throughout the world; the first in literature, the second in warfare, the third in navigation; whence have followed innumerable changes, insomuch that no empire, no sect, no star seems to have exerted greater power and influence in human affairs than these mechanical discoveries.[2]

Bacon's understanding was still conventional in 1920, although by then the attention of historians had largely shifted back from inventions to politics and religion ("empires" and "sects") if not to astrology ("stars").

Today the origins of Bacon's three inventions are a great deal less obscure. None of the three was in fact a European discovery. Printing I have already mentioned. Movable clay type was used for inscriptions in Chinese cast bronze vessels of the mid-second millennium B.C. The earliest extant texts printed on paper from carved wood blocks originated in Korea by 751 and in Japan about 770 (nearly 1,000 years after the Chinese invention of paper, and 400 years before its introduction to Europe). We have an account of Chinese ceramic movable-type printing in the 1040's. The oldest such book that still exists (set in wood type) dates from the 1350's. It is not in Chinese but in the Tangut script of the northwest frontier. Metal type was in use still earlier. Printing with it was brought to a high state of perfection through a succession of royally subsidized Korean experiments that culminated in the early fifteenth century.

As printing evolved across East Asia, its potentialities for changing "the face and state" of literature were by no means unexplored. By 1100, for instance, the Chinese government had authorized the printing of standard collections of texts in an attempt—often repeated but always with limited success—to control education. This was the case not only in the humanistic classics but in such fields as mathematics and medicine.

As for gunpowder, formulas for flare mixtures appear in Chinese alchemical books of the ninth century A.D. or a little earlier. The proportions that make gunpowder explosive were known by 1050, and up-to-date commanders over the next two centuries had at their disposal increasingly destructive flame-throwing devices, bombs hurled by trebuchets, and rocket weapons. Between 1270 and 1320, metal-barreled cannon appeared on the

[2] Francis Bacon, *The New Organon and Related Writings*, Library of Liberal Arts, 97 (New York: Liberal Arts Press, 1960), p. 118. Needham, *The Grand Titration: Science and Society in East and West* (Toronto: University of Toronto Press, 1969), pp. 62–76, has several interesting observations on this passage.

Mongols' main battlefronts in Europe, the Islamic west, and China. Although it is not beyond doubt that the inventions that constitute the cannon were finally assembled in Cathay, the prehistory of the crucial propellant and of weapons that used it to fire projectiles was clearly Chinese.

Magnetic attraction was not a matter of recent knowledge in Bacon's time. It is clear that he was speaking of its use in the navigational compass, for the polarity and directive property of lodestone were known and applied to steering by Europeans late in the twelfth century. In China it is likely that a pivoted lodestone spoon was used for divination in the first century A.D. Well before 1100, the true compass—apparently developed earlier for geomantic siting—was being used in navigation, and the declination of the needle was known to steersmen.

Bacon's statement about the springs of change in the West reminds us, then, that for many centuries Europe was primarily a beneficiary of technology transfers. Through the long centuries in which Europe gradually relearned the arts of civilization after the Roman Empire collapsed, China continued to build upon its high culture. Many important inventions that apply the power of water or wind to machines are first found in Europe more than ten centuries later than in China. A few Chinese priorities of other types are efficient harness for draft animals, the drawloom, deep borehole drilling, the segmental arch bridge, rigging that allowed ships to sail to windward, the axial rudder, porcelain, and cast iron.

A Chinese visiting one of the great cities of Europe in 1400 would still have found it backward. In Bacon's lifetime some Occidental cities could no longer be considered backward. By 1840 what appeared glorious about China to its more sophisticated visitors were mainly vestiges of the past.

Some of the reasons for the reversal in technological preeminence are internal to China: centuries of disastrous fiscal and other administrative policies, the remorseless pressure of increasing population, and a large measure of social stability and cultural homogeneity that left traditional values and forms practically unchallenged as the creativity behind them was sapped by intellectual orthodoxies. Other reasons for the reversal arose in Europe, above all a universal quantitative and logical approach to empirical knowledge and practice that gradually redefined nature, reshaped society, and remade human consciousness. In the final reckoning the predominance of Western science and technology cannot be explained by contributing factors entirely within Europe or outside it, for it was built on the intercourse of civilizations.

## *CHINESE AND WESTERN, TRADITIONAL AND MODERN*

It is almost impossible to think about traditional Chinese science and technology without comparing them, explicitly or implicitly, with their present-day analogs. Otherwise it would be difficult even to identify what past activities are of technical interest. Nevertheless, the transforming influence of the scientific and industrial revolutions was so great that the earlier sciences of China and Europe resemble each other more than either resembles the modern variety. It is important, if one is to think clearly about science and technology as worldwide phenomena, to avoid confusing differences between China and the West with differences between traditional societies and societies that have become essentially modern.

Certain aspects of today's technical activity that may appear to be universal are in fact peculiarly European or peculiarly modern. We normally think, for instance, of a sequence of sciences derived from the classical Greek schemes of knowledge but determined by the gradual spread in modern times of system and quantification, beginning with physics and moving through chemistry, the life sciences, and ultimately (or so Comte believed) through psychology, the yet-to-be-achieved exact science of the human mind. In Europe this scheme and its precursors did not, when they arose, describe established structures of knowledge, but justified projects to bring them about by creating new institutions and habits of thought.

In China, for example, there was no biology. Observations and theoretical perspectives on the manifestations of life were scattered through a very different grouping of knowledge that I will sketch out below. Alchemists, East and West, set down a great deal of information about chemical processes—how much they learned from physicians and illiterate craftsmen we do not know—but the designation of Chinese alchemy as early chemistry ignores the fact that alchemical goals were not cognitive but spiritual. Most topics of modern physics had not been imagined before the demise of traditional Chinese science, and the more old-fashioned subjects—for example, heat, sound, and magnetism—were a matter of dispersed experiences and reflections rather than coherent disciplines.

Another practically universal modern assumption is that technology is applied science, that technological progress and economic benefit are the natural end of new scientific knowledge. Before the mid-eighteenth century, emerging modern science did not affect technology in Europe. Craft traditions were developed by people who, even as literacy diffused through society, had little access to science and little reason to use it. The old certainty that the industrial revolution in Western Europe and the United States grew out of the direct application of modern science has come so seriously into question since the 1960's that we are prepared to see a much subtler influence operating in the nineteenth century. Coming out of the new Western mentality shaped by theoretical science, this influence led the educated to take an active interest in manufacturing, and artisans (along with everyone else) to begin reasoning impersonally and abstractly about facts, processes, commodities, and labor to an extent unprecedented in human history. The design of economic systems to exploit new scientific knowledge is, at least outside the chemical industry, mainly a twentieth-century phenomenon. Familiar though the direct linkage of science to technology may be to readers of this book, it would have appeared wildly exotic to all except a visionary few 200 years ago in Europe or Asia.

Since China's industrial revolution began less than two generations ago, what we find in the early record—as in early Europe—are sciences and manufacturing techniques that had little to do with each other. The sciences reflected the concerns of the tiny literate elite, their cosmologies, and the managerial problems they encountered in their careers and recorded in their writings. Technology was on the whole a matter of craft traditions, passed down privately from father to son or from master to apprentice. It was on these mainly oral and manual traditions rather than on cumulative science, recorded in writing, that the technological preeminence of China was built.

What we know about the early industrial arts comes not from those who did the work, but from scholars writing for fellow dilettantes or from officials writing for their peers who happened to be curious about the work of the lower orders. As the learned compiler of the great technological encyclopedia of 1637 put it, "While the best rice is cooking fragrantly in the palace kitchen, perchance one of the princes would wish to know what farming implements look like; or, while the officers of the Imperial Wardrobe are cutting suits of brocade, another of the princes might wonder about the techniques of silk weaving." Those were the only circumstances in which a civil servant might actually feel a need for the book. "Let the ambitious scholar toss this book onto his desk and give it no further thought; it is a work that is in no way concerned with the art of advancement in officialdom."[3]

Thus in both China and Europe before modern times science and technology went their own ways. Natural philosophers learned from the techniques being practiced about them, at least to an extent that their practical curiosity dictated, but what they read in books counted for a great deal more. The achievements of artisans did not depend on the enhanced scientific understanding of their betters.

## SCIENCE AND SCIENCES IN CHINA

In Europe since classical times the various sciences were part of a single structure which included all systematic rational knowledge. They were part of *scientia,* the part that has to do with nature. When in the late sixteenth century Francis Bacon in England began imagining a physics to which experiment was the essential means to discovery, and in the early seventeenth century Galileo Galilei in Italy invented our world full of physical motions that could be measured and computed, these two styles of science—the experimental and the mathematical—developed separately for a while. But they were still seen as parts of one endeavor.

[3] Song Yingxing [Sung Ying-hsing], *T'ien-kung K'ai-wu: Chinese Technology in the Seventeenth Century,* tr. E-tu Zen Sun and Shiou-chuan Sun (University Park, Penn.: Pennsylvania State University Press, 1966), pp. xiii–xiv (translation modified).

In China there was no single structure of rational knowledge that incorporated all the sciences. Knowing was an activity in which the rational operations of the intellect were not sharply disconnected from what we would call intuition, imagination, illumination, ecstasy, esthetic perception, ethical commitment, or sensuous experience.

The various sciences were neither circumscribed by the philosophies of their time nor subordinated to theology (which did not exist in East Asia). The sciences developed a great deal more independently of each other than in the West. The practitioners of each science extended and revised the concepts and assumptions about physical reality with which it began (the particular sciences emerged between the second century B.C. and the second century A.D.). Over the centuries there was relatively little direct response to contemporary philosophic innovations. For example, Zhu Xi (Chu Hsi; 1130–1200), perhaps the most influential moral philosopher of the last 2,000 years, was intensely interested in astronomy and cosmology, but what he knew of these fields was grossly outmoded; astronomers returned the compliment by ignoring his frequently odd opinions about the sky. They were free to exert that autonomy because there were no institutions able to enforce the authority of Zhu Xi over them. There is an obvious contrast with the educational institutions of Europe, from Plato's Academy and Aristotle's Lyceum to the medieval and early modern universities, in which the natural sciences were kept subordinate to philosophy.

Although in China the sciences were relatively autonomous provinces of knowledge, there were definite limits to what was expected of them. Scientists were well aware of the growth of understanding and of increasing ability to predict, but we do not find the conviction that in the fullness of time all phenomena would yield their ultimate secrets. The typical belief was rather that scientific explanation merely expresses, for finite and practical human purposes, limited aspects of a pattern of constant relations too subtle and too multivariant to be understood completely by what we would call empirical investigation or mathematical analysis. This point was made with great clarity by Shen Gua (1031–95), a polymath who served as astronomer-royal and who made lasting contributions to practically every science.

> Those in the world who speak of the regularities underlying the phenomena, it seems, manage to apprehend their crude traces. But these regularities have their very subtle aspect, which those who rely on mathematical astronomy cannot know of. Still even these are nothing more than traces. As for the spiritual processes described in the *Book of Changes* that "when they are stimulated, penetrate every situation in the realm," mere traces have nothing to do with them. This spiritual state by which foreknowledge is attained can hardly be sought through traces, of which in any case only the cruder sort are attainable. What I have called the subtlest aspect of these traces, those who discuss the celestial bodies attempt to know by depending on mathematical astronomy; but astronomy is nothing more than the outcome of conjecture.[4]

[4] Hu Daojing, ed., *Mengqi bitan jiaozheng* (Brush Talks from Dream Brook), ca. 1090, modern variorum ed., 2 vols. (Beijing: Zhonghua, 1960), item 123. For a more extensive discussion

This understanding did not diminish astronomy in Shen's eyes; he devoted his best energies to improving the ephemerides. His perspective, far removed from that of the modern scientist, would have been quite comprehensible to his contemporaries in eleventh-century Europe, although he was applying it to a level of scientific knowledge much more sophisticated than theirs.

To sum up, the Chinese sciences were able to attain a high standard—at times the highest in the world—without the overarching structure of natural philosophy that subsumed science in Europe and without the naive claims to universal knowledge that modern positivists have sometimes attempted to read back into the Western tradition.

## *THE CHINESE SCIENCES: QUANTITATIVE*

It will be clear by this point that a catalogue of priorities torn out of context can only convey a distorted impression of what is worth knowing about science in China. Regardless of how clearly certain features of early science may appear to anticipate today's knowledge, their contexts had little in common with those of modern science. In describing the sciences below, therefore, context will receive as much attention as accomplishment. The Chinese sciences will be portrayed, in other words, not as a succession of triumphs of certain knowledge abstracted from a morass of superstition, but as one important thread firmly woven into the fabric of culture, only forming its own distinct pattern, isolated from the rest, in very recent times.

The sciences described will be those the Chinese defined by applying their concepts of order to various areas of experience—the sky for astronomy and astrology, the bodies of humans and animals for medicine, and so on. As I have already pointed out, this is a very different demarcation from that of modern science.

### *Mathematics*

It is an old Western habit (which those mathematical pioneers, the ancient Mesopotamians, assuredly would not have shared) to think of mathematics as a quintessentially pure science dedicated to the fundamental exploration of number and extension, with practical applications a matter of little interest to the most esteemed sort of mathematician. In China mathematics was not the queen of the sciences, but their servant.

Nearly 1,000 mathematical treatises, beginning in the second century or a little earlier, survive from previous centuries. The great majority have to

---

of Shen's attitude, see Sivin, "Shen Kua," *Dictionary of Scientific Biography, s.v.* An earlier astronomical passage of similar purport is translated in Sivin, *Cosmos and Computation in Early Chinese Mathematical Astronomy* (Leiden: E. J. Brill, 1969), pp. 61–62. Some scientists believed that the sages of ancient times had attained a complete understanding, but that subsequently it had been lost; see, for instance, Ulrich Libbrecht, *Chinese Mathematics in the Thirteenth Century: The Shu-shu Chiu-chang of Ch'in Chiu-shao,* MIT East Asian Science Series, 1 (Cambridge, Mass.: MIT Press, 1973), pp. 57–62.

do with practical matters of the kinds officials, their clerks, and landowners (and increasingly, in the last 1,000 years, merchants) would encounter: surveying; determining areas and volumes; calculating exchange rates and taxes payable in money and commodities; and figuring costs of transportation, materials, and labor. Techniques remained numerical and algebraic, with trigonometric approximations that made it possible from the eleventh century on to do much of the work of geometry with no more spatial visualization than one would need to solve simple problems on a present-day calculator.

A severely logical and axiomatic corpus of proofs like that of Euclid never emerged to provide a unifying standard of rigor in the quantitative sciences. One cannot argue that national character ruled such rigor out, for we find the beginnings of a universal set of definitions, spanning science, language, ethics, and other fields, in the surviving writings of the abortive Mohist school (ca. 300 B.C.). That some Chinese were capable of constructing geometric proofs is clear from surviving examples beginning in the third century, but this remained as minor a concern as numerical procedures were in Europe before the introduction of algebra from the Orient beginning in the thirteenth century.

Chinese mathematics was instrumental from the start. At first, problems were set up with computing rods on a board ruled off in squares like a checkerboard. As the rods were manipulated in appropriate squares, the counting board could represent a two-dimensional array of numbers (see Fig. 1). The potentialities of this device were gradually exploited until by ca. 1300 it was being used to record the coefficients of equations in several powers of up to four unknowns. But by that time the counting board was being replaced by the abacus, an instrument roughly as speedy as but no more flexible than an early twentieth-century adding machine, extremely well suited to the routine needs of the growing urban merchant class. Because it could only represent a dozen or so digits in a linear array, it was useless for the most advanced algebra until it was eventually supplemented by pen-and-paper notation. There were few important innovations at the highest level of mathematics from the mid-fourteenth century until the seventeenth century, when the Jesuit missionaries prompted an efflorescence of interest in European geometry, true trigonometry, logarithms, and so on. This hiatus may have been part of the price paid for the convenience of the abacus.

The predominantly practical orientation of Chinese mathematics made it neither inferior nor superior to the Western tradition. Its lack of development at the abstract geometric level was balanced by its strength in numerical problem solving. Algebra reached Europe ready-made after it had emerged from a long process of convergent discovery and interaction in China, India, and the Islamic world. Many comparatively late European "discoveries," such as Pascal's triangle of coefficients (published in 1665, but known in the sixteenth century) and Horner's method for solving nu-

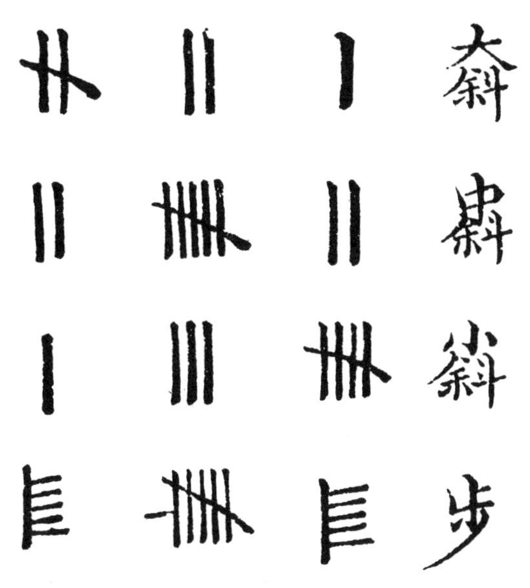

FIG. 1. Example of Matrix Algebra Notation from a Textbook of 1303. The leftmost of these three simultaneous equations corresponds to $-2x + 2y + z = 15$; they are written from top to bottom, and a negative coefficient is indicated by a diagonal slash. (This and other illustrations in this chapter are courtesy of the East Asian History of Science Library, Cambridge, England.)

merical higher equations (1819), were, we now know, named in ignorance of their Chinese origins (the former ca. 1100 or earlier, the latter ca. 1245).

There was a theoretical and speculative side to Chinese mathematics that has generally been ignored by modern historians, at some cost in our understanding of what the art meant to its practitioners. The senses of both words used for "mathematics" before modern times, *shu* and *suan,* include "numerology." They refer as well to a variety of divination techniques that identify regularities and constant patterns—not necessarily quantitative—underlying the flux of natural phenomena.

Prognosticating the future and divining the hidden often, especially in the early centuries of mathematics, were thought to be among the powers of master computators. There is an obvious parallel with the complementarity of mathematical astronomy and astrology, which we find not only in China, India, and Islam, but in the West as late as the time of Kepler. The form of the relationship of course varied with the intellectual and social character of the two activities in each culture.

### *Mathematical Astronomy*

According to the Chinese theory of monarchy, the ruling dynasty remained fit to rule because of the accord the emperor maintained with the cosmic order. This accord depended upon his personal virtue and his correct performance of certain rituals. His special status in the order of nature enabled him to maintain an analogous order in the political realm, for the state was a microcosm. If the emperor lacked virtue or was careless in his duties, disorderly phenomena would appear in the sky or elsewhere in nature as a warning of potential disaster in the political sphere.

This theory divided celestial phenomena into those that were regular and could be computed and those that were irregular and unpredictable and thus omens. Astronomers had two tasks. First, mathematical astronomy (*li*) was to incorporate as many phenomena as possible in a correct calendar—actually an ephemeris that included, in addition to days and lunar months, predictions of planetary phenomena and eclipses. Second, astrology (*tianwen,* which in modern Chinese has come to mean "astronomy") was to observe unpredictable phenomena and to interpret their political meaning. Thus, the emperor could be warned that all was not well in his realm, so that he could mend his ways and take appropriate administrative measures—or be reassured if the omen was favorable.

The calendar, issued in the emperor's name, became part of the ritual paraphernalia that demonstrated dynastic legitimacy (a function not entirely different from that of economic indicators in a modern nation). Astrological observations could easily be manipulated and thus could be dangerous in the hands of someone trying to undermine the current dynasty (the analogy with economic indicators is again perceptible). It was therefore a principle of state that the proper place to do astronomy was the imperial court, and in certain periods it was illegal to do it elsewhere.

The most sophisticated accomplishments of Chinese mathematics are largely concentrated in astronomical and chronological reckoning. Astronomers' ability to use what amounts to higher-order equations, to deal with apparent rather than mean celestial motions, and to determine astronomical constants with great precision grew steadily until the ephemerides reached their zenith shortly before 1300. Although lunar eclipses could be predicted with considerable accuracy by 100 B.C., the lack of the spherical geometry or trigonometry needed to calculate accurately the intersection of the moon's shadow cone with the earth made solar eclipses an abiding problem. The tendency of the imperial court to look abroad for technicians who could deal with it was fateful for the development of astronomy within China, as we will see below.

Beneath the steady evolution of computational astronomy over 2,000 years in China lay a foundation not only of observational instruments but of data-recording systems for centuries on end. Joseph Needham has summarized what is characteristically Chinese in this foundation:

1. The elaboration of large and complex observational and demonstrational instruments, from at least the second century B.C. to the great bronze armillaries of 1421 still on display at the Purple Mountain Observatory, Nanjing (Fig. 2).

2. The invention of the clock drive, and perhaps of the clock escapement

FIG. 2. "Abridged Armillary Sphere" of 1421, Now at the Purple Mountain Observatory, Nanjing

itself, in a long series of great astronomical clocks culminating in the eleventh century. A water-driven mechanism rotated an armillary sighting tube for observation, a celestial globe, and a variety of time indicators (Fig. 3).

3. The maintenance of accurate, dated records of such phenomena as eclipses, novae, comets, and sunspots over a longer continuous period than in any other civilization.

4. Early star catalogues embodying quantitative positional data (earliest possibly from the fourth century B.C., extant observations from ca. 70 B.C.). No less important, although not as old as Mesopotamian cuneiform records, are a recently excavated second-century ephemeris and a table of planetary motions from 244 B.C. to 177 B.C.

5. A coordinate system mainly oriented on the equator and the equatorial pole, unlike the ecliptic system prevalent in Europe and Islam until the late sixteenth century, when it was replaced with a system of Chinese type.

6. Among several early conceptions of the universe, one in which it was boundless and in which the stars floated in empty space. Perhaps because of its audacity, the details of this cosmic scheme were lost early, and no influence on astronomical practice has been documented.[5]

Despite the limitations of pre-telescopic observation, ancient astrological

[5]Based on Needham, *Science and Civilisation*, vol. 3, p. 458; for a fuller discussion and references, see Leo Goldberg and Lois Edwards, eds., *Astronomy in China: A Trip Report of the American Astronomy Delegation* (Washington, D.C.: National Academy of Sciences, 1979), chap. 2.

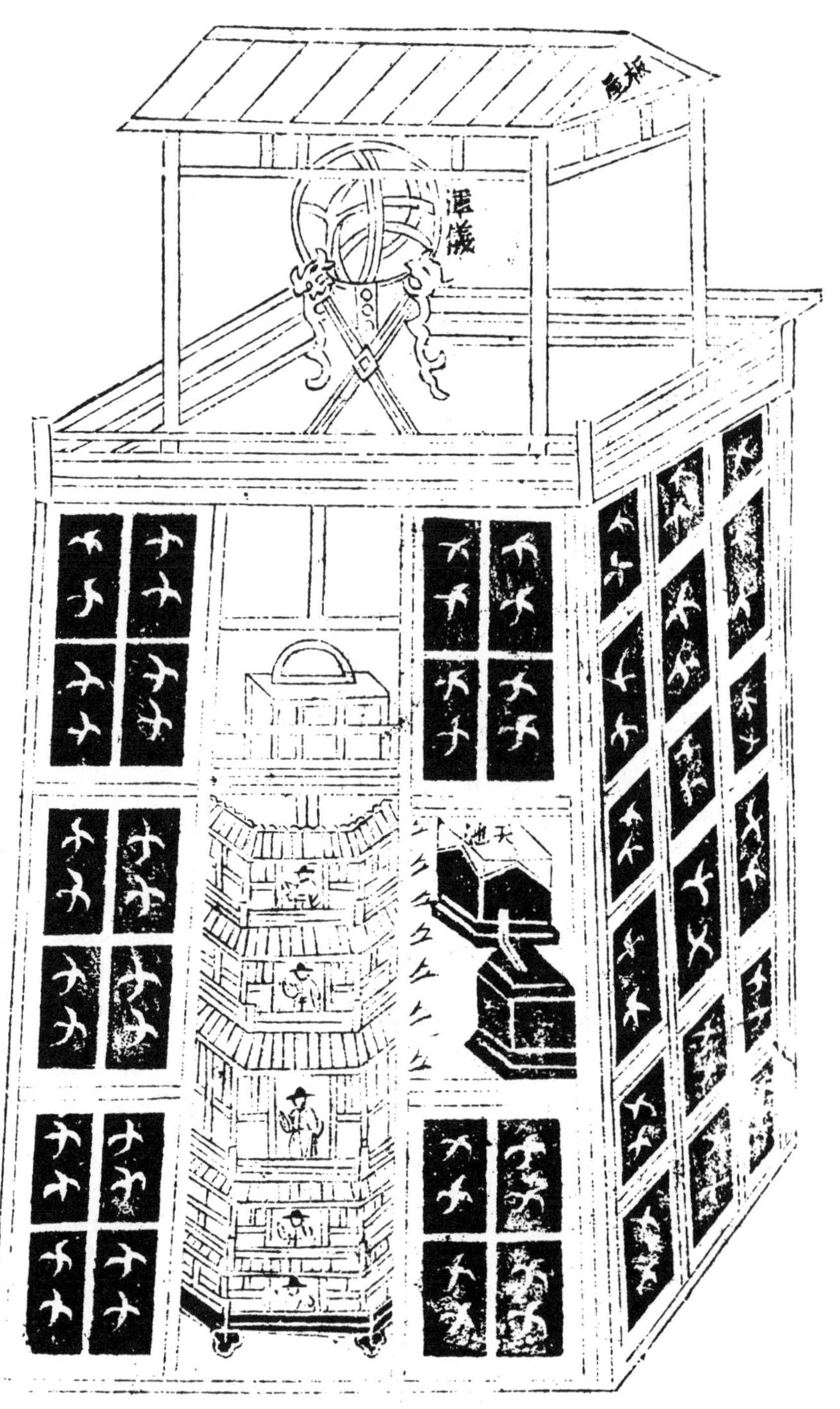


FIG. 3. Water-Driven Astronomical Clock from a Book of ca. 1089

records have proved useful in many ways to modern astronomers. The orbital periods of such celestial bodies as Halley's comet and the frequency of sunspot cycles have been determined with confidence; detailed descriptions of supernova explosions that coincide with today's radio sources have been compiled.[6]

Chinese records have been most productive, supplemented primarily by those of Japan, Korea, and the Islamic world. European records have proved less useful for most of these purposes, since the churchmen and educated laymen who knew astronomy between the later Middle Ages and the seventeenth century generally accepted the Aristotelian dictum that there could be no change in the skies beyond the sphere of the moon. Comets and similar phenomena were considered sublunary (meteorological, more or less) and thus usually not worth noting.

A good part of the historical research under way in the People's Republic of China (PRC) is devoted to exploring early records for information that bears on current scientific problems. Because of the wide scope of recorded astrological portents, the fruit of this labor has benefited other disciplines besides astronomy. For instance, over the past quarter century two major projects have resulted in detailed tables of earthquakes across the breadth of China and through its recorded history.[7]

*Mathematical Harmonics*

The third of the quantitative sciences studied the mathematical relations between sounds and the physical arrangements that produced them. The Pythagoreans in the sixth century B.C., it has often been said, motivated two and a half millennia of scientific exploration with their conviction that number and measure underlay the order perceptible in nature, even those aspects of it that appeared purely qualitative, such as musical harmonies. The Pythagorean faith arose from the portentous discovery that the main musical intervals produced by plucking different lengths of a stopped string are related by the ratios of small whole numbers (a string half as long produces the octave, one two-thirds as long gives the fifth, and so on). Chinese harmonics was a similar blend of mathematics and numerology, probably also initiated by the study of string lengths, but by the second century B.C. concerned with the dimensions of resonant pipes. These were used in turn as pitch standards for ritual bells and stone chimes.

[6] Citations of recent Chinese research are given in Sivin, "Current Research on the History of Science in the People's Republic of China," *Chinese Science*, 3 (1978): 39–58.

[7] Chinese Academy of Sciences, Earthquake Working Committee, Historical Group, ed., *Zhongguo dizhen ziliao nianbiao* (A Chronology of Materials on Chinese Earthquakes), 2 vols. (Beijing: Science Press, 1956); Chinese Academy of Sciences, Institute of Geophysics, ed., *Zhongguo dizhen mulu* (Chronological List of Chinese Earthquakes), 2 vols. (Beijing: Institute of Geophysics, 1970), reprinted in 1 vol. (Washington, D.C.: Center for Chinese Research Materials, 1976). There is a considerable literature on projects of smaller scope. On the use of historical compilations in theoretical studies, see J. Tuzo Wilson, "Mao's Almanac: 3,000 Years of Killer Earthquakes," *Saturday Review*, Feb. 19, 1972, pp. 60–64.

The approaches in East and West were by no means the same. Whereas the Pythagoreans proceeded by subdividing the octave, the Chinese multiplied the string lengths repeatedly by either 2/3 or 4/3 to generate a twelvefold spiral of fifths within the octave. The attempt to draw the most general significances of harmonics was equally ambitious at both ends of Eurasia. To give only one example, the Chinese adapted the pitch pipes as metrological standards of length, volume, and (indirectly) weight.

Harmonics in China is of interest not only for its use in connecting number and nature, but for its application in musical performance—above all the ceremonial music of the court—for music, like astronomy, was cultivated as an aspect of the imperial charisma.

One outcome of Chinese harmonics has survived the end of monarchy. The ability to modulate and transpose freely in contemporary music—and many other aspects of musical innovation in Europe from the time of Bach—depend on equal temperament (the equal spacing of intervals within the octave). The mathematical basis of the equal temperament system has been attributed to the great Dutch scientist Simon Stevin (1548–1620), although first published by Marin Mersenne in 1636. In actuality the system was worked out in a form somewhat more elegant than Stevin's and tested in practice as well, by a prince of the Ming imperial house, Zhu Zaiyu. He described it in a book that appeared in 1584. Zhu was working entirely within the Chinese tradition. There is no evidence of European influence upon him, nor of influence extending from him to Stevin. Still, as Kenneth Robinson and Joseph Needham have pointed out, it may not be coincidence that Stevin was the "inventor" of Europe's first high-speed vehicle, the sail-propelled carriage, which was also known previously in China.

## *THE CHINESE SCIENCES: QUALITATIVE*

Most of the aspects of natural order studied in early science were not governed by number and measure. The patterns of function and dysfunction that were striking in the human body, for instance, could be accounted for only by qualitative theories. In the early West such theories incorporated a variety of explanatory entities—Empedocles' four elements and schemes of pneumata, exhalations, and humors—used to label the aspects of a given thing or activity and thus divulge what it had in common with, and how it differed from, other things or activities that could be so analyzed.

By A.D. 200 at the latest, Chinese thinkers had elaborated two systems that became the chief resorts of those seeking to distinguish the phases of a process in time or a configuration in space (temporal processes in nature were generally considered cyclical, and configurations thought of as finite). One such theoretical entity was the complementary pair yin and yang, which when applied to processes stood for their taking and giving, abiding and transforming, retracting and expanding, relaxing and stimulating aspects, and when applied to configurations stood for the ventral and dorsal, lower and upper, inner and outer aspects, and analogous functions that could be conceived as feminine and masculine pairs. Yin and yang were without exception relational conceptions. As a recent textbook of traditional medicine puts it, "When speaking of the relations of chest and back, the chest corresponds to yin and the back to yang; but when associating chest and abdomen, the chest, being above, corresponds to yang and the abdomen, being below, to yin."[8]

In addition to resolving any whole into its yin and yang aspects, one could understand it in terms of a similar system of five phases (*wuxing*, often mistranslated "five elements" by specious analogy with the Greek "four elements"). This was not a different understanding, merely a finer textured one. In cycles of change the phases labeled "water" and "fire" stood for the most intensive aspect of yin and yang; "metal" and "wood" stood for the less intensive aspects; and "earth" stood for the aspect in which the opposed tendencies were balanced and in effect neutralized each other. The use of the five phases in the study of spatial relations was analogous. Most commonly four of the five stood for the cardinal points of the compass and the quarters of the sun's annual path corresponding to the four seasons, and Earth stood for the central point on which others pivoted. Thus, for instance, in scientific discourse Earth always implied balance and the neutral center; it did not refer to particles of earth as ultimate constituents.

Yin-yang and the five phases were not primarily technical concepts. Like the European notion of cause and effect, they belonged to everyday language and were likely to be used whenever anyone tried to explain structure and change. On the other hand, like cause and effect they had more specialized meanings in learned discourse. As each of the qualitative sciences assumed its classical form, yin-yang and the five phases were given special definitions related to the subject matter of that field and supplemented with other technical conceptions to provide a language adequate for theory. Because, as noted earlier, the sciences evolved independently, the exact meanings of common concepts tended to differ considerably from one discipline to another.

### *Astrology*

Mathematical astronomy has been described above as the science of celestial phenomena that can be predicted and thus need not be observed. Astrology, its complement, depended on data available only through contemplating the sky and sought to discern its significance for current politics, in part by precedent and in part by theoretical analysis.

In the standard histories published by successive dynasties, we find records of a great variety of phenomena involving solar eclipses, sunspots, the fixed stars and planets, and comets, as well as what would now be con-

[8]Guangzhou Armed Forces, Rear Support Units, Medical Administrative Organization, et al., ed., *Xinbian Zhong yixue gaiyao* (New Essentials of Chinese Medicine) (Beijing: People's Hygiene Press, 1972), p. 2.

sidered atmospheric phenomena and terrestrial prodigies. The character of the record and its contemporary use are perhaps best revealed by example. Here is the account of the great supernova of 1006 from the treatise on astrology in the Song dynasty history. It is one of several accounts in Chinese sources which, when colligated, provide rich data on the supernova.

On the 15th sexagenary day, fourth month, third year of the Jingde reign period [May 6, 1006], a Zhoubo star appeared. It emerged in the south of the lunar lodge [a division analogous to the zodiac] Base, 1° west of the constellation Mounted Guard [27 stars, mostly in Lupus]. Its form resembled that of the half-moon, and it had pointed rays. It glowed so brightly that objects could be distinguished by its light. It passed to the east of the constellation Treasury-and-Tower [10 stars in Centaurus], and in the eighth month, following the rotation of the sky, "entered the turbid zone" [i.e. set with the sun and thus became invisible]. In the eleventh month it appeared once again in Base. Thereafter it was seen regularly heliacally rising in the east in the eleventh month and setting in the southwest in the eighth month.

Earlier in the same treatise, the astrological characteristic of a Zhoubo star, one of five canonical classes of auspicious stars, is noted in long-established language: "A Zhoubo star is yellow in color and brilliant; the state in correspondence to which it appears will greatly prosper." The last phrase refers to a system of interpretation in which the lunar lodge of the manifestation indicates the part of China affected. In this case, the lodge Base (determinative star in Libra) corresponds roughly to modern Henan province.

Astrological interpretations are neither unsuccessful science nor mumbo jumbo. They are best understood, like modern economic indicators, as a technical framework for policy debates, resolved, as often as not, on other grounds. Faith in the validity of astrological categories, like confidence in extensively manipulated statistical figures today, is in no sense compromised by their repeated failure to deliver accurate predictions.

In the biography of Zhang Zhibo, an imperial commissioner in 1006, there is a fragment of the discussion that followed the appearance of the new star. We find Zhang attempting to turn the attention of the court away from the auspiciousness of the omens and the minutiae of interpretation toward the moral vigilance of the monarch, which Zhang urged should not be relaxed at moments of good tiding:

When the Zhoubo star appeared, the astronomer-royal reported it as an auspicious portent. The court officials prostrated themselves at the palace gates to offer congratulations. Zhang expressed the view that the Ruler of Men ought to cultivate his virtue in response to the celestial phenomena, but that the appearances and disappearances of such stars had no particular significance [lit., "were not tied to anything"]. He proceeded to outline the essentials for mastering the Way [of correct government].[9]

[9]The three translations are from *Song shi* (Standard History of the Sung Period) (Beijing: Zhonghua, 1977), pp. 1,226, 1,076, and 10,187. For excellent discussions, based on freer translations, see Bernard R. Goldstein and Ho Peng Yoke, "The 1006 Supernova in Far Eastern Sources," *Astronomical Journal*, 70 (1965): 748–53; and for a conspectus drawing on Goldstein and Ho, but also using European and Islamic materials, David H. Clark and F. Richard Stephenson, *The Historical Supernovae* (Oxford: Pergamon Press, 1977), pp. 114–39.

These essentials, we can be sure, supported the policies that Zhang favored.

The inseparability of astrology from politics should not be taken to imply that the former was less pure a science than mathematical astronomy. Their institutional setting was the same, at least until classicists in the seventeenth century took up the computation of ancient eclipses and other phenomena to understand early technical writings, correct the chronology of the Confucian scriptures, and test their authenticity. The great concentration of effort on solar eclipses in early astronomy, the relative neglect of such interesting topics as latitude theory and apparent planetary longitudes, and the use of gnomon-shadow observations for many centuries after armillary instruments had surpassed the gnomon in accuracy, all represent the direct imprint upon mathematical astronomy of court ritual patterns and the social norms behind them. The difference between astronomy and astrology was a contrast of emphasis on the quantitative as opposed to the qualitative, and on objective motions as opposed to the correlation between celestial and political events.

### *Medicine*

The data collected over the centuries about the body, health, and disorders were structured by the nature concepts described earlier, forming a coherent body of theory applied to diagnosis and treatment of illness.[10] Classical medicine, which deserves the adjective "scientific" as fully as its counterparts in Western culture until recent times, provided health care for a small portion of the Chinese populace. The majority of its patients and its more eminent practitioners belonged to the upper crust of society. Most of the afflicted among the Chinese population over the course of history had no access to the few fully qualified physicians. They depended on a great variety of less educated healers, ranging from herbalists to priests—a situation that would have been perfectly familiar in eighteenth-century France.

What we call medicine incorporated and imposed order on experience related to every aspect of health, disease, and injury. One Chinese scheme of its major divisions includes theoretical studies of health and disorder; therapeutics; the theory and practice of longevity techniques, including sexual hygiene; pharmacognosy; and veterinary medicine. Pharmacognosy, the study of vegetable, animal, and mineral substances used in therapy, brought together so much information on the habitats and characteristics of thousands of drug sources that its literature was studied not only for therapeutic

[10] The general level of Western scholarship on Chinese medicine is extremely low. The only philologically sound book in any European language on the theory of traditional medicine is Manfred Porkert, *The Theoretical Foundations of Chinese Medicine,* MIT East Asian Science Series, 3 (Cambridge, Mass.: MIT Press, 1974). Vol. 6 of Needham's *Science and Civilisation,* on biology and medicine, is not yet in press. Several excellent essays by Needham in collaboration with Lu Gwei-djen have been gathered in *Clerks and Craftsmen in China and the West: Lectures and Addresses on the History of Science and Technology* (Cambridge, Eng.: Cambridge University Press, 1970), in particular "Medicine and Chinese Culture" (pp. 263–93). A history of acupuncture by Lu and Needham, *Celestial Lancets,* was forthcoming from Cambridge University Press in 1980.

purposes, but also as compendia of natural history. Prescriptions made up of both crude drugs and extracts were commonly used in combination with a great variety of other therapeutic means, including acupuncture and moxibustion, dietary regulation, calisthenics, breathing exercises, and massage.

In acupuncture and moxibustion, needles were inserted into the flesh and cones of punk burnt on the skin at certain points to affect what traditional physicians considered a system in which vital energy circulated throughout the body (Fig. 4), and what clinical researchers today in China and abroad are more inclined to see as peripheral nerve endings and receptors. On the whole, acupuncture was classically considered a minor component of therapy, usually effective only when continued over a long period. In modern Chinese clinical practice, it is of course impossible to evaluate acupuncture separately from the other remedies with which it is combined (including both Western and Chinese types).

Acupuncture has been promoted to the status of wonder cure, primarily by the American media in their endless quest for novelty. It has proved attractive out of proportion to the rest of traditional therapy for two reasons. It is exotic to Americans, who were very late in learning about its longstanding popularity in China, Japan, and Europe, and who have forgotten its vogue among physicians in early nineteenth-century America.[11] It is usually confused with acupuncture anesthesia (or analgesia), a new development used in modern surgery since the late 1950's. In traditional surgery (limited to external medicine, amputation, and bonesetting), the only anesthetics known to us were drugs—mainly alcohol, datura, and cannabis.

Classical Chinese medicine has often been represented as an empirical science based on the clinically sound use of effective natural drugs and other remedies. In this view, theories served primarily as mnemonic devices or as mystifications to confuse the untrained, or may be dismissed as futilities of the feudal past. Other authors have portrayed medicine as a remarkable corpus of theory—based on adaptions of the yin-yang and five phases concepts—that succeeded in understanding the body as a many-leveled system and treated its ills holistically. They accordingly recommend it as a corrective to the impersonal, excessively lesion-centered and nihilistic tendencies of modern therapy. The mild and usually nonspecific remedies of traditional medicine are sometimes, but by no means always, expected to contribute to this reform. A closer acquaintance with the literature of classical medicine and the remnants of practice that have survived the demise of traditional China suggests that these are partial pictures of a more complicated reality.

Before 1920, the strengths of medicine everywhere lay predominantly in the care of mild and chronic disorders. There was little physicians could do for most acute emergencies beyond strengthening the patient's defenses and

[11] James H. Cassedy, "Early Uses of Acupuncture in the United States, with an Addendum (1826) by Franklin Bache, M.D.," *Bulletin of the New York Academy of Medicine,* 2d series, 50 (1974): 892–906.

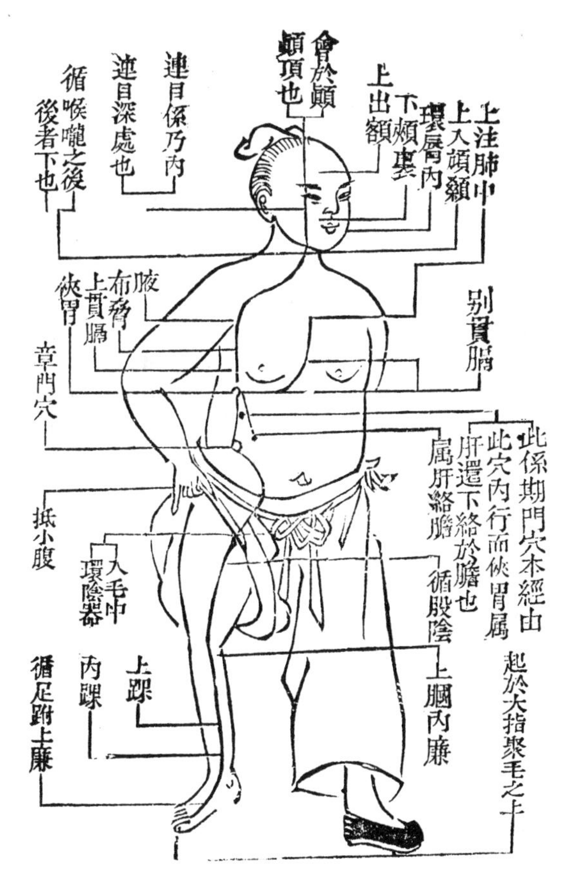


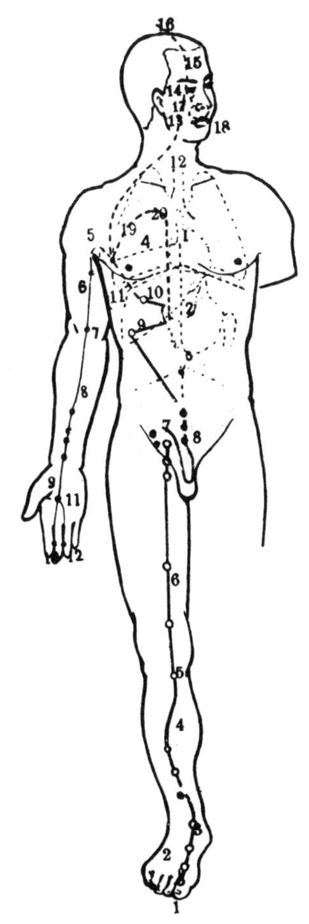

FIG. 4. Similar Diagrams of Acupuncture Loci from an Eighteenth-Century Source and a Textbook of 1960

preparing his family for the worst. With that emphasis on mild and chronic cases in mind, we are better able to judge the importance of gentle, gradual remedies, practices aimed at evaluating and improving the physical, mental, spiritual, and social circumstances of the patient, and theories used to relate a multitude of symptoms to a multitude of therapeutic measures and to make those relations comprehensible to the patient. Theory also maximized control of acute disorders by emphasizing prevention and early treatment.

That Chinese medicine does not give a detailed and accurate picture of anatomy and physiology was not the handicap hindsight makes it. The high state of anatomical knowledge in Europe by the death of Andreas Vesalius in 1564 was the glory of medicine as a branch of learning, but had little application to medicine as the care of living bodies. This situation changed only as disease came to be thought of in terms of localized lesions, as asepsis and anesthesia made local surgical intervention safe, and as the organization of the medical profession imposed a single high standard of qualification. In the United States, for instance, the change was not complete until about 1920.

The strength of classical Chinese medical discourse lay rather in the sophistication with which it portrayed the interrelations of functions on many levels—from the vital processes of the body to the widest conceivable environment of the patient—while never cutting its roots in therapy. The evaluation of Chinese medicine in the light of this strength, rather than according to criteria that could not have been applied anywhere until half a century ago, has hardly begun.

### *Alchemy*

Chinese alchemy used chemical techniques to prepare elixirs, which were perfected substances that brought about personal transcendence and eternal life. Elixirs could also be used for medical purposes and to transform ordinary metals into gold. That is how an alchemist might have defined "external alchemy" (*waidan*); its analogue, "internal alchemy" (*neidan*), used the language of the laboratory to teach meditational (or sometimes sexual) disciplines in which the adept's body was visualized as the reaction vessel and furnace. In the first millennium the two alchemies were regularly practiced together, but after 1200 little activity in the external art was recorded.

The materials and apparatus of alchemy were on the whole the same as those of pharmacology, with some contributions from metallurgy and other practical chemical arts (Fig. 5). Certain developments, such as elaborate distilling vessels, appear so exclusively in alchemical literature that they almost certainly originated there. The same may be said of gunpowder. What may be the earliest mention of flare mixture composition appears, oddly enough, in a list of external alchemists' misguided activities in a tractate on internal alchemy written not later than the end of the ninth century: "There was a case in which sulphur and realgar were mixed with saltpeter and honey, and

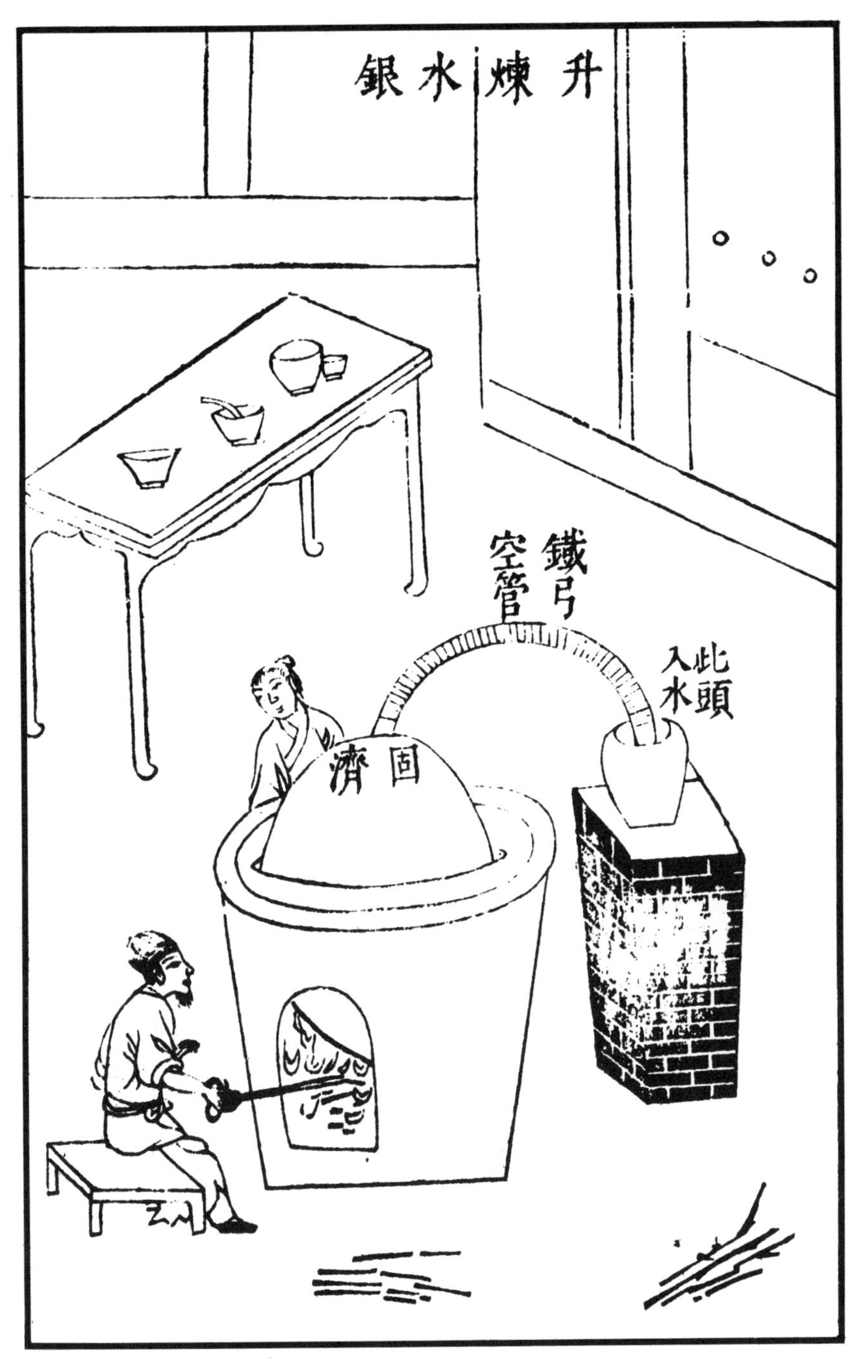


FIG. 5. Commercial Distillation of Mercury, from the Technological Encyclopedia of 1637

burnt. Flames leapt up, burning the alchemist's hands and face and incinerating the building."[12]

The roughly 100 remaining Chinese alchemical books are probably the world's richest source for what was known about reactions and their products up to 1200 and even for a couple of centuries afterwards. They reveal, in fact, that alchemy was not entirely qualitative; some adepts took a lively interest in the gravimetric composition of reagents.[13]

Knowledge of chemical change was a means and by-product, but not the aim of external alchemy. For some practitioners the goal was hardly distinguishable from that of medicine. Others were less interested in a product that would bring health or immortality than in the alchemical process, which they designed to serve as a model of the great cycles of nature, the rhythms of the Tao.

No mortal could experience the cosmic cycles in their millennial sweep. These alchemists accelerated the scale of time, using theories based on yin-yang, the five phases, and numerology, to create, in a laboratory procedure that might require a few weeks to a year, an object of mystic contemplation. Their principle, like that of the Pythagoreans before them, was that to grasp the constant patterns that underlie the phenomenal chaos of experience is to be freed from the bonds of mortal finitude. As in the other Chinese sciences (and less obviously in modern science), the motivation that led to chemical discovery was connected to the deepest values of those who pursued alchemy.

*Siting*

Geomancy, or siting (to use a more informative term proposed by Steven J. Bennett), analyzed topographic features to find balanced land configurations in which to place houses and tombs.[14]

Alchemy, as we have seen, used the yin-yang and five phases concepts to understand and manipulate time as it shaped certain laboratory processes; siting adapted the same concepts to the study of space and form in the landscape. Some practitioners depended on their experience to judge the visual balance of topography, and some used readings taken with the famous geomantic compass. This instrument, which among its 18 or more concentric dials (Fig. 6) incorporated indications of magnetic declination, was apparently a forerunner of the sailor's navigational instrument.

Once its contributions to the evolution of the compass are acknowledged,

[12] *Zhen yuan miao dao yaolüe* (Essentials of the Mysterious Tao of the True Origin) in *Zhengtong dao zang* (Collected Taoist Works), vol. 596: 3a. On this book, see Needham, *Science and Civilisation*, vol. 5, part 3, pp. 78–79.

[13] Sivin in Needham, *Science and Civilisation*, vol. 5, part 4, pp. 300–5; summarized in Sivin, "Chinese Alchemy and the Manipulation of Time," *Isis*, 67 (1976): 513–26, especially p. 521; reprinted in Sivin, ed., *Science and Technology in East Asia*, History of Science, Selections from *Isis* (New York: Science History Publications, 1977), p. 117.

[14] Steven J. Bennett, "Patterns of the Sky and Earth: A Chinese Science of Applied Cosmology," *Chinese Science*, 3 (1978): 1–26.

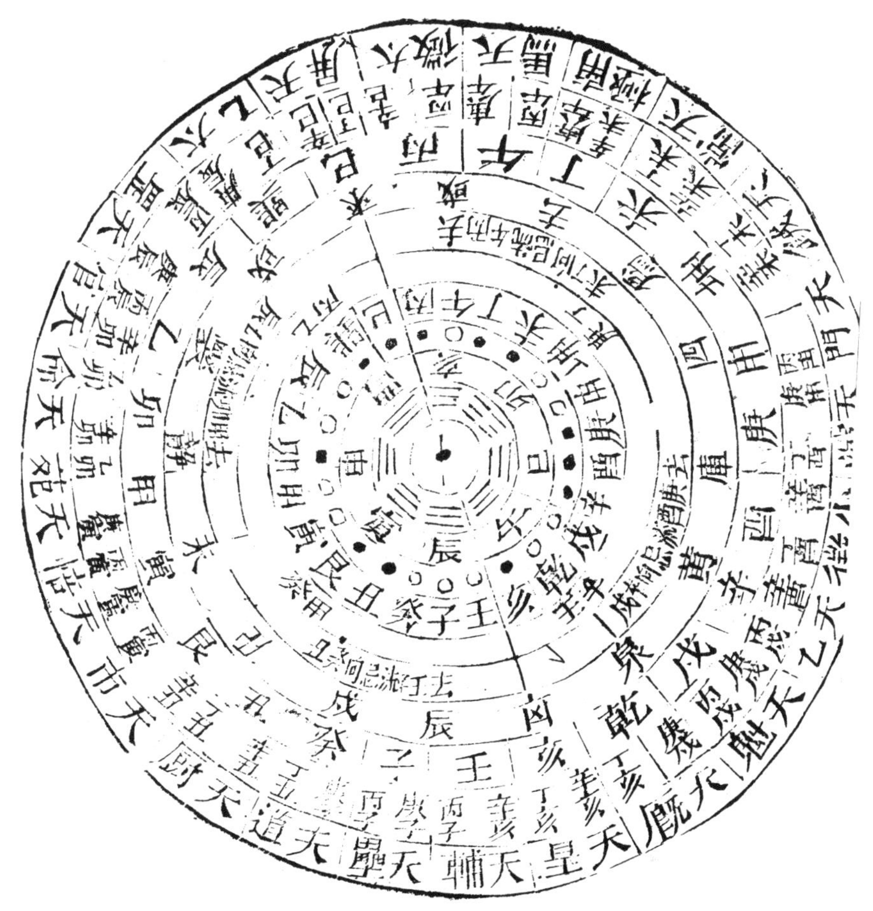

FIG. 6. Woodcut of a Siting Compass, from a Siting Compendium of 1786

siting is sometimes shrugged off as a mass of superstition. There is indeed not a great deal to be said for it if the sole criteria of evaluation are to be those of science today. But those are not the criteria that lead to understanding, since siting was not as narrowly focused as modern science. Siting nevertheless fits the definition of a premodern science, for certain schools of practice worked out a system of abstract and objective reasoning about the natural phenomena that concerned them.

Anthropologists have found siting worthy of study because of the way it was used (and is still being used) to resolve conflicts of status between practitioners' clients. Geographers and landscape architects have expressed interest in siting theory because it accomplished for centuries what modern theorists have been trying to do: provide a systematic framework for reasoning that will reliably yield beautiful sites. Because the balance that siting strove for was dynamic and complex, the benefits of the science are in large part esthetic—not a matter of purely scientific interest, but of no small utility nevertheless. That may explain why siting has proved so durable. It and medicine are the only traditional sciences still widely practiced in Chinese communities today.[15]

*Physical Studies*

This is a general rubric for several traditions that considered a great range of natural phenomena in the light of fundamental concepts. So far these traditions have been little studied, but there is a great deal in their literatures that throws light on early knowledge of change and interaction, whether chemical, physical, or biological.

A few items from the section on trees in one of the most popular treatises suggest the wealth of information to be found there: "If peach is grafted onto persimmon, the fruit will be golden peaches. If peach is grafted onto Japanese apricot, the fruit will be meaty. If male and female gingko trees are planted together, the females will bear fruit. If pomegranate is grafted onto 'flowering cassia' [*Osmanthus fragrans* Lour.], the blossoms will always be red." Because we do not have records left by those who had mastered grafting, it is from sources such as this one, as well as writings on materia medica, that the history of Chinese arboriculture must be reconstructed. An analogously broad range of documents must be studied by those who wish to understand many other ancient activities that required technical knowledge.

It would be misleading to leave the impression that the physical studies literature is an unmixed trove of information pertinent to modern science. The mixed character of the book just cited, for instance, is obvious from two consecutive items in the section on "human affairs": "When a mother cries and her tears fall into the eye of her son, they will cause damage to its pupil and give rise to a cataract. In epidemic warm-factor disorders [a class of febrile diseases], if the clothing of the first person to fall ill is steamed in an earthenware steamer, the rest of the family will not catch the sickness."[16]

The abstract ideas in these extensive compilations also deserve to be investigated. In a study of yin-yang and five phases theory bearing on alchemy in the physical studies literature, P. Y. Ho and Joseph Needham called this theory "a hitherto unrecorded chapter in the prehistory of the conception of chemical affinity," by which they mean that it "seems to take its place in the linear ancestry of the idea that things can be arranged in chemical classes the members of which are susceptible of chemically similar processes."[17]

[15] This is documented only for Taiwan and overseas Chinese communities; the extent to which siting is practiced in the PRC is unknown. On siting in the New Territories of Hong Kong, see Baruch Boxer, "Space, Change and *Feng-shui* in Tsuen Wan's Urbanization," *Journal of Asian and African Studies,* 3 (1968): 226–40.

[16] Zan Ning, *Gewu cutan* (Simple Discourses on the Investigation of Phenomena) ca. 980 (*Baibu congshu jicheng* ed.), 1: 4b and 2: 16b. Cf. the translation of the first passage in Needham, *Science and Civilisation,* vol. 5, part 2, p. 149.

[17] Ho and Needham, "Theories of Categories in Early Mediaeval Chinese Alchemy," *Journal of the Warburg and Courtauld Institutes,* 22(1959): 173–210, especially p. 201; revised version in *Science and Civilisation,* vol. 5, part 4, p. 322.

## INTERACTION OF CHINESE AND FOREIGN SCIENCES

Things and ideas have flowed between China, Europe, and the other great civilizations regularly since the New Stone Age. Whether such characteristics of the earliest Chinese civilization as writing, the manufacture and use of the wheel, and bronze and iron metallurgy diffused into China from abroad, or whether they were reinvented from scratch, is still a matter for heated debate. These two hypotheses need not be considered mutually exclusive, for a mere rumor that something exists vastly simplifies making it. In any case these techniques did not appear in China fully formed, as we would expect had they been imported.

Iron, for instance, appeared in the fifth or sixth century B.C., perhaps eight centuries later than the wrought iron of Anatolia. The Chinese manufacturing process, according to details recorded from the fourth century B.C. on, was for cast iron, which could not be produced at will in the West for another 17 centuries. The production of steel, probably widespread by the third century B.C., was due to the simultaneous availability of efficient furnaces derived from ceramic kilns, reciprocating bellows to provide an air blast, good refractory clays, and phosphorus-rich ores or fluxes to lower the melting point of iron.

Science was part of the flow between East and West from early times. The Greek medical doctrine of the four elements reached China via India by about 500. Physicians from Byzantium or Syria, possibly Christians, are said to have cured a Chinese emperor in the mid-seventh century. Jesuit missionaries in China published a treatise on Western physiology and an anatomical atlas in the mid-seventeenth century, but the two books were very little circulated and received practically no attention. The first European book on Chinese medicine was published in 1682, the first on acupuncture in 1683. Alchemical elixirs of Western origin are noted in seventh-century China, and there is some evidence that Chinese alchemical gold was traded or sold abroad. But the greatest foreign scientific influence on China in early centuries, as in the seventeenth-century heyday of Jesuit missionary activity, was exerted in astronomy. A fuller discussion of that field is therefore in order.

Before the middle of the seventh century, after Buddhism had become rooted in China, Indian astronomers were resident in the Chinese capital. Their techniques were partly derived from Greek astronomy and were more reliable for predicting solar eclipses than those current in China. The political significance of solar eclipses led the Chinese court from the turn of the eighth century to depend on resident foreign astronomers. When the Mongols brought China under their rule in the second half of the thirteenth century, their astronomical functionaries were Islamic, from Persia and Central Asia. Their computational procedures were still more accurate than those of their Indian predecessors. They were still performing the same services to the court when Jesuits began competing with them 350 years later.

The Jesuit missionaries in China in the early seventeenth century were there not to propagate astronomical science but to convert the empire from the top down to Roman Catholicism. The only established access to the top was in the Astronomical Bureau, which had provided court positions to foreigners for nearly 1,000 years. By 1635 the Europeans had gained operational control of the Bureau after submitting to the throne a series of treatises that set out in Chinese the mathematical and cosmological foundations of European astronomy. They maintained their status in the civil service until the middle of the eighteenth century.

The Church's injunction against teaching the Copernican doctrines of the central sun and the planetary earth in 1616, and the condemnation of Galileo in 1633, made it impossible for the missionaries to disseminate the state of the art as it developed over the next century or so or to give Copernicus, Kepler, and Newton credit for their accomplishments without puzzling gaps and circumlocutions. Successive discussions of the new astronomy in Chinese by Jesuit writers before 1760 were thus full of contradictions, which were never explained. What is often considered the great watershed in European scientific consciousness was not revealed in China until late in the nineteenth century, long after it had become commonplace. Nor were the Jesuit writings sufficiently technical and detailed to permit the Chinese to advance world knowledge. Nevertheless, they stimulated a high pitch of astronomical activity outside the court.

The geometric and trigonometric approaches and the cosmological framework of the Jesuit writings in Chinese, obsolete though they became as the seventeenth century wore on, precipitated what can only be called a scientific revolution. The best Chinese astronomers of the time adopted new concepts, tools, and methods. They changed their convictions about what constituted an astronomical problem and what significance astronomy could have for the ultimate understanding of nature.

This metamorphosis of astronomy did not lead to the fundamental changes in thought and society that are naively supposed to be the inevitable outcomes of a scientific revolution. To make a long story short, conceptual revolutions, like political revolutions, occur at the margins of societies. The astronomers who responded to the Jesuit writings were members of the educated elite who above all felt the responsibility for strengthening and perpetuating traditional ideas. They were, in other words, at the center of their society. It is scarcely surprising that they used what they learned from the West to rediscover and master for the first time in centuries the astronomical techniques of their greatest Chinese predecessors.

Only after the Opium Wars of the early 1840's could the Chinese receive a systematic education in the exact sciences as then taught in Europe. This time the educators were not individual priests dependent upon the toleration of their hosts. They were Protestant missionaries exempt from Chinese laws, their right to operate missions and schools guaranteed by imposed treaties and enforced by gunboats. They were no longer appealing, as the

Jesuits did, to an elite intent on adapting new techniques to traditional ends. The Protestant missionaries educated mostly the poor and people of modest means. Even their richer converts came to them because their children had little chance of conventional success in the old society.

The Chinese astronomers trained in Western institutions had no reason even to be curious about what their compatriots in earlier times had done. By 1880 Protestant missionaries, generally working with Chinese, had translated from European languages a number of basic textbooks in astronomy, mathematics, and physics and made them generally available at low prices. Their schools, libraries, bookshops, and other institutions were founded for the purpose of instigating change, not to preserve Chinese civilization.

As the threat of dismemberment by the colonial powers became more imminent, the Chinese government was belatedly persuaded to begin educating modern scientists. In 1866 a department of mathematics and astronomy was added to the Tongwenguan in Beijing, which had previously been a college for interpreters. In 1867 a translation department was added to the Shanghai Arsenal, which had been established two years earlier. There Chinese and foreign employees of the imperial government systematically undertook the translation and publication of modern works in science, engineering, medicine, law, and so forth.

These and the less systematic technical publications of the missionaries were widely distributed and eagerly studied by successors of the amateur groups that maintained the tradition begun by the Jesuit writings. The new translations often played a part in the education of statesmen and reformers, for whom they provided a window on the world. The future lay, however, not with those who saw modern technology as a tool to preserve an empire and a way of life, but with those at the margin of the old society, educated in modern schools and given employment in new institutions.

From about 1900 onwards, Chinese astronomers began to emerge who were fully prepared to benefit from advanced training abroad. They were educated in missionary institutions, as well as in government universities as these appeared. When ten foreign powers extracted heavy indemnities from China after the Boxer Uprising in 1900, the United States used income from its share to support students during technical training abroad.

Scientists trained in the United States, Europe, and Japan, as well as those educated in China, built the first large-scale system of research institutions and an educational system to train scientists. A few of these institutions had been founded by foreigners in the image of their own, but—to redirect the words of a leading historian of China—"on all sides of this gleam of Western light, China was being torn apart by forces so powerful that they made the Westerners' efforts poignantly irrelevant."[18] As the succeeding chapters

[18] Jonathan Spence, *To Change China: Western Advisers in China, 1620–1960* (Boston: Little, Brown, 1969), p. 172.

in this volume will show, China has gradually, since 1949, by fits and starts, invented policies toward education and science that reflect its own priorities rather than the expectations of other nations.

## THE HISTORY OF SCIENCE IN CONTEMPORARY CHINA

In China today it is normal for historical studies to be published in scientific journals, for observatories and other organizations to have research groups for ancient science and technology, and for modern scientists to be knowledgeable about their country's scientific heritage.[19] This situation contrasts so greatly with that of most other countries that it calls for an explanation.

I have already noted that early astronomical and earthquake records are a valuable resource for current scientific research. But this is only one motivation for awareness of history.

Other reasons are related to China's place in the world. For millennia Chinese considered their country the one true center of civilization. Over the past century China has had to make an entirely new place for itself as only one member of a large family of nations. For most of the last 100 years it has been dependent, and looked down upon by foreigners for that reason. The present government is resolved to make it as independent as the imperatives of survival permit, and in doing so has enlisted the full energies of every Chinese citizen.

China's recent policies for technological development have been unique in many respects. They have made unique demands for adaptation on the part of the whole scientific sector of society, which until a couple of decades ago was considered quintessentially Western. Science and engineering were what one learned from foreigners in order to safeguard oneself against them. This view has gradually been changed over the past 20 years by popularizing the history of Chinese science. Children's books, postage stamps, museum exhibits, and school lessons have all carried the message that science is not European but a world enterprise and that over most of history China was one of the great contributors to that enterprise.

It is not difficult to find this point explicitly stated. Consider the afterword to a new book for teenagers on every aspect of technical history: "The achievements of China's ancient science and technology prove that the Chinese people have the ability needed to occupy their rightful place among the world's peoples. These achievements will also encourage our faith and strengthen our resolve, so that in the shortest possible time we may catch up to and surpass the world's most advanced levels of development. China has yet greater contributions to make to humanity."[20] Among scientists this

[19] Recent publications and research are summarized in Sivin, "Current Research."

[20] Chinese Academy of Sciences, Research Institute for the History of Natural Sciences, ed., *Zhongguo gudai kexue chengjiu* (Achievements of Ancient Chinese Science and Technology) (Beijing: Zhongguo qingnian chubanshe, 1978), p. 706. Although written for young readers, this comprehensive work contains contributions by China's best-known historians of science and technology.

consciousness has undoubtedly served to encourage continuity between scientific work and political activity. This continuity was epitomized at a dinner given for the U.S. Astronomy Delegation in 1977 by an official responsible at the provincial level for science and education. When asked how, in view of the nature of science, there could be a characteristic Chinese or Marxist form, he responded that while the truths of science are universal and must be respected, the uses of science are political and must be controlled.

Enhanced consciousness of Chinese scientific history is not entirely an internal matter. The importance and fascination of the Chinese scientific tradition have long been known in Europe, Japan, and the United States. Scholars in many countries have contributed to understanding it, as well as to making the work of many great Chinese historians of science accessible in other languages. Educated people all over the world are now prepared to respond to new revelations about China's scientific tradition—whether it be innovative applications of the ancient art of acupuncture or the unique archaeological finds that have been appearing without interruption since the 1950's. This heightened interest has meant a small but by no means imperceptible rise in the world's esteem. More to the immediate point, it has meant that scientists all over the world are increasingly aware of the special status of science in China and are less inclined than they might be to overlook indications that China is rising quickly in the international scientific community.

*Richard P. Suttmeier*

# *SCIENCE POLICY AND ORGANIZATION*

## *CHINA'S SCIENCE POLICY, 1949–76*

In 1961, John Lindbeck began his introductory essay to the last comprehensive review of Chinese science with these words: "When China's Communists came to power in 1949, they began to outline ambitious plans to thrust China into the company of modern, industrial and scientifically advanced countries."[1] Now, nearly two decades later, a major new direction in Chinese policy can be described in exactly Lindbeck's terms. In 1978 as in 1949, China's scientific development experienced a new beginning. But whereas in 1949 Chinese leaders, according to Lindbeck, were unprepared for the tasks that fell to them, the new directions of 1978 were based on almost 30 years of experience with a variety of successful and unsuccessful experiments in scientific development.

The locus of pre-1949 science in China was the Academia Sinica. In 1950 the Academia Sinica, centered in Nanjing, and the National Academy of Sciences of Beijing were merged to form a new national organization, the Chinese Academy of Sciences (CAS). Starting with approximately 20 institutes in 1950, the Academy grew into an enormous research organization with some 120 institutes in 1966. During the 1950's, it underwent a series of internal reorganizations to bring it more closely in line with the Soviet Union's All-Union Academy of Sciences, on which it was modeled. These changes included the creation of a central research secretariat, a system of research

[1] John M. H. Lindbeck, "Organization and Development of Science," in Sidney H. Gould, ed., *Sciences in Communist China* (Washington, D.C.: American Association for the Advancement of Science, 1961), p. 3.

secretariats within the institutes, and the creation of academic departments for planning and overseeing research in five areas: (1) mathematics, physics, and chemistry; (2) life sciences; (3) earth sciences; (4) technical sciences; and (5) social sciences. Approximately 250 leading scientists were appointed to the quasi-honorific position of "department member," a position roughly comparable to that of academician in the Soviet system.[2]

By the mid-1950's with the CAS well established, attention began to turn to research activities in organizations directed by the central ministries, notably the industrial ministries, the Ministry of Agriculture, the Ministry of Public Health, and the Ministry of Defense. Research in institutions of higher learning was also encouraged, especially in the late 1950's. In 1955–56 a comprehensive 12-year research plan was formulated with Soviet help. It identified 12 research priorities and hundreds of detailed projects.[3] Efforts were made to coordinate Soviet assistance with the plan, and despite the Sino-Soviet break in 1960 much of it was fulfilled by the early 1960's.

In 1956 the Chinese government established two central administrative bodies: the State Technological Commission and the Science Planning Committee. In 1958, during the Great Leap Forward, these two bodies were merged into the Science and Technology Commission (STC), which thereafter served as the nation's supreme agency for science policy and administration. In accordance with the principles of the Great Leap Forward, the Commission was staffed by nonscientist administrators and led by an important Party and military figure, Nie Rongzhen. Control over professional societies was also centralized in 1958, their work being linked to programs for popularizing science; and a Chinese University for Science and Technology was established in Beijing, under the joint auspices of the CAS and the Ministry of Education, to provide advanced training for scientists. Spending on science and technology increased during the Great Leap Forward, as did the number of centrally controlled institutes.

During this period, a number of scientific institutions were established at the provincial and subprovincial levels, among them "branch academies" of the CAS and provincial level branches of the Science and Technology Association, which was formed in 1958 by the merging of the All-China Federation of Scientific Societies and the All-China Association for the Advancement of Scientific and Technical Knowledge. These organizational changes were intended to stimulate scientific and technological activities at the grassroots level, to create a closer relationship between research and the needs of

[2] Chu-yuan Cheng, *Scientific and Engineering Manpower in Communist China, 1949–1963* (Washington, D.C.: National Science Foundation, 1965), pp. 19–20. For a discussion of the organization and purposes of the departments, see Richard P. Suttmeier, *Research and Revolution* (Lexington, Mass.: Lexington Books, 1974).

[3] The 12 priority areas were: atomic energy, radio electronics, jet propulsion, automation and remote control, petroleum and scarce mineral exploitation, metallurgy, fuel technology, power equipment and heavy machinery, problems relating to harnessing the Yellow and Chang Jiang (Yangtze) rivers, chemical fertilizers and the mechanization of agriculture, public health, and basic sciences.

production, and to bring more peasants and workers into science-related activities. Government spokesmen at the time attacked professional values in science and praised the scientific and technological wisdom of the masses.

From the beginning of the new China, professional scientists were subject to political pressures designed to remold them ideologically and to direct their scientific efforts to state needs. Scientists were made to attend political meetings regularly, and to engage in criticism and self-criticism. After 1957 scientists were increasingly criticized for pursuing narrow, specialized research topics, for being excessively preoccupied with publication, and for taking too many of their cues from international science, rather than devoting themselves to the practical problems of Chinese development. During the Great Leap Forward, they were told to "learn from the masses" and forced to reorient their research activities toward the achievement of politically inspired goals.

In the early 1960's, after the failure of the Great Leap Forward and the withdrawal of Soviet assistance, many of the 1958 policies were abandoned. The "professional-mass" contradiction was played down, and many of the low-level science- and technology-related organizations were either consolidated or abandoned. Owing to the break with the Soviet Union, research priorities shifted to the life sciences and defense. Especially high priority was given to the nuclear weapons program, begun in the 1950's, and China's first nuclear test occurred in 1964. It seems that work on missile development also began in earnest during this period. As a further consequence of the Soviet withdrawal, China began to become interested in non-Communist foreign technology.

In general, professional scientists enjoyed a measure of prestige and respect between 1961 and 1966. The research environment was far less politicized than in 1958–60, and the amount of time devoted to attending political meetings was limited by a rule requiring that five-sixths of the scientist's working week be devoted to research.

These favorable conditions for professional science disappeared with the start of the Great Proletarian Cultural Revolution in 1966. The original 16-point charter of the Cultural Revolution was intended to spare the scientific community's work from disruption. But as the Cultural Revolution progressed, scientific activities in most parts of the country were severely disrupted. Most of the nonscientist administrators who had carried out the relatively relaxed policies of the early 1960's were removed from office. The educational system was completely disrupted for approximately three years.

The Cultural Revolution caused a number of profound changes in Chinese science. First, the professional societies were in effect disbanded: their journals ceased to be published, and their professional meetings were discontinued. Second, there was significant decentralization. The STC was abolished, and many research institutes formerly under central control were either closed down or placed under the control of provincial authorities.

Third, the administration of research changed dramatically. Prior to the

Cultural Revolution, research institutes were usually governed by a director (typically a scientist), several deputy directors (including both scientists and nonscientists), and separate committees for academic and administrative affairs. The Cultural Revolution reforms produced a politically oriented administrative structure known as the Revolutionary Committee, which included scientists, Party and administrative officials, younger "revolutionary" scientists, technicians, and workers. Fourth, the conventional boundaries and distinctions between research, education, and production were ignored. "Open-door" research was instituted at research institutes; research institutes and universities were assigned production tasks, and factories were assigned educational and research activities. Perhaps the most radical innovation was opening up higher education on a large scale to workers and peasants, i.e. to people chosen mainly on the basis of sociopolitical criteria.

After the end of the Cultural Revolution in 1969, a number of leaders who had been removed from office began to reappear in public life. Many of them felt that pre–Cultural Revolution policies should be reintroduced; their champions were Premier Zhou Enlai and, after his reappearance in 1973, Vice-Premier Deng Xiaoping. They were opposed by the political elite who had risen to power as a result of the Cultural Revolution, led by what we now know as the Gang of Four. Personnel and policy differences increasingly became entangled in the politics of succession. Zhou Enlai had cancer and Mao Zedong was evidently senile—a change in leadership was in the air.

Science and technology policy became part of the struggle. In 1972 Mao and Zhou became convinced of the importance of redirecting science and reestablishing intellectual standards in the educational system, and Mao reportedly instructed Zhou to pursue the matter. At Zhou's request, Professor Zhou Peiyuan, vice-president of Beijing University and vice-chairman of the Science and Technology Association, produced proposals for strengthening basic research and university scientific training. But these proposals were not implemented, allegedly because of opposition from the Gang of Four.[4]

The debate about science and modernization intensified in 1975 following the Fourth National People's Congress, China's nominal legislature, in January of that year. At that time Zhou Enlai enunciated the "four modernizations," in which he called for the comprehensive modernization of agriculture, industry, defense, and science and technology by the year 2000. Deng Xiaoping, reinstated as vice-premier, was entrusted with implementing the new program and approached his job by initiating a series of studies and reports on the state of the Chinese economy, Chinese industry, and Chinese science and technology. These reports began to appear in the summer of 1975. That fall the press, which was largely under the control of the Gang

[4] Ministry of Education, Mass Criticism Group, "A Political Struggle Around the Question of Basic Theories of Natural Science," *Guangming ribao,* in *Survey of the People's Republic of China Press (SPRCP),* 77–4; and Zhou Peiyuan, "What Is the Intention of the Gang of Four in Obstructing Research on Basic Theories?," *Renmin ribao,* Jan. 13, 1977, in *SPRCP,* 77–4.

of Four, began to attack those who would "reverse the verdicts of the Cultural Revolution," and with Zhou Enlai's death in January 1976 the succession issue intensified. By April Deng had again been removed from office, and it appeared that the Gang of Four was consolidating its position; but soon after Mao Zedong's death in September 1976 its members were removed from office and their followers purged from positions of authority.

By 1977 the "four modernizations" were once again at the center of government policy, and the modernization of science and technology was increasingly being seen as the key to the other three. The mobilization of political support for the new directions in science and technology began in earnest in May 1977 with Party Chairman Hua Guofeng's call to step up the development of science and technology.[5] Ten months of active planning and rethinking culminated in the announcement of important policy and organizational changes at the National Science Conference held in March 1978.[6]

## *THE DEVELOPMENT OF SCIENCE POLICY SINCE 1977*

In 1977 China's scientific leaders formulated a set of plans for the comprehensive development of science and technology in such a way as to support the modernization of agriculture, industry, and national defense. The strategy for scientific development was to lay a foundation for a highly sophisticated research system by 1985, so that China would be able to reach world levels in science and technology by 2000. There was an eight-year plan for the first stage and a 23-year plan for the second.

A broad outline of China's economic plans for industry and agriculture is contained in Party Chairman Hua Guofeng's report to the Fifth National People's Congress in February 1978.[7] Hua spoke of the need to increase grain production by two or three times within the next eight years in the main grain-producing areas, and on all state farms. Areas with low yields and grain deficiencies were to aim at achieving self-sufficiency in two to three years; to this end additional effort would be put into land reclamation, soil improvement, and water control. In addition to local projects, there were to be major state-supported projects on the Yellow, Chang Jiang (Yangtze), Huai, Hai, Liao, and Bo rivers. Grain production increases would also require progress in seed improvement, farming methods, fertilizers, mechanization, and insecticides. The quality of fertilizers and insecticides was to be improved and their cost reduced. Hua talked about mechanizing at least 85 percent of "all major processes of farm work" within ten years. It was hoped that by these measures the value of agricultural output would increase by from 4 to 5 percent annually between 1978 and 1985.

[5]*Peking Review,* 1977, 40 (Sept. 30): 6.

[6]For a useful summary of these events, see Genevieve Dean and Thomas Fingar, eds., *Developments in PRC Science and Technology Policy, October–December, 1977,* S & T Summary no. 5 (U.S.–China Relations Program, Stanford University).

[7]Hua Guofeng, "Unite and Strive to Build a Modern, Powerful Socialist Country," *Peking Review,* 1978, 10 (Mar. 10): 18–26.

China's plans for the modernization of industry were even more ambitious. From 1978 to 1985, the value of industrial output was to increase by 10 percent each year. Light industry was to make available abundant, attractive, and reasonably priced consumer goods. In heavy industry, Hua mentioned the importance of adopting new techniques in the metallurgical, fuel, power, and machine-building industries and in the newer industries, such as petrochemicals and electronics. Furthermore, China's sadly neglected transportation and communications systems were to receive priority attention.

Hua spoke of building up six major economic regions in the southwest, northwest, central south, east, north, and northeast. All the plans for industrial development focused on basic industries. Some 120 large-scale projects were set out, including the building or completion of 10 new iron and steel complexes, 9 nonferrous metal complexes, 8 big coal mines, 10 oil and gas fields, 30 power stations, 6 new trunk railways, and 5 key harbors. The successful completion of this program, it was claimed, would give China "14 fairly strong and fairly rationally located industrial bases." Hua identified the fuel industry, the electric power industry, various raw and semifinished materials industries, and transport and communications as areas where contributions from science and technology were expected to make a particularly significant difference.

A month later, plans for scientific development were announced at the National Science Conference of March 1978. They called for comprehensive preparations for research in 27 different "spheres," with 108 specific research topics being given top priority. Eight "comprehensive scientific and technical spheres, important new technologies, and pace-setting disciplines with a bearing on the overall situation" were singled out for special attention: agriculture, energy, materials science, computer science, lasers, space science and technology, high energy physics, and genetic engineering. Three of these areas are fundamental to the development of the Chinese economy (agriculture, materials, and energy); three others have wide-ranging implications for other fields of science and technology (computers, lasers, and space). The remaining two signify China's new commitment to basic research (high energy physics and genetic engineering).

Details about these eight priority areas were offered by Fang Yi, the director of the reestablished STC.[8] Fang's discussion of priorities in agriculture followed Hua Guofeng's remarks noted earlier. In the sphere of energy, he stressed the discovery and extraction of oil and gas, the need for highly mechanized coal mining, the importance of research into the gasification and liquefaction of coal, and noted the urgent need to improve China's long-neglected electric power industry. Research on new sources of energy, including atomic power, solar energy, geothermal energy, wind, tidal en-

[8] "Abridgement of Fang Yi Report to National Science Conference," New China News Agency (NCNA) dispatch, Mar. 28, 1978, in *Foreign Broadcast Information Service (FBIS)*, Mar. 29, 1978.

ergy, and nuclear fusion, was also mentioned. Interestingly, Fang noted the importance of exploring techniques for energy conservation.

Fang stressed the importance of steel and the need for metallurgical research, particularly on raw materials for the steel industry and on copper, aluminum, nickel, cobalt, rare earths, titanium, and vanadium. Fang discussed the importance of both synthetic materials and materials research for the national defense industry. He repeatedly called for the comprehensive utilization of materials and emphasized the importance of transforming waste materials into useful products.

Agreeing that the level of a country's computer development is a good indicator of its overall technological capacity, Fang said that China would pursue research in a whole range of computer-related areas, including the development of giant computers, microcomputers, computer networks, and intelligence simulation. He expected China to be doing comparatively advanced research in computer science by 1985 and to have "a fair-sized modern computer industry." He added that key industries would be using computers for process control and management functions.

Fang noted the application of lasers to materials processing, precision measurement, remote ranging, holography, telecommunication, medicine, and seed breeding, and their potential application in isotope separation, catalysis, information processing, and controlled nuclear fusion. He said that in the following three years attention would be given to the study of laser physics, laser spectroscopy, and nonlinear optics in order to solve basic problems in optical communications, and he stressed the importance of popularizing laser technology both for the economy generally and for national defense.

Fang's remarks on space technology suggested that China was planning to initiate a comprehensive national effort involving research in space science (including communications and meteorology), the building of modern space centers, and the development of launch vehicles and skylabs.

The inclusion of high energy physics in China's list of priorities is both a curiosity to the foreign observer and, apparently, a bone of contention in China. Mao Zedong reportedly had a particular interest in this field because he felt that its contributions to understanding the structure of matter had significant philosophical or ideological implications. An institute of high energy physics was established in the CAS in 1972, but its work seems to have been stymied by the Gang of Four until after Mao's death. In 1978 Vice-Premier Deng Xiaoping was reported to consider a command of high energy physics essential to a command of various other scientific disciplines. Fang Yi put the case as follows:

> We should step up research in the theory of high energy physics and cosmic rays, consciously promote the interpenetration of high energy physics and the neighboring disciplines, actively carry out research in the application of accelerator technology to industry, agriculture, medicine, and other spheres, and pay attention to the exploration of subjects which promise important prospects of application.[9]

[9] *Ibid.*

China's current plans for high energy physics focus on the construction of a 50-GeV proton synchroton. This work was well under way in Beijing by mid-1978 and was slated for completion in 1982. Reportedly a machine of higher energy is to be completed by 1987.[10]

Finally, Fang Yi said that whereas China was comparatively unprepared to take advantage of the potential applications of genetic engineering, a particularly strong effort would be made to develop this field. He noted the importance of genetic engineering techniques for basic research in cell differentiation and in understanding the growth and development of tumors. He also referred to the potential applicability of genetic engineering in agriculture, medicine, and industry, including pharmaceuticals. Because of China's weakness in genetic engineering, the first three years of the plan were to be devoted to strengthening the organization and coordination of work and the construction of new laboratories. Gradually work in genetic engineering would be integrated with work in molecular biology, molecular genetics, and cell biology.

In addition to the eight priority areas, several other broad areas of science and technology were mentioned as having high priority, one being medicine and public health, another oceanography. It appears that China's priorities in the area of science and technology not only reflect the need to modernize agriculture, industry, and national defense, but also represent various constituencies within the Chinese scientific community. Although physicists and cell biologists are singled out for special attention, the eight priority areas are sufficiently "balanced" to provide opportunities for research support in most fields of science.

### *The Treatment of Scientists*

From the beginning of the Cultural Revolution in 1966 to the end of 1976, scientists—particularly those who were trained abroad or in the preliberation period—were regarded as especially susceptible to the influence of "bourgeois ideology" and were accordingly subjected to various types of harassment. In 1977 this situation changed dramatically. Chinese scientists began to be portrayed in the media as loyal, hardworking servants of the nation, and most were said to have made great progress in acquiring a proletarian outlook. Following this new media treatment came various policy changes affecting the treatment of technical personnel and the quality of professional life. From Fang Yi's report to the Chinese People's Political Consultation Conference in December 1977 we can identify six of these changes.[11]

First, the outstanding achievements of scientists and technologists were to be publicly recognized by returning to the system used in the 1950's of giv-

[10] *Peking Review*, 1978, 23 (June 9): 3–4.

[11] The report, "On the Situation in China's Science and Education," appeared in *Peking Review*, 1978, 2 (Jan. 13): 15–19.

ing cash prizes and bonuses for outstanding work. One purpose of the National Science Conference held in March 1978 was to serve as a national forum for such recognition. Professional ranks and titles were restored (they had been abolished at the beginning of the Cultural Revolution), and long-delayed promotions were made. In addition, efforts were made to rectify mistakes in the assignment of scientists and technologists to production work and other work not suited to their qualifications.

Second, the professional societies were reestablished and encouraged to hold academic meetings in the spirit of "letting 100 flowers bloom and 100 schools contend."

Third, long-term planning for science, technology, and education was reintroduced, a move that restored stability and predictability to the research environment.

Fourth, the leadership structure of research institutes and universities was to be reorganized. The Revolutionary Committee form of organization was to be replaced by the pre–Cultural Revolution system, in which each institute is run by one director and several deputy directors under the general leadership of the Party Committee. The "academic committee," a pre–Cultural Revolution body composed of 20 to 30 leading scientists that served as an advisory group to the institute director was also reestablished. Deng Xiaoping, speaking to the delegates of the National Science Conference, identified the Party's role in a research institute as twofold: first, to ensure that the institute follows the Party's general policies; second, to ensure that the institute's scientific research needs are met.[12] In short, the performance of Party committees in institutes was to be evaluated according to the institute's research performance rather than its ideological "health."

Fifth, the STC was reestablished and instructed to "take charge of overall planning, coordination, organization, and administration of the country's scientific and technological work." The Commission's director, Fang Yi, a member of the Party's Politburo, was the highest-ranking political official to be given responsibility over science and technology since the inception of the People's Republic.

Sixth, the fundamental criterion for admitting students to universities and colleges was to be academic achievement, especially in the sciences.

As a result of these six changes, scientists in China seem likely to enjoy a status more nearly like that of their counterparts in other countries. Moreover, in the higher interests of the nation, the regime seems prepared to face the intellectual elitism that such policies will inevitably produce.

### *Organizational Structures*

The five main divisions for scientific research and training in China are described in detail below. They are the CAS, higher education, the produc-

[12] Deng Xiaoping, speech at opening ceremony of the National Science Conference, in *Peking Review*, 1978, 12 (Mar. 24): 9–18.

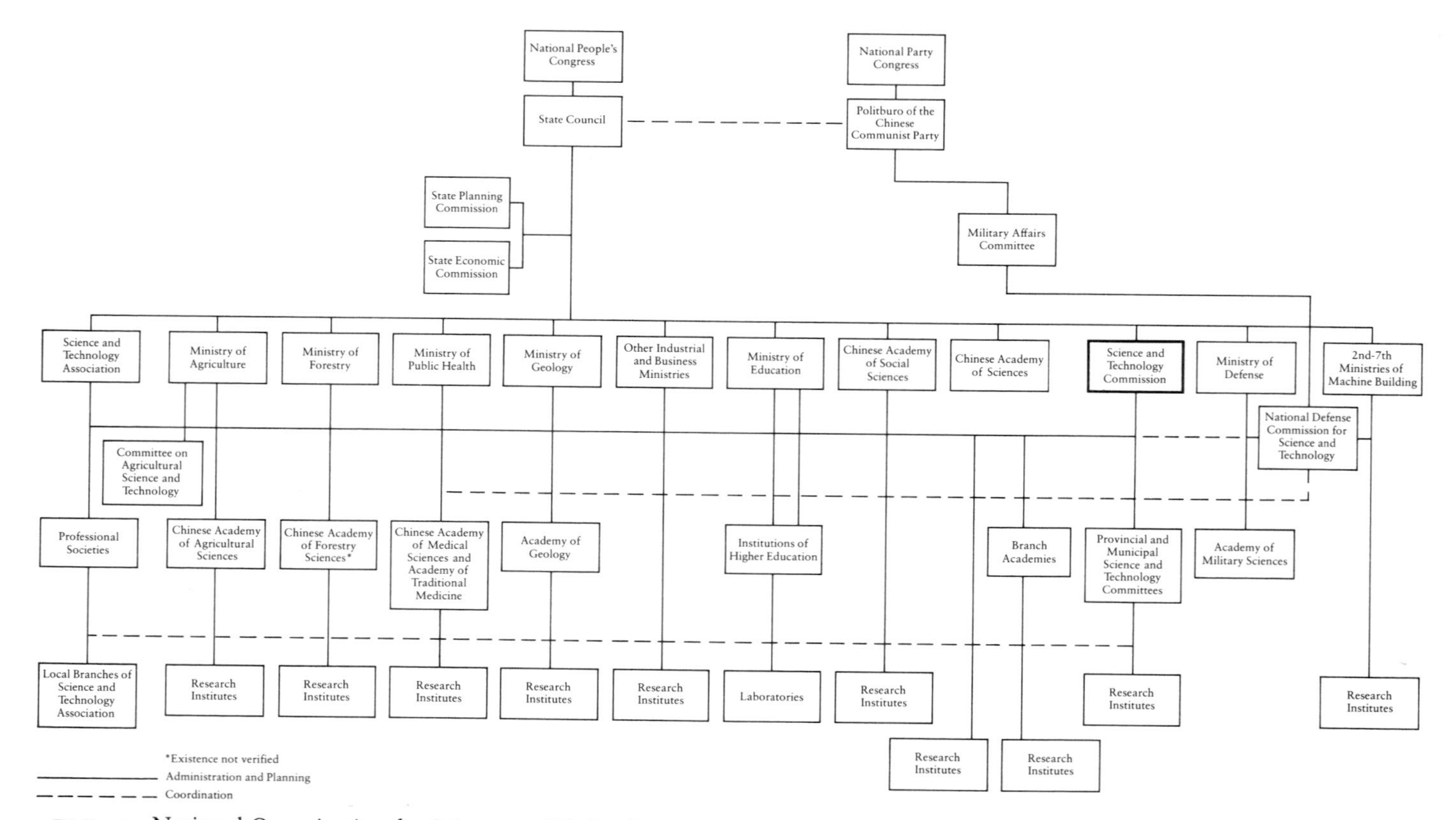


FIG. 1. National Organization for Science and Technology

NOTE: Chinese leaders admit that their rapidly developing scientific establishment continues to be in a state of flux and difficult to chart. Figure 1 suggests one version of the administrative structure of research in China.

SOURCES: *Science and Technology in the People's Republic of China* (Paris: Organization for Economic Cooperation and Development, 1977), pp. 72–73; Harvey Nelson, *The Chinese Military System* (Boulder, Colo.: Westview Press, 1977), pp. 60–62.

tion ministries, national defense, and the "mass scientific network." Policymaking and administration are centered in the STC, the Academy, and the Science and Technology Association. Figure 1 illustrates the overall organization.

*The Science and Technology Commission (STC).* As we have seen, the State Technological Commission and the Science Planning Committee, both established in 1956, were merged in 1958 to form the STC, which became, prior to the Cultural Revolution, the supreme policymaking body for science and technology. During the Cultural Revolution the STC, like other such institutions, seems to have been dissolved. By the end of the Cultural Revolution, therefore, there was no central coordinating body to bring a measure of national coherence to research and development. The first post–Cultural Revolution attempt to provide such direction was the establishment of the "Science and Education Group" under the State Council in 1971. This body was headed by Liu Xiyao, who had pre–Cultural Revolution science administration experience with the STC and who later became minister of education. The Science and Education Group, however, did not seem to possess the broad-ranging authority enjoyed by the pre–Cultural Revolution STC.

The various measures of the new science policy of 1977–78, notably administrative recentralization, the reintroduction of long-term planning, and the overall strengthening of a scientific infrastructure, necessitated the reestablishment of the STC. Fang Yi, who had previously been appointed a vice-president of the Academy, was appointed director, with the rank of minister of state. Among his assistants were such other pre–Cultural Revolution science administrators as Yu Guangyuan, formerly chief of the Science Office of the Party's Propaganda Bureau.

The STC had approximately 200 personnel in 1978, reportedly organized into ten bureaus. It is thought to contain a planning bureau, a basic research bureau, and mission-oriented applied research bureaus in such areas as energy, petrochemicals, materials science, and machinery. As in 1958–66, the STC has major responsibility for coordinating China's program in scientific information, scientific instruments, and technical standards, for promoting innovations, and for implementing the broad policy goals announced at the National Science Conference. Much of its work is carried out by science and technology committees at the provincial and subprovincial levels. In many parts of the country, the director of the provincial STC is a Party secretary. Thus, although the STC network is ostensibly part of the state, it is closely associated with the Party as well.

*The Chinese Academy of Sciences (CAS).* As we have seen, the Academy has long held a central role in Chinese science. In 1978 it consisted of central administrative offices in Beijing and over 100 research institutes in various parts of the country. (This figure includes those jointly run by the Academy and a unit of local government.) The Academy's institutes typically have

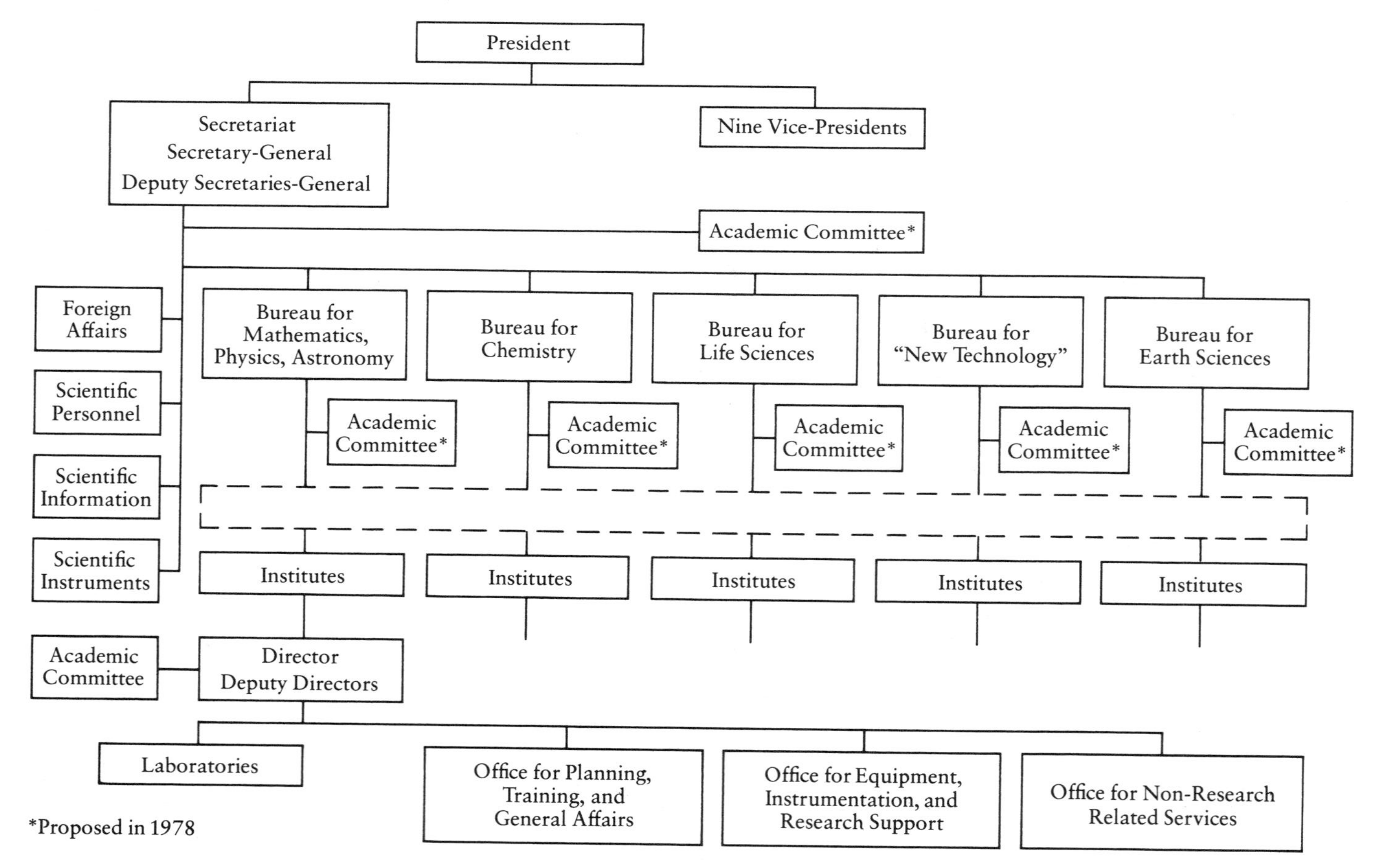


FIG. 2. Organization of the Chinese Academy of Sciences, 1978

anywhere from 200 people to well over 1,000 on the payroll. As many as one-third of these people, however, are support personnel who have nothing to do directly with research.

The organization of the Academy is shown in Figure 2. At both the central and institute levels, administrators are a mix of institute directors and deputy directors, Party officials, administrative and planning officials, and representatives of working scientists. The Academy in Beijing is headed by a president—Guo Moruo (Kuo Mo-jo) until his death in June 1978—and nine vice-presidents, including prominent scientists and nonscientist administrators. Since the 1950's, when the organization of the Academy was influenced by the Soviet system, the Academy has had a powerful administrative secretariat. This office is led by a secretary general, usually not a scientist, and several deputies. As Figure 2 shows, the secretariat's research administration was performed by five discipline-oriented bureaus plus bureaus for such support work as information services and scientific instruments. Academy activities in the social sciences were transferred in 1977 to a new Chinese Academy of Social Sciences.

Since the beginning of 1978, administrative "branches" of the Academy have been reestablished in Shanghai and many of China's provinces. Branch academies were first established during the Great Leap Forward, but declined in importance during the early 1960's. With the increasing number of institutes under Academy jurisdiction since 1976, a trend that can be thought of as a type of recentralization, there appears to be a need to simultaneously decentralize administration *within* the Academy. The branch academies are administrative mechanisms intended to facilitate decision making at the institute, city, and province levels and thus avoid the need to refer all decisions to Beijing.

The administrative staff of an institute is typically divided into three main offices: general affairs, research planning, and personnel training; equipment, instrumentation, and other forms of research support; and non-research support functions. The general affairs and research planning office appears to have the main responsibility for developing research programs at the institute level. Its proposals reportedly are reviewed by the "academic committee" discussed above, which is also responsible for evaluating research work. When the academic committee and the general affairs and research planning office disagree, the matter is referred to the institute director and the Party committee.

Under the new policies of 1977–78, the Academy was expected to be the leading organ in support of basic research, although it retained important roles in applied research, in scientific information activities, and in the development of scientific instruments. Its scientific information services were centered in the Institute of Scientific and Technological Information.[13] Beginning in 1978, major new efforts were made all over China to provide

[13] Cheng, *Scientific and Engineering Manpower*, p. 344.

scientists with improved scientific information services. The Institute of Chemistry in Beijing, for example, was made responsible for improving the country's chemical information services; and steps were promptly taken to coordinate its activities with those of the Academy library, the Institute of Scientific and Technological Information, and the National Library in Beijing. All indications are that Chinese policymakers are interested in creating a sophisticated machine-based national information system.[14] The Academy's contributions will be crucial to the development of such a system.

*Higher education.* Three kinds of institutions offer higher education in science and technology: "comprehensive" universities, offering a broad range of programs in the arts and sciences; technical universities, of which perhaps the best known is Qinghua (Tsinghua); and Soviet-style specialized institutes for training in selected areas of technology, such as chemical engineering. In keeping with the 1977–78 stress on the value of expertise and the need for high-quality training, the government designated 88 universities and colleges (and a number of primary and secondary schools) as "key point" centers of excellence. The Ministry of Education shares the responsibility for administering some of these schools with local government bodies and coadministers the specialized institutes with the relevant production ministries.[15]

Until 1957 research in universities was generally not stressed, since most attention was being given to the development of a strong CAS. During the Great Leap Forward interest grew in university research, chiefly applied research, and this emphasis was emphatically reinforced following the Cultural Revolution. But budgets for university research remained meager, with the result that universities supplemented them where possible by doing contract research with industry, a tendency that reinforced the emphasis on applied research. Moreover, as a result of the disruptions in higher education in 1966–76, many universities failed to sustain even an applied research program. Beginning in 1977, however, university research was encouraged. Graduate programs were introduced, and the emphasis was switched to basic research, with priorities probably to be determined jointly by the university and the Academy.

National responsibility for the administration of university research seems to be vested in an office for science and technology within the Ministry of Education, but at least three other bureaucracies may also be involved:

1. Provincial governments. As a result of Cultural Revolution reforms, provincial governments seem to have some measure of control over most universities. Most provinces have an education bureau, which in turn has an office for science and technology. In addition, provincial science and tech-

[14] Fang Yi called for the establishment of such a system in his report to the National Science Conference.

[15] Dean and Fingar, *Developments in PRC Science and Technology Policy*, p. 28.

nology committees affiliated with the STC also have an interest in research in institutions of higher education within their provinces.

2. Production ministries. The specialized technical institutes for higher education maintain close ties with the relevant production ministries as well as with the Ministry of Education. Since such ministries are the most likely source of employment for graduates of technical institutes, they can be expected to exert some influence on the education and research that goes on there.

3. The CAS. There are four institutions under the joint control of the Academy and the Ministry of Education: the Chinese University of Science and Technology (now at Hefei in Anhui province), Zhejiang University (Hangzhou), the new University of Science and Technology (Haerbin, Heilongjiang province), and the Shanghai University of Science and Technology. In addition, there was formerly a fair amount of interchange between CAS institutes and universities in large cities where both exist; thus staff of the CAS institutes in Changchun sometimes lectured at Jilin University, and professors from the University sometimes worked at CAS laboratories. Although this kind of interaction was disrupted during 1966–76, it has presumably been resumed.

Funds for research in institutions of higher education have come from a variety of sources, notably the research budgets of the Ministry of Education, discretionary funds in the university budgets, and funds from provincial-level educational bureaus and from production units near the university (these last administered by the local science and technology committee). If the universities are to play a larger role in basic research, it seems likely that a greater share of research funding will come from the Ministry of Education.

*The ministerial sector.* Another sector of Chinese scientific research and development consists of the production ministries and the Ministry of Public Health. In a recent study of research and development organizations in China, Susan Swannack Nunn identified some 2,165 research institutes, most falling within the ministerial sector.[16] They seemingly report to their parent central ministry through its office of science and technology, but are also typically linked to design institutes, factories, and other enterprises under the ministry's jurisdiction.

Such institutes may also maintain working relationships with institutions of higher education (for training programs as well as research) and with institutes of the Academy system. For instance, when basic research and exploratory applied research on new catalysts performed at the Academy's Institute of Chemical Physics in Dalian reached the stage of potential applicability, the results were conveyed to the Beijing Institute of Petro-

[16] Susan Swannack Nunn, *Directory of Scientific Research Institutes in the People's Republic of China* (Washington, D.C.: National Council for U.S.–China Trade, 1977–78), 3 vols.

chemical Research, an institute of the Ministry of the Petrochemical Industry, for further applied research and development. The Institute of Petrochemical Research then worked with various of the Ministry's design institutes, refineries, and petrochemical complexes in developing commercially usable catalysts.

There are also differentiated research academies. For example, the Academies of Agricultural and Forestry Sciences under the ministries of those names have a number of institutes under them.[17] The Ministry of Public Health has both the Chinese Academy of Medical Sciences and the Chinese Academy of Traditional Chinese Medicine under it, which between them control approximately 20–30 important institutes. Just as the institutes under the Academy of Medical Sciences maintain active contacts with hospitals, so the institutes under the Academy of Agricultural Sciences are linked to the national system of agricultural research.

Like the CAS and higher education generally, the ministerial sector has experienced political and administrative decentralizations as a result of the Cultural Revolution. Many institutes previously under central ministerial control were placed under provincial administrative jurisdiction, and some under joint central and local jurisdiction.

*National defense.* Although one could include national defense in the ministerial sector, its importance warrants separate treatment, and indeed the Chinese themselves regard it as a separate sector. Not surprisingly, we know very little about research and development in this area. It is believed that the chief suppliers of defense equipment are the Second to Seventh Ministries of Machine Building; if these ministries follow the usual patterns, they probably maintain institutes for research, development, and design. In addition, the work of some CAS institutes is thought to be heavily defense-oriented. Prior to the Cultural Revolution, a separate Science and Technology Commission for National Defense, which was directly under the Party's Central Military Commission, coordinated planning and research in the ministerial and Academy sectors and operated some 14 relatively large research institutes. This Commission ceased to operate during the Cultural Revolution, but was reestablished in 1977 or 1978, presumably with the same function as before.[18]

*"Mass science."* Grass-roots research conducted by workers and peasants is recognized by the Chinese as a separate and important sector of scientific and technological activities.

The idea of "mass science" predates the Great Leap Forward, but first became prominent during that period as part of China's development strategy of "walking on two legs" (professional and mass). Mass science pro-

[17] *Ibid.*, vol. 1, p. 15. After being combined in the early 1970's, Agriculture and Forestry are again separate ministries.

[18] See Harvey W. Nelsen, *The Chinese Military System* (Boulder, Colo.: Westview Press, 1977), pp. 60–62, 109–10.

grams encouraged workers and peasants to participate in scientific activities and promoted the establishment of local research institutes, experiment stations, and centers for popularizing science. Such activities were de-emphasized in the early 1960's, only to be re-emphasized during the Cultural Revolution. As of 1978, what might be called the "mass scientific network" was most prominent in the Chinese countryside, where from the province down, and particularly within the commune and subcommune units, research and experimentation activities were institutionalized as shown in Figure 3. In 1978 there were an estimated 14 million people involved in mass science activities in the countryside.[19] Mass science programs exist in urban areas as well.

During periods of radical science policy, peasant and worker scientists were urged to go to research institutes, and established researchers were sent to factories and farms to "learn from the masses." During periods when radicalism was checked, the mass scientific network was more an instrument for speeding the implementation of research-based innovations and of popularizing innovations. The mass scientific network includes institutions for training workers and peasants, such as the May 7th cadre schools and July 21 universities.[20] It also includes centers for scientific exchange between workers, peasants, and trained technical personnel.

Mass science activities are oriented toward practical production problems. They are designed to exploit what is seen as a vast fund of knowledge and a vast potential for creative innovation among workers and peasants, not to mention the availability of innumerable observers for gathering data. The fields where the mass scientific network has been used include meteorology, seismology, geology, medicine, and, of course, agriculture. In industry, mass science activities include worker-initiated technical innovations and mechanisms for sharing knowledge of new technical innovations among factories. From time to time "worker scientists" and professional scientists and engineers are brought together to explore the applicability of professional knowledge to production problems identified by workers.

The mass scientific network has also functioned as a mechanism for recruiting exceptionally talented workers and peasants into professional roles. Thus Zai Zezhuan, a worker in an electric-lamp factory, displayed such extraordinary innovative abilities in the early 1960's that he was appointed to the faculty of Shanghai's Fudan University, where he was still teaching in 1978.

The policy statements of 1977–78 clearly acknowledge the value of the mass scientific network and the importance of popularizing science. Al-

[19] Jon Sigurdson, "Technology and Science in the People's Republic of China: An Introduction," draft manuscript (Research Policy Program, University of Lund, Sweden), p. 149.

[20] These institutions take their names from directives issued by Mao Zedong on May 7, 1966, and July 21, 1968. Reportedly, in the summer of 1976, there were some 15,000 July 21 and May 7th universities, enrolling 780,000 students (Byung-joon Ahn, "China's Higher Education and Science in Flux," *Contemporary China,* Mar. 1977, p. 22).

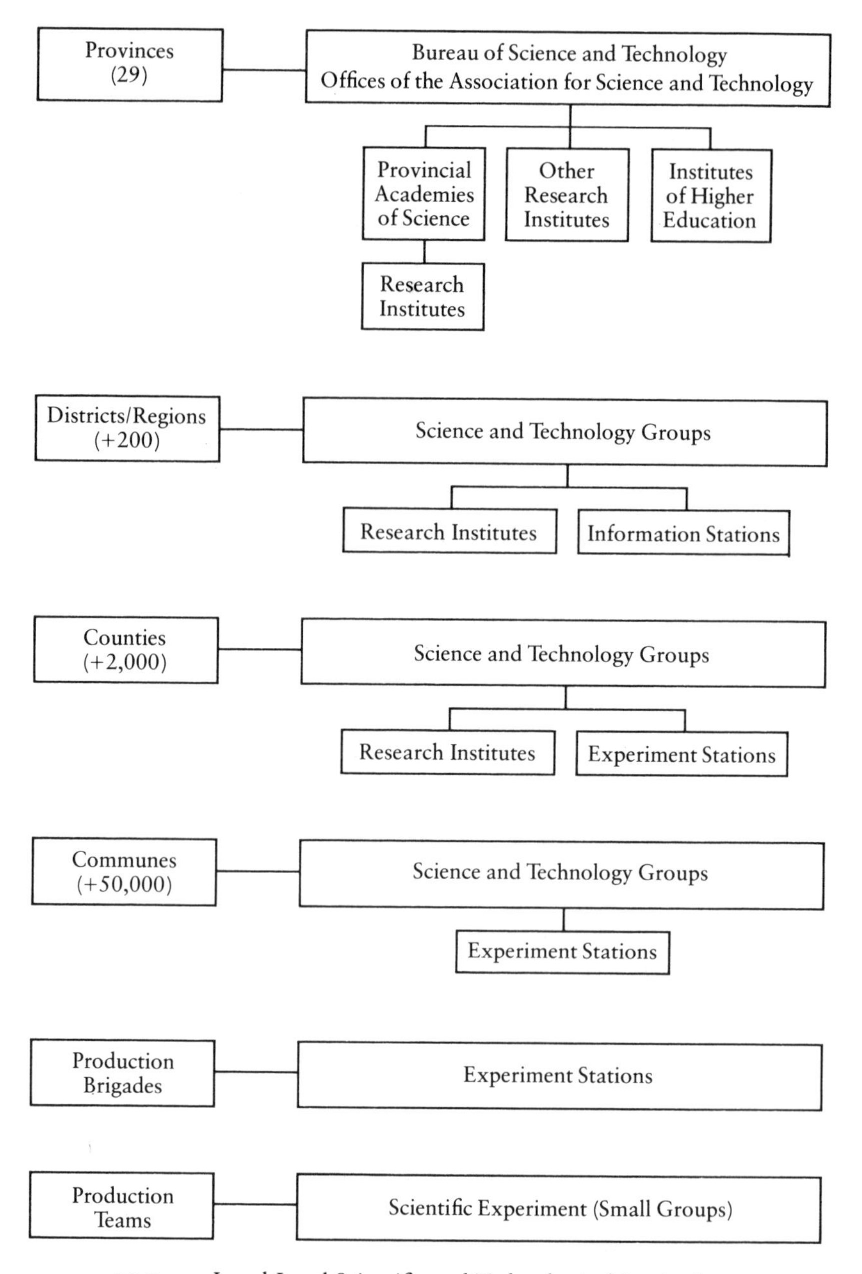


FIG. 3. Local-Level Scientific and Technological Institutions
SOURCE: *Science and Technology in the People's Republic of China* (Paris: Organization for Economic Cooperation and Development, 1977), p. 60.

though the mobilization of untutored workers and peasants into science- and technology-related activities has its costs, it does provide China with a readily available means for the rapid popularization of new scientific knowledge and the diffusion of technological innovations. It may be China's way of successfully overcoming the serious problems that have affected other developing countries in this regard.

*The Science and Technology Association (STA).* As we have seen, the Science and Technology Association was formed in 1958 as a result of the merger of the All-China Federation of Scientific Societies and the All-China Association for the Advancement of Scientific and Technical Knowledge. From its beginning, therefore, it had a dual orientation toward professional societies on one hand and popularization and mass science on the other. Like most organizations, the STA was disrupted during the Cultural Revolution. Though it remained in existence, its member scientific societies ceased operations altogether and only in 1977 or 1978 resumed holding meetings and publishing journals. Until the United States and China established diplomatic relations, the STA served as the main liaison organization for scientific exchange.

### *Education, Professional Manpower, and the Budget*

A further component of China's post-1977 science policy is to improve higher education and thereby increase China's pool of professional manpower. In his report to the National Science Conference, Fang Yi set a goal of increasing the number of professional research workers from an unknown number to 800,000 by 1985.[21]

When the PRC was established in 1949, China's scientific and technological manpower supply was extremely limited. Although the educational system expanded enormously from 1950 to 1966, producing well over a million scientists and engineers, there was not sufficient stability to allow for the emergence of a new cadre of scientists capable of leadership. The serious disruptions of higher education caused by the Cultural Revolution made a bad situation considerably worse, leading some to suggest that a whole generation of scientists was lost.

The Chinese themselves lack comprehensive manpower data and in 1978 initiated a national manpower survey to remedy this situation.[22] The last careful Western analysis of China's manpower supply was the estimates prepared by Leo Orleans in 1975.[23] Orleans concluded that there were approximately 575,000 scientists and 725,000 engineers in that year, with

[21] "Fang Yi Report to National Science Conference."

[22] This survey is being conducted with the participation of the State Planning Commission, the STC, the Ministry of Civil Affairs, and the State Statistical Bureau (NCNA dispatch, June 23, 1978).

[23] Leo A. Orleans, "Scientific and Technical Manpower," in *Science and Technology in the People's Republic of China* (Paris: Organization for Economic Cooperation and Development, 1977).

180,000 of the scientists specializing in agriculture, 240,000 in medicine, and 155,000 in other fields. In qualitative terms, Deng Xiaoping in 1975 categorized China's scientific manpower as follows:

1. Senior scientists, including those trained abroad or in China before 1949, who in spite of advancing age remained the leaders of the scientific community.
2. Scientists trained abroad after 1949, whom Deng called the mainstay of work in science and technology.
3. Scientists trained in China after 1949. Deng noted that their quality was rather uneven and that their training had been subject to serious interruption by political meetings and periods of physical labor.
4. Worker/peasant "scientists," whose competence was generally rather low.

Given the heavy demand for trained manpower for research and development, education, engineering and design, and administration, it becomes clear why China's leaders have broken so dramatically with past policies by aggressively seeking training opportunities abroad. Highly trained, experienced scientists capable of leadership will become even scarcer when the senior, Western-trained generation of scientists disappears.[24] Foreign training will be extremely important in expanding the leadership corps.

In addition to sending students, teachers, and researchers abroad for training, steps were taken to improve China's own education system, especially in science and technology. Entrance examinations were reinstated in 1977, and those of 1978 reportedly placed even greater emphasis on intellectual achievement than those of 1977.[25] Approximately 578,000 students were admitted to institutions of higher education in 1977 and 1978. If pre–Cultural Revolution patterns are any guide, approximately 70 percent of these were in science and engineering.

The new education policy also encouraged special national searches for extraordinarily gifted children to be admitted to universities below the normal university age. For example, in 1978 some 91 gifted children, ranging from ages 11 to 16, were admitted to the Chinese University of Science and Technology in Anhui province.[26] Graduate training in higher education institutions and in the CAS was also reintroduced. By the fall of 1978, the Chinese expected, to have some 9,000 graduate students enrolled in 207 institutions of higher education and 144 research institutes.[27] Finally, steps

[24] I have attempted a quantitative assessment of China's professional manpower problems with particular reference to the corps of leaders needed in research and education in *Science, Technology, and China's Drive for Modernization* (Stanford, Calif.: Hoover Institution Press, forthcoming).

[25] Pierre M. Perrolle, "Engineering Education in China: A Report on Observations of the U.S. Engineering Education Delegation to China, September 8–October 2, 1978" (Paper presented to the Workshop on the Development of Industrial Science and Technology in the PRC: Implications for U.S. Policy, St. George, Bermuda, Jan. 3–7, 1979).

[26] *Ibid.*

[27] *Ibid.*

were taken to strengthen the July 21 universities, which have a potentially important role in training lower-level technical manpower. Run by factories and other production units, these schools offer an abbreviated and specialized university curriculum focused on the sponsor organization's technological needs.

However we interpret Fang Yi's target of 800,000 technical personnel by 1985, it will be difficult to reach. There were not enough experienced personnel available as of 1978 to handle China's ambitious programs in research and development and in engineering, even without considering educational functions. The shortage was certain to be felt in undergraduate education, in graduate programs, and in the July 21 universities.[28]

It will be necessary for China to increase its expenditures for science and technology, not only for the expansion of the educational system to meet projected manpower needs, but also for research and development. "Big science" projects, such as the 50-GeV accelerator at the Institute of High Energy Physics and the new institute for nuclear fusion research in Sichuan province, will be particularly expensive. Moreover, there is China's wage bill to consider. Long-denied promotions for research workers and university faculty have been made; the resulting salary increases will increase total spending on science and technology. Accordingly, the 1979 budget for science and technology showed a 10 percent increase over that of 1978 to 5.87 billion *yuan* or approximately 5.2 percent of the state budget.[29]

A shortage of trained manpower looms as the major obstacle to Chinese scientific development over the short run. In assessing the prospects for Chinese science, however, it is important to consider the broader historical context. For over a century China's efforts to develop modern science and technology have been hampered by recurring periods of political instability. It remains to be seen whether these have come to an end. If they have, China's new science policies offer a good prospect for creating the scientific and technological prerequisites for achieving the nation's modernization goals.

[28] This subject is explored in greater detail in Suttmeier, *Science, Technology, and China's Drive for Modernization.* For the situation in July 21 universities specifically, see Perrolle, "Engineering Education in China."

[29] Zhang Jingfu, "Report on the Final State Accounts for 1978 and the Draft State Budget for 1979," *Beijing Review,* 1979, 29 (July 20): 17–24.

*Saunders Mac Lane*

# *PURE AND APPLIED MATHEMATICS*

### *The Background In Mathematics*

The various recent developments of mathematics in the PRC exhibit many interesting special features. In general, the high development of mathematics and the direction of that development in any country or region depends on several factors, which include:

1. The presence of potentially talented mathematicians
2. Effective stimulus from knowledgeable older mathematicians
3. Social and political demands or conditions on mathematical activity

The recent situation in China presents a fascinating interplay of these three influences, which have all been present—and this, despite the occasional insistence in 1976 in China on the political line that talent is equally shared by all. As will appear, there are clear examples, not just in the West but also in China, that some young people have an especial talent for mathematical insights and discoveries. Since 1978, this important point is recognized in China, especially in statements at the 1978 National Science Conference and in the recent nationwide mathematics examinations to determine future entrance into universities.

In this article, we will endeavor to summarize in the light of these three factors the present situation of mathematics in China, with especial attention to a few striking recent mathematical discoveries. A good part of the basic information was collected during the 1976 visit to China of a Delegation of Pure and Applied Mathematicians, under the auspices of the Committee on Scholarly Communication with the People's Republic of

China.[1] The report of that Delegation is used here repeatedly. One must recall always that our information about science in China today is quite limited, and that much of the recent development of mathematics there has been carried on in deliberate isolation from the West under the slogan "by our own efforts."

The Chinese people have long displayed a talent for mathematics. Centuries ago, they discovered independently the theorem of Pythagoras, which states that the square on the hypotenuse of a right triangle is the sum of the squares on the legs of that triangle. However, they did not embody this and other geometric theorems in a sophisticated deductive system such as that developed by the Greeks. In A.D. 500, Zu Chongzhi knew the value of $\pi$ to seven decimal places, and the formula $V = (4/3)\,\pi r^3$ for the volume $V$ of a sphere of radius $r$. The Chinese also knew the binomial expansions for $(x + y)^2, (x + y)^3, \ldots$ and understood the behavior of the coefficients in these expansions, as they are arrayed in a triangle ("Pascal's" triangle, at right). They knew the solution of the quadratic equation. Western textbooks of number theory still solve "simultaneous congruences" by using the "Chinese Remainder Theorem." A simple sample is this one: The smallest odd number which leaves a remainder 6 when divided by 7 is the number 13. The Chinese remainder rule for solving such problems testifies to an early interest in number theory—an interest which we will see continued in the present. All this indicates an early and active development of mathematical ideas in China. However, with the Ming dynasty (1368–1644) the practices of Chinese civilization shifted; mathematical discoveries ceased, and soon the older mathematical results were forgotten.

$$
\begin{array}{ccccccccc}
 & & & & 1 & & & & \\
 & & & 1 & & 1 & & & \\
 & & 1 & & 2 & & 1 & & \\
 & 1 & & 3 & & 3 & & 1 & \\
1 & & 4 & & 6 & & 4 & & 1
\end{array}
$$

In the seventeenth century, the Jesuits brought some Western science and mathematics to China. However, it was not until the 1930's that there was a substantive development of modern mathematics in China, sparked by a number of Chinese mathematicians who had received some training abroad. For example, both the geometer Su Buqing and the analyst Chen Jiangong had been trained in Japan and then taught at Zhejiang University (Hangzhou), where they built up effective schools of their students.

### *Talent and Stimulus*

In the 1930's a number of young Chinese mathematicians went abroad to study, usually for a Ph.D. In such cases it seems that the impact of up-to-date knowledge on able young scholars can lead to striking developments in a variety of fields of mathematics. Thus Shiing-Shen Chern studied geometry in the European centers of Paris and Hamburg, and has gone on to become an influential world figure in both topology and geometry, especially

[1]Committee on Scholarly Communication with the People's Republic of China, *Pure and Applied Mathematics in the People's Republic of China,* Report no. 3 (Washington, D.C.: National Academy of Sciences, 1977).

in the study of complex manifolds. Wei-Lang Chow studied in Chicago and then with the German algebraic geometer B. L. van der Waerden in Leipzig; he has gone on to do decisive work in algebraic geometry (the "Chow points" of an algebraic variety bear his name). P. L. Hsu and later K. L. Chung studied probability and statistics, and have contributed extensively to these fields. (Hsu studied in London in 1936–40 and taught in the United States in 1945–1947, then returned to China.) C. C. Lin came to the West in 1940, studied at the Center of Applied Mathematics at Toronto and the California Institute of Technology, and has stayed in the United States to become a leading figure in applied mathematics and mathematical aspects of astrophysics. These five cases—plus a number like them—all illustrate the way in which talented young Chinese mathematicians were able to learn the most sophisticated and advanced aspects of Western mathematics and to carry these ideas further. For example, the pregnant ideas of the great French geometer Elie Cartan on infinite continuous groups were for some time hardly understood in Paris, while Chern has been one of the few who grasped their significance. Note also that four of these five mathematicians left China and stayed in the West (in these cases in the United States). There was, in effect, a considerable brain drain for China. However, this was not peculiar to the situation in China. In the early 1960's a count indicated that about two-thirds of the 30 leading Japanese mathematicians held positions in the United States!

Often the stimulus of foreign training works through several generations. One case of importance for the development of number theory in China is worthy of note. In the 1920's K. C. Yang came to study in the United States. He obtained a Ph.D. at the University of Chicago writing his thesis (on the Waring problem) under the direction of the noted American algebraist and number theorist L. E. Dickson. Yang then returned to China, where he stimulated interest in number theory. His son is the well-known physicist C. N. Yang. Also in the 1920's Hiong King-lai (also written Hsiong Ching-lai) studied in Montpellier in France. On his return to China he was the teacher of S. S. Chern. He was a professor at the Qinghua (Tsinghua) University in Beijing. There in 1931 he noticed some strikingly interesting articles published in a Shanghai science magazine by one Hua Luogeng. He proceeded to hunt up Hua, who turned out to be just 20 years old, with no formal university training, simply working for his father as an accountant in a small village. Hiong brought Hua to Qinghua University and encouraged his interest in number theory. Hua soon became a lecturer in the university. Then in 1936 Hua in his turn went to Cambridge, England, on a fellowship from the China Cultural Foundation. Cambridge was at that time the focus of active developments in number theory, especially in the use of analytic methods to solve problems of number theory. Hua immediately was in the middle of these developments, making a number of contributions to the same Waring problem which appeared in Yang's thesis. (This is the problem: every integer can be written as a sum of at most four squares or as the sum of at most nine cubes of integers. Similarly, how many $k^{th}$ powers does it take to represent every integer as a sum of $k^{th}$ powers?) In Cambridge, Hua not only contributed to the attack on this problem, but also widened his knowledge of other parts of number theory, including especially the so-called "geometry of matrices." In 1938 Hua returned to China to become a professor at the Southwestern Associated University at Kunming. His talent and his mastery of number theory will play a considerable role in our subsequent story.[2]

In 1946 a mathematical institute was established in Nanjing, with the aim of bringing modern mathematics to China. S. S. Chern was the director. Many talented college graduates in mathematics were brought from all over the country to this institute for intensive training. Because of the disruptions of the civil war, the program lasted only two years, but it had a long effect. At least eight of the young graduates went on to get doctorates in the West; they include Liao Shantao, now a professor at Beijing University, specializing in topology, and Wu Wenjun, now a permanent member of the Mathematics Institute in Beijing.

### *Organization and Direction*

With the liberation in 1949 a new pattern was set for the development of sciences in China. Initially the Russian influence was strong, as it was in the country at large. Following the Russian model, much of the direction of scientific research was located in the numerous institutes of the Chinese Academy of Sciences (CAS), in Beijing. Various Chinese scholars returned from abroad to help in the development. In particular Hua Luogeng, who had spent the period 1946–50 in the United States, returned to China and became director of the Institute of Mathematics of the CAS. This Institute succeeded in assembling a number of able mathematicians. Hua's wide knowledge of mathematics, ranging from number theory and algebra to analytic function theory, served to focus many of these interests. There was also considerable attention to promoting work in the application of mathematics. Su Buqing had many students in classical differential geometry, and Wu Wenjun (who had studied in France) contributed penetrating research in topology.

There are at hand a number of summaries of Chinese research in mathematics during the period 1949–60: a bibliography[3] by Tsao Chia-kuei, an article[4] by M. H. Stone in an A³S volume, a translation of a report[5] by Hua Luogeng on mathematics in China, plus the translation of two articles[6] on

[2]See Stephen Salaff, "A Biography of Hua Lo-keng," *Isis*, 63, 217 (1972): 143–83.

[3]Tsao Chia-kuei, *Bibliography of Mathematics Published in Communist China During the Period 1949–1960* (Providence, R.I.: American Mathematical Society, 1961).

[4]M. H. Stone, "Mathematics, 1949–1960," in Sidney H. Gould, ed., *Sciences in Communist China* (Washington, D.C.: American Association for the Advancement of Science, 1961), pp. 617–29.

[5]Hua Lo Keng, "Report on Mathematics in China," *Notices of the American Mathematical Society*, 6 (1959): 724–30.

special aspects of mathematical research by Chi Keng Look and by Su Buqing and Ku Chao-hau. The latter two articles are rather special. For example, the article by Look records some remarkable contributions to the geometry of matrices by a few Chinese experts; there are listed ten articles by Hua, six by Look, six joint articles by Hua and Look, plus three by other combinations of authors. The more general summary article by Hua enthusiastically records many contributions to a variety of fields of mathematics, emphasizing practical applications and the inspiration coming from the Great Leap Forward. The bibliography by Tsao lists some 1,271 titles of mathematical research articles by Chinese authors. Stone's summary observes that this listing includes a small number of really significant contributions (for example those of Hua already noted above and those by Wu) but that most of the work was concentrated on a few sharply defined aspects of analytic function theory, differential geometry, classical analysis, algebra, and topology—and that much of the work is concerned with detailed studies of problems already posed outside China. Stone also observed that "the growing attention [in China] to applied mathematics has so far been more apparent at the relatively routine levels associated with the solution of specific practical problems than at the level of fundamental mathematical research into the underlying physical phenomena."

The period 1960–66 appears to represent similar mathematical emphasis. In this period the *Acta Mathematica Sinica (Shuxue xuebao)* was probably the principal journal publishing mathematics in China; it was issued by the CAS in Beijing. The American Mathematical Society published a cover-to-cover translation of this journal under the title *Chinese Mathematics,* up to the interruption of publication in China in 1966. An examination of volume 8 of that translation (chiefly covering the 1966 volume 16 of the Chinese publication) reveals a variety of sound but not spectacular articles on standard subjects of pure mathematics; I counted six articles on algebra and number theory, 13 on analysis, plus eight on Fourier analysis and six on differential equations, 14 on geometry and topology, six on Lie groups, seven on probability, with one each on combinatorics and on mathematical logic. There is evidently an emphasis on Lie group theory, geometry, topology, and Fourier analysis, but the general balance is not very different from what would be found in a contemporary Western journal of pure mathematics. There is a noticeable absence of applied mathematics. There is only one article on combinatorics, a field with useful applications, while the articles on differential equations and on probability theory appear to be more theoretical than applied. In other words, this 1966 sample of Chinese mathematics, which was taken just before the Cultural Revolution, shows mathematical research closely following the model of such research in the West.

[6]Chi Keng Look, "A Study of the Theory of Functions of Several Complex Variables in China During the Last Decade," *Notices of the American Mathematical Society,* 7 (1960): 155–63.

*Number Theory*

An outstanding example of the work of the Mathematics Institute in Beijing is given by developments there in number theory. Since the director, Hua Luogeng, had wide interests in number theory and wrote an effective textbook on the subject in 1957, he encouraged a number of other Chinese mathematicians, notably Wang Yuan and Chen Jingrun. This has resulted in some notable progress on the "Goldbach conjecture."

Some centuries ago, one Goldbach observed that every small even number could be written as a sum of two prime numbers, thus:

$$4 = 2 + 2,\ 6 = 3 + 3,\ 8 = 3 + 5,\ 10 = 3 + 7,\ 12 = 5 + 7, \ldots$$

and so on. He conjectured that *every* even number could be so written. To this day, no one knows whether or not this is so, but the attempt to find out has required more and more sophisticated methods, involving the use of calculus and other analytic tools which comprise the subject of analytic number theory. In 1930 the Russian mathematician I. M. Vinogradov had considered the case of odd numbers; if the Goldbach conjecture is true, every odd number ought to be a prime or a sum of *three* primes. Using elaborate analytic methods, Vinogradov proved that every *sufficiently large* odd number is a prime or the sum of three primes. It was quite typical of his methods that such a result is established not for *all* odd numbers, but only for those which are "sufficiently large" in the sense that they are all the numbers above some very large bound.

This striking result of Vinogradov immediately attracted attention. Hua, during his stay in Cambridge in 1936–38, provided some improvements in one of the lemmas used by Vinogradov. The Russian mathematician Y. V. Linnik in 1946 obtained a new and considerably simpler proof of the Vinogradov result for odd numbers. However, no one succeeded in using these methods to get the original Goldbach conjecture for (sufficiently large) even numbers. Then in 1948 the Hungarian mathematician A. Renyi took a different tack. He found an integer $k$ such that every sufficiently large even number is the sum of a prime and a second number which is a product of at most $k$ primes. If $k$ were 1, this would be the Goldbach hypothesis; unfortunately, in Renyi's proof the number $k$ was quite large.

This result, however, immediately stimulated number theorists in many countries to get the same conclusion with a better value of $k$. There resulted a sort of international competition. In 1958 Wang Yuan in Beijing made considerable progress. In 1963 the Russian B. V. Levin showed that $k = 4$ would do (and that even assuming the famous but as yet unproved Riemann hypothesis, he could get only $k = 3$). In 1965 Chen Jingrun improved slightly on Levin's result. Also in 1965 the Russians A. A. Buhštab and A. I. Vinogradov independently proved that $k = 3$ would do (*without* using the Riemann hypothesis!).

Finally, in 1966 the Beijing mathematician Chen Jingrun announced[7] that $k = 2$ would do. That is, he proved that every sufficiently large even number can be written as the sum of a prime and a second number which is either a prime or a product of two primes. Chen's detailed proof of this result was not published until much later,[8] in 1973—probably because of the interruption of scientific publication in China during the Cultural Revolution. His result stands today as the best result so far on the elusive Goldbach conjecture. A group of his colleagues at the Mathematics Institute in Beijing—Ding Xiaxi, Pan Zhengdong, and Wang Yuan—in 1975 found a considerable simplification of his proof. The whole sequence of discoveries does serve to illustrate the international exchange of ideas and methods in mathematic—and the fact that Chinese mathematics is at the forefront of this aspect of analytic number theory.

*Operations Research*

From number theory we turn to the Chinese development of a type of immediately applicable mathematics. In the Second World War, British and American scientists found that they could use analytical and mathematical methods to handle a wide variety of practical problems of military tactics, such as the problem of finding the best method to search for a submerged submarine last located at some known point and time. In peacetime, this field of activity, called operations research, operations analysis, or sometimes systems analysis, proved to have a number of useful applications, both to problems of military systems and to problems of business tactics, as well as to the efficient execution of manufacturing operations. Many of these problems depend on choosing parameters so as to achieve an optimum of some desired result—as for example with the important process of "linear programming."

In 1959, after the failure of the Great Leap Forward, there were substantial problems with the wheat harvest in China. Some of these problems came to the attention of Hua Luogeng. He prepared for the *Guangming ribao (Guangming Daily)* a long article on the "science of operations and programming," with application to the improvement of particular problems of transportation and distribution. Some of these problems have an immediately practical and quite simple character. For example, given two wheat fields located at different points along the same (straight) road, at what point should one locate a common threshing floor for these two fields in such a way as to minimize the quantity: weight of wheat times distance transported? The answer is that the threshing floor should be located right at the larger of the two wheat fields. Indeed, were this floor moved any dis-

[7] Chen Jingrun, "On the Representation of a Large Even Integer as the Sum of a Prime and the Product of at Most Two Primes," *Kexue tongbao* (*Science Bulletin;* foreign-language ed.), 17 (1966): 385–86.

[8] Chen Jingrun, "On the Representation of a Large Even Number as the Sum of a Prime and the Product of at Most Two Primes," *Scientia Sinica,* 16 (1973): 157–76.

tance along the road toward the smaller field, this would entail the transportation of a heavier load of wheat (from the larger field) over that distance. There is a similar solution for the more complex problem of locating a threshing floor for several wheat fields. Hua's wording of this situation, when the several fields are joined by roads without loops, is that the best place for the threshing floor is where the quantity of wheat brought in along each road is less than half the total amount of wheat—and this again because, were the quantity so brought in more than half, a move of the threshing floor along that road would reduce the total weight times distance transported. This solution, with others, was presented by Hua and his collaborators in a mathematics paper[9] published in 1961.

Since that period, Hua has taken the lead in training many Chinese in the practice of operations research. A group of experts (at least 30) was assembled at the Mathematical Institute in Beijing. Hua and some of his collaborators traveled widely to consult and lecture. At one point, Hua is said to have lectured (by telephone network) to an audience of about 100,000 people. The intent of his efforts is to make these methods accessible to all.

Various developments published in the Western literature on operations research have been studied and widely used in China. The favorite methods are the *optimal seeking method,* called OSM, and the *critical path method,* CPM.

In OSM, the problem is that of finding approximately the largest possible (maximum) value of some (empirically) given function. In the usual case, the function has a single maximum at some unknown position; one wishes to calculate a few values of the function in such a way as to make a close estimate of the real maximum value. In 1953 the American statistician Jack Kiefer published a suitable "Fibonacci" method of search for such a maximum; his method is an optimum method, in a certain "minimax" sense (i.e. for a given number of trials, one minimizes the maximum error). This method, with others, is incorporated in the Chinese OSM procedure, which has the advantage that it can be explained to workers who are unfamiliar with algebra.

> OSM thus fits well with the Chinese doctrine that workers are capable of almost any achievement. In some of the uses of OSM that we encountered, polynomial interpolation would almost certainly have been more efficient, though mathematically somewhat more demanding. However, great progress has undoubtedly been made through the use of OSM, and perhaps promulgation of the idea of systematic optimization has been even more important than OSM itself.[10]

CPM is a method of approaching production problems by drawing a diagram with arrows which show the expected duration of each task and the necessary order of precedence of the tasks (when there is such an order).

[9] Hua Luogeng et al., "Application of Mathematical Methods to Wheat Harvesting," *Acta Mathematica Sinica,* 1961, 1/2: 77–91.

[10] CSCPRC, *Mathematics in the PRC,* p. 23.

From such a diagram (which is often called a PERT chart, for "project evaluation review technique"), one can calculate the time required to complete the entire production project and also the latest time at which each component task must be started if the whole project is not to be delayed. Those particular sequences of tasks whose timely completion is essential are called "critical paths"—hence the name CPM and the utility of the method.

The use in China of these OSM and CPM methods indicates that in such questions Chinese science has been quick to pick up and put to use practical devices first developed in the West (the Mathematical Institute in Beijing maintains a substantial library of Western mathematical journals). The mathematicians working on operations research have not only popularized these two methods, but have also made use of a variety of more sophisticated procedures to handle queuing problems, decision problems, quality control, and the like. The visit of the 1976 Delegation of Pure and Applied Mathematicians did not turn up any Chinese developments of striking new methods of these types, but did note several impressive and elegant applications—one to transportation problems in the logging industry, another very ingenious one dealing with the optimum way of switching the cars of a freight train in a freight yard.[11]

### *Social and Political Influences*

The emphasis on mathematicians doing operations research in China is a considerable contrast to the situation in the United States, where the relatively few experts on the subject are more likely to be present in consulting firms, business schools, and engineering schools than in standard departments of mathematics. This emphasis in China is part of the way in which the pressing social needs there have conditioned the directions of scientific development. During the visit of the 1976 Delegation, the speakers at the Institute of Mathematics in Beijing emphasized the idea of "open-door" scientific research. By this they meant that the doors of intellectual institutions should be open to deal with the problems of production and agriculture, and that "scientific research should serve proletarian politics, serve the workers, peasants and soldiers and be integrated with productive labor." The same idea was put forward at other institutions. At the various universities, professors and students often made extensive visits to (sometimes distant) factories and communes to study practical problems there. In 1976 many students in their third year of study prepared specific projects or essays dealing with such problems. It now appears that such preoccupation with practical problems will not be so dominant.

Political aspects also play a role. In 1976 at most institutes, including that in mathematics, the scientific staff was expected to spend at least one day a week in careful discussion of the current political situation. One gathers the impression that each staff member was expected to be able to state the cur-

[11] *Ibid.*, pp. 42–43.

rent political "line" in his own words. All told, this created for mathematical and other scientific research a situation very different from that in Western universities or institutes. From this perspective, the continuation in China of deep mathematical discoveries (as in the case of number theory) is even more striking.

These political conditions vary considerably from time to time. The "line" changes, but there always seems to be a line which is to be prominent in scientific activity. Currently, it asserts that the Gang of Four had denied the role of science in promoting production. There is now a five-sixths rule, which asserts that scientific workers are to have five-sixths of their working time to devote to science itself. The previous fraction was presumably lower. This change, and others, promise well for the future of Chinese science.

In the past, individual scientists have been put at times under considerable social pressure. For example, Hua Luogeng, despite his extensive practical activity in developing operations research, may have been under such pressures at the time of the antirightist campaign in 1957. He was also severely criticized by the Red Guards at the time of the Cultural Revolution in 1966. In August of 1968 the worker-army team entered the University of Science and Technology in Beijing (where Hua was then teaching). Soon thereafter Hua himself published a striking article of self-criticism in the *Renmin ribao (People's Daily)* of June 8, 1969. Parts of this article were republished in English in November 1969 in the magazine *China Reconstructs.*[12]

The Cultural Revolution evidently involved a major disruption of science. Publication ceased, universities were closed, and many mathematicians (including many of those at the Mathematics Institute in Beijing) were temporarily assigned to work in factories. This must have had a considerable influence on the course of mathematical research in China.

Our information on these influences is quite fragmentary. The principal journal, *Acta Mathematica Sinica,* ceased publication in the middle of 1966, to resume publication in 1974 (volume 16 was published in 1966, volume 17, no. 1, in 1974). Some publication of mathematical articles in English recommenced at least a year earlier in the journal *Scientia Sinica* (volume 16, beginning February 1973). For example, the topologist Liao Shantao in 1966 submitted a two-part paper on ergodic properties of systems of differential equations. The first part[13] was published in 1966; the second part, with some revisions, was published in 1973 in *Scientia Sinica.* The striking result of Chen on the Goldbach conjecture, as described above, was announced (without detailed proof) in 1966; his full proof was published in

[12] See *Renmin ribao,* June 8, 1969; and *China Reconstructs,* Nov. 1969.

[13] Liao Shantao, "Applications to Phase Space Structure of Ergodic Properties of the One-Parameter Transformation Group Induced on the Tangent Bundle by a Differential System on a Manifold," *Acta Scientiarum Naturalium, Universitatis Pekingensis,* 12 (1966): 1–43 (in Chinese); and idem, "An Ergodic Property Theorem for a Differential System," *Scientia Sinica,* 16, 1 (Feb. 1973): 1–24.

1973, also in *Scientia Sinica.*[14] Thus we can conclude that in mathematics there was an interruption in the publication of results of about six and a half years (1967–73).

There was a similar interruption in the training of mathematicians. Most universities were closed in 1966 or soon after that; when they reopened, mathematics was not one of the first subjects to appear. Thus the American Delegation visit in 1976 found the first class of newly trained mathematicians about to graduate from Beijing University since 1966. This gap had the striking effect that the Delegation found no active research mathematicians much younger than 35. This is striking because Western mathematicians are accustomed to the great stimulus of the arrival each year of one or two brilliant and iconoclastic young mathematicians—hence the belief that progress depends on the stimulus from the young.

These particular disruptions are now past. Moreover, since 1977 the situation in education has changed dramatically. The science curriculum has been altered, and it seems that there is now (as there was not before) a considerable emphasis on examinations at all levels of education. The recent extensive arrangement to send Chinese scientists to the United States and other Western countries for further training is another very useful move to help the development of science within China. It would also appear that within China the emphasis on following the "correct line" in scientific research has now been eliminated. In 1978 there was a national mathematics competition for middle-school students; the top 50 or so of the students who took part were allowed to enter any university they chose. All told, all the Western mathematicians and mathematical educators who have visited China since 1977 emphasize that all these developments amount to a drastic change in attitude.

The recent (March 1978) National Science Conference bids fair to open a new stage in China's socialist development of science and technology. Chairman Mao's 1956 slogan of "Letting One Hundred Flowers Bloom, Letting One Hundred Schools of Thought Contend" is said now to be used to justify scientific debate among intellectuals. Reports indicate that in mathematics this has led to a greatly increased emphasis on proving new theorems, with considerable competition in this endeavor. At the 1978 National Science Conference, Deng Xiaoping emphasized the "four modernizations": agriculture, industry, national defense, and science and technology. He emphasized that science is part of the production forces, and that of course "there are now and there will be many theoretical research topics with no practical application in plain sight for the time being." These slogans, if followed through, will again allow for the progress of all aspects of mathematics.

[14] Chen, "Representation of a Large Even Integer," *Science Bulletin;* and idem, "Representation of a Large Even Integer," *Scientia Sinica.*

*Probability and Statistics*

The development of statistical methods in China is closely related to the work in operations research. There is widespread emphasis on the use of the method of orthogonal experiments; there is the standard method of "block designs," initiated in the West in connection with agricultural experiments and extensively used since then. The 1976 Delegation saw several enthusiastic applications of this method; for example, to the improvement of gummed tape, of the output of a brewery, and of the design of the insulation used in refrigeration cars for railroads. The Delegation also obtained a copy of a textbook on orthogonal experimental designs, published in 1975 by the Shanghai Normal University. This text covers its subject quite thoroughly, at a practical level.

Another standard method of statistics is the use of regression analysis, intended to fit a straight line to points representing given data. A Chinese textbook on this subject was published by the Mathematical Statistics Group at the Institute of Mathematics.[15] This book treats the standard techniques for regression but does not discuss the theory of statistical distributions. In the West, regression analysis is often used as a test of hypotheses. The techniques for such testing depend essentially upon the theory of distributions, hence are not covered in this book. However, the 1976 Delegation did hear of an interesting application of regression analysis to the manufacture of steel.[16]

Part of the intellectual foundations of statistical methods lies in probability theory. An interesting textbook on elementary probability was published in 1975 by the Applied Mathematics Group of the Shanghai Normal University. The contents of the book are not much different from the initial chapters of the standard Western book (W. Feller's 1950 *Probability Theory*),[17] except that the Westerners' use of drawing balls from urns and of models from gambling (probably the origin of probability theory) is replaced in the Chinese text by models of real industrial problems. The parallel between the 1975 Chinese book and Feller's book is not surprising; it does turn out that Feller's book was translated into Chinese in 1964.[18]

The foundations of probability do still involve substantial intellectual disagreements, such as those between the frequentists and the Bayesians. Roughly speaking, the frequentists define the probability of an event as the fraction of favorable cases (in an infinite string of trials), while the Bayesians start from an *a priori* probability (½ for a fair coin to come up heads) and modify this in the light of subsequent evidence, using a famous formula due to the Reverend Bayes. The issues involved in this divergence of methods are

[15] Institute of Mathematics, Mathematical Statistics Group (Beijing: Institute of Mathematics, 1975).

[16] CSCPRC, *Mathematics in the PRC*, p. 47.

[17] William Feller, *An Introduction to Probability Theory and Its Applications, Vol. 1* (New York: Wiley & Sons, 1950).

[18] William Feller, *Kailülun ji qi yinyong, ce 1*, tr. by Hu Dihe and Lin Xiangjing of *An Introduction to Probability Theory and Its Applications, Vol. 1* (Beijing: Kexue chubanshe, 1964).

quite subtle, as may be seen from a current cogent explanation by Bradley Efron.[19] In China, it appears that no such controversy grows, since the underlying Marxian doctrine decides at once in favor of the frequentist approach. This is illustrated by the following abstract of an article written by a teacher from the Jiangsu Teachers College:

> ABSTRACT: The major difference between the two definitions of probability is explained. One is the classical definition based on the "frequency of events" viewpoint; the other is the subjective definition based on the "degree of belief" concept. The author emphasized that the subjective view must be rejected in favor of the materialistic viewpoint. Teachings of Engels and Mao are quoted to illustrate the objective process of statistical testing. In particular, the question of properly analyzing a random event and deducing its statistical properties is discussed. The concept of infinite number of trials as used in probability theory and the approximate method of estimating probability with finite samples are also discussed.[20]

Except for such philosophical influences, the Chinese scientists appear to have made considerable use of the elementary statistical techniques developed in the West, although they have not made many striking theoretical contributions of their own. After all, the direction and thrust of the social pressure in China has been almost entirely toward the practical applications of known techniques.

One striking recent result concerns what is called a Markov process—a continuous process which is determined by giving the matrix $P$ which consists of the probabilities of a change during time $t$ from one state $i$ to another state $j$. This "transition" matrix in its turn determines another "density" matrix $Q$, which describes the process more concretely. Long ago there arose the question: does $Q$ in its turn determine $P$? In 1940 Feller found a positive answer for a special class of these Markov processes—those called "conservative." Finally in 1974 Hou Zhending from the Changsha Institute of Railways found necessary and sufficient conditions that a given density matrix $Q$ arises from a unique transition matrix $P$ of probabilities. His abstract said: "Since the Great Proletarian Cultural Revolution, we have gone out of the school gate and come to the practical field of railroads; we have thus gained the same [sic] experience and impression in practical work."[21]

In May 1978 the Statistical Laboratory of Cambridge University awarded its Rollo Davidson prize to this paper by Hou. This is one of the first examples of the award of an international prize to a Chinese mathematician.

### *Analytic Function Theory: The East Advances Beyond the West*

One of the recent striking results found by Chinese mathematicians lies in the study of the "distribution of values" of a meromorphic function. This

[19] Bradley Efron, "Controversies in the Foundations of Statistics," *American Mathematical Monthly*, 85 (1978): 231–46.

[20] Du Wuchu, "A Look at Subjectivism and Metaphysics in Mathematics from the Definition of Probability," *Shuxue de shijian yu renshi* (Mathematics: Its Cognition and Practice), 1974, 3 (July): 10–13.

[21] Hou Zhending, *Scientia Sinica*, 17 (1974): 141.

goes back to one of the great themes of the development of Western mathematics in the nineteenth and early twentieth centuries. In the nineteenth century it gradually appeared that the calculus for functions of an ordinary (real-number) variable $x$ had much more profound properties where applied to functions of a complex variable $z$. As usual, such a complex number $z$ has the form $z = x + iy$, where $x$ and $y$ are (ordinary) real numbers and $i = \sqrt{-1}$. These complex numbers $z$ can be effectively pictured as the points with coordinates $x$ and $y$ in the usual plane, and this geometric view is important to the understanding of functions of $z$. Thus a function $w = f(z)$ can be considered as mapping the points $z = x + iy$ of one (complex) plane onto the points $w$ of another plane. Such a function $f$ is *analytic* if for each value of $z$ it has a first derivative, found just as in the calculus. Because of the special properties of the complex plane, the existence of a derivative is a very strong property, and the investigation begun in the nineteenth century by Cauchy in France and Weierstrass in Berlin of the remarkable consequences of this property constitutes the subject of analytic function theory. It was probably the dominant and most profound branch of mathematics in the West in the period 1880–1920.

One of the great theorems in this subject was Picard's theorem (1880). He considered *meromorphic* functions $w = f(z)$ which were defined for all complex numbers $z$, and which were analytic everywhere except for certain singular values of $z$, when the function has what is called a pole (at which it behaves like the function $w = (1/z)^n$ at the origin $z = 0$). Such a function need not take on *all* possible complex numbers $w$ among its values. Picard in 1880, however, proved that such a meromorphic function, provided it is not an algebraic function, must take on all possible values with *at most* two exceptions (two exceptions are possible; the exponential function

$$w = e^z = 1 + z = z^2/2 = z^3/6 = \dots$$

never takes on the values $w = 0$ or $w = \infty$).

This famous Picard theorem led to many further investigations and refinements, especially by other French mathematicians, such as H. Poincaré, J. Hadamard, E. Borel, and G. Valiron. Then in 1925 the Finnish mathematician R. Nevanlinna made decisive advances[22] in this subject of "value distribution." The subject as developed represents the beauty and subtlety of mathematical ideas in an exemplary fashion.

In 1932–34 Hiong King-lai (the teacher of Chern and Hua, mentioned above) went on sabbatical leave to Paris to study with Valiron on value distributions. Hiong's student Zhuang Jidai also went to Paris in 1936–38 to work with Valiron. Both Hiong and Zhuang returned to China and trained Yang Le, Zhang Guanghou, and other Chinese in this specialty. Valiron had been concerned with the existence of a "Borel direction" for a meromorphic

[22] For a recent discussion, see Rolf Nevanlinna, *Analytic Functions*, tr. of 2d German ed. (New York: Springer-Verlag, 1970).

function. Such a direction is a half-line starting at the origin near which the function $f$ takes on almost all values (with two exceptions, as usual) in the expected way. Nevanlinna, however, was concerned with certain "deficient values" of such a function. In 1973 and 1974 the Chinese mathematicians Yang Le and Zhang Guanghou (who both graduated from Beijing University in 1962) established a surprising and beautiful connection between these two concepts. If $p$ is the number of deficient values of $f$, and $q$ the number of Borel directions for $f$, then necessarily $p = q$. At the same time they established a number of related results in this field. For example, they showed that any appropriate set of directions could be the set of Borel directions for a suitably chosen meromorphic function.

This is a good example of a deep contribution to a central field of mathematics. Its development illustrates the way in which mathematical ideas grow with the successive contribution of many people for different countries. The Chinese ascribe this success to the use of philosophic concepts:

> For the past 50 years mathematicians have studied deficient values and singular directions, but always in isolation. Guided by their study of Chairman Mao's philosophical work, *On Contradiction,* Yang and Zhang used the dialectical method in a survey of foreign work on value distribution theory. They concluded that deficient value is a global concept reflecting the fact that a function infrequently assumes the deficient value and changes slowly. Singular direction, on the other hand, is a local concept reflecting the fact that a function assumes numerous values and changes rapidly. The two concepts are not isolated but interdependent and interrelated. There is an organic link between them and the two are united in a single contradiction. So they dialectically linked their research in the two fields which had long been studied separately.[23]

### *Computers and Computer Science: The East Copies the West*

The Chinese have been fascinated and attracted by the Western development of high-speed computers. Recognizing their importance for modern technology, they have learned many Western techniques; many Chinese institutions have constructed computers for special or general use. These computers are displayed with pride to Western visitors; for example, the 1976 Delegation of Pure and Applied Mathematicians was shown computers at Beijing University, at Qinghua University, at Fudan University, and elsewhere.

A number of American computer scientists have been invited to visit and to lecture in China; often such visitors have given systematic courses of lectures, for example, at the Institute of Computer Technology in Beijing (one of the institutes of the CAS). This emphasis on hearing foreign lectures is one indication of the current Chinese desire to learn from the West, especially of computers and similar practically effective devices. This Beijing Institute of Computer Technology is a large organization which currently employs somewhat over 1,000 people—though about 500 of these belong to

[23] *China Reconstructs,* June 1977, p. 42.

the manufacturing division, which is in effect a small factory attached to the Institute (this is a quite typical Chinese arrangement, to attach a factory or shop to a university or institute). The Institute also has a number of other divisions, including a division of some 90 persons working on software and theory.

The Institute has recently completed a new computer, the 013 computer, which is the most powerful yet announced in China. It has rated speed of 2 mips and a word length of 48 bits. Only one copy of this machine exists; before bringing it into manufacture, a few years of experimental use and software system development are intended. However, the Shanghai Radio Factory Number 13 seems to be manufacturing another machine, with rated speed of 1 mip. (A mip is one million instructions per second.)

A recent visitor reports the following overall impression of this machine:

> The 013, which is 10 times as fast as the fastest computer shown to us during our 1973 trip, obviously represents significant Chinese progress in hardware technology. On the other hand, some of its architectural details show the unfortunate effect of isolation from world trends in computer architecture. We find the instruction set somewhat baroque, the programmable 512 word cache difficult to use effectively (particularly in compiled code), and the absence of any hardware relocate especially unfortunate. The use of paper tape equipment to the exclusion of punched cards is bound to cause inconveniences. The disc storage system provided is rather minimal, and will make it difficult to develop an adequate disc oriented operating system, to say nothing of a more advanced file system. (It may be that magnetic disc technology will prove more difficult to master than all-electronic CPU technology.)

With this machine, the Chinese use a variety of compilers including one labeled BCY, which employs language very similar to the Western language ALGOL 60. It is noteworthy that Chinese chain printers carry roman characters only, so that it is at present impossible to comment programs adequately in Chinese characters. This problem may exercise a very harmful effect on the development of Chinese programming. (This is so far the only case known to me where the complexities of Chinese script may have a real effect on the progress of science. In mathematics papers proper, the text is written in Chinese characters, but the equations are in Latin symbols.)

Theoretical computer science is a subject which has developed very rapidly in the last ten years in the West. The group of Chinese students of this subject in Beijing have taken up the study of a number of topics popular in the West: grammar and grammar transformation, cross-compiling, software portability, algorithm analysis, etc. In the West, there has been a very active development of complexity theory, in which one endeavors to determine theoretically the *minimum* number of steps which is absolutely necessary to carry out a given type of computation. (For example, at the minimum, how many multiplications are needed for the product of two $9 \times 9$ matrices?) This is a subject which turns out to involve great subtleties and conceptual difficulties. The Institute of Computer Technology has recently formed a small group to specialize in this subject. However, some other parts of Western computer science have not yet been developed very far in China; for example, there is little work being done there on computer graphics.

The relationship to mathematical logic is also of interest. In the West, mathematical logic developed as a separate discipline in the 1920's and 1930's. It was stimulated by interest in questions of the foundation of mathematics, and it was generally regarded as a very pure and highly abstruse branch of mathematics—one shunned with suspicion by most mathematicians. One of the major topics in logic was the study of recursive functions—those functions of whole numbers which could be defined from the successor function (the function "add 1") by a suitable finite system of equations. This class of functions came to prominence in connection with Kurt Gödel's 1931 analysis of his famous incompleteness theorem, which stated that if the standard formal systems of logic were consistent, one could always formulate in them a statement or potential theorem which could be neither proved nor disproved. This study of recursive functions led the British logician Alan Turing in 1936 to analyze the theoretical character of all possible computations which could be made by a deterministic computing machine. He showed that the functions of whole numbers which could be calculated by such a "Turing machine" were exactly the recursive functions introduced by Gödel.

When, shortly after this, the actual computing machines began to be used, the connection between the highly theoretical studies of the logicians and the practical needs of computer scientists was rapidly made. Mathematical logic in the West, originally the purest part of pure mathematics, found numerous cogent applications of its methods.

In China, according to the various reports, it was quite otherwise. In the 1920's, the philosopher Jin Yuelin had started research in mathematical logic. One of his students, Hu Shihua had trained quite a few students in this subject, and in the years before the Cultural Revolution there was an active group of about 20 younger workers built up at the Computer Institute and at Beijing University. The Cultural Revolution ended most of the work of this group and seems to have put an end to instruction in symbolic logic throughout China. However, a number of former experts in symbolic logic came to work in aspects of computer science at the Computer Institute in Beijing! In other words, the effects of the Cultural Revolution in China suppressed the subject of formal logic (which now may return to some extent), but did not suppress its application to computer science. It is interesting to note that the difficulties with formal logic arose from the accusation that logic lacked practical utility, rather than from any more specific conflict with Hegelian or Maoist principles. (The situation may be different in other Eastern countries. For example, in Poland, scholars with inclinations toward philosophy found it appropriate to become specialists in the "safe" field of mathematical logic—which was very highly developed in Poland, beginning in 1920. In the Soviet Union, symbolic logic has occasionally been criticized as un-Hegelian.)

## *The Finite Element Method: The East Competes with the West*

In the West, the capacities and availability of computers have been an effective stimulus to numerical analysis—the study of effective methods of numerical calculation of all sorts, ranging from inverting a matrix to finding solutions for specified equations. Relatively little activity in this field has been observed in current Chinese mathematics, except for one topic: the finite element method.

This topic refers to a method of approximating the solution of (elliptic) partial differential equations. Such equations arise in potential theory (e.g. for electric potential) and in many other branches of classical mathematical physics. In many cases, the desired function $\phi = \phi(x,y)$ satisfying the partial differential equation can also be described as the function which minimizes a certain expression $M(\phi)$ involving $\phi$. Here we can think of $\phi$ as a function of two real variables $x$ and $y$, pictured as the coordinates of a point in the plane, so that $\phi = \phi(x,y)$ is defined for all points $(x,y)$ in some region $\Omega$ of the plane—possibly a quite irregular region. Usually the values of $\phi$ on the boundary of the region are specified in advance—a "boundary value problem." Now the problem of minimizing the expression $M(\phi)$ amounts to picking all the values $\phi(x,y)$ so as to get the smallest possible value for the resulting $M(\phi)$. Since there are an infinite number of choices of values of $\phi$ to be made, this is a minimum problem in an infinite dimensional space. The idea of the finite element method is to approximate this by a finite problem, where only a finite number of values need be chosen to get the minimum. One proceeds by cutting the region $\Omega$ up into pieces—for example, into triangles (see p. 71)—and assumes that within each triangle the function has some especially simple form (e.g. it is a linear function). This linear function is then completely determined by its values at the vertices of the triangles. Thus one has to choose not *all* the values $\phi(x,y)$ of the function, but only the values at the vertices of the triangles, so the original infinite problem has been approximated by a *finite* one.

This finite element method was long implicit in mathematics. Its modern form was introduced by Richard Courant in 1943, but was neglected until a number of structural engineers about 1955 independently suggested the method on empirical grounds. By this time, it could be used effectively in computations. What was needed was a justification of the method by a convergence theorem, a proof that with finer and finer triangles the approximate solution found for $\phi$ by triangles did come closer and closer to the "real" solution. This type of theory was developed in the West, beginning about 1965. At the same time Feng Kang in China worked on this theory, publishing in 1965 a paper[24] giving a general convergence theory, together with applications to elasticity problems. This was clearly a case of a theo-

[24] Feng Kang, "Finite Differences Schemes Based on Variational Principles," *Chinese Journal of Applied Mathematics and Numerical Mathematics*, 2 (1965): 238–62.

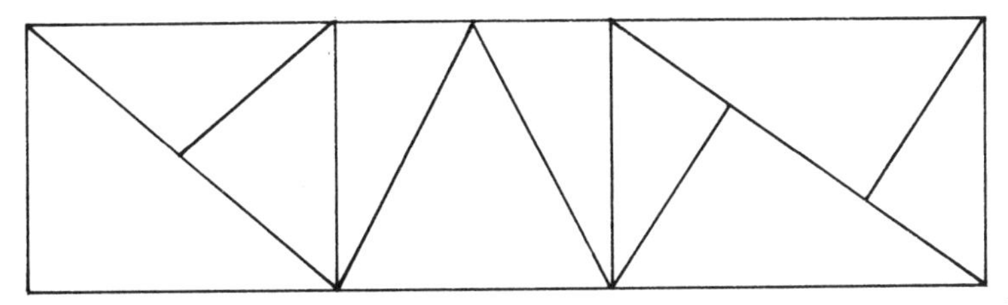

retical discovery motivated by practical problems, and made simultaneously in China and in the West.

Feng's work remained unknown and hence unrecognized in the West, where very extensive further research in such theoretical problems continued. The reason is probably that Feng's paper was published in Chinese, and to boot in a new journal (the *Chinese Journal of Applied Mathematics and Numerical Mathematics*) which was not then regularly translated and which in fact promptly ceased publication with the Cultural Revolution. Here (and in other cases) the effects of this revolution have cost Chinese scientists some international recognition.

The Chinese scientists continue to be aware of Western work in this field and do now make extensive use of the finite element method. Among the lectures given by Chinese mathematicians to the 1976 Delegation, at least four were concerned with the finite element method—including one by a student who used it to calculate the stress on the connecting rod of a specific diesel engine (i.e. in the picture above, the rectangle in the figure is replaced by the diagram of a connecting rod, suitably subdivided into triangles, with especially small triangles being used for those parts of the rod where the stress is greatest or most variable). Once such a problem is reduced to finite elements, calculating the minimum desired is essentially a problem in elementary calculus—for functions of several variables. In several Chinese universities, it appears that the standard calculus course is almost immediately focused on carrying out such very practical many-variable problems in the finite element method. This is an example of the principle that "education must serve proletarian politics and must be combined with productive labor. Students must take an active part in revolution and construction."

The finite element method, since it involves finite choices, appears recently to involve some questions of philosophy. In *Scientia Sinica* for January 1978 (vol. 21, no. 1), pp. 19 and 20, there is published an article entitled "The Infinite Similar Element Method for Calculating Stress Intensity Factors." After an introduction, the article goes on to comment as follows:

> F. Engels, the proletarian revolutionary teacher, has profoundly explained the dialectical relationship between finiteness and infinity: "Infinity is a contradiction, it is full of contradictions. It is already a contradiction that an infinity is composed of purely finite terms, and yet this is the case." (*Anti-Dühring*) According to this point of view, we realize that it is a pair of contradictions in itself to use the finite element method to calculate an elastic body which has an infinite degree of freedom. It reflects our cognition that infinity is composed of finite terms, and that infinity can

only be cognized through finite terms. However, there is qualitative difference between infinity and finiteness. The former is the summation of finiteness. For example, at the crack tip, it is impossible to reflect completely the singularity of stress by using a finite number of elements. Instead, it is necessary to develop the finiteness into infinity. The "infinite-similar element method" is just propounded on this basis. Of course, this "infinite element" has still to be cognized through finiteness in the end, and we complete this leap in cognition by using the method of combined elements which is still to "transform infinity into finiteness."

The work of Feng on the finite element method was not known in the West because of publication difficulties. There is now good news about the prospective removal of some of these difficulties. A new series of scientific books entitled "Pure and Applied Mathematics" was announced in Beijing in November 1978. All of these books are to be published in Chinese and in an English translation. The first volume is entitled *Applications of Number Theory to Numerical Analysis* and is written by Hua Luogeng and Wang Yuanhe. It will present the extensive work of these authors on numerical approximations to integrals. What is surprising in the work is the idea of using number theory and properties of algebraic numbers (traditionally very pure mathematics) to solve problems of computation. Specifically, they consider integrals $\int f$ of a function $f = f(x_1, \ldots, x_n)$ of $n$ variables, the integral being taken over a unit cube. The approximation succeeds by the careful choice in this cube of points whose coordinates involve "algebraic" numbers such as the 7th or 11th complex roots of 1 (technically, numbers in cyclotomic fields).

This new science series of books is planned to include a number of other major contributions of Chinese mathematicians, such as Chen on the Goldbach conjecture, Yang and Zhang on value distribution theory, and Feng Kang and Shi Zhongzu on the mathematical theory of elastic structure.

*Topology: The East Joins with the West*

The work of Yang Le and Zhang Guanghou in meromorphic functions, as discussed above, is a striking example of a Chinese contribution to a classical nineteenth-century field of mathematics. Topology is one of the newer (twentieth-century) fields; here Wu Wenjun has made major contributions. Topology has to do with the qualitative properties of geometric objects, such as manifolds (the sphere, the torus . . .) and fiber bundles, like the well-known Mobius strip (take a long strip of paper, twist through 180° and paste the ends together—the result is a surface with just one side). In order to distinguish one such fiber bundle from another qualitatively different one, topologists use certain algebraic invariants called "characteristic classes." The first important such classes were discovered by the American Hassler Whitney, the Swiss Eduard Stiefel, and the Chinese S. S. Chern.[25] Wu Wenjun, after studies in Nanjing with Chern, went to France. There he

[25] Shiing-Shen Chern, "Integral Formulas for the Characteristic Classes of Sphere Bundles," *Proceedings of the National Academy of Sciences USA*, 30 (1944): 269–73.

discovered[26] another family of such characteristic classes, now called the Wu classes and usually denoted by $W_i$. Since then Wu has gone on to publish many other incisive results on topology and to stimulate several other Chinese topologists (Wang Qiming, Zhang Sucheng, Shen Xinyao, Yu Yanlin, Dai Xinsheng, and others). In 1975 he traveled with Wang to Paris to attend a celebration for the French topologist Henri Cartan. There Wu learned of some of the spectacular recent work by the American topologist Dennis Sullivan, who had introduced certain new invariant "functors" in topology. Wu followed up this work and contributed a number of new insights[27] to the discussion of these $I^*$ functors. His insight and grasp of mathematics have enabled him in this way to work effectively in one of the most active current fields of mathematical research. Chern reports that currently Wu is working on algebraic groups—another subject of lively current interest in the West.

*From Geometry to Ship-Lofting: The East Redirects Its Own*

We now consider an example of the way in which pure mathematicians were directed by the Cultural Revolution to undertake some very applied work which turned out, after all, to make use of some of the pure mathematics.

Differential geometry is a subject in which the methods of calculus (differentials, derivatives, and integrals) are applied to the understanding of curves and surfaces, by measuring their tangents, their curvature, and other related quantities. The curves and surfaces in question may be located in ordinary Euclidean space, or possibly in a projective space. (For example, one may add to the ordinary Euclidean plane a new line "at infinity" so arranged that pairs of parallel lines, not meeting in the Euclidean.plane, do meet at a point "at infinity" in the projective plane.) Application of calculus to configuration in these projective spaces led to the subject of projective differential geometry, emphasized by E. Wilczynski at the University of Chicago and active generally in the West in the period 1900–30. After that time, most of the decisive results under this particular label were already at hand, so the subject disappeared in the West—or more exactly, was absorbed and transformed in other geometrical studies.

Su Buqing, as already mentioned, had studied mathematics in Japan in the 1920's. There he became interested in projective differential geometry. On his return to China, he continued to work and teach actively in this field, publishing over a hundred research papers—as an example we list one published in 1966.[28] At this time, Su was a professor of mathematics at Fudan

[26] Wu Wenjun, "On the Product of Sphere Bundles and the Duality Theorem Modulo Two," *Annals of Mathematics,* series 2, 49 (1948): 641–53.

[27] Wu Wenjun, "Theory of $I^*$-Functor in Algebraic Topology (Effective Calculation and Axiomatization of $I^*$-Functor on Complexes)," *Scientia Sinica,* 19 (1976): 647–64.

[28] Su Buqing, "Contributions to the Theory of Conjugate Nets in Projective Hyperspace VI," tr. from *Acta Mathematica Sinica* 16 (1966): 528–36, *Chinese Mathematics* 8 (1966): 550–59.

University in Shanghai. At this university the Cultural Revolution struck soon and vigorously, and it is evident that Su's views of his work were considerably reformed. Western reference journals give no further indications of work by Su on projective differential geometry. In 1976, the American Delegation met Su, still or again a professor, at Fudan University. For this Delegation he gave a brief talk on the contributions of the old, the middle-aged, and the young. He said:

> I started research in mathematics in 1920's in differential geometry. I wrote more than 100 papers having nothing to do with practice. I did not know what was useful. What I knew was of no help in practice. It is all out of date now. What you need to do is to begin with a practical problem, dig deeper, and bring theory back to practice. Our country is not advanced enough; if I do this kind of work, it is of some help to our country.[29]

The next day, Su gave a second, technical talk, "On Some Problems in the Design of the Shapes of Ships."[30] This dealt with ship-lofting. Shanghai is near the coast; many ships are built there. The traditional way was to lay out the actual shape of a vessel in a full-sized loft. Now, instead of doing this full scale, it is more efficient to calculate ahead of time how to cut those curves at the edge of each plate. This involves finding suitable approximations to the curvature of the desired curve—a task evidently related to the measurement of curvature by differential geometry. It was to this task that Su had turned his talents, once he was blocked from teaching by the Cultural Revolution and encouraged to attend to real problems.

This work has been noted widely in China. On October 2, 1977, a Hsinhua dispatch from Beijing was headed "Mathematicians Score New Research Achievements." The dispatch cited the work of Chen Jingrun on the Goldbach conjecture, that of Hua Luogeng on operations research, joint work of Hua and Wang Yuan in number theory, the results of Yang Le and Zhang Guanghou on meromorphic functions, and the work of Su Buqing, in the following words:

> Su Buqing, a veteran professor of mathematics in Fudan University of Shanghai, went with young teachers to a shipyard to cooperate with the workers and technicians. They renovated lofting, the first process of shipbuilding, applying differential geometry of curves and surfaces in laying off a full-sized working drawing of the lines and contours of a ship's hull. Their research result on the smoothing and fairing of the form of ship bow curves is up to advanced world levels and helpful to achieving greater, faster, better, and more economical results in China's shipbuilding industry.

### *Mathematics Variously Applied to the Problems of China*

The results of Professor Su and his students on ship-lofting, the studies of orthogonal experimental design, and the work of Hua and his numerous

[29] CSCPRC, *Mathematics in the PRC*, p. 95.
[30] *Ibid.*, p. 36.

collaborators on operations research are chosen samples of the many ways in which Chinese mathematicians have been encouraged, cajoled, and prodded to do work directly helping the socialist reconstruction of China. Here are a few other samples.

The problem of designing gear teeth came to the attention of the Higher Mathematics Research Group at Xian Jiaotong University. They wrote on the "Calculation of Normal Tooth Shape of the Roller Cutter for Cycloid and Planetary Gears."[31] Huang Lili and Zhong Hangchun of a Tools Research Institute in Chengdu wrote on " 'Integrated Error Curve Families' of Gears and Measuring Methods";[32] they observed that the introduction of these new concepts and measuring methods effectively eliminate a series of defects.

A team from Nankai University used elementary calculus to solve two practical design problems suggested by workers in the Tientsin No. 1 Machine Tool Factory—calculating an angle required in machining and designing a pair of elliptic gears.[33] A group from the Yangzhou Teachers College made a mathematical analysis of a clamping device which had been developed by shop workers in order to effectively machine certain polygonal shapes.[34] A practice team from the Department of Mathematics at Beijing University joined with technicians from a tractor plant to design piston rings, using suitable differential equations.[35] There are many similar examples; it is not surprising that one of the members of the 1976 Delegation claimed that in three weeks in China he saw more machine shops than in his previous 60 years in the United States.

That Delegation heard a careful analysis by two students at the Shanghai Normal University of the rather complex sequence of motions executed by the new Chinese automatic rice planter.[36] The next day, the Delegation visited an agricultural commune and saw the rice planter in action—they saw it get stuck in the mud. Such difficulties aside, it is clear that such an automatic way of transplanting rice shoots would enormously decrease the amount of back-breaking labor previously required in Chinese agriculture.

The 1976 Delegation also heard lectures dealing with the analysis of the flow in a network of pipes, turbocompressors, and the vibration of turbine blades. They also heard a variety of reports on the use of mathematics in related sciences: earthquake prediction,[37] meteorology, oceanography (Chinese oceanography tends to concentrate on phenomena in the Yellow Sea), magnetic exploration. There is also some effort in China in more classical types of applied mathematics: work on hydrodynamics involving what is

[31] *Shuxue de shijian yu renshi,* 1973, 3 (Sept.): 11–21.
[32] *Scientia Sinica,* 1973, 4 (Nov.): 582–611.
[33] *Shuxue de shijian yu renshi,* 1974, 2 (Apr.): 15–20.
[34] *Ibid.,* 1973, 4 (Dec.): 14–18.
[35] *Ibid.,* 1974, 3 (July): 5–9.
[36] CSCPRC, *Mathematics in the PRC,* p. 38.
[37] *Shuxue de shijian yu renshi,* 1973, 4 (Dec.): 1–13.

called "homogeneous isotropic turbulence,"[38] plasma physics, quantum field theory, and relativity.

In the West, there has been an elaborate and necessary development of the theory of linear control circuits, such as the electronic circuits used to control satellites. A decisive advance in this theory was the discovery by R. E. Kalman of a new "filter" for feedback now called the Kalman filter. The Institute of Mathematics in Beijing organized in 1973 a seminar course on the properties of the Kalman filter. Subsequently An Hongzhi of this academy devised a new filtering method.[39] Previously, with Yan Jiaan, he had written a paper extending the results of some Western research on "limited memory" filtering methods.[40] This development again illustrates the Chinese activity in adapting to its uses various ideas developed in the West.

The balance of these various activities in applied mathematics does lie in the direction of immediate applications, such as might be needed for rapid industrial development. The 1976 Delegation, after considering the state of classical applied mathematics, drew the following conclusion:

> Since the Cultural Revolution, mathematics, like most sciences in China, has been re-oriented to serve production and to serve society. Accordingly, mathematics is now taught from an applied point of view and research is now directed toward practical problems.
>
> This policy constitutes a nearly complete reversal of the earlier Chinese attitude, in which attention had been directed almost exclusively to pure mathematics. There was no tradition of research in applied mathematics on which to build, and there were practically no experts to lead in this new direction. The consequences are evident in the present state of applied mathematics.
>
> First, there appears to be little research on the invention and development of new techniques for solving differential and integral equations, on novel variations of existing techniques, or on the fundamental problems associated with such techniques. These activities constitute an essential part of classical applied mathematics in the West and are of great importance for science and technology. Second, there seems to be no general awareness of the great variety of methods of classical applied mathematics as developed in other parts of the world during the last 30 years. These include such methods as singular perturbation theory, boundary layer analysis, ray methods, the method of matched asymptotic expansions, the two-time and multiscale techniques, turning point theory, and the method of uniform asymptotic expansions. Although these methods dominate current Western work in classical applied mathematics, they seem to be completely absent from the Chinese work we saw.*
>
> Third, applied mathematics in China is generally directed at very specific technological problems and is not addressed to the general properties of the solutions of

*We remind the reader that what we saw may represent only a small fraction of the Chinese effort. Indeed, from scientific publications (e.g., current issues of *Scientia Sinica*) and other sources of information, it appears that among the scientists and engineers there may be more competence in classical applied mathematics than we saw. Perhaps we failed to meet these people because they are not members of institutions of instruction and research in mathematics.

---

[38] CSCPRC, *Mathematics in the PRC*, pp. 32, 34.

[39] *Shuxue de shijian yu renshi*, 1973, 4 (Dec.): 47–56.

[40] *Ibid.*, 1973, 3 (Sept.): 28–38.

large classes of problems. Instead it seems to aim for the particular solutions of particular problems, often found by numerical methods using electronic computers. This gives much of the work the character of engineering analysis rather than that of applied mathematics. Fourth, many of the older standard techniques of applied mathematics are employed. These include such techniques as separation of variables, Fourier and Laplace transforms, and eigenfunction expansions. When these methods are inapplicable, there is no resort to modern analytic methods. Instead numerical methods are employed on electronic computers. Thus, either exact, explicit solutions are obtained for simple problems or numerical solutions are obtained for particular complex problems.

### *Publication and Politics*

One might venture to measure the effects of the Cultural Revolution on mathematical research by the balance of articles published in China (though this will manifestly miss both research which has not been published and research which involved only practical advice and consultations, with no written record).

Above, we summarized the contents of the Chinese journal *Acta Mathematica Sinica* for the (partial) year 1966. The journal reappeared in 1974; here we compare the 1966 counts with corresponding counts covering the 1974 and 1975 issues:

| | *1966* | *1974–75* |
|---|---|---|
| Algebra and number theory | 6 | 4 |
| Analysis | 13 | 3 |
| Combinatorics | 1 | 0 |
| Differential equations | 6 | 6 |
| Fourier analysis | 8 | 0 |
| Geometry and topology | 14 | 5 |
| Lie groups | 6 | 0 |
| Logic | 1 | 1 |
| Probability | 7 | 0 |

We note that every topic has diminished except for the relatively practical topic of differential equations. This, however, is only half the story, because in 1974 and 1975 there are a number of new topics not represented at all in 1966 (except perhaps that statistics might be regarded as a sort of continuation of the 1966 interest in probability theory). Here are the new topics:

| | *1974* | *1975* |
|---|---|---|
| Approximation methods | 6 | 7 |
| History and philosophy | 1 | 5 |
| Mechanics | 3 | 1 |
| Operations research | 4 | 5 |
| Statistics | 4 | 3 |
| Theoretical physics | 2 | 1 |

All these added topics (except perhaps for history) belong strictly to the applications—and these publications on the applications are in addition to those in the new journal *Mathematics: Its Cognition and Practice*. The contents of the latter journal in 1975 and 1976 are strictly immediate-type applications[41] plus some book reviews, expository articles, and the inevitable lead articles on political-historical topics.

It is apparently primarily in these lead articles that mathematical research makes explicit contact with Hegelian dialectics and with Marxism–Leninism–Mao Zedong thought. Some of the titles in *Acta Mathematica Sinica* are:

"Minutes of a Forum Studying Chairman Mao's Important Directive on Theoretical Problems"[42]

"Notes on the Study of the History of Chinese Mathematics—on the Obstructiveness and Destructiveness of the Reactionary Confucianists in the Ancient Development of Mathematics in Our Country"[43]

"On the Struggle Between the Legalists and the Confucians and the Development of Ancient Mathematics in China"[44]

"Karl Marx's Mathematical Manuscripts (Selections): III. On the History of Differential Calculus"[45]

"The Brilliant Victories of Materialist Dialectics—Notes on Studying Karl Marx's 'Mathematical Manuscripts' "[46]

Note especially the titles dealing with Marx. He had a lifelong interest in mathematics and left over 1,000 pages of unpublished manuscripts, chiefly dealing with the foundations of calculus. These manuscripts had little influence at the time; also they did not use the discoveries made in his lifetime on the foundations of calculus. It would therefore seem that the careful study of his manuscripts by Chinese mathematicians is primarily an act of revolutionary piety.

### *Education in Mathematics*

The most striking fact about the present Chinese educational system is that so few mathematicians and other scientists have been educated recently. Their number, in a country with a population over 700,000,000, is very small by any standard—and certainly too small for the prospective needs of socialist development in China. Part of this situation is due to the way in which the Cultural Revolution interrupted the normal educational process. When the universities were reopened about 1972, the standard course of study lasted only three years, and this on top of a cut in the length of the middle-school education. There seems also to have been essentially no

[41] CSCPRC, *Mathematics in the PRC*, pp. 101–5.
[42] *Acta Mathematica Sinica*, 18 (1975): 75–80.
[43] *Ibid.*, 17 (1974): 227–33.
[44] *Ibid.*, 18 (1975): 81–85.
[45] *Ibid.*, pp. 1–17.
[46] *Ibid.*, pp. 149–56.

graduate work in mathematics—and little in other sciences. All pupils graduated from middle schools were expected to go to the farm or the factory for several years of work. Their subsequent (possible) admission to a university was then based on recommendation from the commune or the factory where they worked and not on entrance examinations of any sort. When they did arrive at the university some years after their last active contact in the middle school with mathematics and other science, it took them some time to refurbish their elementary knowledge.

These conditions and arrangements had some social justification but, taken together, they do not provide circumstances conducive to training scientists for today's complex studies. Since October 1976, however, the circumstances have again changed. There is a new enrollment system, with some college entrance examinations so as to ensure the quality of the new students, and some admission of students directly from middle schools. There is also a clear intent to introduce graduate work in the sciences. Finally, there is now a definite recognition of the role of talent. In his speech at the opening ceremony of the National Science Conference on March 18, 1978, Deng Xiaoping said:

> The discovery or training of talented people by our scientists and teachers is itself an achievement and a contribution to the state. . . . A number of outstanding mathematicians in China today were discovered in their youth by older generation mathematicians who helped them mature. Some of the newcomers may have surpassed their teachers in scientific achievement, but the teachers' contributions are indelible, nonetheless.

With this recognition and these changes, perhaps plus others, it is likely that China will be able to discover and educate the needed numbers of mathematicians and scientists. In Chinese terms, there will be new people who are both "red and expert."

Specific information about the present process of education in mathematics is sparse. In 1976 in the universities, the education was structural so as to "combine theory with practice." For example, the first-year calculus course at Beijing University was illustrated throughout by the problem: Where does one support a horizontal laser tube so as to minimize the maximum deflection? Similarly, the third-year course on the subject of complex variables (i.e. holomorphic functions, as above) was introduced and motivated by the fact that an irrotational electromagnetic field in the plane exhibits the properties of a holomorphic function. In other words the mathematical object, a holomorphic function, was not defined directly but only indirectly, through the corresponding properties of one of its physical interpretations. Every student in China spent substantial amounts of time visiting factories in order to learn how to make practical applications. At the end of the course, a student might spend up to four months on a "graduation project," which is likely to be such a practical application.

As for secondary education, the 1976 Delegation had a fascinating visit with five members of a team that plans the mathematics curriculum and writes the texts for the primary and middle schools of Shanghai.[47] The team explained the philosophical background of their work: Learning proceeds from practice to general knowledge and then back to practice. Engels and dialectical materialism were amply quoted. The team exhibited the dialectic relation between pairs of opposites with several mathematical examples: positive and negative numbers, constants and variables, powers and roots, and differentiation and integration. However, when we pressed them as to how they would teach the use of differentiation in constructing the tangent line at a point of a curve, their explanation was much like that in a standard intuitive Western text on calculus.

From the meeting the Delegation also developed an outline of the middle-school curriculum in mathematics, as recommended by this team:[48]

*First two years.* Basic facts from algebra and geometry. Parallel lines introduced intuitively; similar triangles introduced as a means of measurement and then applied to actual measurements in surveying, etc. Sometimes, projects carried out with a factory (example: a program for computer control of a milling machine, involving some use of Boolean algebra).

*Third and fourth years.* Polar coordinates, spirals and their use for the design of cams; the cycloid and its use in production problems. Logarithms, and the basic concepts of calculus, based on geometric intuition.

*Fifth year.* More calculus, some statistics and operations research, including orthogonal experimental design and the optimal seeking method (OSM).

Thus in Chinese terms, it is emphasized that mathematics originates from the real world and is a product of the needs of the people. In the West we might say: The mathematics is well and thoroughly motivated.

### *Conclusion: Stimulus from Below Versus Direction from Above*

Let us try to summarize the accomplishments of Chinese mathematics in the last dozen years (i.e. since the start of the Cultural Revolution).

In pure mathematics there have been two striking and important contributions to classical mathematical problems—the result of Chen on the Goldbach hypothesis and the theorems of Yang and Zhang on the value distribution of meromorphic functions. Parallel with these is the substantial contribution of Wu to currently active questions in algebraic topology. There have been a number of other results on a less spectacular level; in fact, considering the small number of active research mathematicians, the level is impressive. However, there have been in China no major breakthroughs involving new mathematical methods or new conceptual approaches. In the West, the vitality of mathematics research does depend on the repeated cases of such advances. They are often made by young mathematicians as an

[47] CSCPRC, *Mathematics in the PRC,* pp. 64–66.
[48] *Ibid.,* pp. 65–66.

immediate result of their graduate training. In China, for a dozen years, there has been no such regular process of graduate training.

Some topics of central importance in recent Western developments simply do not appear in China. Algebraic geometry is the study of the properties of curves, surfaces, and higher-dimensional configurations defined by polynomial equations. It originated in the nineteenth century with projective geometry, was advanced by German mathematicians, and then by the Italians. In this century it has gone through three successive revolutions in the United States and in France. Today it stands in a central role—but it does not seem present in Chinese mathematics—nor does it appear to have any practical applications, although it does of late have vital connections with current problems of field theory in physics. As already noted, mathematical logic is hardly present in China. Other inactive subjects there include numerical analysis, Lie group theory, and many parts of algebra, both concrete (ring theory) and abstract (homological algebra and category theory). Such gaps and omissions are not surprising, given both the small number of mathematicians in China and the general observation that different countries often show special emphasis and favoritism in choice of research topics in pure mathematics.

In applied mathematics, the Chinese emphasis on solving the practical problems of farms and factories has borne fruit. Western techniques of operations research and orthogonal experimental design have been widely disseminated by Hua and others. Practical problems ranging from the shape of gear teeth to procedures for ship-lofting have been tackled with enthusiasm by Chinese mathematicians. Computers and calculation by the finite element method have been pressed hard, in intended competition with the West. At the present stage of the technological development of China the solution of such immediate problems was clearly a matter of pressing and practical interest. Thus the idea, arising from the Cultural Revolution, of deploying the scarce stock of Chinese mathematicians in these directions has had the intended and desirable effect.

For the long pull, however, the development of applied mathematics is not complete; the immediate use of older techniques is not balanced by a deeper study of modern sophisticated methods of applied mathematics, and there does not seem to be an adequate stock of trained, younger applied mathematicians. The new plans for education may rapidly change this situation, and the prospective further training of Chinese scientists in the West will help.

Chinese mathematics today is not culturally or fundamentally different from Western mathematics. The topics, the terminology, and the theorems of mathematics in China are among the topics, the theorems, and the terminology of Western mathematics. This is in part because the Chinese have very reasonably wanted to exploit for their own purposes the helpful Western ideas on computers, operations research, or number theory. More fundamentally it is because the results and theorems of mathematics are indeed international and independent of cultural setting or philosophical position.

Occasionally, it has been said of Chinese mathematics progress that results were achieved by the applications of Marxism-Leninism or of Mao Zedong thought. Now while we can admire and marvel at the great accomplishments of Chairman Mao in pulling China into the twentieth century, we cannot agree that his thought—or any such system of philosophical principles—can have utility in mathematical investigations beyond good feelings and general inspiration. The attempt to apply Hegelian dialectics in teaching mathematics is downright ridiculous. Thesis, antithesis, and the resolution of contradictions may be useful slogans for some purposes and consequently have some philosophic impact (though I doubt that). They do not serve in any real way to explicate pairs of opposites like positive and negative numbers or differentiation and integration. There is a simple and direct mathematical formulation of the notion of an inverse or an opposite. Specifically a function $g$ is *inverse* to a function $f$ if and only if both the composite $g \circ f$ ($g$ following $f$) and the composite $f \circ g$ are the identity. There are more general extensions and modifications of this simple notion of an inverse—the most sophisticated being perhaps the notion of "adjoint functor," a new mathematical concept not yet much known in China. However, these exact mathematical formulations owe nothing to and profit in no way from the speculative ideas of Hegelian or Marxist dialectic. The use of philosophical formulas and slogans by Chinese scientists is, under the circumstances, natural and inevitable. The use of such explanations by Maoists in the West is less defensible.

At the start of this chapter, I observed that the progress of a theoretical science like mathematics depends on talent, on stimulus from other scientists, and on social encouragement, including direction from above. All three of these elements have been present in China. As our discussion shows, there is no doubt as to the presence of many talented mathematicians there. There has been real stimulus both from older generations of Chinese mathematicians and from Western mathematics. This latter has acted both by direct contacts and through the literature. In China there is much more direction from above than in the West. The direction is concerned both with an insistent emphasis on the importance of immediate problems and with slogans stating the political line. There is no doubt that the emphasis on practical problems has been effective, though some considerable part of the effect had depended on the applications of ideas worked out in the West. The role of the political slogans and philosophic emphasis is less clear; it may have been necessary in the political circumstances of a revolutionary China. It is probably *not* necessary for the structure of mathematics, which should be independent of all politics and all philosophy, resting only on logic and the truth.

This is not to deny the direction of science by society—which exists in the West too. We are perhaps fortunate that for us this direction is expressed indirectly, by variations in the financial support of science and not by slogans or "lines."

*Beauty or Utility*

The 1976 Delegation on Pure and Applied Mathematics eventually became bored with the fact that every visit to every facility began with a presentation of the political line—essentially the same line at every facility. Hence, in the last visit, at the Shanghai Normal University, the Delegation requested (and received) a much abbreviated briefing. Then later, upon request, one of the Delegation members gave a talk about the form of graduate education in the United States. The talk emphasized the great freedom of choice in the United States—choice of students by the graduate school, choice of courses by the student, choice of research topics by the professors. The talk also brought out the American belief in the existence of mathematical talent and the importance of developing this talent as a natural resource. This emphasis was so sharp that the responsible member subsequently assembled the whole Delegation and the local mathematicians again at the end of the visit—and this time the Delegation received a full description of the current line. I suppose it must have been evident that this description was needed.

In the subsequent discussion session, one member of the Delegation asked about one of the units in the normal college—the unit dealing with international education. In response we were told that the unit dealt almost exclusively with education in Russia and with the weakness of that education. For good measure, we were also reminded that a *New York Times* reporter, in a dispatch from Moscow, had recently reported that Russian education was now almost exactly like American education. The Delegation was clearly in the doghouse.

A different issue then came up. One of the Delegation members then brought up the beauty of mathematics and the role of this beauty in motivating research. His Chinese counterpart responded with an emphasis upon the utility of mathematics—it originates in practical problems, and its beauty lies in its use for those problems. This dialogue, as recorded in the report of the Delegation, never did reach a conclusion between beauty and utility. Neither was explained in depth. As for beauty, it is the innate symmetry and balance of mathematical results which attract thoughtful attention and which often suggest generalizations—as in the case of Clark Maxwell, who predicted radio waves without any experimental evidence, just the symmetry of his equations for the electromagnetic field. As for utility, von Neumann and others have observed that some of the best inspirations of modern mathematics have originated in the natural sciences; moreover, in the recent development of China, these origins and the prospective application must play a permanent role.

China has emphasized utility, the West beauty. Both are there, and the answer may lie not in the dialectic contradiction between them, but in the balance. Perhaps both parties can approach a better balance!

*Nicolaas Bloembergen*

# PHYSICS

## PHYSICS AND NATIONAL GOALS

Physics played a prominent role at the National Science Conference held in Beijing in March 1978. The speeches of Premier Hua Guofeng and Vice-Premier Deng Xiaoping provided ideological justification and political approval at the highest level for the new science policy. The new policy was further elaborated by Fang Yi, a member of the Politburo of the Central Committee of the Communist Party of China, vice-premier of the State Council, and minister in charge of the State Science and Technology Commission. Fang outlined an eight-year plan for science and technology and enumerated eight comprehensive areas of "important new technologies and pace-setting disciplines that have a bearing on the overall situation; namely, agriculture, energy resources, materials, electronic computers, lasers, space science and technology, high energy physics, and genetic engineering."

Physics is intimately, or even exclusively, involved in all except the first and last of these fields, as the following excerpts show.

*On energy:* "Atomic power generation is developing rapidly in the world, and we should accelerate our scientific and technical research in this field and speed up the building of atomic power plants. We should also step up research in solar energy, geothermal energy, wind power, tide energy and thermonuclear fusion."

*On materials:* "We must solve the key scientific and technical problems in producing special purpose materials, structural materials and compound materials necessary for our national defense industry and new technology and evolve new materials characteristic of China's resources."

*On electronic computers:* "In the next three years we should rapidly develop basic research on computer science and related disciplines, lose no time in solving the scientific and technical problems in large-scale integrated circuits."

*On laser science and technology:* "We will study and develop laser physics, laser spectroscopy and nonlinear optics in the next three years. We should solve a series of scientific and technical problems in optical communications, raise the level of routine lasers quickly and intensify our studies of detectors. We expect to make discoveries and creations in the next eight years in exploring new types of laser devices, developing new wavelengths of lasers and studying new mechanisms of laser generation, making contributions in the application of lasers to studying the structure of matter. We plan to build experimental lines of optical communications and achieve big progress in studying such important projects of laser applications as separation of isotopes and laser-induced nuclear fusion. Laser technology should be popularized in all departments of the national economy and national defense."

*On space science and technology:* "We should attach importance to the study of space science, remote sensing techniques, and the applications of satellites."

*On high energy physics:* "We expect to build a modern high energy physics experimental base in ten years, completing a proton accelerator with a capacity of 30–50 BeV in the first five years, and a giant one with a still larger capacity in the second five years. Completion of this base will greatly narrow the gap between our high energy accelerators and advanced world levels and will stimulate the development of many branches of science and industrial technology. . . . The high energy physics experimental base is a key project on the nation's list of scientific research centers, and it is necessary to organize the forces of all quarters to build and use it jointly."[1]

These extensive quotations from the highest science policymaker in China demonstrate the importance currently attached to physics. Physics, however, has played a significant role in the PRC since 1949, especially in such strategic areas as atomic weapons, aeronautics, ballistic space missiles, and computers.[2] The accompanying table compares China's performance in these strategic domains with those of other major powers. It is clear that even during the height of the Cultural Revolution research in physics continued in quasi-military establishments. (No information on classified research institutes in China has been published, and all reports on science and technology in China underestimate the state of the art to an unknown degree.)

Apart from military requirements, China's planned technological devel-

[1] "Abridgement of Fang Yi Report to National Science Conference," New China News Agency, Mar. 28, 1978; in *Foreign Broadcast Information Service,* Mar. 29, 1978.

[2] Susan B. Rifkin, "The Development and Use of Nuclear Energy in China," *IR & T Nuclear Journal,* 1, 4 (Apr. 1969): 1–22, and 1, 5 (May 1969): 1–13; M. S. Minor, "China's Nuclear Development Program," *Asian Survey,* 16 (June 1976): 571–79; and Manfredo Macioto, "Scientists Go Barefoot," *Successo,* Jan. 1971, pp. 232–38.

*Comparative Timetable of Strategic Achievements of China and Other Major Powers*

| Achievement | Year | | | | | |
|---|---|---|---|---|---|---|
| | U.S. | U.S.S.R. | Britain | France | Japan | China |
| First reactor | 1942 | 1946 | 1947 | 1948 | — | 1956 |
| First A-bomb | 1945 | 1949 | 1952 | 1960 | — | 1964 |
| First H-bomb | 1952 | 1953 | 1957 | 1968 | — | 1967 |
| First satellite[a] | 1958 | 1957 | — | 1965 | 1970 | 1970 |
| First jet plane | 1942 | 1945 | 1941 | 1946 | — | 1958 |
| First Mach 2 jet | 1957 | 1957 | 1958 | 1959 | — | 1965 |
| First prototype computer | 1946 | 1953 | 1949 | — | 1957 | 1958 |
| First commercial computer | 1951 | 1958 | 1952 | — | 1959 | 1966 |
| First transistor | 1952 | 1956 | 1953 | — | 1954 | 1960 |
| First integrated circuit | 1958 | 1968 | 1957 | — | 1960 | 1969 |

SOURCE: Adapted from Manfredo Macioto, "Scientists Go Barefoot," *Successo*, Jan. 1971, pp. 232–38.
[a] Weights in kg were as follows: U.S., 14; U.S.S.R., 83; France, 42; Japan, 23; China, 173.

opment clearly implies a major effort in applied physics. A key ideological point in China is the extent to which basic physics should be involved in this effort. Clearly the technological development of the Western world and the Soviet Union would not have been possible without substantial activity in basic physics. But "to catch up" and reproduce this technology does not require the same amount of basic scientific research. The table vividly illustrates this point; China has been able to compress the amount of time required significantly.

## HISTORIC ROOTS OF PHYSICS IN CHINA UNTIL 1970

Until A.D. 1300, China was scientifically and technologically ahead of the West. Such devices as the magnetic compass and the mechanical clock were Chinese inventions. But from 1300 to 1900 scientific activity virtually halted. The socioeconomic reasons for this have been discussed by many scholars and need not concern us here.

One feature—the separation of mental and manual activities—was, however, particularly detrimental to the development of the physical sciences. Scholars, who constituted the highest social class and the administrators of the empire, concentrated on humanistic studies, the Chinese classics, and philosophical and legal issues. In contrast, craftsmen belonged to low social strata. The mistrust of an elitist intelligentsia that has surfaced on many occasions in the PRC is deeply rooted in the social practices of Imperial China.

Toward the end of the nineteenth century, China's scientific and technological backwardness became so apparent that some reforms—with inevitable Westernizing influence—were reluctantly instituted. Beijing and Fudan Universities were founded in 1898 and 1908, respectively. The Science Society of China was established in 1914 and the Academia Sinica in 1928.

But opportunities for physics research were severely restricted by difficult socioeconomic conditions and internal political strife. Later the Japanese war and occupation eliminated the opportunity for research. A substantial number of students were nevertheless trained in the sciences at the undergraduate level, and many of the best went abroad in the 1930's for training in Japan, Europe, or the United States. After the end of the Japanese occupation when the revolutionary struggle in China reached its climax, many intellectuals left the mainland. At the end of 1949 over 3,500 Chinese were enrolled in American colleges and universities. Although this influx of Chinese talent enriched physics in the United States, many outstanding physicists with Western educations remained in or later returned to China.

In 1950 the Chinese Academy of Sciences (CAS) was established by merging the Academia Sinica and the National Academy of Sciences of Beijing. The CAS was patterned after the Soviet model of a central science organization and soon became the center of research activities and the main recipient of state funds for science in the nascent PRC. The Institute of Physics of the CAS was founded in 1950 and started with a group to study magnetic materials. Until 1959, the Soviet Union exercised a dominant influence on science in the PRC. The primary consideration in founding new research institutes was their relevance to the development of heavy industry. Higher education was reorganized along Soviet lines. Soviet experts taught in China, and increasing numbers of Chinese students went to the U.S.S.R. for advanced physics training. The most important subject was nuclear physics. Many students were trained at the Soviet counterpart of CERN, the Joint Institute for Nuclear Research in Dubna. A substantial number of Chinese authors contributed articles to the *Soviet Journal of Solid State Physics* in the late 1950's. The Soviet Union also provided China's first experimental nuclear reactor and cyclotron.

Before 1949 physics research in China was virtually nonexistent and did not lead to significant publications, but by 1961 an earlier survey, *Sciences in Communist China,* conducted by the American Association for the Advancement of Science, could indicate the regular publication of research articles in nuclear physics, solid-state physics, fluid dynamics, acoustics, optics, and electronics.[3] Most of these papers were in the area of applied physics and often related to instrumentation, improvements in techniques, or the application of physical techniques to problems in other areas of science. The published papers are of good quality, and the majority would have been accepted for publication in Western journals of applied physics. They also demonstrate that Chinese scientists followed world developments in physics.

The Soviet pattern for research and development was discredited during

[3]Sidney H. Gould, ed., *Sciences in Communist China* (Washington, D.C.: American Association for the Advancement of Science, 1961). See especially Robert T. Beyer, "Solid State Physics," pp. 645–58; and T. Y. Wu, "Nuclear Physics," pp. 631–43.

the Great Leap Forward (1958–59) and ended with the complete Soviet withdrawal of personnel and material aid from China. During this brief period, basic research was not in favor; its administration was decentralized and "popularization of science" was promoted. But between 1959 and 1966, while Liu Shaoqi was chairman of the PRC, modern research had the opportunity to prove itself. Chinese physicists showed their "self-reliance." They not only maintained, but actually significantly enlarged the level of research activity previously established with Soviet assistance, including additional facilities for nuclear research. Most significantly, a program for enrichment of weapons-grade materials was started and carried to quick successes (see table) without outside assistance.[4] Work in hydrodynamics and jet propulsion blossomed. The Chinese also started their own laser research and development program. An outstanding accomplishment in x-ray analysis by any international standard was the structure determination of insulin (1965). China had established a national, independent, and viable activity in physics.

But the Great Proletarian Cultural Revolution (1966–69) interrupted these activities. Universities were closed, and research activities were severely curtailed in the CAS. Of the 120 institutes in the CAS, only 17 remained under its immediate jurisdiction; 19 were under "dual level of leadership, with the locality in charge"; and the remainder were placed under direct control of local authorities or industrial enterprises. Basic research was discouraged, intellectual elitism eradicated, and the close integration of science with production processes emphasized.

Research in some institutions, especially those important to national defense, was apparently shielded from the Cultural Revolution's full impact. The programs under direct military control in nuclear weapons and missiles and a number of local or CAS institutes with work relevant to national security in electronics, optics, and communications enjoyed some measure of protection. The universities, however, were the main battleground of ideological conflict. They reopened gradually after 1971, but their role in research, never large, had to start from scratch.

Physics manpower at the beginning of the 1970's consisted of three categories. (1) The "old guard," who received doctorates from European and American universities before 1950. They number fewer than 50. About half hold top positions of technical leadership; they must devote most of their efforts to administrative and political matters and are not active in research. (2) A larger contingent of physicists trained by the Soviets during the 1950's. Many of these provide scientific leadership for various research projects. (3) Those physicists who received advanced training inside the PRC between 1956 and 1966 and are now active in research. There was no graduate training of any kind in physics between 1966 and 1974.

[4] Rifkin, "Nuclear Energy"; Minor, "China's Nuclear Development Program"; Macioto, "Scientists Go Barefoot."

## THE ORGANIZATION OF PHYSICS RESEARCH IN THE 1970's

### The Major Institutes of Physics

The research institutes of the CAS, including those that were partially or wholly transferred to local control in the 1960's, are the primary research-oriented organizations. The main task of universities is education, and the primary objective of industrial institutes is production.

*Institute of Physics (CAS, Beijing).* Nearly all visiting foreign physicists have seen this Institute, which lies on the northwest side of Beijing. The trip reports of visitors from different countries and from different specializations agree remarkably well, although a three-day visit by a delegation[5] can obviously acquire more information than a half-day visit by an individual. The director of the Institute, Shi Ruwei (Ph.D., Yale), leads a staff of 700, of whom 90 percent work in the laboratories. There are eight departments in the Institute: Magnetism, Crystallography, Low Temperature, High Pressure, Acoustics, Lasers, Plasma Physics, and Theoretical Physics.

The magnetism group, one of the first groups begun after the Liberation, was started in 1950. In 1975 this group was concerned with magnetic bubbles, microwave ferrites, magnetic thin films, and theory.

The crystallography group has about 50 workers and has been involved in crystal growth since 1958. This group also studies crystal structure by means of x-ray diffraction and measures physical properties of the crystals being grown. Materials of current interest are yttrium aluminum garnet–neodymium, lithium iodate, and gadolinium gallium garnet.

The low-temperature group has a Collins-type helium liquefier, built in 1964. It has been improved periodically and can now produce 20 liters/hour. The physicist who produced the first liquid helium in China, Hong Zhaoshen, studied at the Massachusetts Institute of Technology and the Kamerlingh Onnes Laboratory in Leiden.[6] Work on the first dilution refrigerator in China is in progress. Research on various aspects of superconducting devices is also carried out.

The high-pressure group has a static high-pressure system with pressures up to 45 kilobars. Shock waves for generating dynamic pressures are also available. Both methods are used for producing artificial diamonds.

The acoustic department has done ultrasonic work on high-power transducers, surface waves, and holograms.

The laser group was started in 1970. Its subgroups are active in color holography, optical processing, semiconductor lasers, and (since 1974) nonlinear optics.

The plasma group was begun in 1973, after a preliminary study of the

[5] Anne Fitzgerald and Charles P. Slichter, eds., *Solid State Physics in the People's Republic of China* (Washington, D.C.: National Academy of Sciences, 1976).

[6] C. K. Jen, *Science and Public Affairs*, 30 (Mar. 1974): 18.

Western literature of one to two years carried out by five workers, two of whom had obtained advanced degrees in the U.S.S.R. None, however, had previous experience in plasma physics. In 1975 the group contained 20 workers, scientists, and technicians. A small Tokamak machine was completed in 1974. The cleaning of the vacuum components with hydrofluoric acid was done by carrying them outdoors on a windy, wintry day to provide adequate ventilation. The group has detected neutrons, but was doubtful that they were of thermonuclear origin. A glass laser system for plasma target interaction was started in 1972 and was capable of delivering 36 joules in a 6-nanosecond pulse in 1975. A plasma pinch experiment was also being conducted.

The theory group consisted of about ten physicists, working in such areas as relativity and quantum field theory. In addition, each of the Institute's other seven departments had a few theoretical workers. The magnetism group, for instance, had five. Thus, the total theoretical staff of the Institute numbered several dozen. Some topics studied by the theory department were superconducting tunnel junctions, especially those involving niobium; spin-wave modes in the range of conditions where magnetic dipolar interactions are important; relativistic moving-medium electrodynamics; the gauge theory of gravitation; optical information processing and analog-type optical simulation of computers; crystal growth; the not-yet-understood phenomenon of the increase in neutron-scattered intensity in a static electric field; and the scaling law for magnetic phase transitions.

*Institute of Semiconductors (CAS, Beijing).* In 1978 the director of this Institute was Huang Gun (Ph.D., Bristol). Semiconductor research in China was started in 1956 under the auspices of the Institute of Physics, and the first semiconductor devices were made in 1958. The Institute of Semiconductors was established as a separate entity in 1960. Its staff grew from an initial 100 to 900 in 1976. Of these, 600, including 400 university graduates, were involved in scientific research. The Institute is housed near the city center in four buildings of traditional Beijing architecture with a total area of 10,000 $m^2$. Security is tight. City dust and crowding present major problems. Growth of silicon crystal and production of silicon devices began in 1963. The first integrated circuits were made in 1965.

As of 1978, the four major areas of activity were:

1. Semiconductor materials. Studies of gallium arsenide (GaAs) for use as laser substrate, liquid and vapor phase epitaxial GaAs for making microwave devices, and epitaxial silicon for integrated circuits.

2. Integrated circuits technology. Computer-operated mask making, MOS circuits stability and control, and passivation and reliability studies.

3. Microwave devices. Silicon impatt diodes as well as GaAs Gunn devices and avalanche oscillators.

4. Semiconductor lasers. Emphasis on GaAs and gallium aluminum arsenide (GaAlAs) heterojunction lasers.

*Institute of Biophysics (CAS, Beijing).* This Institute was established in 1958 and is located near the Institute of Physics. The director is Bei Shizhang (Ph.D., Tübingen), who heads a staff of 400 scientific and technical workers. Of the seven divisions, the following four are most closely involved with physical techniques: Radiobiology, Crystal Structure, Physics of Receptors, and Experimental Technology, which includes magnetic resonance spectroscopy and other probing techniques. The Institute appeared to be active in many areas, but discussion of these topics clearly belongs more appropriately in the chapters on other sciences. The Institute had an electron microscope imported from Japan, nuclear and electron magnetic resonance spectrometers, equipment for optical and spectroscopic analysis, x-ray diffraction apparatus, and nuclear radiation detectors, all with associated electronic equipment, as well as chemical and biological facilities.

The Institute's outstanding work on the structure of insulin, finished in 1965 just before the Cultural Revolution, has not been followed with related work. As of 1979, there was little activity in the structural analysis of biochemical compounds.

*Institute of Electronics (CAS, Beijing).* The director of the Institute, Gu Dehuan, heads a staff of 1,100 people, 600 of whom are research personnel.

Medium- and high-power traveling wave tubes (1 kw at 5-cm wavelength) have been developed. Pulsed klystrons at 15-megawatts peak power at 250,000 volts are also at an advanced stage of development. High-power transformers, vacuum tube rectifiers, and alnico permanent magnets have all been made in China.

Millimeter tubes including reflex klystrons operating at 250 milliwatts and carcinotrons (millimeter wave tubes) at 1 kw were manufactured at the Institute. Since 1970 there has been an active program in high-power gas lasers, notably carbon dioxide with cold cathodes and argon ion lasers.

*Institute of Atomic Energy (CAS, Beijing).* The director of this Institute is Jian Sangjiang (Ph.D., Paris). One of the deputy directors is the theoretician Peng Huangwu (Ph.D., Edinburgh). The Institute, which was started in 1958, consists of two separate laboratories. The main experimental facility is 50 km southwest of Beijing, and there is a cosmic ray station at 3,220 m altitude, located in the western mountains of Yunnan province. The Institute's personnel number about 1,000, of whom 400 are university graduates, but some of these may have been transferred to the Institute of High Energy Physics.

The principal facilities are a heavy water reactor, supplied by the Soviet Union in 1958; a small cyclotron started by the Soviets, but completed by methods of self-reliance after their departure in 1960; a 2.5-MeV Van de Graaff electrostatic generator; and a homemade swimming pool reactor.

The 1.2-m cyclotron has been changed to an isochronous ridge-focused machine. In 1979 its performance was a current of 80 microamperes with variable energy: for protons 3–20 MeV, for deuterons 3–14 MeV, and for alpha-particles 6–28 MeV. The relative spread in energy of the extracted beam is 1 percent with a 40 percent extraction efficiency. External beams are focused with a quadrupole doublet and bending magnets.

The Soviet heavy water reactor was redesigned in 1962 and went critical in 1964. After the Cultural Revolution, it was upgraded from 7 to 10 megawatts. The thermal flux was $3.5 \times 10^{13}$ neutrons/cm$^2$ sec, and the fast flux was approximately the same. Total $U^{235}$ inventory was 5 kg. Reprocessing of the aluminum-clad fuel rods with 1.2 percent enriched $U^{235}$ was being done elsewhere. The reactor design appeared primitive but reliable. The small swimming pool reactor was completed in 1972 and has a power output of 3.5 megawatts. The reactors were used primarily for material testing and isotope production. There was a "hot lab" with eight manipulators for chemical separation. The waste treatment reduces radioactivity to $10^{-11}$ curie/liter before dumping.

The home-built detection equipment included a lithium-drifted germanium detector with a resolution of 4 keV at the $Co^{59}$ $\gamma$-ray line and a Charpak-type proportional chamber. Other detectors, including plastic scintillators and Geiger-Müller and Cerenkov counters, were being built by industry.

The 2.5-MeV Van de Graaff generator was being upgraded to a proton current of 50 microamperes. It was used for nuclear reactions with three-body final states.

Some work on isotope separation by old-fashioned electromagnetic techniques was also carried out, but other techniques, including lasers and centrifuges, were being contemplated.

*Institute of High Energy Physics (CAS, Beijing).* This Institute was separated from the Atomic Energy Institute in 1974 for "budgetary reasons." Perhaps a growing commitment to a high energy physics program was anticipated. Such a program clearly could not be justified on programmatic grounds of immediate relevance, which characterizes most Chinese work in nuclear physics. This Institute has conducted theoretical studies in elementary particle physics, design studies for accelerators, and a limited amount of experimental work on wire chambers and detectors for very high energy particles and cosmic rays.

Fang Yi revealed plans for the construction of a 50-GeV proton accelerator in his March 1978 policy speech. (For details about the site and organization of this accelerator, see the technical note on high energy physics by L. C. L. Yuan in this volume.) It appears reasonable to assume that the decision to build an accelerator was partly based on studies carried out at the Institute of High Energy Physics and that the Institute will be of growing importance in the 1980's.

This type of high energy accelerator can be constructed in China on a self-reliant basis at relatively modest risk and cost. It is not clear whether it will contribute significantly to the world's understanding of basic physics. However, the accelerator will enable the PRC to catch up on several important high-technology developments. It will also provide a solid base for training in experimental high energy physics and make future exchanges with CERN, the Fermi National Accelerator Laboratory, and other institutions more meaningful. China has been realistic and not succumbed to the temptation of leapfrogging and attempting an ambitious super-accelerator-cum-colliding-beam facility.

*Institute of Nuclear Physics (Shanghai; CAS and local control).* Established in 1959, this Institute is located about 35 km northwest of Shanghai. Everything is Chinese made. The cyclotron, with a 1.2-m pole diameter, was started in 1963. The 120-kw RF generator was surplus equipment from the Beijing Broadcasting Corporation. The magnetic field is 14.6 kilogauss. The internal beam current is 2.8 milliamp of 27.2-MeV protons, 13.6-MeV deuterons or 27.2-MeV alpha-particles. There is a "rabbit" target arrangement for irradiation. It is a classical cyclotron setup.

There is a 0.12-megacurie cobalt source, supplied by Canada, in a water pool 5 m deep in a heavily shielded vault.

The neutron generator uses an eight-step cascade three-phase transformer. The energy of the 3–5 milliampere deuterium ion beam is 200 kilovolts. The total ($4\pi$ steradian) neutron production is $10^{10}$/sec. The work is concerned mostly with applied nuclear physics—isotope production, radiation damage, irradiation of biological objects (seeds), and detector development. The Institute collaborates with the Physics Department of Fudan University.

*Institute of Optics and Precision Instruments (Shanghai; local control).* This Institute was founded in 1966 and is under local control. The director, Lu Ming, heads a staff of 1,000, of whom 400 are actively engaged in research. The Institute is housed in a number of large, well-equipped buildings on the outskirts of Shanghai. Security is tight. A prominent member of the laser group is Lin Zunji.

The Institute has a well-engineered YAG-Nd system and excellent crystal-growing facilities. The Institute produces its own laser-quality YAG and other crystals for electro-optic devices.

There is also a neodymium-glass laser system with eight glass rod amplifiers and two Faraday isolators. A disk laser system with three glass slabs of 10-cm aperture each was under construction in 1975. The glass was fabricated in ceramic crucibles to avoid platinum inclusions. There was close cooperation between the glass-manufacturing facility and the Institute's laboratory so that high-quality laser glass was available. The Institute also has a high-power cw $CO_2$ discharge laser facility of unknown size. Unfortunately it has not been visited by many foreigners. Since it has probably the most advanced optical and laser facilities available in China today, further information would be most desirable.

*Institute of Ceramics (Shanghai).* This Institute is also known as the Institute of Silicates. It is under the direction of Zheng Xiquan and has 900 workers. It was separated from the Institute of Metals and Ceramics in 1960. Its six departments are concerned with crystal growth (diamond, quartz, niobates, tantalates); special glasses (chalcogenides, high silica, photochromic glass); special ceramics (organic carbon-fiber reinforced ceramics, amorphous semiconductors, high-temperature anticorrosive coatings); new analytic methods (chemical); instrumental analysis (optical, physical); and pilot plant and workshop techniques. Leading solid-state physicists here include Tian Kewen, Yu Jinliang, and Zhang Shouqing.

*Physics Research Facilities at Universities*

The primary mission of the universities is education, and until recently, there was no graduate education. This situation may change rapidly under the new eight-year plan for scientific development, announced at the National Science Conference. As I discuss more fully in the next section, the universities often have rather extensive training and production facilities for physics devices. Facilities for crystal growth, semiconductor device technology, and laser fabrication exist at several universities. As part of this training program, more advanced developmental work that can be designated as applied research may be carried out. It should be kept in mind that this work started from scratch in 1971 at the end of the Cultural Revolution. Research programs at universities were often reactivated in collaboration with CAS institutes. Although every province in China now has its own university or equivalent educational institution, only the major physics departments of national universities that have been visited by Western scientists are mentioned here.

*Beijing University.* Beijing University has beautiful grounds in the northwest part of the city. The vice-chancellor is Zhou Peiyuan (Ph.D. in hydrodynamics, California Institute of Technology), and the head of the Physics Department is Wang Zhuxi (Ph.D., Cambridge). Another senior member of this department is Huang Gun (Ph.D., Bristol), well known in the West for his book on lattice dynamics, coauthored with Max Born during Huang's stay in England in the 1940's and early 1950's. The Physics Department has four specialties: theoretical physics, laser physics, low-temperature physics, and magnetism. Research activities are on a much smaller scale, but are rather similar in nature to the corresponding activities at the CAS Institutes of Physics and Semiconductors. The department has a helium liquefier with a capacity of 20 liters/hour, facilities for growth of electro-optic crystals, construction of helium-neon lasers, and somewhat dated spectroscopic analytical and electromagnetic instruments.

*Qinghua (Tsinghua) University (Beijing).* The Physics and Electronics Departments are two of the 11 departments in this technical university, which is sometimes referred to as the M.I.T. of China. There are elaborate production facilities for semiconductor devices and for electronics and electric instrumentation. Zhang Wei, the vice-chairman of the University, obtained a Ph.D. in civil engineering at the University of Berlin. The head of the Physics Department until 1976, Ma Wenzhong, has apparently been removed from office.

*Fudan University (Shanghai).* The Physics, Optics, and Atomic Energy Departments are the three departments involved with physics at Fudan University. They have close relations with a number of factories and the Institute of Nuclear Physics. Lu Hefu is professor of theoretical physics, Yang Fujia is a section head, and Ms. Xie Xide is professor of semiconductor physics. Production facilities associated with the University include an electronics plant that grows silicon and GaAs single crystals and makes semiconductor components, silicon-integrated circuits, and electronic instruments. The optical plant makes electro-optic sources and gas lasers. A faculty member built the first dye laser in China in 1975 without ever having seen one in operation before. The design was copied from the literature, with rhodamine 6G pumped by a pulsed nitrogen laser. There is an active group in theory, which instituted one of the first graduate seminars after the Cultural Revolution.

*Nanjing University.* The Physics Department of Nanjing University covers a rather wide range of topics and has seven specialties: low temperature, crystallography, semiconductors, magnetism, radio physics, acoustics, and nuclear physics. The facilities were similar in nature to those seen in Beijing and Shanghai.

The acoustic lab has a large anechoic chamber, 11.5 × 10.5 × 9.7 m, with good measurement facilities, which is also used by outsiders. It is headed by Professor Wei Yongjue. The nuclear physics group, led by Professor Shi Shiyuan, has a small, disk-loaded linear accelerator, powered by a 230-kw magnetron. The 0.7-MeV, 60-milliampere pulses of 1-microsecond duration, repeated 300 times per second, are used for irradiation effects in plastics. Detection equipment and pulse-height analyzers for Mössbauer effect experiments are also in the lab.

The leading member of the Physics Department is Professor Bao Jiashan, and the rather large crystallography group is led by Professor Feng Duan.

*Zhongshan University (Guangzhou).* The Physics Department of Zhongshan University is active in lasers and nuclear physics. Nonlinear electro-optics effects and holograms are studied. The nuclear physics section has built semiconductor radiation detectors and an x-ray crystal spectrometer.

*Jiaotong University (Xian).* This is a technical university with 25 departments, including mechanical, electrical power, and radio engineering. There is no physics department proper. But there are elaborate mechanical test facilities for the strength of materials and for electronic semiconductor components. Harvard-educated Huang Chang is one of several professors in radio communications.

## *RESEARCH, EDUCATION, AND COMMUNICATION IN PHYSICS*

### *Status of Research*

The reader with technical expertise will already have noticed from this description of China's research facilities that they lag considerably behind those of the West. The PRC cannot participate in advancing the frontiers of knowledge in physics at the moment—a fact well recognized by China's scientific and political leaders. Their primary aim is to catch up in technologically important areas; hence a base in applied physics is needed. This section attempts to assess the breadth and quality of the base in applied physics on which the eight-year plan for science and technology is grounded.

Usually the primary basis for such an evaluation would be an analysis of publications in research journals. In the case of China such an approach is not feasible since much research does not get published. During the Cultural Revolution publication was discouraged in order to deter personal glorification and prevent intellectual elitism. Regular publication of two physics journals, *Wuli* (Physics) and *Wuli xuebao (Acta Physica Sinica),* has been resumed under the auspices of the Chinese Institute of Physics in Beijing. A review of the complete contents of both publications for the year 1976 reveals the applied nature of much of the work. Some articles are of a review nature; others are nonscientific, i.e. quasi-popular, educational, philosophical, ideological, or blatantly political. It appears that the more innovative articles are preferentially published in *Acta Physica Sinica.* Its editorial organization and referencing system is similar to that of Western research journals. This journal also contains more theoretical papers, without immediate applications, dealing with statistical mechanics, elementary particles, cosmology, and gravitation. In 1977 a new journal, *Gaoneng wuli yu hewuli (Physica Energiae Fortis et Physica Nuclearis),* on nuclear physics and elementary particles appeared, and it can be expected to take over part of the function of *Acta Physica Sinica.* A significant increase in published work may be expected under the auspices of the eight-year plan. In addition, there are a number of journals that span a variety of scientific fields. Some physics articles also appear, for example, in *Zhongguo kexue (Scientia Sinica).*

The published papers tend to confirm the more detailed observations that have been made by many visiting Western scientists. A large proportion of the experimental research is concentrated in four major areas of activity: semiconductors, lasers and electro-optics, low temperature, and nuclear physics. Other fields with considerable activity include magnetism, acoustics, high pressure, cosmic rays, and theory. The following evaluation is based on recent reports by U.S. scientists, a detailed account of solid-state physics in 1975,[7] and Panofsky's report on his visit in October 1976.

*Semiconductors.* This field has received the most emphasis in the general area of solid-state physics. The main goal is to advance semiconductor technology for the electronics industry. The activities are phenomenological in approach, developmental in nature, and applied in orientation. The main emphasis is on integrated circuits, injection lasers, microwave devices, and amorphous semiconductors. The Institute of Semiconductors covers all four of these areas. Work on injection lasers is also carried out at the Institute of Physics, the Institute of Optics and Precision Instruments, and Beijing University. Most universities not only engage in work on integrated circuits, but actually have production lines on campus. The foremost educational objective is training of technicians with manual skills.

The production of integrated circuits is based on silicon. Pure single crystals are produced in China and are commercially available with typical wafer diameters of 1.5 inches (3.8 cm). Complete production lines for MOS circuits at Qinghua University and for bipolar circuits at Fudan University include substrate preparation, photolithography, oxidation diffusion, epitaxy, metallization, and bonding. Conventional techniques, adapted from the Western literature, are followed. There is some research on building a block library for mask generation and on methods to determine sodium contamination of silicon dioxide in MOS devices by neutron activation analysis and capacitance measurements. Chinese scientists have established a basic silicon technology in a relatively short time, with only indirect outside communication and no outside technical assistance. They have the capacity to build large computer systems, as well as all the less sophisticated products of solid-state electronics.

Injection lasers are based on GaAs-GaAlAs. Research in this area includes studies of the phase diagram of the GaAlAs system at low aluminum concentration, the influence of heating and thermal resistance on emission characteristics, and optimization of laser characteristics by the use of an orthogonal matrix method.

Microwave devices are based on a variety of structures: p-n junctions, Schottky barriers, tunnel diodes, impatt diodes, and bulk negative-resistance devices. A rather advanced research effort at the Institute of Semiconductors was concerned with avalanche relaxation oscillations in planar Gunn diodes. A clear observation in bulk structure has been made, and an analysis of the associated recombination radiation was in progress.

The amorphous semiconductor work at the Institute of Ceramics deals with switching and memory properties in chalcogenides, specifically the GeTeSSb system. These effects were shown to be mostly thermal in origin, and no important applications have been developed. The Chinese appeared

[7] Fitzgerald and Slichter, *Solid State Physics.*

not to perceive the diminishing interest in the West for this type of switching device.

In view of the interest in devices and relatively short-term applications, it is understandable that there are no research programs on spectroscopy of semiconductors at any frequency, no band structure determinations, etc. There is also little investigation of materials other than the silicon and GaAs systems mentioned. It is less understandable that little emphasis is placed on infrared photodetectors, solar cells, window materials, fibers and thin-film structures, and light-emitting diodes. These devices have important applications in integrated optics and optical processing. The nearly exclusive emphasis on GaAs injection lasers, with noticeable duplication of effort in different institutes, amounts to a serious imbalance of effort. It is, of course, possible that activities not shown to visiting scientists partially redress this shortcoming.

As of 1978, China had a semiconductor technology base that in some major areas was only five or six years behind Western technology. The gap could be narrowed rapidly during the eight-year plan.

*Lasers.* It was already apparent in 1975 that considerable emphasis was placed in China on laser development. This was reaffirmed in the eight-year science program, where lasers were explicitly mentioned as one of the eight thrust areas. The Shanghai Industrial Exhibition exhibited a YAG-Nd laser range finder with a digital readout meter for the general public. It also displayed an argon-ion laser coagulator for ophthalmic applications and a 100-watt cw $CO_2$ discharge laser for welding, cutting, and scribing operations.

In 1975 laser technology at the Institute of Optics and Precision Instruments lagged only three to five years behind the Western state of the art. Workers here grew YAG-Nd crystals of high quality and had eliminated problems of coring. Thermal and elasto-optic coefficients were measured. The crystals were used in a 20-megawatt oscillator amplifier system and efficiently generated the fourth harmonic at 264 nm. This radiation was used for spontaneous parametric down conversion of visible light in an ADP crystal. A striking demonstration of generation of blue and red light was observed.

Good neodymium-glass laser systems with rod and disk geometry were also available. Systematic investigations of self-focusing and damage thresholds had been carried out. The results would have been publishable in a Western journal on applied physics.

In comparison, the glass laser system at the Institute of Physics in Beijing used for laser-plasma target interactions was more primitive. The nonlinear optics investigations and the argon-ion laser setup in this Institute reminded visitors of the apparatus and setups used in the early 1960's in American laboratories. Here the technology gap was ten years or more.

The first Chinese dye laser was constructed at Fudan University in 1974–75. There have been problems with obtaining suitable dyes for laser and saturable absorbers; no materials have been imported from Japan or other countries. It is not known how much the dye laser program has developed since this self-reliant start in 1975. It is clear that rapid strides can be expected in the early 1980's.

The university workshops at several universities have a production line of helium-neon lasers. Vacuum- and gas-handling techniques are taught. Optical coatings, optical alignment, and interferometric quality control are also part of this technical training.

At Beijing University a cadmium-ion laser with an output at 441.6 nm was used to detect nitrogen dioxide by red fluorescence of its dissociation products. This university also developed a range finder with spherical infrared-emitting GaAs-diodes and silicon-photodiodes as detectors. This is one of the few examples of another semiconductor application in optics besides the intensive program on injection lasers discussed previously.

No pulsed carbon-dioxide, gas-dynamic, chemical, or electron-beam lasers have been reported by Western visitors. These may well exist, however, since Chinese scientists have clearly demonstrated their competence to copy any existing laser technology.

Extensive work on electro-optic crystal growth is conducted in many institutes. The work everywhere is competent and professional, with clean laboratories and well-built equipment. The following materials have been produced: alkali halides, diamond, sapphire, quartz, silicon, gallium arsenide, gallium phosphide, lithium iodate, lithium niobate and tantalate, ammonium dihydrogen phosphate, bismuth germanate and silicate, yttrium aluminum garnet, and yttrium aluminum perovskite with neodymium.

Holography is used in character-recognition studies. Automated typesetting of Chinese characters is a possible and an important application. Work on three-dimensional storage of optical information in chalcogenide glass, $Al_2S_3$, is pursued at the Institute of Ceramics. Holographic interference patterns produce a temperature grating that is converted to a phase grating in the glassy semiconductor.

Notable by its absence is work on laser spectroscopy, nonlinear spectroscopy, or laser light scattering in solids, liquids, and gases. Basic atomic mechanisms have not received attention. The coverage in the laser field is uneven and probably historically determined by somewhat arbitrary initial choices. The Western literature is generally followed closely and with understanding. Scientists are eager to hear about current developments from visiting Western experts. In 1975 I observed that "laser technology in China will undoubtedly be ready for rapid expansion, if and when the political timetable for China's economic development deems this desirable. The groundwork for this technology has already been laid." The new eight-year plan will probably verify this prediction.

*Low-temperature research.* The low-temperature work is narrowly focused, and a great amount of overlap exists among institutions, notably the

Institute of Physics in Beijing, the Institute of Metals in Shanghai, and Beijing and Nanjing Universities. The emphasis is on superconductivity, magnet wire made of $Nb_3Sn$ and NbTi, and on Josephsen junctions of the lead-oxide-lead type. At Beijing University a 100-kilogauss superconducting solenoid with a 1.4-cm bore diameter has been constructed. (In comparison, 160-kilogauss solenoids with a 4-cm bore are available in the United States.)

Three topics with a different slant are the construction of a dilution refrigerator and a superconducting gravity meter at the Institute of Physics and a broad-band millimeter wave detector at Nanjing University. The latter was to be used in astronomy, and a sensitivity of $10^{-14}$ watts/$cm^2$ at millimeter wavelengths was planned. The gravity detector used the Josephsen effect in a Dayem bridge arrangement.

Notably absent from the low-temperature program is basic work on liquid and solid helium, specific heat of alloys, Kondo effect and other properties of magnetic materials, Fermi surface work, and thermal and electrical transport properties.

It is rather puzzling that a relatively large amount of effort is devoted to superconducting technology. Its promise for power and signal applications in electrical engineering is not close to realization, even in Western economies. The students, of course, get training in gas handling and liquefaction, which is undoubtedly useful in rocketry and the steel-making industry. Perhaps the field was started on the basis of overoptimistic statements of several European low-temperature physicists who visited in the 1960's. The capability of producing superconducting magnets may be relevant for the construction of the high energy accelerators that are now being planned. As of 1975, the time lag in this particular technology was perhaps five to ten years. The effort in low temperature appears unbalanced both in its internal focus and in relation to other efforts in physics.

*Nuclear physics.* Activities in this area are also application oriented. The cyclotrons, reactors, and neutron generators are used for isotope production and for irradiation experiments on materials and living objects. Ion implantation is used in semiconductor device technology. The Mössbauer effect is used in the structure determination of magnetic materials. There are some nuclear-scattering experiments with three-body final states. Fewer publications in experimental nuclear physics have appeared than in solid-state physics. The equipment seen by Western visitors in the nuclear institutes appeared to lag behind that of corresponding Western organizations by about 20 years.

The Institute of Nuclear Physics in Shanghai has a cooperative program with Fudan University. Basic training in nuclear physics instrumentation, construction of particle detectors (including lithium-drifted germanium), and associated electronic counting equipment is available. Nuclear physics is used in scattering from solid surfaces and channeling in crystals. Thickness gauges and liquid-level gauges based on radiation detectors have been developed. Angular correlation and Mössbauer experiments have been used to determine magnetic structures, for example, of bismuth calcium vanadium iron garnet. Two multichannel analyzers, of 4,096 channels each, are available, one imported from France, the other a Chinese product. Rapid counting of $C^{14}$ activity for biological assays is carried out with two large liquid-scintillation counters. Fluorescent x-radiation is used to test calcium content in cement. The 3-MeV Van de Graaff machine at Fudan University is used for elastic and inelastic neutron scattering with a time-of-flight spectrometer. The $C^{12}$ $(p,\gamma)$ $N^{13}$ reaction is used for determining carbon content in various materials. Theoretical work at this University is concerned with nuclear fission.

At the Atomic Energy Institute in Beijing, elastic deuteron scattering from carbon $C^{12}(d,d)C^{12}$ has been investigated and compared with theory. Sodium-iodide and cesium-iodide crystals for scintillation counters have been developed; so have plastic scintillators, lithium-drifted germanium detectors, and photomultiplier tubes. The manufacture of these devices was transferred to industry in 1975.

It is clear that the main emphasis in nuclear physics is on applications. The Chinese have demonstrated the capability of developing nuclear weapons; they have the instrumentation for monitoring radiation and weapons tests. They conducted their twenty-second test in September 1977. The previous test had been a hydrogen device exploded in the atmosphere in November 1976. There are at least seven nuclear reactors. The gaseous diffusion complex at Lanzhou produces an estimated 300 kg of $U^{235}$ per year. There are persistent rumors of a second facility. China has the required supporting base of nuclear instrumentation.

*High energy physics.* The PRC was scheduled to embark on the construction of a 50-GeV proton accelerator. (For further details about this, see L. C. L. Yuan's technical note in this volume.) This requires coordination in experimental work in magnetism, electronics, low energy accelerators, and nuclear physics instrumentation.

The Chinese have conducted design studies and built some wire chambers for detection of high energy particles at their cosmic ray station. They reported the detection of a very heavy particle in 1972, but no further confirmation has since been obtained.

This theoretical activity in fundamental particles and cosmology is remarkable because it does not further the people's material welfare. It has been justified on philosophical grounds—for example, Mao's principle that matter is infinitely divisible.

*Other fields of physics.* There is little work in atomic and molecular physics in the basic sense of spectroscopic or collisional probing. This illustrates once more the lack of interest in understanding basic atomistic mechanisms. It is possible that some molecular spectroscopy occurs in chemistry departments. Perhaps the field of fluid dynamics is involved in such studies, but if so, it was not shown to Western physicists. Fudan University has a program on high-intensity discharge lamps, and there are numerous programs on gas laser development. Spectrometers for various frequency ranges, from microwaves to ultraviolet rays and x-rays, are available, but they are used primarily as service instruments.

There is more activity in acoustics and magnetism. In the former, work in acoustic holography and surface acoustic wave generators and detectors is being pursued. Interdigital electrodes on ferroelectric materials are used. As of 1975, the techniques were perhaps only four years behind those used in America. Acoustic impedance matching of ultrasonic cleaning baths and anechoic and reverberation chamber studies of noise pollution and architectural acoustics are more mundane applications. The Cultural Revolution diverted a group at Nanjing University to this last project from acoustic probing of Fermi surfaces.

The effort in magnetism concentrates again on device applications. Ferrite-core memory elements for computers, studies on bubble transport and switching in garnet films, microwave ferrite devices, and thin metallic magnetic films of CoGd for use with bubble devices are examples. Properties of magnetic films were studied by Faraday rotation; microwave resonance was used in the study of low-anisotropy ferrites. Routine modern equipment, x-y recorders, and oscilloscopes of Chinese origin, as well as East German microscopes, are readily available. An ion microprobe of French origin is also used.

One magnetism project appeared to address a physics question of more fundamental interest. It studied the creep of domain walls in Permalloy films in terms of the dynamics of the Bloch walls driven by an applied high-frequency ac magnetic field.

There are theorists attached to most experimental groups. In addition there is a separate theoretical department, with about ten members, in the Institute of Physics. They concentrated on quantum field theory, general relativity, theoretical astrophysics, and nuclear physics, and are now probably closely associated with the Institute of High Energy Physics. Theoretical work on domain wall motion, tunneling in Josephsen junctions, optical information processing, and the theory of crystal growth are closely connected with experiment. The last project attempts to explain a surprising increase in the neutron-scattering intensity when $\alpha$-$LiIO_3$ is subjected to a static electric field. Although most of these theories are preoccupied with classical or semiclassical phenomenology, some rather sophisticated work is going on in critical exponents of certain continuous phase transitions. It is based on renormalization group theory and diagrammatic analysis.

The state of research in the PRC in the 1970's may be summarized as follows. Little basic physics research in the Western sense of the term was carried out, with the notable exception of some theoretical work in high energy physics. Most research had a device- and application-oriented flavor. The Chinese were following the international state of the art with a lag of five to ten years. Coverage was uneven, and activities in many specialties were lacking. In selected high-priority areas (semiconductors and lasers), the lag was reduced to three years.

### *Status of Education and Communication*

It is difficult to understand the most populous nation and one of the world's great twentieth-century powers shutting down all academic education. But this was precisely what happened during the Cultural Revolution, and it was not until 1971 that a new freshman class was admitted to Beijing University. Other universities slowly started to reopen around the same time. But by 1975 academic buildings were still only about half full of students and faculty. This hiatus in China's higher education followed the "brilliant directive" issued on July 21, 1968, by Chairman Mao:

> It is still necessary to have universities; here I refer mainly to colleges of science and engineering. However, it is essential to shorten the length of schooling, revolutionize education, put proletarian politics in command and take the road of the Shanghai Machine Tools Plant in training technicians from among the workers. Students should be selected from among workers and peasants with practical experience, and they should turn to production after a few years' study.

This directive established the factory-run July 21 universities (which still exist) and also imposed the following strict conditions on university curricula (which for the most part no longer apply):

1. Undergraduate education was shortened from the five-year, pre–Cultural Revolution, "bookish" education. In practice, this resulted in a three-year engineering program and a three-and-one-half year program in undergraduate physics (stretched to four years for theorists).
2. The curriculum was closely tied to practical objectives, the student devoting one-third of his time to practical work.
3. The student's preparation for university study consisted mostly of ten years of primary and secondary school and about two years of work in the countryside, in a factory, or in the armed forces.
4. During each undergraduate year, the student had to serve one month in the army and devote two weeks to agricultural production.

*Physics curricula in 1975.* Because the curriculum was closely tied to practical objectives, there was no general physics, and the field was divided into subspecialties. The training in each of these was heavily applied and device oriented. Any student admitted to a particular subfield found it nearly impossible to switch to another. Typical subfields offered at various universities were semiconductor physics, laser physics, low-temperature physics, nuclear physics, acoustics, magnetism, microelectronics, radiophysics, optics, and crystal physics. No single university offered all of these. The students spent a third of their time in a short training period in different production workshops on campus. The final semester was devoted to a graduation practice project at the university, a factory, or a research institute.

The training in physics departments was narrowly oriented, reminiscent of that in engineering colleges in early twentieth-century America. A broad education was regarded as an undesirable luxury. The aim of China's higher education system was to supply technicians for the country's industry and agricultural communes. It was well suited for this purpose. There was very intense student-faculty contact, and the student got exposed to "real-life problems." The course material was carefully delineated, and mimeographed lecture notes were prepared for each course. There was individual tutoring for students who needed special assistance. No one was allowed to fail. Further technical training in two-year programs at selected factories was provided.

*Student selection and placement in 1975.* In 1975 Beijing, Nanjing, and Fudan Universities each had about 500 physics students. Their numbers were planned to double by 1978. In 1974, the first class to graduate after the Cultural Revolution was too small to meet China's needs. So over 100,000 students were trained in university-sponsored short courses and in correspondence courses. In addition, the July 21 universities gave instruction in basic mathematics, physics, and mechanics, combined with more specialized on-the-job training.

The university students were initially selected by "peer youth groups" while doing their two-year stints as peasants-workers-soldiers in commune, industry, or military service. The most important criteria were correct political outlook, class consciousness, and the capability of doing good manual work. The system clearly had no use for extremely bright, bookish types or for socially maladjusted and nonconforming youths. After this initial selection by the "masses," applications were reviewed by commune and local municipal or provincial authorities, who set quotas according to their perceived needs. The universities made a final selection, taking about half of the candidates chosen for them.

On graduation, most students returned to their former commune, factory, or other organizational unit. At Fudan University, 80 percent of the physics majors returned in this manner. But 10 percent were selected to remain at research institutes, and another 10 percent were retained as "junior faculty" by the University. Beijing University, which may be exceptional, retained 20 percent of its students and sent 40 percent to research institutes and 40 percent into industry.

Despite this, the 1975 educational system was not geared to physics research or to provide faculty for academic teaching. The broad view required for scientific and technical leadership was not provided. A number of university graduates who were retained as "junior faculty" or accepted on the staff of research institutes received additional training equivalent to graduate study. Some may have broadened their outlook and experience through self-reliance, private study, and innate intellectual power. There was, however, a serious lack of accomplished and broadly trained physicists in the

25- to 35-year-old age group. Innovation in physics, as well as in physics-based technology, was heavily dependent on the leadership of the Western- and Soviet-educated generations, whose training in physics was completed before the Cultural Revolution. This older generation of physicists has recently been called on to help in carrying out the new eight-year program in science and technology.

*Exchange of personnel.* Transfers of people between institutions and visits of up to a year in different localities have been the most important mechanisms for diffusing awareness and know-how in the scientific community. The interchange of students and faculty between universities, research institutes, and industry has also been widespread. Both students and faculty come in direct contact with practical problems; in return, industry is able to obtain advice from universities and research institutes. For example, the Institute of Physics has transferred the production of diamond, quartz, and ruby to industry, and the Institute of Optics and Precision Instruments collaborates closely with the glass industry in the manufacture of platinum-free laser glass. The Institute of Semiconductors transfers knowledge on reliability and techniques of failure analysis to the semiconductor industry.

*Seminars and conferences.* Intraspecialty seminars are frequent for universities and research institutes working in the same city in the same field. For example, theoretical seminars on quantum field theory and relativity bring together workers from the Institute of Physics, the Institute of High Energy Physics, and the Beijing Observatory. But there are no general colloquia for cross-fertilization between specialties, although national conferences on specialized topics are not uncommon.

*Research journals.* Journals play a relatively minor role in comparison with personal communications—a marked contrast to the rest of the world. The major physics research journals are *Wuli, Acta Physica Sinica,* and the new journal on high energy and nuclear physics, *Physica Energiae Fortis et Physica Nuclearis.* The time delay between submission and appearance is often more than a year. Many research results do not get published, even if they are not considered classified or privileged. Limited reports are sometimes circulated to a list of "grapevine" connections.

No publications in important foreign physics journals of work carried out in China have taken place. But foreign journals do play an important role in informing Chinese physicists about work abroad. They are centrally reprinted for low-cost circulation. Although students are required to study one foreign language, generally they cannot read foreign literature with ease unless it is translated into Chinese. A few physics books of especially wide use have been translated from English or Russian into Chinese. Some texts have also been written in China. In May 1976, for example, the Physics Department at Beijing University and the Foi Research Institute in Guangzhou jointly published a comprehensive book on lasers entitled *Principles of Lasers.*

*Libraries.* The major universities and institutes have an impressive collection of foreign physics books. Separate catalogues are kept for Chinese-, Russian-, Japanese-, and Western-language publications. The coverage of important physics books and journals is about as complete as that of major institutions in America. In several universities the physics books are placed in a large central collection, which impedes use by faculty and students, but departments often have their own collections. The institutes, of course, have collections relating to their specialties. There is an interlibrary loan system, and microfilm copying is used extensively. Photocopying is not generally available.

The language barrier severely curtails the use of the library collections. Many important foreign books and periodicals have not been checked out for years. Abstract journals, all of foreign origin, are used even less by Chinese physicists.

## *PHYSICS PERSPECTIVES IN CHINA*

The basic question to be addressed in this concluding section is whether the state of Chinese physics is such that the goals set in the eight-year plan can be met. The answer hinges largely on how quickly more physicists can be educated and the training of existing physicists upgraded. Drastic changes in the educational process, which will undoubtedly affect physics, took place in 1977–78. For example, a four-year curriculum now appears to be the norm. The first three years may be common to all physics students. The program thus provides a broader base of general principles. In the fourth year, differentiation takes place between different specialized subfields of physics.

Scientific and technical personnel have been sent abroad to study and attend international science conferences. There have been student-exchange agreements between China and major universities in Europe and the United States, and foreign scientists have been invited to serve as lecturers and research advisors.

It is clear, however, that China's senior physics professors, though few in number, can take care of a fundamental graduate curriculum in physics at the major universities. They can give the required lectures in theoretical physics, covering electromagnetic theory, mechanics, and statistical and quantum mechanics, as well as topical courses in solid-state, elementary particles, and quantum field theory. But the backup staff of junior faculty and teaching fellows could present a problem at first. The brightest graduate students could, however, be readily selected, and in one or two years the problem of formal class training could be solved.

The major thrust areas of the eight-year program relating to physics are enumerated in the first section of this chapter. They are realistic in that the stress falls on applied physics. There is no mandate to start doing a lot of basic physics research in the Western sense of the term. Catching up with the West can be achieved by following foreign developments, reading the international journals, attending some key international conferences, and copying or buying newly developed equipment. The PRC already has shown its ability to do this, notably in the field of semiconductor and laser materials.

But what about "catching up" in the various subfields? One area in applied physics that was to receive major emphasis was computer science, especially the problems associated with large-scale integrated circuits and peripheral equipment and software. Here, performance of China's semiconductor technology to date indicates that the goals in computer science should be realizable. A second major area was laser technology. More work was to be done in laser physics, laser spectroscopy, and nonlinear optics. The power output of the standard lasers was to be raised and studies of detectors intensified. Experimental lines of optical communications were planned, and studies of laser isotope separation and laser-induced fusion plasmas were to be stepped up. Laser technology was to be popularized in all departments of national economy and national defense.

All these are feasible goals; they aim at introducing technology into China that already exists elsewhere. The Chinese have shown their ability to construct all types of lasers and to grasp the basic literature in the field of quantum electronics. The question of how fast they will be able to catch up depends largely on how much instrumentation they are willing to import and how large a fraction of their precious trained manpower they are willing to devote to this subject.

High energy physics is one major point in the eight-year program that, at first sight, does not appear to fall in the category of applied physics. "The high energy physics experimental base is a key project on the nation's list of scientific research centers, and it is necessary to organize the forces of all quarters to build and use it jointly." Fang Yi's statement about the planned 50-GeV proton accelerator suggests that the machine is seen as a test of the nation's technological organization. When it is completed in the mid-1980's, its capability will be the same as that of comparable machines existing at Brookhaven and CERN 20 years earlier. The accelerator technology will be applied to industry, agriculture, medicine, and other subjects. The Chinese have wisely ignored overenthusiastic advice from some Western visitors to build more ambitious machines and jump to the frontiers of high energy physics research. Their program is likely to succeed and will undoubtedly enhance China's standing in international physics.

One can have nothing but admiration for Chinese physicists who have struggled through such a difficult period in history and yet have built a base in applied physics from which the new eight-year plan can be launched. Although the present number of highly trained physicists is small, their dedication, their willpower, and that remarkable capacity of the Chinese people for hard work should be able to rapidly develop the intellectual talents that are now hidden in the younger generation of this populous nation.

China will probably substantially increase its imports of know-how from

Europe and the United States, but at the same time remain self-reliant. Although its professed aim is (or was) to surpass the West by the year 2000, its more immediate and realistic ambition is to catch up with Western science and technology. A substantial effort in basic physics research, in the Western sense, is not part of the eight-year plan, and its objectives do appear realistic. I am optimistic about the successful completion of this plan, especially in view of recent political developments, which encourage vastly increased scientific exchange and technological imports.

*L. C. L. Yuan*

# *A TECHNICAL NOTE ON HIGH ENERGY PHYSICS*

On March 28, 1978, at the National Science Conference in Beijing, Fang Yi, vice-premier of science and technology, announced an eight-year development plan for science and technology.[1] Top priority was given eight areas: agriculture, energy, materials, computers, lasers, space science and technology, high energy physics, and genetic engineering. The inclusion of high energy physics may seem surprising, since it is an intrinsically fundamental and highly prestigious field with little application at present. Its development will be expensive and will drain resources and manpower from other areas.

The decision to develop high energy physics is, however, both courageous and wise. It is, moreover, a decision that has the support of Chinese leaders, both past and present. Mao Zedong and Zhou Enlai believed in the importance of fundamental research for the development of science and technology. (Mao, particularly, was fascinated with elementary particle theory. In the early 1950's, he had a long discussion with the Japanese physicist S. Sakata on the possibility of discovering an ultimate elementary particle.)

The information in this note is based on my visits to China in 1973, 1977, and 1979; private communications with members of the Beijing Proton Synchrotron project; and various publications, including Anne Fitzgerald and Charles P. Slichter, eds., *Solid State Physics in the People's Republic of China* (Washington, D.C.: Committee on Scholarly Communication with the People's Republic of China, 1976); W. K. H. Panofsky, *Observations on High Energy Physics in China,* United States–China Relations Report no. 3 (Stanford, Calif.: U.S.–China Relations Program, 1976); H. Y. Tsu, "The Preliminary Program for High Energy Physics in China," in *Proceedings 1978 International Conference on High Energy Physics* (Tokyo, 1978); and *Summary of the Preliminary Design Study of Beijing Proton Synchrotron* (Beijing: Institute of High Energy Physics, 1978).

[1]See Appendix A for the text of major speeches at this conference.

Furthermore, high energy physics is now a tremendously active science, and highly significant breakthroughs have occurred fairly frequently in recent years. If China is to contribute to this field, it must develop its own high energy research facilities. Among the ancillary benefits of this effort will be the increased knowledge in China of the large-scale mobilization and organization of advanced technology and the administration of a large number of active scientists—skills that China needs to become a modern nation.

In the fall of 1973 my wife, Professor Wu Chien-shiung of Columbia University, and I returned to the PRC for the first time since we left China in 1936. We spent seven and a half weeks visiting scientific institutions, universities, and industries in 18 different cities. At a banquet in Beijing, given for us by Premier Zhou Enlai and attended by many of China's leading scientists and officials, Zhou demonstrated that he knew most of the leading scientists intimately and was familiar with their fields of specialization. He posed most of the questions and led the discussion. The subjects discussed ranged from general scientific and technological developments to such matters as the urgent need for birth control in China in order to raise living standards. Zhou emphasized the importance of fundamental research in science and its possible long-term benefits to the Chinese people and the world. He touched on the development of China's high energy physics program and said that he strongly supported it, as did Chairman Mao. He was interested in the possibility of Chinese scientists participating in work at CERN (European Organization for Nuclear Research) after I forwarded such an invitation from the CERN authorities. It was at this dinner that Zhou gave me formal permission to visit the Cosmic Ray Research Station at Dongzuan near Kunming (see below). I was the first non-Chinese scientist to see it.

In the autumn of 1977, not long after the fall of the Gang of Four, we again visited China, this time for a month. The people we met seemed to be more relaxed and much happier than in 1973 when the Gang's influence was strong, and they spoke more openly. In Beijing we met, for the first time, with Fang Yi, the executive vice-president of the Chinese Academy of Sciences (CAS), at a special meeting in honor of our visit to the PRC. The late Professor Wu Youxun, who knew us well, made the formal introductions and praised the tremendous reorganization work carried out at the Academy by Fang Yi. Later at a sumptuous banquet with many prominent scientists and old friends present, the main topic of discussion again focused on China's high energy physics program. Fang was especially interested in learning about high energy physics in the United States. I extended an invitation to Chinese high energy physicists to come and work at the Fermi National Accelerator Laboratory from Dr. Robert R. Wilson, then Fermi director. Fang acknowledged this with appreciation. A team of eight Chinese physicists and engineers headed by Professor Xie Jialin subsequently came to the United States in May 1978 for a three and one-half month stay, first at Fermi and later at Brookhaven National Laboratory.

During our 1977 visit we were invited by Vice-Premier Deng Xiaoping to a discussion and dinner at the Great Hall of the People in Beijing. I was deeply impressed by Deng's dynamic, forthright personality and his strong support of the development of science and technology in China. He professed himself to be a military leader, with little knowledge of science, but said he would serve as an ordnance officer supplying the logistical requirements of China's scientists.

Deng's comments on the status of science and technology in China were extremely frank. He indicated that science and technology had made steady progress since China's liberation in 1949, and that the gaps between some branches of science and technology in the West and in China, such as biological research and solid-state physics, had closed gradually. However, he said that China's progress in most areas of scientific research and education had been stopped cold by the Gang of Four. He added that in science and technology China lagged behind the West by at least ten years, and estimated that it would probably take 20 years for China to catch up and be in a competitive position with the rest of the world. At this point, the late vice-president of the CAS, Wu Youxun, interrupted to say that he felt Deng was perhaps overpessimistic on this point, and that in his opinion the gap could be closed in 15 years. We then discussed science education in China and possible improvements to the present system. Deng seemed quite receptive to our suggestion that China establish special science-oriented middle schools for training talented or gifted science students like the Bronx High School of Science in New York, which has produced many well-known American scientists.

In March 1979, owing to the overenthusiastic response of various industrial and scientific sectors to the four modernizations program, the desire to order Western machinery and equipment was so overwhelmingly strong that if these desires were fulfilled, the Chinese economy would be greatly strained. Hence, the PRC government ordered a thorough reexamination of the whole modernization program to reassess the allotted priorities of various projects.

In my latest trip to the PRC, during June and July 1979 to participate in the first meeting of the U.S.–PRC High Energy Physics Cooperation Committee in Beijing, I learned that some major readjustments to the four modernizations program had been decided on. Top priority in this program was given to agriculture and the light industry, whose modernization will immediately benefit the masses of the country, whereas the development of heavy industry was to be severely slowed down. During this period of economic readjustment, top governmental leaders conceded that there was great opposition to the high energy program from many sides, but these leaders stood firm and the program came through unscathed, except that the completion date for the Beijing Proton Synchrotron (BPS) was wisely extended from 1982 to 1985. The great advantage of this decision is that the extended time schedule will enable the PRC to build its own magnets and most other components of the BPS.

Chinese research on high energy physics emphasizes two areas: cosmic rays and elementary particle physics employing very high energy accelerators.

*Cosmic Ray Research*

In 1952 the Institute of Modern Physics in Beijing established a cosmic ray station near the top of a mountain northeast of Kunming, the capital of Yunnan province, with a 50-cm multiplate cloud chamber to measure high energy shower events. Unfortunately, the station was built below the mountain peak and was not free from obstruction to cosmic radiation from all directions.

A new station was built at Dongzuan, about 220 km northeast of Kunming, 3,220 m above sea level. It was operating by 1965, but experimental work was interrupted during the Cultural Revolution (1966–69). The station has a permanent staff of 60 scientists, technicians, and engineers, divided into two groups, one working at the station, the other at the Institute of High Energy Physics in Beijing. Every six months the two groups change places to prevent prolonged high-altitude exposure.

The primary research objectives at the station are to study cosmic ray showers in the vicinity of the shower axis and to look for heavy mass particles in high energy interactions (>100 GeV).

In 1973 the experimental setup consisted of a 1.8 by 1.8 by 0.72 m cloud chamber placed in a large magnet with a magnetic field of $\simeq$7.5 kilogauss. Field strength of $\pm$3 percent was uniform over the entire illuminated region of the cloud chamber. A 2-m high multiplate cloud chamber was placed underneath the cloud chamber, and a similar multiplate chamber was placed above it. A 1-m thick iron plate calorimeter was placed beneath the chambers. Layers of Geiger-Müller counters were employed as triggers. The electronic control system consisted of transistorized, homemade circuitry.

Slow expansions were used in the cloud chamber to obtain clear pictures, and the chamber was cleaned five times after each trigger expansion to ensure a clean picture for the next expansion. The cycling time of 11 minutes was consequently quite long.

The station had a well-equipped machine shop with a milling machine, four lathes, drilling machines, and a scraper. There was also a preliminary analyzing laboratory with high-powered precision microscopes.

In 1972 an extremely interesting shower event was observed in the cloud chamber pictures. Its mass was determined to be 10 GeV by means of the inherent accurate momentum determination of a cloud chamber system. This observation was reported at the 1973 International Conference on Cosmic Rays held at Denver, Colorado, and at CERN. Detailed results were published in *Scientia Sinica* the same year. Unfortunately, no further events of a similar nature have been observed.

In 1977 a second cosmic ray observation station was established in the Himalayan mountains near Lhasa. A large number of emulsion chamber stacks placed under lead plates were set up over a wide area (over 200 $m^2$) of a mountain peak at an altitude of 4,500 m. Again, the main objective was to study high energy interactions in extensive showers. The first batch of exposed emulsion chambers from this station had been developed, and the results were being analyzed in 1979.

*Elementary Particle Physics Research*

Most of this research takes place at the CAS Institute of High Energy Physics in Beijing, established in 1974 and headed by Professor Zhang Wenyu. The Institute has a staff of 1,000 members (300 scientists, 500 technicians and engineers, and 200 administrative and support staff). One-quarter of the staff are women, mostly in technical positions.

The Institute has eight departments besides the Cosmic Ray Department discussed above.

The Physics Research Department has a staff of over 100 people and comprises many laboratories, including an electronics laboratory and a detector laboratory that works on multiwire proportional chambers and other high energy detection devices.

The Accelerator Department has about 200 staff members. It is designing and constructing the 50-GeV alternate gradient proton synchrotron and planning the second step of the Chinese high energy physics program (see below). Modules for a proton linear accelerator (linac) are being constructed. A half-MeV module has been successfully completed and tested. An accelerator theory group studies the theoretical aspects of the accelerator under design, such as the optimum parameters obtainable, the beam injection and extraction, and problems concerning beam instability.

The Experimental Facilities Department deals with the construction of accelerator foundation buildings and water and power facilities. Since the Institute is in the midst of a substantial accelerator-building program, this division is expanding rapidly.

The Elementary Particle Theory Department has about 50 members and is headed by Zhu Hongyuan. This group has worked on various aspects of elementary particle theory, theories of magnetic monopoles, and nuclear structures at medium energies ($\simeq$few hundred MeV). Some sample topics are listed below to show the group's range of interests:

Wave functions of mesons in the straton model
Production and decay of the new particles in the $SU_3^{(1)} \times SU_3^{(2)}$ model
U(6) symmetry and the straton model
Field-current relations in the new particles
Hadron symmetries
Hyper nuclei
Gravitational waves
Electromagnetic decay of the new particles $\Psi'$ and J
Interaction of electrons, magnetic monopoles, and photons

The New Techniques Department deals with new technologies, such as superconductivity and other cryogenic applications, and studies new principles for particle acceleration.

The Applied Physics Department deals with the practical aspects of high energy physics in industrial, medical, and other applications and has constructed small particle accelerators for them. A 30-MeV electron linac is used to test for fractures and other failures in heavy materials, such as steel plates. This division also has a Van de Graaff accelerator to investigate Mössbauer effects and other low energy physics problems. In 1978 there were about 100 people in this division.

The Computer Department has a Nr 320 computer, which is manufactured in China using transistorized circuitry and has a speed of 300,000 operations per second. Computer programming for various applications is being studied.

The Technical Safety Department is new and has a staff of about eight persons. In addition to the Institute's nine departments, there are a large machine shop with a staff of about 200 and several other administrative and service sections.

*The 50-GeV Alternating Gradient Proton Synchrotron*

The construction of a 50-GeV alternating gradient proton synchrotron is the first step in China's ambitious high energy physics program. This accelerator, announced in 1978, was originally scheduled for completion within five years, and a much higher energy accelerator was expected to be built within five years of the completion of the first. However, as mentioned above, the completion date for the BPS has been extended from 1982 to 1985. The two accelerators will be built 40 km from Beijing, near the well-known Ming Tombs. The new Beijing High Energy Physics Experimental Center, as the site will be called, was expected to be in full operation by the late 1980's.

The 50-GeV accelerator is designed to provide a beam of protons with an intensity of up to $1 \times 10^{13}$ protons per pulse. The cycling time of four seconds includes a flat-top feature with a duration of up to one second. The flat-top feature of the beam is an essential requirement for carrying out practically all experiments using electronic detector systems. The purposes of this accelerator are to provide a high energy source for performing particle physics experiments and to serve as an injector for the next, much higher energy accelerator.

A 750-keV Cockcroft-Walton generator will be used as a preinjector, but $H^-$ ions will be employed as the ion source instead of $H^+$ ions. The $H^-$ ion source has many advantages, as has been demonstrated at Fermi National Accelerator Laboratory and Brookhaven National Laboratory. For example, a considerably lower, but continuous current of only 60–120 ma is required, instead of the usual 200 ma from a conventional positive ion source.

The 750-keV $H^-$ ions are injected into a linac that accelerates the $H^-$ beam continuously to an energy of 200 MeV. The original design called for a 200-MeV beam to be injected into the 50-GeV main ring, which has a magnet gap of 8.6 cm. However, I was informed by Lin Zongdang, who was the head of the BPS project when I was in Beijing in June 1979, that an alternative design of the BPS had been adopted. In the new design, the magnet gap is reduced from 8.6 to 5 cm, and a 2-GeV booster synchrotron is added. The additional booster requires only a 100-MeV linac injector instead of the 200-MeV one originally required. The much-reduced size of the main-ring magnet gap drastically reduces the size of the magnets, and such a change probably enables substantial savings in the overall construction cost of the BPS. Even more important, this change provides a much better machine with uniquely higher beam intensity at 50 GeV, as well as a high-intensity 2-GeV proton beam and secondary beams for possible early physics experimentation and practical applications. The linac consists of a low energy beam transport system and nine accelerating tanks. The radio frequency (RF) power is supplied to each tank from a separate RF station with 5-Mw output. The overall length of the linac is approximately 160 m. A beam transport system will guide the 200-MeV $H^-$ ion beam into the main ring of the accelerator where the $H^-$ beam is stripped by a thin metal foil, converting it into a proton beam. The proton beam is then accelerated in the main ring. This has a radius of 216 m and consists of 28 RF stations that accelerate the beam up to an energy of 50 GeV. After the beam reaches the required energy for experimental purposes, it is ejected either by fast extraction with a beam duration of several microseconds or by slow extraction with a beam duration of up to one second.

*Plans for Other Accelerators*

The BPS and the higher energy accelerator planned to follow it will take at least ten years to complete. In the meantime, Chinese scientists want to continue conducting high energy physics research and other biophysics, solid-state physics, and atomic structure research using a synchrotron radiation source. This is a powerful and versatile facility, since it has tunable wavelengths from vacuum ultraviolet to hard x-rays, light intensities of three or four orders of magnitude greater than those of the most powerful x-ray tubes, and highly collimated and polarized synchrotron radiation beams of only a small fraction of a milliradian. With such a radiation source, a wide-ranging program of extremely interesting research can be pursued, including studies of ultraviolet and x-ray photoelectron spectroscopy, extended x-ray absorption fine structures, x-ray diffraction on biology systems for experiments in structural biology (for example, with proteins, enzymes, muscles, and membrane systems), x-ray–induced luminescence, and extended crystallography. A variety of technological applications, such as writing compact integrated circuits and investigating solar energy processes, is also possible.

Chinese physicists at the Chinese University of Science and Technology in Hefei, Anhui province, are planning to build a synchrotron light source to carry out such research work. Since Chinese physicists have had considerable experience in constructing linacs, they plan to build an electron linear accelerator with a 400-MeV injector and an electron synchrotron that will accelerate the electron beam up to 800 MeV. As of 1979, the final design of the accelerator system had not been decided. The high-duty-cycle linac will probably also be used for nuclear physics experiments.

The Chinese University of Science and Technology has played a crucial role in China's high energy physics programs because it has trained over half of the technical staff of the Institute of High Energy Physics and still maintains a close association with it. Professor Zhang Wenyu serves as both director of the Institute of High Energy Physics and head of the University's Modern Physics Department.

The 50-GeV proton synchrotron at the Beijing High Energy Physics Experimental Center and the synchrotron radiation light source at the Chinese University of Science and Technology are not the only high energy research facilities under design and construction in China. Leading nuclear physicists at research institutes in Lanzhou, Shanghai, and Beijing are highly enthusiastic about studying heavy-ion nuclear interactions and want to obtain heavy-ion accelerators to carry out their research objectives.

The CAS Institute of Atomic Energy in Beijing has already ordered a 13-MeV tandem Van de Graaff accelerator from the High Voltage Engineering Corporation of Boston, Massachusetts, to carry out experiments using both heavy and light ions. The machine includes a 150-kv negative-ion injector with a direct extraction off-axis duoplasmatron, a lithium charge exchange source, a sputter source, and a 90° inflection magnet capable of handling up to mass 240 at 40 keV. A nanosecond pulsing system for light ions is also provided, and a laddertron charging system is used to generate the terminal high voltage. Operating at 13 million volts, the beam performance includes 5 pμA protons at 26 MeV, 0.25 pμA chlorine, 0.2 pμA bromine, 0.1 pμA iodine, and 0.5 pμA helium. This machine will probably be used to investigate high energy deep-inelastic processes, high-spin states of rare nuclei, and their decay processes. Physicists at Lanzhou University in Gansu province are contemplating building a heavy-ion tandem machine themselves, since they have had some practical building experience. However, as of 1979, no definite plans had been made public.

Chinese leaders have chosen to emphasize high energy physics in their modernization program for two reasons. First, high energy physics is at the frontier of science. As a result of fundamental research carried out by high energy particle accelerators, many important contributions have been made to man's understanding of matter. Second and more practically, the technological developments required in building high energy accelerators, sophisticated detectors, and data processing and analysis systems provide many valuable spin-offs for other industries. The United States and member nations of CERN have made outstanding progress in such areas as fast electronics, sophisticated electronic circuitry, superconducting magnet development, fast computers, computer software and computer control systems, high-voltage high-power supplies, high-power radio frequency electronic tubes, klystrons, large-area sensitive detectors for medical diagnosis, sophisticated electronic display devices, and precision measuring instruments. The Chinese hope that their high energy physics program will improve their industrial performance by providing similar benefits.

I think the Chinese will succeed in implementing their high energy physics program and will make substantive scientific contributions in the long run. The timing of the program's completion will depend on the extent and speed of the training of technical personnel in the United States and Europe, the amount of technical help China receives from technologically advanced countries, and the country's economic condition.

*John D. Baldeschwieler*

# *CHEMISTRY*

The United States Delegation in Pure and Applied Chemistry, of which I was a member, visited China in May and June 1978. The trip began in Beijing with visits to the Institute of Chemistry of the Chinese Academy of Sciences (CAS), as well as to other leading research institutes in chemistry, physics, environmental chemistry, pharmaceutical chemistry, and biophysics. The Delegation then divided into two groups, one visiting Haerbin and the Daqing oil field complex in northern Manchuria, while the second visited the Institute of Chemical Physics and nearby industrial installations in Dalian. The Delegation reassembled to visit the Jilin Institute of Applied Chemistry and Jilin University in Changchun and a number of industrial sites close to Shenyang and Fushun. The Delegation also saw several research institutes and industrial activities in chemistry and chemical engineering in the Shanghai area, Northwest University at Xian, and research institutions at Lanzhou.

This chapter is based on the Delegation's trip report.[1] It should be stressed that any attempt to review the status of chemistry and chemical engineering in a country as varied and complex as China is necessarily incomplete.

## *THE BEGINNING OF MODERN CHEMISTRY IN CHINA*

Although the history of science in China is long and impressive, the history of modern chemistry begins only with China's interactions with the West in the nineteenth century. It was not until the twentieth century that organized chemical research started in earnest. An initial impetus was the

[1] John D. Baldeschwieler, ed., *Chemistry and Chemical Engineering in the People's Republic of China* (Washington, D.C.: American Chemical Society, 1979).

formation in 1914 of the Science Society of China, organized by students who had studied in the United States.

During the Republican period (1911–49), new research organizations and professional societies were established. Notable among these were the Academia Sinica in Nanjing and the National Academy of Sciences of Beijing, which in 1949 merged to form the CAS. Within the Academia Sinica, an Institute of Chemistry was established in 1928 in Shanghai. Similarly, another Institute of Chemistry was established at the Beijing Academy in 1929. The Chinese Chemical Society, which published a journal of chemistry, was founded in 1934 and had approximately 2,000 members by the late 1930's.

## *THE INSTITUTIONAL STRUCTURE OF CHEMICAL RESEARCH*

China's centrally controlled institutional structure for research and development consists of the CAS, the universities under the Ministry of Education, and the production ministries. The three sectors are controlled and coordinated by the Science and Technology Commission, which was reestablished in October 1977.

The central administrative offices of the CAS are in Beijing, but it operates 14 chemical research institutes in different parts of China:

Beijing: Institutes of Chemistry, Environmental Chemistry, Photography, Chemical Engineering, and Metallurgy.

Shanghai: Institutes of Organic Chemistry, and Silicate Chemistry and Technology.

Dalian: Institute of Chemical Physics.

Lanzhou: Institute of Chemical Physics.

Changchun: Jilin Institute of Applied Chemistry.

Fuzhou: Institute of Structural Chemistry.

Chengdu: Sichuan Institute of Chemistry.

Taiyuan: Institute of Coal Chemistry.

Anhui province: Institute of Rare Earths.

The CAS also has an Institute for Scientific and Technological Information and a scientific instrument design facility in Beijing.

China's institutions of higher education can be divided into four categories:

1. Comprehensive universities offering courses in a broad spectrum of disciplines in the arts and sciences; Beijing, Jilin, Fudan, and Northwest Universities fall into this category.

2. Universities specializing in science or engineering, such as Qinghua (Tsinghua) University.

3. Specialized institutes for particular areas of technology, such as the Shanghai Institute of Chemical Engineering.

4. Workers' universities and schools run by production units, an innovation of the Cultural Revolution.

Most institutions of higher education are under the direction of the Ministry of Education, but many have strong ties with local governments. The University of Science and Technology (Hefei, Anhui province) and Zhejiang University (Hangzhou) are unique in being operated jointly by the Ministry of Education and the CAS.

## *RESEARCH IN CHEMISTRY*

### *Organic Chemistry*

*Institute of Organic Chemistry, Shanghai.* China's leading program in organic chemistry is at the Institute of Organic Chemistry. As of 1978, work was under way on polypeptides, insect hormones, prostaglandins, nucleic acids, and steroids.

An example of the work on polypeptides was a structural study on trichosanthin, a protein isolated from a plant used as an abortifacient in Chinese folk medicine, which has shown anticancer activity. The compound has molecular weight 240,000, and its isolation by gel electrophoresis and gel chromatography was followed by a sequence determination using a polypeptide sequencer manufactured in the United States. Crystals had been obtained, and an x-ray study was under way.

In the insect hormone area, the Institute was engaged in synthetic work on preparing mimics for the silkworm juvenile hormone and some work on the chemistry of ecdysone. A pheromone of the piny bollworm had been identified as the structure:

$$C_4H_9CH=CH-(CH_2)_2-\overset{H}{\overset{|}{C}}=\overset{H}{\overset{|}{C}}-(CH_2)_6CO_2CH_3$$

The insect requires a one-to-one mixture of the *cis* and *trans* isomers at the double bond on the left to make the compound effective.

A new sesquiterpene derivative called arteannuin (see Fig. 1) had been isolated from artemisia. Its unusual structure was demonstrated by a combination of physical methods, including $C^{13}$ nuclear magnetic resonance (NMR) done at the Institute of Photography in Beijing, the only laboratory in China where the necessary instrumentation was available. The x-ray crystal structure determination was performed at the Institute of Biophysics in Beijing.

Work on prostaglandins began in 1971. Some compounds were being prepared by biosynthesis and others by total chemical synthesis. The routes involved were relatively sophisticated and involved the use of up-to-date synthetic methodology.

In nucleic acids, the most significant effort was a cooperative project with the Institute of Biophysics to achieve the first synthesis of a transfer RNA, in

Fig. 1

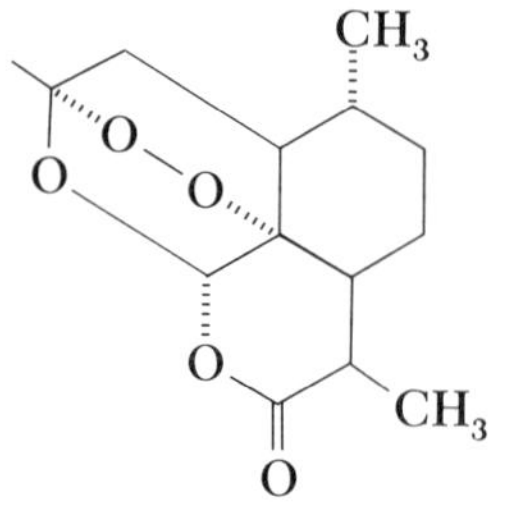

particular of yeast alanine t-RNA. The Institute of Biophysics was doing the enzymatic studies by splicing polynucleotides to produce larger pieces, while the Institute of Organic Chemistry was studying the synthesis of smaller segments. Elegant, modern techniques were used, including the use of $P^{32}$ with enzymatic ligase reactions in the synthesis of dodeca- and hexadecanucleotides. This project, which involved the active collaboration of the Institutes of Experimental Biology and Biochemistry in Shanghai, represented an attempt to match in the nucleic acid area the insulin synthesis achieved in 1965.

There was also considerable effort to produce much-needed medicinal steroids, including birth control agents, from available materials. Some work was in progress on total synthesis, using a variant of the Torgov scheme. Transformations starting from the available disosgenin and the increasingly available hecogenin and togogenin were also under way.

The Institute's work in organo-element chemistry—so called following Soviet practice—included organometallic chemistry and synthesis and characterization of organofluorine, organoboron, and organophosphorus compounds, as well as of compounds containing tin and arsenic. For instance, studies were being pursued of the reactivity and synthetic applications of arsenic ylides. These are more reactive than the corresponding phosphorus compounds and thus lead to significant yields in cases where the analogous phosphorus reagents are inert. For example, in the following reaction:

$$(C_6H_5)_3As = CHCO_2Me \; + \; (C_6H_5)_2C = O \rightarrow$$
$$(C_6H_5)_2C = CHCO_2Me \; + \; (C_6H_5)_3AsO$$

a 77 percent yield was obtained with the arsenic ylide, but the yield is 0 percent under the same conditions for the phosphorus analog. Work was also under way on polymers of fluorinated olefins and on the preparation of organophosphorus-chelating reagents for selective extractions of metals. All fluorinated intermediates had to be prepared as well, and there were extensive studies on mechanisms of fluorination of organic compounds. A class of stable fluorinated surfactants, whose general structure is indicated below, had been prepared as stable materials for applications in extreme conditions.

$$CF_3(CF_2)_nO(CF_2)_mSO_3^{-}K^{+}$$

A few specific projects were under way in physical organic chemistry. One involved detailed studies of liquid crystals aimed at developing appropriate technology in this area. The Institute had explored the properties of patented liquid crystal materials used in display devices. Considerable expertise had been developed in evaluating these materials and was being applied to the development of new liquid crystal substances and mixtures. Another project was the study of free radical addition to unsymmetrical fluorinated olefins and the role of polar effects in these reactions. There was also work on the thermal dimer of phenyltrifluoroethylene, whose structure had originally been misassigned by Soviet workers, but was corrected and reported at the same time as similar American work. Spectral and NMR studies were being pursued on some small fluorinated compounds, and stereochemical and structural factors involved in mass spectrometric fragmentation patterns and NMR coupling constants were being analyzed.

The Institute's instrumental support was modest. Many chemicals were prepared and instruments constructed at the Institute. For example, the Institute machine shops had made a high-pressure liquid chromatography equipment. Commercial instruments included standard infrared and ultraviolet spectrometers, a mass spectrometer, a gas chromatography/mass spectrometer (GC/MS) instrument, and a high-resolution mass spectrometer. But only a relatively simple 60-MHz proton NMR machine was available; there was no electron spin resonance (ESR) or Raman instrumentation.

*Institute of Materia Medica, Beijing.* The program of the Institute of Materia Medica included synthetic organic chemistry, phytochemistry, antibiotics, analytical chemistry, pharmacology, and the cultivation of medicinal plants. There were some partial synthetic approaches to harringtonine (see Fig. 2), an anticancer compound in clinical use. The Institute's work has also resulted in the identification of a series of compounds related to the structure shown in Figure 3. This compound is in active clinical use in

Fig. 2

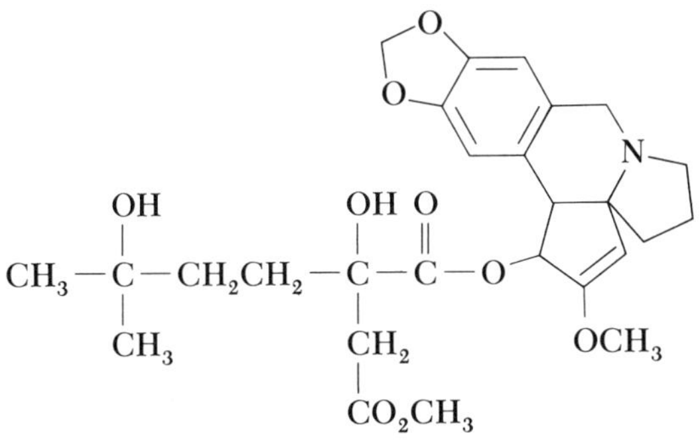

O
O
N
OH OH O
$CH_3-C-CH_2CH_2-C-C-O$
$CH_3$ $CH_2$ $OCH_3$
$CO_2CH_3$

China, and for some forms of chronic viral hepatitis is apparently the best drug known.

Fig. 3

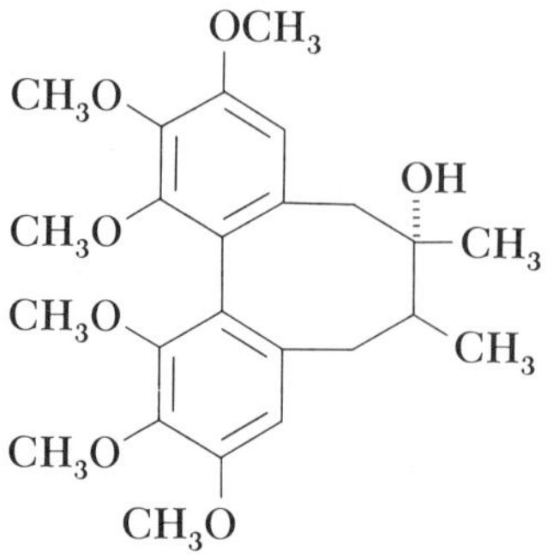

Fig. 4

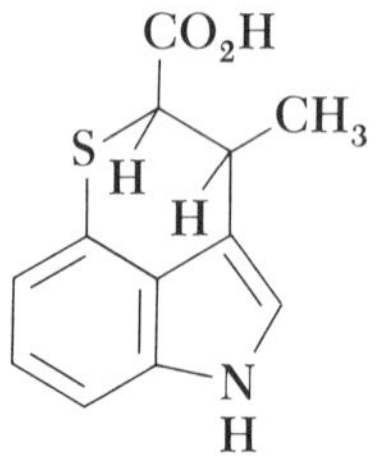

The structure of zhuangxinmycin (see Fig. 4), a new antibiotic active against gram positive and negative organisms and useful in treating urinary infections, has been determined on the basis of chemical and spectroscopic evidence.

Finally, a compound has been isolated from *Artabotrys uncinatus,* a traditional medicinal herb, and its structure determined by a combination of chemical and spectroscopic studies (see Fig. 5). Apparently it is useful in combating protozoa.

Although the spectroscopic instrumentation in the Institute of Materia Medica was routine, with only a rather ancient 60-MHz NMR machine, the Institute's location in Beijing made it possible to use the 250-MHz NMR machine at the Institute of Chemistry and the XL-100 for $C^{13}$ NMR at the Institute of Photography.

*Institute of Materia Medica, Shanghai.* Work on camptothecin (see Fig. 6) and various derivatives at the Institute of Materia Medica in Shanghai was extensive. Workers have found that 10-hydroxycamptothecin and 12-chlorocamptothecin have much-improved pharmacological properties over the parent compound, which is a highly toxic anticancer substance. Because of increased interest in some of these derivatives, they have worked

Fig. 5

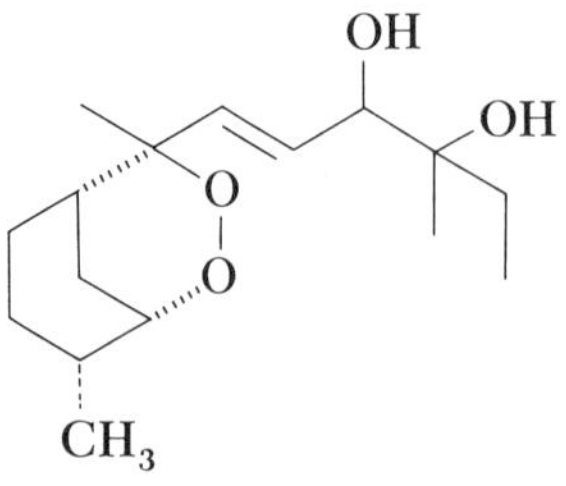

Fig. 6

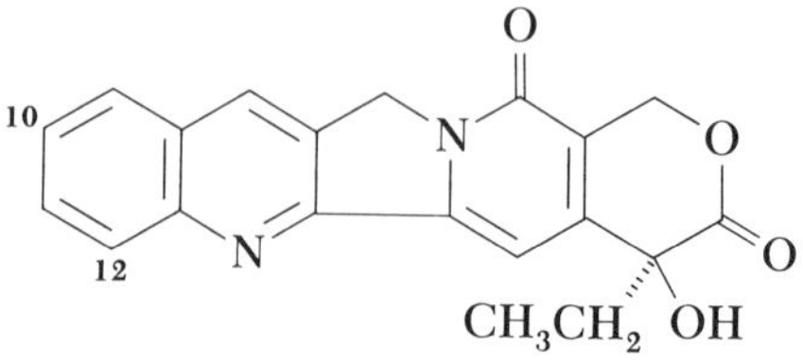

out a clean 10-step synthesis of camptothecin itself. This proceeds in an overall 18 percent isolated yield and seems to be broadly applicable to a variety of derivatives. Work was also under way on derivatives of lycorine and harringtonine.

The Institute's work in phytochemistry can be illustrated by the recent isolation of rorifone ($CH_3SO_2\ (CH_2)_9C = N$) from a traditional Chinese medicinal herb. Rorifone is useful in the treatment of bronchitis and is now used in the synthetic form to avoid side effects from the herb's other components. Rorifone was also the subject of the usual studies of structure-activity relations in this series. A new antihelminthic, quisqualic acid (see Fig. 7) had also been isolated from traditional medicinal plants.

Fig. 7

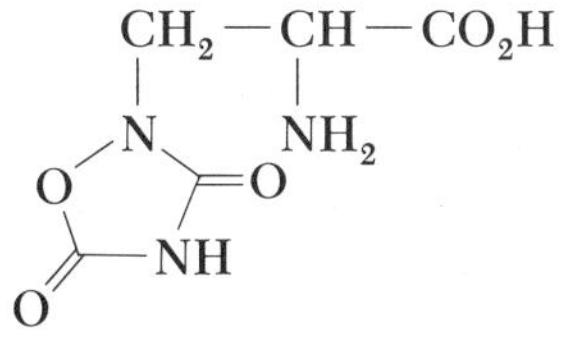

$CH_2 - CH - CO_2H$

N, $NH_2$, O, =O, NH, O

One of the unique opportunities available to Chinese chemists is the wealth of medicinally active substances found in their traditional medicines. There are approximately 5,000 widely used Chinese medicines, and only about 10 percent of these have been examined seriously by chemists. The exploration of these traditional Chinese medicines is one of the most exciting aspects of Chinese chemistry.

*Institute of Biochemistry, Shanghai.* The Institute of Biochemistry has made significant efforts in protein and peptide chemistry. The synthetic strategy involved conventional methods of preparing peptides with seven or eight amino acids. These peptides were carefully purified, and polypeptides were assembled using an insoluble supporting resin. As a result, many fewer reactions were actually performed on the polymer, and the probability of error was correspondingly smaller. Glucagon was synthesized by this strategy; its 29 amino acids were assembled by building a tripeptide on a solid support and adding a pentapeptide, a hexapeptide, a nonapeptide, and a hexapeptide. The overall yield was 83 percent. Syntheses of oxytocin and vasopressin developed at this Institute are now the bases for commercial processes.

### *Inorganic Chemistry*

*Rare earth chemistry.* If the class of elements called "rare earths" had been discovered in China, they would not carry that name. These elements are abundant, and almost all chemical research laboratories are looking for applications of rare earths and rare earth compounds.

The Jilin Institute of Applied Chemistry has worked on separation, extraction, and purification of rare earths, as well as on the synthesis of rare earth compounds and alloys, particularly semiconductors. It has made compounds of selenium and gadolinium on a one-to-one basis and produced rare-earth-doped laser materials.

Workers at Fudan University were looking for rare earth compounds that, when stimulated either by ultraviolet radiation or by low energy x-radiation, would convert this radiation to visible light for use in fluorescent lighting.

*Organometallic chemistry.* Research at the Institute of Organic Chemistry in Shanghai involved the production of molecular nitrogen compounds of the type $Mo(N_2)_2L_2$ where L is a diphosphine. When some of these complexes are acidified with sulfuric acid, ammonia is generated. The work has involved $L = Ph_2PCH_2CH_2PPh_2$, which does not yield $NH_3$, and the unsymmetrical $Ph_2PCH_2CH_2PEt_2$, which does show production of ammonia on acidification. A crystal structure determination was under way to decide which isomer the researchers had made in the unsymmetrical diphosphine.

Research in organometallic chemistry in China is still quite limited. Organometallic chemistry is often more difficult than organic synthetic chemistry because in many cases it requires the handling of extremely air-sensitive compounds, either in Schlenk ware or in dry boxes. Chinese efforts in organometallic chemistry are likely to lag until they succeed in coming up with an important new compound or homogeneous catalytic process.

*Chemical crystallography.* The most advanced work in crystallography was at the Institute of Biophysics in Beijing. Bovine insulin was synthesized in Shanghai in 1965, and crystallization and research on its structure were subsequently undertaken there. Workers had reached a 2.5-Å resolution by 1971 and a 1.8-Å resolution by 1973.

The Chinese have demonstrated a solid understanding of the solution of protein structures, including data collection and the preparation of heavy atom derivatives. However, they have not made the kinds of model-refinement calculations pioneered in the West.

The equipment in use was a Phillips diffractometer made around 1972. This was the only four-circle, automatic diffractometer in China, and it was kept very busy. Although primarily used by the Institute of Biophysics, it was scheduled by other institutes.

Although structural chemistry was reportedly one of the main tasks at the Institute of Chemistry in Beijing, the crystallographer there only had film cameras, although he did have access to the diffractometer at the Institute of Biophysics in Beijing. Work focused on a 23-atom alkaloid with four molecules in the common noncentrosymmetric space group $P2_12_12_1$. A whole year was spent on finding a trial structure, but no refinement had been initiated. Such a structure would usually be solved in a routine fashion in the United States, using one of several direct methods.

The x-ray equipment at the Shanghai Institute of Organic Chemistry included a power diffractometer, a unit for low-angle x-ray scattering and a Nonius Weissenberg camera. In the particular case of the Mo-dinitrogen complex mentioned earlier, the unit cell data had been collected in Shanghai, but the diffraction data had been collected at the Institute of Biophysics in Beijing. The space group of this particular material could be either $P2/c$, which is a centrosymmetric group, or $Pc$, which is noncentrosymmetric. The group at Shanghai claimed that the structure was $Pc$ because an optical second harmonic generator (SHG) test was positive. Second harmonic generation is a new technique involving laser frequency doubling, and it is interesting to learn that it has been adopted in China. It has been employed at the Silicate Chemistry and Technology Institute for screening compounds that might have piezoelectric and other properties dependent on the absence of a center of symmetry.

*Physical Chemistry and Chemical Physics*

Catalysis and polymer science dominated the programs in physical chemistry at almost every institute.

The effort in heterogeneous catalysis was highly applied and was concerned with studying process variables of existing catalytic systems, improving catalysts for existing processes, and developing catalysts for commercial processes. Standard instrumentation for heterogeneous catalyst surface characterization was evident at many institutes: BET mercury porosimeters, x-ray diffraction, and scanning electron microscopy. Such advanced tools as electron spectroscopy for chemical analysis (ESCA), Mössbauer, low energy electron diffraction (LEED), Auger, and extended x-ray absorption fine structures (EXAFS) were not used. The lack of emphasis on detailed reaction kinetic studies involving model compounds and the limited effort devoted to mathematical modeling of catalytic reactions were surprising.

Polymer efforts in China were comprehensive, but not particularly novel. Molecular characterization of polymers was performed through the usual spectroscopic techniques of infrared and NMR, using primarily imported instrumentation. Solid-state studies involving x-ray diffraction were also undertaken, again mostly with imported equipment. There was little activity in theory, chain configuration studies, or thermodynamics.

As of 1978, many institutions had begun or were planning to start programs in laser chemistry. Most of the laboratories were at the stage of laser construction rather than at the stage of active research on laser chemistry problems.

*Institute of Physics, Beijing.* A $CO_2$ TEA laser was in use with an output of several joules/pulse, and multiphoton dissociation of $BCl_3$ was under investigation. The observation and analysis was based entirely on chemiluminescence. The spectra of instantaneous and delayed fluorescence were being analyzed in an attempt to understand dissociation mechanisms. This work repeated that carried out by Rockwood and Rabideau at Los Alamos in 1973.

*Jilin Institute of Applied Chemistry, Changchun.* A relatively short, cylindrically shaped, metal-enclosed $CO_2$ TEA laser with an output of 300 mjoule/sec (line-tuned) was used to study the multiphoton dissociation of $CF_3I$ for the separation of $C^{13}$. This program seemed quite successful. The Chinese have enriched $C^{13}$ as Paul Houston did in 1977 at Cornell University.

A homemade argon-ion laser with an output of 7 w was used in this Institute to pump a dye laser (jet arrangement). Using homemade Rhodamine-6, the conversion efficiency was reported to be 10 percent to approximately 15 percent. Spectroscopy of highly excited atoms, two-photon absorption, multiple-photon absorption, opto-acoustic detection, and laser Raman spectroscopy were the scheduled types of work. In many of the laboratories, it was evident that the Chinese ability to get lasers running was excellent.

*Fudan University, Shanghai.* In 1978 the Chemistry and Optics Departments were planning to initiate a sophisticated laser chemistry program in the near future. The Optics Department had several lasers including:

1. Subnanosecond $CO_2$ TEA laser system under testing. This laser system followed the design of Yablonovitch (Harvard). Subnanosecond pulses (approximately 500 psec) were to be used for the study of the multiphoton excitation and dissociation of polyatomic molecules. This system was a spirally arranged pin electrode laser.

2. Nitrogen laser pumped dye laser. Two nitrogen laser pumped dye lasers were in operation. Researchers were attempting line narrowing by etalon and power amplification using a second dye cell pumped by the same nitrogen laser. The peak power of the small nitrogen laser was reported to be $\approx$500 kw.

There were many other institutions interested in setting up laser chemistry groups, notably the Institute of Chemistry in Beijing, the Institute of Chemical Physics in Dalian, and the Chinese University of Science and Technology in Hefei (Anhui province). The scientists in these institutions, although they had not worked on laser chemistry research, were well informed on the current literature.

As of 1978, there were no molecular beam research programs in China. But the Beijing Institute of Chemistry, the Dalian Institute of Chemical Physics, and the Chinese University of Science and Technology were planning programs in microscopic chemical kinetics.

Active research programs in molecular spectroscopy for the determination of molecular structure involving microwave, infrared, visible, and ultraviolet spectroscopy in the gas phase were largely absent. There were high-quality programs in molecular orbital calculations at a number of institutions, but theoretical chemistry, particularly activities involving large-scale molecular calculations, were not part of China's basic research effort.

Nor was there research on statistical mechanics. Apparently this field is considered a subdiscipline of physics in China, rather than an interdisciplinary field in both chemistry and physics.

### *Nuclear Chemistry*

Research in nuclear chemistry covering low energy nuclear physics and radiochemistry was conducted at the Institute of Atomic Energy, Beijing, the Institute of Nuclear Physics, Shanghai, and the Institute of Chemical Physics under the Lanzhou branch of the CAS. Each Institute had a cyclotron, other low energy accelerators, and radiochemistry laboratories.

*Institute of Atomic Energy, Beijing.* The major facilities at the Institute of Atomic Energy were a 1.2-m diameter cyclotron for the acceleration of protons and deuterons, a 600-kv Cockcroft-Walton D-T accelerator to produce 14-MeV neutrons, a Soviet-built 7-mw heavy water reactor containing 2 percent uranium-235, and a light water swimming pool reactor containing 3 percent uranium-235.

The chemistry program included radiochemistry, analytical chemistry, waste-water treatment, fuel reprocessing, and plutonium chemistry. In analytical chemistry, there were programs in neutron activation analysis, charged particle and fast-reaction activation analysis, and proton-induced x-ray fluorescence. Work on activation analysis included the study of rare earths in meteorites. There were also programs in transplutonium chemistry, high-pressure ion exchange, production of radio pharmaceuticals, and a limited number of tracer studies in medicinal chemistry.

The transplutonium program consisted of a study of americium-curium separation. Samples of americium-241 were irradiated in a neutron reactor to produce curium-242 at a multimicrocurie level. The separation method under study involved cation ion exchange resins. The gamma spectra were analyzed on an 800-channel analyzer manufactured by Intertechnique in France. All other equipment used for gamma-ray counting was made in Beijing. The sodium iodide detector was operated on-line. Alpha spectra were also measured on-line by gold-silicon solid-state detectors.

The radioisotope facility was the major producer of radioisotopes in China. The principal production facilities were for the production of $Au^{198}$, $Cr^{51}$, $I^{131}$, $I^{125}$, $Cs^{131}$, and $P^{32}$ (carrier free). The principal production was $I^{131}$ and $Au^{198}$. Each of these isotopes was processed in a batch process of about 40 curies. The facility was constructed in 1970 and in full operation by 1972. Some of the equipment was built in the U.S.S.R., but the more recent equipment was Chinese. The Institute of Atomic Energy also had facilities for preparing $C^{14}$- and $T^3$-labeled compounds. $C^{14}$ was prepared by reactor irradiation of $Ba(NO_3)_2$; the final product was barium carbonate, $Ba^{14}CO_3$. The specific activity of the $C^{14}$ was 20 millicuries per millimole $BaCO_3$. $C^{14}$-labeled sodium benzoate and sodium cyanide were also prepared, but there was no complicated organic synthesis with $C^{14}$.

The Institute of Atomic Energy also had a significant program of tritium-labeling the active ingredients of Chinese herbs by a modification of the Wilzbach method. Compounds labeled included scopolamine, avacoline, and harringtonine.

*Institute of Nuclear Physics, Shanghai.* The Institute of Nuclear Physics was concerned primarily with nuclear reactions and the study of fission. Isotope research dealt mainly with radioisotopes produced in the cyclotron and tritium labeling. The Institute had a Canadian $Co^{60}$ source of 120,000 curies for radiation research, as well as a 1.2-m diameter cyclotron built at the Institute. A Tandem accelerator to produce 12-MeV protons, similar to the FN accelerator marketed by High Voltage Engineering Corporation, was planned.

The Institute had a facility for producing semiconductor detectors, including Ge-Li coaxial-type detectors for gamma rays and Si-Au surface barrier detectors for charged particles. The germanium was obtained from the Institute of Nonferrous Metals in Beijing. The Institute was producing plug board integrated circuits and had already built 512-channel analyzers.

There were three types of perturbed angular correlation experiments in progress. One involved measuring the magnetic moment of the first excited state of cadmium-111 produced from indium-111. Another used cadmium-111, apparently implanted by recoil, to measure local magnetic fields in solid-state materials. The third and perhaps most interesting experiment was the application of perturbed angular correlation to study the enzyme nitrogenase, which is responsible for nitrogen fixation. The indium-111 isotope was bound to adenosine triphosphate (ATP), which is directly associated with the enzyme. It was not clear whether this rather nonspecific binding of indium could provide much useful information on the rotational correlation time for the enzyme.

*Analytical Chemistry*

Examples of programs in analytical chemistry in progress at several research institutes included:

*Institute of Chemical Physics, Dalian.* Work in gas chromatography began in 1954. By 1960 research workers had done a significant amount of work with glass capillary columns, and thermal conductivity and flame ionization detectors. These units were put into production, and cuts from petroleum fractionation were studied. The researchers had established the relationship between the boiling point of various petroleum fractions and the retention times on the glass capillary columns and simultaneously worked on the dynamic theory of chromatography.

Microanalytical methods for organic compounds that include nitrogen, chlorine, sulfur, or phosphorus were subsequently developed using gas chromatography. An important problem has been elimination of adsorption on the chromatographic bed. In some cases, the compound for chemical analysis was converted to another form to minimize adsorption. Workers at this Institute have made major advances in developing flame ionization and electron capture detectors, including new detectors that can discriminate selectively for nitrogen and phosphorus in organic compounds. These detectors have a selectivity against hydrocarbons of the order of $10^4$, and the sensitivity is approximately one picogram per second. The detectors' response was optimized by detailed study of the effects of the flow rates and composition of hydrogen, oxygen, and the carrier gas (nitrogen, prepurified by molecular sieve, activated charcoal, and silica gel).

*Fudan University.* Work has concentrated on polarography for dissolved cations in electrochemical analysis and anodic stripping voltametry (ASV) to determine dissolved $Cu^{+2}$, $Zn^{+2}$, $Pb^{+2}$, and $Cd^{+2}$ in waste waters. The general level of detection achieved for these ions was of the order of 0.1 parts per billion by weight. The various reagents used in the analyses were purified by ASV.

The group at Fudan University has also devised field instruments that use a gold film electrode for measuring arsenic, tellurium, and selenium by ASV. These rugged instruments, which were mass-produced by an instrument factory in Shanghai, could make measurements in the range of 0.1 to 10 parts per billion.

## *APPLIED RESEARCH IN CHEMISTRY AND CHEMICAL ENGINEERING*

### *Chemical Engineering*

There was a great deal of applied research and development in chemical engineering on processes that were already under way, but little in support of longer-range goals. Thus, there was great interest in catalytic techniques and polymers, but little in kinetics, applied thermodynamics, reaction engineering, surface and colloid chemistry, transport phenomena, fluid mechanics, separation processes, mathematical modeling, process dynamics and control, optimization, and computational techniques.

### *Petroleum and Petrochemical Research*

The history of petroleum refining in China is long and impressive. Research has been carefully coordinated, and plants and refineries are almost entirely of Chinese design and construction. The petrochemical area, in contrast, has not had a long history, and the Chinese have imported technology in an effort to catch up as rapidly as possible.

Research is concentrated in institutes of petroleum and petrochemistry at refinery and petrochemical complexes and in the provinces in order to maximize the efficiency with which local problems are identified and solved.

The institutes were generally well equipped. Research equipment ranged from small-scale microreactors to rather large-scale pilot plants, for example, a 10-liter-per-hour riser cat cracker at Beijing Petrochemical Research Institute and a 1,300-liter polypropylene polymerization autoclave at Fudan University. Computerization and on-line analytical instrumentation, which are so evident in U.S. research laboratories, were the most obvious gaps. There was heavy reliance on Chinese-made GC units (including capillary column and stationary phase). But some foreign equipment has been bought, and the institutes were generally well equipped with infrared and mass spectrometers, x-ray diffractometers, DTZ and DSC, x-ray fluorescence, and atomic absorption spectrometers. Examples of research programs at the more capable petroleum and petrochemical research institutes are given below.

*Institute for Petrochemical Research, Beijing.* This Institute had a leading role in introducing fluid catalytic cracking to China. It worked out manufacturing procedures for amorphous silica/alumina catalysts and developed sieve-type catalysts and a riser cracking unit, which were commercialized in 1975. It also developed a platinum/alumina fixed-bed reforming process based on catalyst research carried out at Dalian Institute of Chemical Physics. The scaling work and process commercialization were done at Daqing in 1965. The Institute began research on bimetallic reforming systems in 1969 and 1970 and claimed to be producing semicommercial quantities of bimetallic reforming catalyst. It appeared to be concentrating on a $Pt/Sn/Al_2O_3$ system.

This Institute, in collaboration with other petroleum institutes, has produced commercial hydrogenation catalysts for coker gas oil, coker gasoline, and lubricating oil hydrogenation (cobalt and nickel-molybdenum). In 1965 it developed a two-stage propane-deasphalting process with supercritical solvent recovery, which was marketed in 1967.

*Institute of Petrochemistry, Haerbin.* The Institute of Petrochemistry has developed adhesives for use in cementing wires to strain gauges. They included phenolformaldehyde, polyimides, and silicones. The Institute was also working on structural adhesives, including epoxies, silicones, nylons, and polyimides, and had developed over 20 for commercial use in automobiles, agriculture, and the petrochemical industry. One interesting use lay in the cementing of alloys to the ends of drill bits used in Daqing oil field. Apparently, the tungsten carbide alloys could not be welded to the steel bit ends because the welding temperature (over 1,000° C) destroyed the useful alloy properties; so structural adhesives developed at the Institute were used instead.

*Lanzhou Institute of Petroleum Research.* The Lanzhou Institute of Petroleum Research was principally engaged in research on lubricating oil additives, catalytic cracking and hydrofinishing catalysts, new oil products, and refining processes. The main problems were related to production technology, but the Institute also worked on technical issues assigned by the state. Since 1968 it has cooperated with schools, colleges, and other institutes to develop and manufacture metal organic salt detergents/dispersants, oxidation/corrosion inhibitors, extreme-pressure lubricating oil, and rust additives. One detergent-type additive produced was polybutene succimide of molecular weight about 1,000. Polyisobutylene was also being polymerized to a molecular weight of several thousand for use as a viscosity index improver for lubricating oils. The Institute developed and manufactured bead catalysts for TCC units and microspheroidal catalysts for fluid-bed catalytic crackers and catalysts used in lubricating oil hydrotreating. Institute workers had developed a Dill Chill process based on a U.S. journal article (Exxon process). They were also examining primary amines as ashless additives for lubricating oils. The work was essentially applied, with a small basic research component.

*Institute of Organic Chemistry, Shanghai.* The Institute of Organic Chemistry was studying the production of petroprotein from yeast. The Chinese started research in petroprotein production in 1971 and claimed in 1978 to be piloting two different processes. One used normal paraffins as a feedstock. In the other, a gas oil was used as feed to produce petroprotein and dewaxed oil suitable for use as a transformer oil. The process involved heating (for sterilization) the hydrocarbon feed and added nutrients, passing them through a micropore filter into an air-lift fermentor, and leaving them there for five hours at a pH of 3.5–4.0. The product of the first fermentor went to a second fermentor to lower residual hydrocarbons and then to a gas separator. The product was water-washed several times and centrifuged. The centrifuged material was then treated with 0.5 wt. percent NaOH for nucleic acid removal without destruction of the cell walls, resulting in 60–70 percent nucleic acid reduction. The petroprotein proceeded to a thin-film evaporator and then to a drum drier. Trace hydrocarbons were removed by hexane or alcohol (50 ppm level), and the protein was filtered. The nucleic acid products were finally film-evaporated and drum-dried and used for agricultural purposes, such as plant growth regulation.

As of 1978, the petroprotein was undergoing tests as an animal feed supplement. It was being fed to pigs and chickens in partial replacement of fish meal. The Institute had a joint program with the Shanghai Agriculture Science and Animal Husbandry Institute that included an examination of long-term genetic effects. An increase in the cost of Peruvian fish meal used as a supplement to supply the necessary amino acid balance in admixture with cereal grains provided the economic incentive for the project. The major problem still unresolved was the nucleic acid by-product. Nucleic acids constitute 6–7 percent of the biomass on a dry basis, and the Institute was searching for economic uses.

### *Catalysis*

The heterogeneous catalysis effort was highly applied. It concerned process variable studies on existing catalytic systems, the improvement of catalysts for existing processes, and the development of catalysts for processes that the Chinese wish to commercialize. There were numerous studies at various institutes on butadiene production via butene oxydehydrogenation. These were essentially process variable studies with microreactor and small-flow reactor equipment using standard catalyst systems. Catalytic studies involved the addition of metal oxides to the base system and a gross evaluation of yields and products. Studies of the development of microspheroidal fluid-bed zeolitic cracking catalysts and fixed-bed platinum and multimetallic reforming catalysts were also common. The Chinese displayed considerable knowledge of Western literature and patents.

China's abundance of rare earths has led several institutions to try using them as catalysts. Their programs have involved screening various rare earths alone and as additives to existing catalyst systems to determine what reactions they catalyze.

The standard heterogeneous catalyst surface characterization tools available at the different institutes included BET, mercury porosimeters, and x-ray diffraction. There were no advanced tools such as ESCA, Mössbauer, controlled atmosphere electron microscopy, LEED, Auger, or EXAFS. There appeared to be little, if any, solid-state chemistry devoted to synthesizing new materials with potential catalytic value, with the exception of rare earths. Detailed reaction kinetics involving model compounds was not evident, and only limited effort was devoted to mathematical modeling of catalytic reactions.

*Jilin Institute of Applied Chemistry, Changchun.* The catalysis laboratory of the Institute of Applied Chemistry was engaged in structural analysis and characterization of heterogeneous catalysts. The laboratory did surface area, porosity, and high-pressure porosity studies as well as BET measurements, x-ray diffractometry, differential thermal analysis, and electron spin resonance.

The reaction of ammonia with propylene to form acrylonitrile was a specific area of interest. A catalyst that contains P, Mo, Bi, Ni, Co, Fe, and Ca was being used at the Shanghai refinery (probably the Sohio catalyst). Institute workers were attempting to substitute rare earths for nickel and cobalt in the hope of improving the catalyst.

The production of methanol from water gas at low pressure ($\simeq 50$ atm) using the ICI methanol process involving copper on alumina and chromia was also being studied. The Institute was attempting to reproduce this catalyst system from information available in the literature. A final area of interest was the oxidation of methanol to formaldehyde using a FeMo oxide on silica in a fluidized bed arrangement. As of 1978, this process had not been commercialized.

*Dalian Institute of Chemical Physics.* Workers at this Institute were working on catalytic reforming reactions. They were studying multicomponent reforming catalysts and investigating the bimetallic catalysts Pt-Au and Pt-Ag. They were particularly interested in the reaction mechanisms of hydrogenolysis and isomerization and postulated the cyclopropyl radical as an intermediate for both types of reaction. This hypothesis agreed with molecular orbital calculations.

One project in heterogeneous catalysis was connected with catalyst evaluation and catalyst kinetics. Workers were studying the dehydrogenation of long-chain paraffins by bimetallic catalysts containing platinum, as well as hydro-forming reactions. These catalysts were characterized by x-ray studies, electron microscopy, and hydrogen-chemisorption. For dehydrogenation, the particle radius should be greater than 200 Å. Apparently, workers were successful in developing a reforming catalyst that gives 55 percent aromatics from Taqing oil, compared with 45 percent aromatics using conventional catalysts.

The laboratory's high-vacuum equipment was outstanding and consisted of Kovar glass seals, excellent flexible metal bellows to join glass to metal systems, Varian high-vacuum valves and flanges, and Balzer high-temperature ultra-high-vacuum valves.

*Fudan University.* Fudan University's catalysis group did applied, industrially sponsored work for polymer plants in the Shanghai area, much of which involved screening and developing catalysts. Studies included Ziegler-Natta $TiCl_3$-Al alkyl catalysts for polypropylene polymerization. Batch polymerizations were carried out in 2-, 30-, and 1,300-liter autoclaves, the latter being a very large apparatus for a university laboratory. The group was also trying to develop a solvent-free, solid-phase catalyst system that would not require washing residual catalyst from the polymer.

The laboratory had good instruments, including a Rigaku rotating anode, an x-ray powder diffractometer and low-angle camera, a large number of high-pressure liquid chromatographic units, a Mettler T160N balance, and a large number of semiworks scale autoclaves.

## *Polymers and Synthetic Scientific Fibers*

Work in China includes efforts to synthesize polymers by the usual radical and ionic routes, make block copolymers by the living polymer technique, and prepare condensation polymers of polyesters, polyamides, and polyimides. There were major efforts to find new polymerization catalysts and to produce specialty polymers with unique functions, such as polymers containing crown-ether for analytical separations, blood plasma substitutes, and pharmacologically active polymers.

The usual spectroscopic techniques of infrared and NMR were used for molecular characterization of polymers. Molecular weights were generally determined using viscosity, imported vapor phase osmometers, homemade membrane osmometers, and homemade or imported light-scattering photometers. An appreciable effort had gone into making gel permeation chromatographs, which had just begun to be manufactured in China, and the development of packings for GPC columns was strongly emphasized.

Solid-state studies involving x-ray diffraction and scattering were usually undertaken with imported equipment. These involved degree of crystallinity, orientation, or crystal or particle size measurements. Little work on polymer crystal structure was evident. Solid-state infrared spectroscopy was generally undertaken for analytical purposes, and there was some use of dichroism for orientation studies. Solid-state NMR studies were not common, and there were no pulsed NMR studies on solids. Laser small-angle light-scattering studies were in progress for investigating spherulitic polymers and liquid crystals. Electron microscopy was used in the principal laboratories; transmission electron microscopes were usually Japanese, but scanning microscopes made in Shanghai were also used.

Apart from routine tensile testing, mechanical property studies were not extensive. Several laboratories used homemade torsion pendulums, and one had a fairly sophisticated digitalized Japanese viscoelastic spectrometer. Conventional dielectric loss equipment was available in several laboratories. Solution and solid-state characterization procedures were thus routine. The Chinese working in polymer science were conversant with Western developments and well prepared to make significant advances with the development or acquisition of suitable techniques.

There was little activity in such areas of theory as chain configuration thermodynamics, chain dynamics, glass transitions, or crystallization.

*Institute of Chemistry, Beijing.* There was an appreciable polymer effort at the Institute of Chemistry, dating back to 1953. Solution studies in the 1970's primarily involved the use of GPC. An instrument developed at the Institute used both differential refraction and ultraviolet-absorption-type detectors. Emphasis had been placed on developing new column-packing materials, which included polymers of divinyl benzene cross-linked polystyrene with ion exchange resins. The column resolution was determined by interaction between groups on the polymer and the substance studied. Highly cross-linked columns with small pore size had been used for steroid and amino acid analyses.

Solid-state and polymer morphology studies were conducted using a photographic small-angle light-scattering apparatus (SALS) to study liquid crystals of aromatic polyamides in $H_2SO_4$ solution. A thermotropic nematic to cholesteric transition was observed, analogous to what is observed with low-molecular-weight cholesterol esters. At this transition the $H_v$ light-scattering pattern changes from one having cylindrical symmetry to a four-leaf clover type, suggesting nonrandom orientation correlations. This transition was also studied by depolarized light intensity techniques. The change in the scattering pattern on deforming the solution was seen upon microscopic examination by the appearance of a streak-type pattern with a periodicity corresponding to diffraction from striations. The study of this system was obviously of interest from the point of view of spinning high-modulus fibers of the "Kevlar" type.

Dynamic mechanical spectroscopy was being studied with an up-to-date

Japanese instrument manufactured by Uamoto Company, which had been automated by the Chinese to give digital data printouts. The instrument had been used to determine the anisotrophy of the viscoelastic spectrum of polyethylene terephthalate produced by stretching. A homemade torsion braid analysis apparatus and a Japanese thermogravimetric analysis apparatus were also available.

Some work was in progress on composites involving polyamides. The only work on inorganic polymers was on silicones and represented pioneering work for China's silicone industry. Some work on doped unsaturated polymers as semiconductors was in progress. Photoconductive polymers for holography, involving charge transfer complexes with polyvinyl carbazole, were being developed. These work by causing thickness changes when the charged polymer film is heated above its glass temperature.

There were attempts to develop organic sulfur compounds (e.g. thiobisphenol) as melt stabilizers. Oxidation mechanisms were being investigated. Attempts were also being made to modify polypropylene fiber to improve aging, making it more dyeable, and therefore better for clothing purposes. This was done by copolymerization with 5 percent of monomer units (see Fig. 8). This copolymer was blended with polypropylene.

Fig. 8

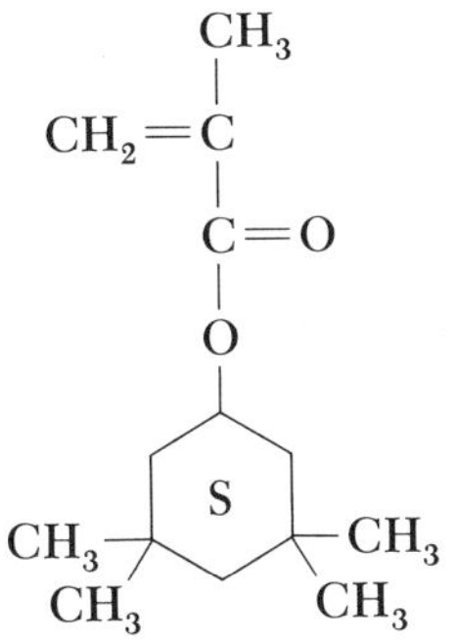

Fig. 9

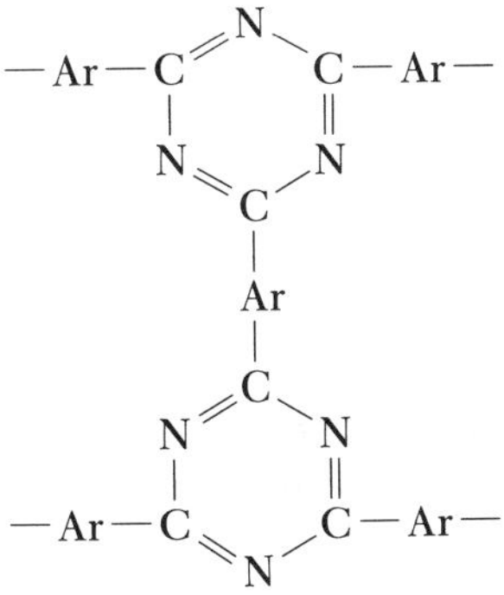

There were also studies of end-capped polyimides with improved solubility, intended for wire coating. Polymerization of dinitriles and of polymers containing triazine rings was also investigated in an effort to produce polymers of improved temperature and thermal deformation stability. A typical dinitrile polymer is shown in Figure 9.

*Jilin Institute of Applied Chemistry, Changchun.* One polymer effort, at the Institute of Applied Chemistry, was concerned with cis-1,4 polybutadiene. Prepared with a nickel catalyst, it had been characterized by studies of microstructure, morphology, gel content, intrinsic viscosity, molecular weight distribution, branching, Mooney viscosity, cold flow index, density, stress-strain curve, and stress relaxation. The polymer had a gel content of less than 1 percent and a molecular weight distribution broader than polymers obtained abroad. Mooney viscosity was apparently not adequate for characterizing processability.

Another polymer chemistry program was concerned with polymerization of isoprene and butadiene using rare earth catalysts. Synthetic zeolytes were produced to serve as catalyst supports. Catalytic activity among the 14 rare earth elements was studied and found to vary markedly. (Samarium and europium had exceptionally low activity.) This variation did not correlate with the valence state or with changes in the infrared or Raman spectra of the metal complex with tributyl phosphate, presumably due to differences in *f* orbital hybridization. The polybutadiene's microstructure was found to differ slightly, ranging from 94–96 percent cis-1,4 in going from the lighter to the heavier rare earths. In a manner similar to that of alkyl-lithium catalysts, the molecular weight distribution was found to shift toward higher values.

The laboratory at Jilin Institute of Applied Chemistry was equipped with a dielectric loss apparatus, a torsion pendulum, and a GPC, all homemade. There was no equipment for melt rheology. A homemade DSC was under construction in 1978, and preliminary scans were available for melting low-density polyethylene.

*Institute of Organic Chemistry, Shanghai.* This Institute's work on natural polymers included studies on blood plasma substitutes, such as carboxy methyl amylose prepared from cornstarch. This was somewhat reminiscent of U.S. work on "Dextran" in the late 1950's. Solutions exhibited typical polyelectrolyte behavior. Samples were characterized by paper chromatography using $C^{14}$ tracer and by molecular weight distribution studies involving fractionation and osmotic pressure measurements. Enzymatic degradation rates were appreciably suppressed with increasing degrees of substitution. By 1978 synthetic plasmas had been animal tested and were commercially available for clinical use.

Another major effort was concerned with fluoropolymers, including FEP copolymer ($C_2F_4$-$C_3F_6$) and fluoroplastic 40, an EPTE ($C_2H_4$-$C_2F_4$) copolymer. These were of interest for corrosion-resistant metal coatings and for wire insulation. (Work on fluoropolymers started at the Beijing Institute of Chemistry, but was transferred to Shanghai because of a Freon plant there.) The fluoropolymers were characterized by electron microscopy (a Shanghai-made SEM, 100 Å resolution). Correlations were made between spherulite morphology, composition, and stress cracking. Compositions were established by infrared spectroscopy, laser pyrolysis, DSC, and DTA. Melting points were found to be linear functions of the composition of the copolymer. The heat of fusion depended upon composition, suggesting the inclusion of $CF_3$ in the crystals as point defects. Effects of cooling rate and annealing on this phenomenon were being studied. Crystallization kinetics were studied by DSC. The Chinese believed that fractionation occurs upon crystallization from the melt. X-ray analysis of crystal imperfection was planned.

Although organoelement chemistry was an area studied at this Institute, no work on inorganic polymers (e.g. polyphosphazines) or silicones was in progress in 1978. Silicone studies were said to be done at a university in Guangzhou and a laboratory of the Ministry of Chemical Industry.

## SUPPORTING INFRASTRUCTURE

### Instrumentation

It is evident that in high-priority areas, particularly in applied research closely related to major national priorities, Chinese research institutions in chemistry and chemical engineering are able to obtain the facilities required to support their missions. Simple laboratory instruments, such as gas and liquid chromatographs, electrochemical instrumentation, high-vacuum hardware, pumps, flanges, recording instruments, and meters in refinery control rooms, were all designed and manufactured in China. Machined parts for high-performance requirements using unusual materials were well made and comparable in quality to similar components in the United States. A variety of electronic test instrumentation of Chinese manufacture, such as scopes and meters of all types, was also available, manufactured in cities such as Shanghai, Jilin, Changchun, Haerbin, Lanzhou, and Anshan. More complex instrumentation, such as NMR, mass spectroscopy, automated x-ray diffraction, and electron microscopy, was imported, mainly from Japan.

Sophisticated instruments with integrated microprocessing, available in the United States, Western Europe, and Japan, are missing in China. For example, in 1978 there was only one Fourier transform NMR spectrometer in China, a Varian XL-100 at the Beijing Institute of Photography. There was no instrumentation available for microwave spectroscopy, Mössbauer spectroscopy, ESR, ESCA, LEED, or Auger spectroscopy, although homemade ESR spectrometers were reported at the Beijing Institute of Chemistry and the Dalian Institute of Chemical Physics, and the Shanghai Institute of Organic Chemistry had an imported ESCA spectrometer. With a few exceptions, most instrumentation operated without the benefit of computer interfaces.

Those institutions with adequate instrumentation had high-priority programs with applied emphasis, such as the Silicate Chemistry and Technology Institute and Fudan University's Catalysis Research Institute. Institutes with a basic research mission had limited or inadequate instrumental facilities by U.S. standards.

### Instrument Development

The development of simple instrumentation in China seems to depend on the initiative of individual scientists working in collaboration with instrumentation factories. Instruments developed in this way include gas chromatography, high-pressure liquid chromatography, polarography, vapor phase osmometry, and instrumentation for measuring boiling point elevation.

Prototypes are generally first constructed in the extensive shops associated with each research institute. They are then produced in small quantities by an instrumentation factory and distributed to research institutes for testing and improvement. Research institute shops and instrumentation plants are normally responsible for making almost all the components, as well as for assembling the final instrument.

More sophisticated instruments, such as scanning electron microscopes, GC/MS instruments, and infrared spectrometers, are developed by the CAS Institute for Scientific Instrumentation, which has electronic and mechanical design capability for larger systems, as well as facilities for prototype manufacturing. There have been several instruments produced in this way.

### *Scientific Communication*

The technical societies, meetings, and journals that were suspended during the Cultural Revolution (1966–69) have resumed gradually since 1972. The best scientific work, including that in chemistry and chemical engineering, is usually published in *Scientia Sinica,* which has an English-language edition. *Kexue tongbao (Science Bulletin),* which covers all branches of science, resumed publication in 1973, as did *Huaxue tongbao* (Chemistry Bulletin). The *Journal of Analytical Chemistry* resumed in 1974, *Huaxue xuebao (Acta Chimica Sinica)* in 1975, and *Polymer Communications* in 1978.

The Institute for Scientific and Technological Information carries out research on information technology and plans to set up a national information network. It maintains liaison with the central CAS library in Beijing, as well as with various institute libraries, and publishes an abstracting journal, *Zhongguo huaxue huagong wenti* (Problems of Chemistry and Chemical Engineering in China). This abstracts important foreign journals for Chinese readers who do not understand English.

Research institute libraries have good selections of Western journals, distributed from a central library in Beijing, four to six months after initial publication date. These include *Chemical Abstracts, Science,* and *Chemical and Engineering News.* Collections are fairly complete and were not interrupted by the Cultural Revolution. But collections of Western books are poor. Many books were outdated, and recent publications were scarce. Patent literature is said to be available in microfilm form.

Scientific societies also discontinued activities during the Cultural Revolution and were only just resuming operations in 1978. A general meeting of China's Chemical Society was held in September 1978 in Shanghai with 500 to 600 people attending.

## *EDUCATIONAL PROGRAMS IN CHEMISTRY AND CHEMICAL ENGINEERING*

Students who intend to enter a university in a science program usually study chemistry and physics for two years at secondary school. In 1978, the undergraduate curriculum in chemistry was as follows. First year: General chemistry, mathematics, physics, English, and political science (Marxism and Maoism). Second year: Analytical chemistry, organic chemistry, physics, and English. Third year: Physical chemistry, structural chemistry, industrial chemistry, and organic chemistry. Fourth year: Specialized courses with particular emphasis on catalysis and structural chemistry.

The development of a curriculum approaching Western standards will take some time. Most university faculties were depleted after the Cultural Revolution, the teaching laboratories were antiquated, and modern Chinese textbooks were in short supply. Since 1978 students have been admitted by competitive examination to graduate programs in chemistry, among other disciplines, for the first time since the Cultural Revolution. Both university departments and CAS institutes accepted graduate students. It is significant that the CAS institutes have primary responsibility for basic scientific research under the National Science Plan; they suffered much less during the Cultural Revolution than the universities and were in a better position to reestablish graduate research programs.

Arrangements for chemical engineering education in China are varied. Among the best departments are those of Shanghai Chemical Engineering College, Zhejiang University at Hangzhou, Tianjin University, Qinghua University at Beijing, and the Dalian Engineering College. Other centers often mentioned as very good include Chengdu and Sichuan Engineering Colleges, the Guangdong Chemical Engineering Institute in Guangzhou, Northwest University in Xian, and Beijing Chemical Engineering Institute. There are two specialized petroleum engineering institutes at Taqing and Shengli oil fields.

The Shanghai Institute of Chemical Engineering was established in 1952 by combining the chemical engineering departments of five different universities in East China. Zhekiang University has colleges of both science and engineering, while Tianjin, Qinghua, and Dalian Universities only have engineering programs. Northwest University has one engineering department—chemical engineering. The petroleum engineering institutes cover every aspect of the petroleum industry.

A typical curriculum, such as that at Zhekiang University, consists of a series of required courses, including two terms each of physics, inorganic chemistry, organic chemistry, analytical chemistry, and physical chemistry (all with laboratory work), and mathematics. There are also courses in thermodynamics, transport phenomena, reaction engineering, polymer chemistry, physics and technology, petrochemical engineering (which is mostly applied thermodynamics), equipment, design principles, and unit operations. The curriculum includes politics (Marx and Mao) and six semesters of English, physical education, mechanical and electrical engineering, and drawing. A particularly interesting section in chemical engineering colleges is "chemical machinery," which is seldom found in the United States today. Most departments have simple computers for research and none for teaching.

*

The level of Chinese effort and achievement in chemistry and chemical engineering is substantial, but uneven. There are areas of obvious high priority, but also some significant gaps.

An area of particular interest is the high-quality effort in pharmaceutical chemistry, particularly development of pharmaceuticals from traditional Chinese medicinal materials, which is represented by programs at the Institutes of Materia Medica in Beijing and Shanghai and at the Institute of Organic Chemistry in Shanghai. Another significant project in organic chemistry is the yeast alanine t-RNA synthesis at the Institute of Biophysics in Shanghai.

Polymer chemistry is a more applied area of intensive Chinese effort. China produces most of the conventional polymers and has a broadly based research program with emphasis on applications and short-range goals. The strongest efforts focus on the development of new polymerization catalysts. Another applied area of considerable interest is the development of solid-state materials for electronic, electro-optical, and acousto-optical use, including piezoelectric materials. Chinese work in gas chromatography is extensive and uniformly good.

China is developing an extensive and modern petroleum and petrochemical industry, including the required infrastructure and research base. One area of particular interest is China's experience in the operation of full-scale shale oil production facilities. The Chinese effort in catalysis is extensive and appears to be well coordinated with the chemical process industries. The most obvious deficiency is the lack of on-line analytical techniques and microprocessors for process control.

In contrast to these areas of strength, modern organometallic chemistry appears to be practically nonexistent in China. Although Chinese organic chemistry is strong in some areas, the invention of new synthetic reactions, new methods for structure determination, and the development of physical organic chemistry have not been sufficiently emphasized. There is also little crystallographic work on automated equipment for x-ray diffraction studies. Furthermore, except for the strong program in molecular orbital theory, there was in 1978 no quantum and theoretical chemistry, statistical mechanics, molecular spectroscopy, molecular beam research, or microscopic kinetics, although serious programs were planned in some of these areas.

Although many research institutes have computers, chemists do not always have ready access to them, and they have not been aggressive in developing applications for computers in their work, such as simulating spectra and fitting kinetic data.

Chemical research in China is highly application oriented. It is mainly concerned with day-to-day problem solving, generation of information geared to improving existing processes, improvement in catalysts, and development of improved products. Little research is aimed at understanding underlying phenomena or attempting to recast problems in terms of basic scientific questions. Separately managed teams of scientists devoted to long-range research aimed at pioneering new products or new processes are almost nonexistent. It is evident that so far the Chinese have decided to emphasize short-range research to maximize the short-term development of their productive capacity. But interest in American methods of conducting long-range research and coordinating it with existing programs is high. As China continues on its rapid industrialization program and its scientists develop a better understanding of critical technical issues, the emphasis on longer-range research may increase.

Scientific manpower will remain a critical issue in the future development of Chinese chemistry and chemical engineering. Most leading scientists are in the 55-to-70 age range and were trained in the West. Measures must be taken to retrain and upgrade the age cohort from 30 to 45 and to graduate larger numbers of well-trained chemists and chemical engineers. This will require improved graduate training and better coordination between the universities and academy and ministry institutes.

The major question for the future is whether the Chinese will be able to absorb and make the most of high technology. Given the essential elements of political stability, good leadership, adequate resources, and a functioning infrastructure of research and training institutions, the results could be dramatic.

*Leo Goldberg*

# *ASTRONOMY*

## *ASTRONOMY IN CHINA*

China's most important contributions to astronomy were made early in its history during the thousand-year period that began with the Han dynasty (206 B.C.–A.D. 221). Aided by the invention of large and intricate astronomical instruments, notably an assortment of massive armillaries dating back to the second century B.C., the Chinese measured and recorded vast amounts of data on the motions of the moon and planets, eclipses, sunspots, comets, aurorae, meteors, and the famous supernova of A.D. 1054 in the constellation of Taurus. The existence of these accurate records has been a boon to a number of subfields of modern astronomy. It has, for example, been deduced that comets recorded in China in A.D. 989, 1066, 1145, and 1301 were earlier appearances of Halley's comet. The dating of the supernova explosion of 1054, the remnants of which are observable today as the pulsar-centered Crab nebula, has been of great importance in elucidating the physics of exploding stars. More recently, the analysis of ancient sunspot records dating back 20 centuries is helping astronomers decide if the 11-year sunspot cycle is a permanent or transient phenomenon.

As interpreted by Nathan Sivin,[1] the decline of Chinese astronomy after the thirteenth century A.D. is a fascinating story. In ancient China, the appearance in the sky of unpredictable phenomena was viewed as evidence of incompetence or lack of virtue on the part of an emperor. Therefore, the calculation of an accurate calendar and an ephemeris for the prediction of planetary phenomena and eclipses was a vital activity of the imperial court.

[1] Nathan Sivin, *Cosmos and Computation in Early Chinese Mathematical Astronomy* (Leiden: E. J. Brill, 1969); and idem, "Current Research on the History of Science in the PRC," *Chinese Science*, 3 (1978): 39–58.

Lunar eclipses were predicted routinely as early as 1000 B.C., but solar eclipses remained essentially unpredictable and hence dangerous to the emperor. To solve this problem, the Chinese court looked abroad, first to India and Islam and later, in the early seventeenth century, to the Jesuit missionary Matteo Ricci and his colleagues. For over one hundred years, the Jesuits effectively controlled astronomy in China. The new ideas and techniques they spread were widely adopted by the Chinese astronomers, but the Jesuits also effectively suppressed the propagation of Copernican astronomy. Not until the nineteenth century did the Chinese become fully aware of the great scientific revolution that had been under way in Europe for two centuries.

In the aftermath of the Opium Wars, Protestant missionaries, who came to China in great numbers after 1850, educated relatively large numbers of people of all classes in modern scientific disciplines. Threatened by foreign conquest, the Chinese government embarked on a major program of advanced training in science and technology, and at the turn of the century Chinese astronomers began graduate studies abroad. On returning home, they built China's modern observatories. Even now they play a prominent part in Chinese astronomy and are involved in China's campaign to upgrade and expand astronomical research and education.

Unfortunately, while the foundations of modern astronomy were being laid in the West in the first half of the twentieth century, China endured a succession of devastating civil wars, foreign invasions, and other disasters that prevented Chinese astronomers from partaking in the exciting developments that followed the introduction of large reflecting telescopes and the breakthroughs in atomic and nuclear physics. It was not until 1934 that China's first modern observing facility, Purple (and Gold) Mountain Observatory, was completed on a hill overlooking Nanjing. Following the Japanese occupation of the city in 1937, the observatory's astronomers were forced to move to Kunming in Yunnan province, where they established a solar observing facility. This installation was maintained after the war as an outpost of the reoccupied Purple Mountain Observatory, but in 1972 became the Yunnan Observatory, an independent institute governed jointly by the Chinese Academy of Sciences (CAS) and the provincial government. In 1974 Yunnan Observatory initiated an expansion program evidently intended to make it the leading center for optical astronomy in South China.[2] The corresponding center for North China is Beijing Observatory, which was constituted in its present form in 1958 during the Great Leap Forward. Shanghai Observatory, the oldest contemporary observatory in China, has been run by the Chinese since 1950. It is headquartered in Shanghai and has an observing section at Zo-se not far from the city. Once called the Zikawei (Xujiahui) and Zo-se Observatories, they were founded by French Jesuits in

[2] Leo Goldberg and Lois Edwards, eds., *Astronomy in China: A Trip Report of the American Astronomy Delegation* (Washington, D.C.: National Academy of Sciences, 1979), p. 62. Interested readers may refer to this report for further information.

1872, principally for transmitting time signals to French warships.[3] Shaanxi Observatory, in the province of that name, was founded in 1972.

Virtually all growth in Chinese astronomy has occurred since the Communist government took power in 1949. But since the degree of "redness" or political reliability of astronomers has at times been considered more important than professional expertise, the pace of astronomical progress has been somewhat uneven. By the late 1970's, the possibilities for rapid growth seemed very bright. After the death of Mao Zedong and the arrest of the Gang of Four, the new government announced its determination to transform China into a powerful, modern country by the year 2000, largely through massive applications of science and technology. Astronomy can be expected to benefit from this major policy change for several reasons, not the least of which is its importance in space research and applications. At the National Science Conference in March 1978, Fang Yi announced that space science and technology were to be one of the eight sectors of activity receiving high priority in resource allocation during the following eight years.[4] This was to include remote-sensing techniques as well as the manufacture and launch of a variety of scientific and applications satellites. We should also not overlook China's pride in its ancient astronomical tradition and its apparent recognition that solutions to fundamental problems of the nature and behavior of matter may come from astronomy.

Until 1950 relations between American and Chinese astronomers were close. A number of Chinese astronomers studied and worked at American observatories and later facilitated exchanges of information between the two countries. But with the Korean War, direct communication ceased altogether, and after 1960, when the PRC withdrew from the International Astronomical Union, contact virtually ended. For a time, it was possible to keep in touch with astronomical research in China by reading *Acta Astronomica Sinica,* but even that source of information was lost between 1966 and 1974, when the publication of research results was suspended.

Since 1973, reports on visits to China by astronomers from the United States and other countries have indicated a high level of astronomical activity. These reports were confirmed when a delegation of Chinese astronomers spent one month visiting observatories in the United States in November and December 1976. A return visit by an American delegation in October 1977 provided much of the material for this chapter.[5]

## *ORGANIZATION AND ADMINISTRATION OF RESEARCH AND TEACHING*

Astronomical research in China is conducted at the five independent observatories mentioned above, as well as at a number of other CAS institutes and several universities.

[3] G. K. Miley, "Astronomy in China Today," *Sky and Telescope,* Mar. 1974, p. 148.
[4] *Peking Review,* Apr. 1978, p. 10.
[5] Goldberg and Edwards, *Astronomy in China.*

The observatories are operated by the CAS, but those outside of Beijing are managed jointly with provincial or municipal authorities, presumably to make them responsive to local needs and problems.[6] Basic research is not regarded in China as an end in itself, and therefore the programs and budgets of the CAS are reviewed and approved by the State Science and Technology Commission as part of the total economic plan for the nation.

According to policy statements by Deng Xiaoping and others,[7] the scientific direction of research institutes is to be entrusted to competent scientists; but as a safeguard against elitist control, observatory directors are required to report to Party Committees, whose members are appointed by the CAS from lists of nominations proposed by the employees of the observatories. Committee members are drawn more or less equally from scientists, administrators, and workers. An observatory director administers affairs much as an executive officer or deputy director of an American laboratory does, but the Party Committee controls the observatory's policies and budgets. The chairman of the Party Committee is usually the leading administrator of the observatory, and his function is both to provide support to the technical staff and to enforce the policies of the Communist Party. An observatory scientist is expected to qualify as both "red and expert," but in the new climate of enthusiasm for science and technology, he may, if sufficiently expert, be tolerated even though he may not be an active Party supporter.

Beijing Observatory has no central headquarters; its five research departments—Solar Physics, Stellar Physics, Radio Astronomy, Astrometry and Time Service, and Latitude Measurement—conduct their work at four field stations. The oldest station, built at Tianjin in 1957, is the principal latitude station for China. The Shahe station, 20 km from Beijing, was begun in 1958. Its staff of 100 research workers and assistants, supported by 80 technicians, conducts research in theoretical astronomy, solar physics, and astrometry. The Xinglong station, 150 km northeast of Beijing, was constructed between 1964 and 1968. Its staff of 25 astronomers and 13 technicians engages in stellar physics and theoretical extragalactic research. The Miyun station is located 100 km northeast of Beijing near the great Miyun reservoir, and its instrumentation for radio astronomy research is the most advanced of the four stations. Construction began in 1967, and the staff numbers 80 astronomers and technicians.

Beijing Observatory is directed by Zheng Maolan, a well-known stellar spectroscopist who had previously worked at Haute Provence Observatory in France.[8] The chief administrator is Yu Jiang, and the observatory secretary is Hong Siyi, who was trained as a scientist and serves as spokesman in

[6]Susan Swannack Nunn, "Research Institutes in the People's Republic of China," *U.S.–China Business Review,* Mar./Apr. 1976, pp. 39–50.

[7]*Peking Review,* Apr. 1978, p. 8.

[8]Since this essay was written, Zheng has died. The new director is Wang Shouguan, a well-known radio astronomer who has been responsible for building the Miyun station. Hong Siyi is now deputy director.

dealings with foreign astronomers. The observatory is administered by a seven- to nine-person Party Committee, with Yu as chairman.

Yunnan Observatory is organized into five research sections: Solar Physics, Nonsolar Astrophysics, Astrometry, Celestial Mechanics, and Technical Support (engineering, optics, computers, data processing, etc.). In 1979, construction was under way on a new building with 30,000 sq. ft. of floor space for offices, laboratories, and a telescope. When all planned facilities are completed they are expected to house 500 research and support personnel. The observatory's director, Wu Minran, also participates in the solar research program.

Shanghai Observatory, with a staff of about 100, is organized into three research sections: Time Service, for which it is the national headquarters; Time Standards; and Photographic Astrometry and Astrophysics. The director, Ye Shuhua, is noted for her work in astrometry.

Purple Mountain Observatory is the oldest observatory in modern China. Before 1950 the staff consisted of only ten astronomers. After 1949 the observatory became an institute of the CAS and now has a greatly augmented staff of 150 research workers in addition to support personnel. The work is organized among eight research sections: Solar Physics, Stellar Physics, Radio Astronomy, Planetary Astronomy, Ephemeris, Time Service, History of Astronomy, and Satellite Tracking. The director, Zhang Youzhe, is one of China's leading astronomers and is well known for his work on binary stars, solar system astronomy, and the history of astronomy.

Very little is known about the size and capabilities of Shaanxi Observatory. It is China's newest observatory, and its work seems to be more applied than basic.

Although not strictly a scientific research institute, the Nanjing Astronomical Instruments Factory should be mentioned; it is operated by the CAS and works closely with the astronomical community. The factory began its work in 1958 with a staff of 70–80 employees, which has since increased to 300. It is engaged in the design and construction of complete astronomical instruments, including those for observing artificial satellites, the sun, planets, and stars for time service and latitude measurement. Work is carried on in ten separate, modern buildings housing optical, mechanical, and electronic workshops. Over 20 types of large and small instruments were produced in recent years. For the observation of artificial satellites, factory workers built a 43/60-cm Schmidt-type telescope, optical printing theodolites, and a large satellite-tracking camera. For solar work, they produced spectrographs and two 40-cm horizontal solar telescopes for the Yunnan and Purple Mountain Observatories and chromospheric telescopes equipped with narrow-band filters. For time service and latitude determination, they built several novel types of photoelectric transit instruments and a photographic zenith tube. For astrophysics, they produced a 60-cm reflecting telescope and in 1979 were engaged in a major effort to produce a 2-m telescope, which will be the largest in China, for the Xinglong station. When the American Astronomy Delegation visited the factory in October 1977, a large area had been cleared for an addition that could double the size of the factory. No indication was given of the specific projects that would be undertaken in the new space, except that they would be "for astronomy."

### *University Research and Teaching*

Apart from the observatory-institutes, significant research in astronomy is carried on at three universities: Beijing University, Nanjing University, and the Chinese University of Science and Technology at Hefei in Anhwei province. In the past, research was given a lower priority in the universities than in the institutes, but following the restoration of postgraduate education, great emphasis was placed on linking research and advanced training.

Nanjing is the only university with an independent department of astronomy; its chairman and most distinguished member for many years was Professor Dai Wensai, who died in 1979. The department was formed in 1952 from the Astronomy Departments of Guangzhou and Shandong Universities, which were moved to Nanjing to be near Purple Mountain Observatory. About 50 faculty members teach approximately 120 undergraduate students, of whom 30 to 40 are graduated each year. Within the department, the four principal divisions are stellar and solar astrophysics, radio astronomy, astrometry, and celestial mechanics. Within these general divisions, the faculty teach most subjects of current interest in astronomy. For the first year and one-half students pursue a common course, which includes mathematics and physics. Thereafter they specialize in a particular branch of astronomy in which they write a thesis during their last year. A fair number of the departmental staff are active in research and interact closely with the Purple Mountain Observatory. The department began accepting postgraduate research students in the late 1970's for the first time since the beginning of the Cultural Revolution.

Astronomy in Beijing University is administered as a section of the Department of Geophysics, with a faculty of 17 and a student enrollment of about 70. Students specializing in astronomy are required to take courses in mathematics, physics, electrical engineering, and English. Courses concentrate on solar physics, radio astronomy, and stellar physics. About half of the faculty is active in research, chiefly in solar astronomy. Beijing University was expected to enroll 1,500 students in three-year postgraduate programs, but it is not known whether astronomy was included. University-observatory collaboration in the Beijing area is not as effective as it is in Nanjing.

The Chinese University of Science and Technology was founded in 1958 in Beijing and moved to Hefei during the Cultural Revolution. At least eight members of its Department of Physics engage in research on high energy astrophysics, density-wave theory of galaxies, cosmology, and other topics in theoretical astrophysics. In 1978 the university embarked on an intensive graduate-training program that will probably include astronomy.

## RESEARCH PROGRAMS

In the West, roughly half of the scientists engaged in astronomical research were trained in other disciplines, principally physics, and many of them still identify themselves as physicists, chemists, geophysicists, etc. The pattern in China is similar, although not yet as well developed. In China as in the United States, problems disclosed through observational astronomy have captured the attention of an increasing number of theoretical physicists, the most active of whom are found at seven CAS institutes in Beijing and at the Chinese University of Science and Technology. Recent astronomical visitors to China have reported the attendance of considerable numbers of Chinese physicists at their lectures and seminars on astronomy given at Fudan (Shanghai), Yunnan (Kunming), Beijing, and Nanjing Universities.

It should be emphasized that the following discussion of research programs refers to late 1977. Until now Chinese astronomers have selected their subfields for two reasons: potential practical applications and limitations in telescope size.

### *The Sun*

Observational solar research in China is strongly utilitarian and devoted principally to studies of solar active regions and solar flares. One of the goals—if not the primary motivation—of these studies is the prediction of flares, especially those that might generate earth-bound streams of high-speed particles (so-called proton events). Observations of optical radiation from the sun are made at three locations: the Shahe station of Beijing Observatory, Purple Mountain Observatory, and Yunnan Observatory. Solar images are routinely photographed in white light and in the red light of hydrogen with small patrol telescopes, and spectra with good resolving power are recorded with multichannel spectrographs at Yunnan and Purple Mountain Observatories. The spectrographs are sometimes used to construct monochromatic images and maps of solar magnetic fields. (The horizontal telescopes and spectrographs and a new generation of patrol instruments were produced by the Nanjing Astronomical Instruments Factory.) These three observatories and Shaanxi Observatory also monitor the sun at microwave radio wavelengths. Solar patrol observations from the four observatories are sent to Beijing for predictions of solar activity; Purple Mountain Observatory also makes predictions from its own data. A much more advanced radio telescope for solar observing, an interferometric array of 16 9-m dishes, operates at 67-cm wavelength at the Miyun station of Beijing Observatory.

Since 1968 the Chinese have successfully conducted three solar-observing expeditions despite formidable logistical difficulties. The first was a coordinated optical, radio, geomagnetic, and ionospheric observing program during the total solar eclipse of September 22, 1968, in Xinjiang province by a group representing more than eight observatories, institutes, and universities.[9] The second was an expedition to observe the annular solar eclipse of April 29, 1976, from a snowy plateau 5,500 m above sea level in the Karakoram Mountains near Ürümqi in western China, a desolate region marked by severe cold and high winds.[10] In addition to studying the so-called gravitational anomaly during the eclipse, the group exposed 2,700 feet of motion picture film for teaching and popularization purposes. Finally, in 1968 Beijing Observatory astronomers took measurements of the solar-spectral energy distribution in the 0.60–1.20 $\mu$ range with a monochromator at an altitude of 5,000 m in the Himalayan Mountains of southern Tibet.[11] The availability of such high-altitude sites could be an important asset to the Chinese should they wish to establish a coronagraph station, which would be of considerable practical value, for example, in the forecasting of geomagnetic disturbances.

Solar astronomy in China is devoted to classical topics and does not as yet embrace the newest developments in the field. For example, there has been no work on such problems as global solar oscillations, the origin and evolution of large- and small-scale magnetic fields, differential rotation, or coronal and interplanetary dynamics. These gaps are due primarily to the lack of sophisticated equipment, but theoretical investigations of these subjects are also neglected.

### *Solar System*

Solar system research is a small and relatively specialized activity in China, but its quality is high. It is conducted at Nanjing under the leadership of Zhang Youzhe at Purple Mountain Observatory and until his death in 1979 by Dai Wensai at Nanjing University. The observational work at Purple Mountain Observatory is devoted to improving knowledge of orbits of faint comets and asteroids, to studying changes in the light curves of Eros and other asteroids, and to observing eclipses of Jupiter's satellites and occultations of stars by planets. The multiple rings of Uranus were discovered in March 1977, independently of the same discovery in the West. In all their observational work, the Chinese have shown considerable skill in extracting weak signals from noisy data.

The Nanjing astronomers have a good grasp of modern dynamical astronomy, as is shown by their theoretical and computational work on the orbits of comets and asteroids, the shapes of comet tails, and the orientation of the pole of Eros.

[9] Solar Eclipse Group for Coordinated Observations, "A Summary Report on the Coordinated Optical, Radio, Geomagnetic and Ionospheric Observations of the Solar Eclipse of 1968 September 22 in Sinkiang," *Chinese Astronomy,* 1 (1977): 105.

[10] *Peking Review,* Sept. 1976.

[11] Lin Yuanzhang, Hu Yuefeng, and Shen Longxiang, "Determination of Spectral Distribution and Solar Radiation in Near Infrared Beyond Earth's Atmosphere," *Acta Astronomica Sinica,* 15 (1975): 149.

At Nanjing University, the late Professor Dai was engaged in a careful review of existing theories of the origin of the solar system and had concluded, in line with prevailing opinion in the West, that the solar system evolved from a huge, slowly rotating cold nebula in which such mechanisms as the formation of planetesimals from dust rings and ice particles and the accretion and collapse of gas played a crucial role.

Finally, the recovery of fallen meteorites is a rich field of investigation in China, involving the organized pursuit of meteorite falls, mineralogical and chemical analysis, and dating from potassium-argon ratios. On March 8, 1976, the spectacular fall of a shower of stone meteorites in northeastern Jilin province resulted in the recovery of more than a hundred meteorites, the largest of which weighed 1,770 kg.[12]

*Time-Service and Positional Astronomy*

The determination of time and such related work as measurement of latitude variation, the motion of the poles, and irregularities in the earth's rotation are among the largest astronomical endeavors in China. Relevant observations are carried out at all five observatories, and both the instruments used and the results obtained are of high quality. Work in this field is closer to world standards than in most other branches of Chinese astronomy.

Each of the five observatories operates a transit telescope made by Zeiss in Jena, East Germany, but converted by the Chinese from photographic to photoelectric recording. But the most impressive device employed at most observatories is the impersonal astrolabe,[13] the original form of which was invented by A. Danjon in France a few years before World War II. The astrolabe then represented a great advance over the conventional type of transit instrument because it eliminated personal error in the determination of time and latitude by the observation of fundamental stars. It could also be used to refine the positions of the fundamental stars and to determine the fundamental astronomical constants. In its original form, the astrolabe was used visually, but after 1971 the Nanjing Astronomical Instruments Factory designed a series of greatly improved devices employing automatic photoelectric recording. The optical design was also greatly improved, and in the newer versions the optical parts were enclosed in an evacuated chamber, which nearly eliminates atmospheric refraction as a source of error.

Most observatories used quartz crystal clocks as local time standards, but a variety of much more accurate atomic clocks were also available or under development. At Shanghai Observatory, the headquarters of the National Time Service, five rubidium clocks served as local standards, and hydrogen maser clocks were being developed and assembled. The rubidium clocks

[12] Chinese Academy of Sciences, Jilin Province, and Jilin City: Joint Survey Team, "Lively Practice of Mass Participation in Scientific Research: On a Survey of the Meteoric Shower in Jilin," *Hongqi*, 1976, 6 (May 23).

[13] Photoelectric Astrolabe Research and Production Group, "The Photoelectric Astrolabe Type II," *Chinese Astronomy*, 1 (1977): 79.

were said to have a stability of one part in $10^{12}$, which is comparable with similar clocks in the United States. The hydrogen clocks may eventually be ten times more stable. Cesium clocks were under development at Beijing's Institute of Metrology, which was responsible for developing standards and calibrating instruments for laboratory and industrial use. About ten cesium clocks had been ordered from Oscilloquartz in Switzerland.

Chinese work on the motion of the earth's polar axis, while still unpublished, was expected to appear soon. From detailed statistical studies of their own and foreign measurements, the Chinese have concluded that a considerable part of the apparent motion of the pole is caused by local variations in the direction of gravity at the observatory locations.[14] This is because all instruments used to date are oriented with respect to gravity.

Work on improving the precision of time determinations is centered at Shanghai Observatory and emphasizes continued experiments on hydrogen maser clocks and the use of a long-baseline radio interferometer.

### *Satellite Tracking and Celestial Mechanics*

As in other countries, the need for accurate determinations of the orbits of artificial satellites has given a new importance to celestial mechanics. The observation of artificial satellites is carried out at Purple Mountain, Yunnan, and Shaanxi Observatories, and apparently at a number of other tracking stations as well. Earlier, satellites had been tracked with small tracking theodolites of 15-cm aperture and a Schmidt-type telescope of 60-cm primary mirror and 43-cm aperture, made in China in 1964 and installed at Purple Mountain Observatory. But these instruments are now being replaced by an array of large, high-quality satellite-tracking cameras produced in quantity by the Nanjing Astronomical Instruments Factory. The camera has Schmidt-type optics with a primary mirror 85 cm in diameter and a doublet correcting lens of 60-cm aperture. The field of view is 5° by 10° with the longer dimension in the tracking direction. The mounting has four axes, which facilitates tracking when one axis is pointed in the direction perpendicular to the orbital plane. By 1978 at least one such telescope was already in operation, and there were indications that a network of four or five additional telescopes was to be constructed in the near future. With respect to satellite-tracking cameras, it appeared that China had already caught up with the most advanced countries.

Two principal groups are engaged in work on the determination of the orbits of artificial satellites and in related theoretical studies: a well-established group of about 30 people at Purple Mountain Observatory and a newer group at Yunnan Observatory, which will probably become the leading center for work on satellite orbits and cataloguing. Celestial mechanics is one of the five divisions of this expanding observatory, which has apparently been given an assignment similar to that of Smithsonian Astrophysical

[14] Goldberg and Edwards, *Astronomy in China*, p. 39.

Observatory in the United States, namely, to monitor the whereabouts of orbiting satellites on a continuous basis.

*Stellar Astrophysics*

Studies of the physical properties of individual stars are carefully tailored to match the relatively modest capabilities of Chinese telescopes and auxiliary instruments. The largest telescope in operation in 1977 had an aperture of 60 cm, although the full 90-cm aperture of the primary mirror of the Schmidt-type telescope at the Xinglong station could also be used for spectroscopy. Most work in stellar astrophysics is conducted at Purple Mountain Observatory and at Xinglong, but Yunnan Observatory planned to initiate a full-fledged program in stellar astrophysics once it acquired a 1-m general purpose reflecting telescope from Zeiss Jena (as it probably did before the end of 1978). The Zeiss telescope was to be operable at f/13.0 Cassegrain and f/45 Coudé foci, where spectrographs ordered from Zeiss were to be installed. A Cassegrain camera for direct photography was to be available, in addition to a photometer made in China.

Of particular interest is a long-range and intensive search for flare and other types of variable stars in the Milky Way's dark clouds. The program is a cooperative one between Purple Mountain Observatory and the Xinglong station of Beijing Observatory, both of which use identical double-astrograph telescopes manufactured by Zeiss Jena. The double astrograph consists of two identical refracting telescopes of 40-cm aperture mounted together and photographing the same region of the sky, thereby avoiding confusion between real transient phenomena and photographic defects. An important by-product of this program was the detection at Purple Mountain of an unusual class of high-temperature variable stars with ultrashort periods of less than two and one-half hours, both in globular clusters and in the Milky Way generally.[15] Other work at both Xinglong and Purple Mountain includes spectroscopic studies of novae, symbiotic stars, and hot, emission-line (Be) stars.

Research in stellar astrophysics is also conducted at Shanghai Observatory and in the Department of Astronomy at Nanjing University. Shanghai Observatory is studying the proper motions of young stars in the Orion association, using plates made between 1902 and 1916 and in 1975 with the double astrometric refractor of 40-cm aperture at the Zo-se station.

Although Chinese astronomers are doing fine work in stellar astrophysics within the limits of the equipment available to them, they cannot hope to be competitive in world astronomy without bigger telescopes and up-to-date auxiliary instrumentation, including computers, modern photographic emulsions, and electronic detectors. Most telescopes now in operation were begun or finished before the Cultural Revolution. But rapid progress can be

[15] Zhu Youhua, "A Hot Cepheid with an Ultra-short Period," *Chinese Astronomy,* 1 (1977): 302.

expected from now on, assuming that the Chinese government maintains a favorable attitude toward science and technology. The 1-m reflector due to be installed at Yunnan Observatory has already been mentioned. More importantly, the Nanjing Astronomical Instruments Factory has designed a 2-m reflecting telescope, which was being constructed both at the factory and in various other cities. The telescope was to have an English-type mounting and a modified Ritchey-Chrétien (R.C.) optical system with an f/9.0 focus in addition to a Coudé focus at f/45. Computer calculations show that a two-element R.C. focus corrector yields images less than 0.3 arc sec diameter over a 0.9° field for direct photography with 30-cm square plates. The primary mirror blank was made of a Pyrex-like material in the U.S.S.R. and delivered to China around 1963. The mirror was to be figured to f/3.0 at the factory on a machine built in China before 1967, which suggests that the project was postponed during the Cultural Revolution.

The telescope has several interesting design features, including an ingenious arrangement for obtaining a Coudé focus without changing the secondary mirror, and has evidently been designed with care and competence. The telescope was to be completely computer controlled[16] and to be available for use as a national facility by all qualified Chinese astronomers, as was the 1-m telescope in Yunnan. No decision on auxiliary instruments for the 2-m telescope had been announced, but they were likely to include a camera for direct photography, a photoelectric photometer, a classical grating spectrograph for the R.C. focus, and possibly a Fourier transform spectrometer for the Coudé focus.

### *Radio Astronomy*

All astronomical observatories in China are doing some work in radio astronomy, usually with small dishes for patrolling the sun at wavelengths of 3.2 cm and 10 cm, but the Miyun station of Beijing Observatory operates by far the most important and sophisticated instrument. As completed in 1976, it consisted of 20 parabolic antennas each 9 m in diameter, arrayed along an east-west line 2,300 m long and operating as an interferometer with 1 arc-minute resolution at 67-cm wavelength. By the end of 1979, the array was to have been expanded and transformed into an aperture-synthesis array by the addition of 12 9-m reflectors operating at a wavelength of 1.7 m, making it a highly sophisticated and sensitive instrument, which, together with the 2-m optical reflector, will enable the Chinese to carry on first-class research in galactic and extragalactic astronomy. A second aperture-synthesis array under construction in 1979 was to consist of 20 dishes, each 1.5 m in diameter, arrayed along an east-west line about 100 m long. This telescope will be used to observe the sun at 3-cm wavelength with a resolution of about 1 arc minute (1/30 of a solar diameter).

[16] A. H. Cook, *Scientific Visits to China, 10–25 November 1973* (London: Royal Society, 1974).

*Extragalactic Astronomy and Cosmology*

Lacking large telescopes, the Chinese have been unable to engage in modern observational work in extragalactic astronomy or cosmology, but they will have this capability when the 1-m and 2-m telescopes are finished. In anticipation of these new facilities, a considerable number of Chinese astronomers and physicists, principally at Beijing, Nanjing, and the Chinese University of Science and Technology, have been actively analyzing and interpreting the world literature on extragalactic astronomy. Theoretical work on galaxies and cosmology is pursued by bright and capable people, although they are handicapped by lack of close contact with experimental research. There is considerable work both in Beijing and Nanjing on various aspects of the density-wave theory of spiral galaxies, as a result of the influence of C. C. Lin of Massachusetts Institute of Technology, who originated the theory and stimulated the interest of a number of very able young theoreticians during his visits to China. Work on the cosmological implications of quasars is in progress, both by physicists at the Chinese University of Science and Technology and by a collaborative group drawn from the Institutes of Mathematics, Physics, and High Energy Physics in Beijing.

Other work in observational cosmology includes the analysis of data related to predictions of theoretical world models, namely, redshifts, radio fluxes, spatial distributions of galaxies, and clusters of galaxies and quasars. Antimatter models of quasars and active galactic nuclei are being examined at Purple Mountain as they have been in the West (with rather negative results). At Shanghai, there is work on the correlation between the x-ray luminosities of clusters of galaxies and the dispersion of cluster velocities, a topic of interest in the West. Considerable work is also going on within the Beijing institutes on the theoretical foundations of cosmology, including an attempt to generalize the Einstein gravitational-field equations from the viewpoint of gauge theories of gravitation.

Application of this generalized theory to observation has led to some rather unconventional predictions, such as the nonexistence of black holes and the prediction of a nonlinear velocity-distance relation for galaxies in which the velocity increases with the square root of the distance rather than according to the more generally accepted linear relation of Hubble. Although the "big-bang" theory of the origin of the universe has been criticized for conflicting with dialectical materialism,[17] there is no indication that Chinese theorists are pressured to prove the nonexistence of any phenomenon that might be interpreted as an act of creation. In discussions with American visitors, many Chinese scientists demonstrated their familiarity with the "big-bang" theory and were prepared to talk about such central questions as the present lack of homogeneity in the universe in light of the isotropy of cosmic background radiation. The Chinese feel that

[17] Ning Chen, "Comments on the Byurakan Direction in Cosmogonical Investigation," *Acta Astronomica Sinica,* 16 (1976): 93.

current Western work on the early stages of the universe is somewhat oversimplified.

It is clear that cosmology in China has attracted a considerable number of highly talented young theorists who would be at home in large Western observatories. And there is no doubt that they would profit greatly from up-to-date and close contact with the work of large telescopes.

### *Study of Ancient Records*

The study of ancient records of such astronomical phenomena as solar eclipses, aurorae, novae, comets, and sunspots occupies research groups at Purple Mountain, Shanghai, and Yunnan Observatories and at the Research Institute of the History of Science in Beijing. A number of important contributions to modern astronomy have emerged. The most widely publicized and dramatic discovery, by Hubble in 1928, was the identification of the Crab Nebula as a supernova remnant of a star that exploded in A.D. 1054.[18] In 1976 the Group on Arrangement of the Chinese Ancient Sunspot Records at Yunnan Observatory completed a fascinating study of 112 records of observations made between 43 B.C. and A.D. 1638.[19] The sunspots were observed with the naked eye through haze in the early morning or late afternoon. During the past 200 years or so, the number of sunspots appearing on the sun has varied in an approximately 11-year cycle. But recently J. Eddy called attention to the almost total absence of sunspots between 1645 and 1715, which suggests that the 11-year cycle may be a relatively recent phenomenon.[20] However, the Chinese found the 11-year cycle to be quite distinct throughout the period of their investigation—including the 1645–1715 period, when the sunspots continued to follow the 11-year cycle although in greatly reduced numbers. Valuable information on the durability of comets has been provided by Zhang Youzhe, director of Purple Mountain Observatory, from a research project on the orbit of Halley's comet during the past 4,000 years.[21]

### *Practical Applications*

It is evident that past priorities in Chinese astronomy have been given primarily to practical applications; for example, the prediction of solar flares and other solar events from radio and optical observations of the sun, studies of possible correlations between polar motions and earthquakes and between Chang Jiang (Yangtze River) flood levels and sunspot activity, time determinations, and the tracking of artificial satellites. Thus, the solar flares of September 17, 18, and 20, 1977, which interrupted shortwave radio communications for over half an hour, were reportedly predicted by the Bei-

[18] See I. S. Shklovskii, *Supernovae* (London: Wiley/Interscience, 1968), p. 50.

[19] Chinese Academy of Sciences, Yunnan Observatory, Ancient Sunspots Records Research Group, "A Re-compilation of Our Country's Records of Sunspots Through the Ages and an Inquiry into Possible Periodicities in Their Activity," *Chinese Astronomy*, 1 (1977): 347–59.

[20] J. Eddy, "The Maunder Minimum," *Science*, 192 (1976): 1,189.

[21] Zhang Youzhe, "The Orbital Evolution and History of Halley's Comet" (Paper read to American Astronomy Delegation, Oct. 18, 1977).

jing, Purple Mountain, and Yunnan Observatories on September 15. Purple Mountain Observatory and Beijing Planetarium jointly edit a popular almanac that includes tables for the rising and setting of the sun and moon as observed from many cities. According to the Chinese press, the 1977 edition contained useful information for industry, agriculture, fishing, animal husbandry, and civil aviation, and on daylight hours. Such preoccupation with practical applications is natural in a developing country, especially considering the relatively low cost of the necessary equipment. It also accords with China's ancient tradition of supporting astronomy for calendrical purposes.

### *Publications*

The results of astronomical research in China are published mainly in the *Acta Astronomica Sinica* and to a lesser extent in *Scientia Sinica, Kexue tongbao (Science Bulletin)*, and *Ziran zazhi* (Nature Magazine). The *Acta* began publication from Purple Mountain Observatory in the 1950's, but were interrupted between 1966 and 1974 by the Cultural Revolution. *Scientia* is a publication of the CAS and, like the *Proceedings of the National Academy of Sciences,* carries occasional articles on astronomy.

The *Acta* are published in Chinese, except for truncated English abstracts, and therefore have not been accessible to most foreign readers. Thus, the recent initiation of complete translations of the journal by T. Kiang and the Pergamon Press is a welcome development. Manuscripts and duplicated articles are circulated in China, sometimes as reprints and sometimes in lieu of formal publication, but they seldom reach foreign countries.

## *POSSIBLE DIRECTIONS FOR THE FUTURE*

Although astronomical research in China will almost certainly continue to satisfy practical needs, there are signs that increasing attention will be paid to the more basic aspects of astronomical research in the immediate future. The construction of the 1-m and 2-m telescopes, the expansion of the radio astronomy program at Miyun, and the rapid growth of Yunnan Observatory are clear indications of China's intention to become a world leader in astronomy. Although no firm plans beyond these projects have been announced, recent visitors to China reported many discussions with Chinese astronomers about future possibilities for large facilities.

### *Optical Astronomy*

Chinese solar astronomers have recently expressed interest in such matters as comparative image quality in evacuated and unevacuated solar tower telescopes, the factors determining solar seeing, and the design details of such sophisticated instruments as the Kitt Peak solar vacuum telescope and magnetograph. They have also been engaged for some time in testing possible sites for future solar installations.

Stellar and extragalactic astronomers are debating how to proceed with the further development of Yunnan Observatory after the completion of the 1-m telescope. No firm date has been set for the completion of the 2-m telescope for Xinglong, but China's lack of experience in the construction of large telescopes and "the damage done by the Gang of Four" slowed progress on this key project. Having expended so much talent and effort on the telescope, the Chinese are unlikely to stop at building only one, and Yunnan Observatory is the most logical location for the second instrument. The Chinese astronomers realize that telescopes larger than 2 m are required to put them at the forefront of astronomical research. They know that about a dozen telescopes in the 3- to 6-m size range have already been built in other countries. Moreover, design studies in the United States and Europe project a Next Generation Telescope (NGT) as large as 25 m in diameter for the 1990's.[22] The 1- and 2-m telescopes will provide the Chinese with necessary experience in the design, construction, and operation of large telescopes, and they should be capable of completing a 4-m telescope by the mid- to late 1980's.

The scientific value of a large telescope is measured not only by its aperture, but also by the quality of the site on which it is located. A reasonable percentage of clear nights is an important requirement, but even more essential is the quality of the "seeing," which determines the size of a star's image after its light has traversed the earth's turbulent atmosphere. Astronomers speak of the "figure of merit" of a telescope, that is, the ratio of the primary mirror diameter to the image size. As a rule, large telescopes have been cost-effective only when they have been located on sites with good seeing. It is not known whether any sites exist in China where the seeing is good enough to justify installation of a very large telescope. Conventional wisdom in the West holds that first-class seeing cannot be expected within a large continental landmass. It is obvious that this question should be settled by appropriate measurements as soon as possible.

In Western countries, the efficiency of optical telescopes has been increased greatly during the past two decades by advances in such auxiliary instrumentation as spectrographs; photometers and detectors, including both photographic and electro-optical devices; and minicomputers for telescope control and data storage. Chinese instruments adhere to high standards of optical and mechanical design and construction, as is shown by the products of the Nanjing Astronomical Instruments Factory and the Optical Instruments Factory in Shanghai. Good-quality diffraction gratings are ruled at a factory in Changchun in Jilin province. For detectors, the Chinese use single-channel photoelectric photometers and photographic film and plates. Most are made in China, but some are imported from the United Kingdom and East Germany. A few Eastman Kodak plates are also used and were probably obtained from sources in Hong Kong since the United States

[22] "Optical Telescopes of the Future" (Proceedings of conference held by European Southern Observatory, Geneva, Dec. 12–15, 1977).

does not permit the export of photographic emulsions to China. Chinese astronomers are severely handicapped by this lack of access to the newest Kodak plates, which are markedly superior to their competitors in both grain and sensitivity. The revolution in electronic detector technology also has yet to reach Chinese astronomy. If the Chinese are to equal Western countries in observational capability, they must soon acquire such devices as image intensifiers, television scanners, and detector arrays, which surpass the most sensitive photographic plates in efficiency by factors of 10 to 50.

*Radio Astronomy*

The current instrument construction program at the Miyun Radio Astronomy Station was scheduled for completion by the end of 1979. Visiting American radio astronomers reported that their Chinese colleagues were well aware of the opportunities for further development of instruments, but were understandably reluctant to talk about specific plans until this construction phase was completed. The most obvious next step would be expansion of the 1.70-m array by adding a north-south arm to improve beam characteristics at low declinations and/or by lengthening the east-west arm to increase resolution. Since relatively few radio telescopes in the world operate at wavelengths longer than 1 m, the Chinese could, at relatively low cost, fill an important gap in the radio spectrum with an instrument that could be made the best in the world.

Another attractive possibility is the development of instrumentation for millimeter wavelength measurements. Some work at 8-mm wavelengths is conducted at Purple Mountain Observatory. In addition, astronomers at Beijing University designed a novel radio heliograph for observations at millimeter wavelengths. Eight parabolic reflectors, each 40 cm in diameter, are placed on an equatorially mounted beam, 9.2 m long, which tracks the sun while rotating at one revolution every ten minutes in a plane perpendicular to the line of sight. The result is a synthesized map of the sun at time intervals of five minutes, with a resolution of about 1 arc minute at 4-mm wavelengths. The design is said to be purely exploratory, but if implemented, it would be the best instrument of its kind in the world.

As in optical astronomy, progress in radio astronomy instrumentation in China requires sensitive amplifiers and detectors as well as large radio telescopes. As of 1979, China did not have parametric or cooled amplifiers, but it would be surprising if they did not soon acquire them, given their practical importance in modern technology.

*Computers*

The largest computer available to astronomers, at the Institute for Computer Technology in Beijing, was built entirely in China. It has a word length of 48 bits, a multiplicative speed of about 3 microseconds and a main core memory with a capacity of 130K words. Another smaller electronic computer known as the TQ-16, which is also available to astronomers, appears roughly comparable with the American CDC 1604, which went into production about 1960. The TQ-16 has a word length of 48 bits, a core memory of 32K words, and an operating speed of about 30 microseconds. One TQ-16 computer is used at Yunnan Observatory for satellite tracking and celestial mechanics, and another was being installed at Purple Mountain Observatory for the same purpose. A third TQ-16 is used for optical designing at the Nanjing Astronomical Instruments Factory. Minicomputers for telescope control and data acquisition and storage were not in use in China as of 1979.

*Manpower and Communication Problems*

In astronomy, as in all fields of Chinese scholarship, standards for the selection and training of young people for careers in research improved greatly after 1976. Now that China has returned to more traditional methods of selection and training, the country can anticipate a steady output of well-trained and talented astronomers. For the present and immediate future, however, the astronomical community in China confronts some difficult personnel problems. A gap of about ten years in the age distribution of active Chinese research astronomers was created during the Cultural Revolution. Conceivably, the missing numbers could be replaced by an accelerated training program during the 1980's, but in later years there will surely be a period when leadership will be in short supply. Most research astronomers active in the late 1970's were between 35 and 50 years old and had the benefit of a full and well-rounded education. But unfortunately, they have had rather little contact with leading research astronomers in other countries. Although the Chinese scientists make very thorough use of the world astronomical literature, reading scientific papers cannot fully substitute for face-to-face meetings in which new ideas can be exchanged years before they find their way into print. It would seem urgent for Chinese astronomers to begin to enter into normal scientific intercourse with astronomers abroad, by attending international scientific meetings and by work exchanges.

China's continued absence from the International Astronomical Union (IAU) is not only a severe loss to the world community of astronomers, but hinders the development of astronomy within the country. The problem at issue, which also concerns China's relations with most other international scientific unions, is the membership of Taiwan, which the PRC considers unacceptable. (In August 1979 a delegation of six astronomers from China attended a general assembly of the IAU in Montreal.)

The Chinese Astronomical Society, which used to hold both national and regional astronomical meetings at which papers were presented, became inactive about 1972. After the National Science Conference in March 1978 called on societies of the natural sciences to expand their educational activities, the Chinese Astronomical Society was reactivated.

Astronomy offers rich opportunities for international cooperation between China and other countries.[23] In the past, many fundamental astronomical programs have depended on data from a variety of instruments making observations over long periods of time from locations all over the world. Very long baseline interferometry at radio wavelengths requiring simultaneous observations by telescopes as far apart as an earth diameter is a recent example of worldwide collaboration in astronomy. It would be a particularly fruitful one for China, given its geographical separation from the United States and Western Europe. Moreover, both American and Chinese astronomers would undoubtedly welcome opportunities to exchange visits of several months' duration. When China acquires large telescopes, it might be possible to arrange for exchanges of observing time.

Finally, the Beijing area badly needs a central headquarters in which astronomers in all subdisciplines of astronomy at numerous observing stations and research institutes can be brought together.

With a population over 900 million, China has an abundance of talented and enthusiastic young people from which to fashion a large and competent astronomical community despite the handicap of ten years of low standards in the selection and training of scientists. It will not be easy for the Chinese to acquire the large and sophisticated instruments either in operation or in prospect in the most advanced countries, but given the allocation of sufficient resources and a period of political calm, they will probably succeed.

[23] Goldberg and Edwards, *Astronomy in China*, p. 71.

*Clifton W. Pannell*

# GEOGRAPHY

China, in addition to being the most populous country on earth, has the world's third largest area, 9.6 million km$^2$. This vast surface, both land and water, is highly varied and diverse. It is also much modified by the activities and works of man. This human exploitation of surface land and water resources has prompted Chinese leaders and planners to seek out the talents of geographers to better understand and use China's physical environment. Geography has thus become a valuable science in China.

Several sources of data were used to compile this chapter. First is the published scholarly research available in Chinese periodicals and translations. However, between 1966 and 1977, this was very sparse. Second, several scholars have written about geography in China, and these reports and studies have been used.[1] Third, I was a member of the first American delega-

[1] Marwyn S. Samuels, "Geography in China: Trends in Research and Training," *Pacific Affairs,* 50, 3 (Fall 1977): 406–25; Kuei-sheng Chang, "The Geography of Contemporary China: Inventory and Prospect," *Professional Geographer,* 27, 1 (Feb. 1975): 2–6; Pierre Gentelle, "Recherche et enseignement géographiques en Republique populaire de Chine," *Annales de géographie,* 74, 402 (Mar./Apr. 1965): 354–59; Hsieh Chiao-min, "The Status of Geography in Communist China," *Geographical Review,* 49 (Oct. 1959): 535–51; Yi Zhen, "Zai kaimen ban keyan de guanghui da dao shang jixu qianjin" (Advancing on the Road of Open-Door Scientific Research), *Dili zhishi,* 1976, 4 (Apr.): 15–17; Bohdan Kikolski, "Contemporary Research in Physical Geography in the Chinese People's Republic," *Annals of the Association of American Geographers,* 54, 12 (June 1964): 181–89; Kono Michihiro, "Scholars Learn from the Masses: Impressions of the People's Republic of China," *Chiri* (Geography), 11 (1966): 48; S. Leszczycki, "The Development of Geography in the People's Republic of China," *Geography,* 48, 2 (May 1963): 139–54; Laurence J. C. Ma and Allen G. Noble, "Recent Developments in Chinese Geographic Research," *Geographical Review,* 69, 1 (Jan. 1979): 63–78; Allan Rodgers, "Some Observations on the Current Status of Geography in the People's Republic of China," *China Geographer,* 1 (Spring, 1975): 13–24; Editorial Staff, "Xiang kexue jinshi pandeng dilixue gaofen" (Toward Advancing Science, To Climb the Highest Peak of Geographic Study), *Dili zhishi,* 1977, 12 (Dec.): 1–2; Herold J. Wiens, "Development of Geo-

tion of geographers, which visited China in August 1977, and was briefed at the Institute of Geography of the Chinese Academy of Sciences (CAS) in Beijing and other leading centers of geographic work and scholarship.[2] Finally, in September and October 1978, a group of distinguished Chinese geographers headed by Huang Bingwei, director of the CAS Institute of Geography in Beijing, visited the United States. Wu Chuanjun, a leading economic geographer at the same Institute, was the deputy leader. These two individuals were without question two of the most influential geographers in China in the late 1970's. The delegation stopped off at the Department of Geography of the University of Georgia and attended an American-Chinese symposium of geographers at the Racine, Wisconsin, Wingspread Conference Center.[3] I was able to spend several days with these scholars at both places and to hear their papers and discuss geography in China with them.

## *HISTORICAL SKETCH*

Geographical science in China has a long and honorable history, comparable to the descriptive and cartographic traditions established by Anaximander, Eratosthenes, and Strabo in the West. This tradition can be seen in the enormous body of geographical material in China's dynastic histories and provincial, regional, and local gazetteers.[4] The latter, of which a few date as far back as the Tang-Song period (A.D. 618–1127), describe local and regional flora, fauna, topography, hydrology, and natural catastrophes in China and include crude sketch maps. They are roughly analogous in scale and detail to the Domesday rolls of medieval England. Such classics as the *Shu Jing* and *Shui Jing,* which are over 2,000 years old, are even earlier examples of descriptive economic and physical geography texts.

An early work that set the ancient standard for historical scholarship in China, the *Shi Ji* (Historical records), by one of China's greatest historians, Sima Qian (Ssu-ma Ch'ien; died ca. 85 B.C.), contains chapters on hydrology and economic geography. The pattern was followed by other great scholars. For example, Ban Gu (Pan Ku; died ca. A.D. 92) also included a valuable essay on geography in his classic, the *Han Shu (The History of the*

---

graphic Science, 1949–1960," in Sidney H. Gould, ed., *Sciences in Communist China* (Washington, D.C.: American Association for Advancement of Science, 1961), pp. 411–81; and Jack F. Williams, "Two Observations on the State of Geography in the People's Republic of China," *China Geographer,* 9 (Winter 1978): 17–31.

[2] CAS, Institute of Geography, "Presentation to Ohio Academy of Science Geography Delegation," Beijing, Aug. 12, 1977; and Ohio Academy of Science, Geography Delegation to the People's Republic of China, "Oral Presentations at the Institute of Geography," Beijing, Aug. 12, 1977.

[3] Joint United States–China Symposium, Wingspread Conference, "Geographic Perspectives on the Environment: American and Chinese Views," Racine, Wis., Oct. 12–14, 1978.

[4] Joseph Needham, *Science and Civilisation in China* (Cambridge, Eng.: Cambridge University Press, 1959), vol. 3, pp. 497–556.

Han). This genre of dynastic histories set the pattern for recording significant events for future dynasties.[5] China (New York: Columbia University Press, 1958).

The discovery in 1973 of several ancient maps in Han dynasty tombs near the city of Changsha in Hunan province has caused a reappraisal of the history of the development of Chinese cartography.[6] The maps, which date to approximately 200 B.C., include two topographic and military maps and a plan for a prefectural administrative center. The topographic map, which depicts physical and relief features such as streams, lakes, and water bodies, shows that the Chinese had mastered surveying and scaling. The military map, which is also to scale, depicts a variety of natural and cultural features and indicates that Chinese cartography had developed sufficiently to produce thematic maps with easily comprehensible and effective symbols.[7]

Quantitative cartography developed slowly in China, but some of the basic principles were expressed by Bei Xiu in the third century A.D. Bei called for scaling by graduated division and the use of geometric grids for defining location and of triangles for determining distance. Although few Chinese cartographers followed Bei's principles, some important quantitative maps were compiled between the eighth and sixteenth centuries. For example, the eminent cartographer Jia Dan (729–805) continued such work during the Tang dynasty. He was followed by Shen Gua (1031–95) during the Song dynasty.[8] Explorers such as Xu Xiage (1586–1641) and Gu Yanwu (1613–82) maintained the tradition of geographic inquiry, study, and reporting in China through the centuries.[9]

It was not until the twentieth century, however, that geography became established as an academic discipline. The first geography departments were established in 1924 at Nanjing University and in 1928 at Zhongshan University in Guangzhou. Geography in Europe and North America generally developed from geology in the late nineteenth and early twentieth centuries and was mainly oriented to man's adaptations and use of the earth's surface. The first geography departments in China also followed this approach,

[5] John K. Fairbank, Edwin O. Reischauer, and Albert M. Craig, *East Asia: Tradition and Transformation* (Boston: Houghton Mifflin, 1973), pp. 66–68; and Burton Watson, *Ssu Ma Ch'ien, Grand Historian of*

[6] Mei-ling Hsu, "The Han Maps and Early Chinese Cartography," *Annals of the Association of American Geographers*, 68, 1 (Mar. 1978): 45–60; and Chen Cheng-hsiang, "The Historical Development of Cartography in China," *Progress in Human Geography*, 2, 1 (Mar. 1978): 101–20.

[7] Study Group for Han Silk Manuscripts from Mawangdui, "Report on the Ancient Maps Found in Tomb Number Three at Mawangdui, Changsha," *Wenwu*, 1975, 2 (Feb.): 35–42; and Dan Jixiang, "A Map of More Than Two Thousand and One Hundred Years Ago," *Wenwu*, 1975, 2 (Feb.): 43–48.

[8] Hsu, "Han Maps"; Needham, *Science and Civilisation in China;* and Chen, "Cartography in China."

[9] Hsieh Chiao-min, "Hsia-ke Hsu: Pioneer of Modern Geography in China," *Annals of the Association of American Geographers*, 48, 1 (Mar. 1958): 73–82.

which was hardly surprising given the substantial number of Chinese geographers who trained abroad in Europe, Japan, and North America during the period.

The Institute of Geography was first established in 1940 through the support of Boxer indemnity funds returned by the British government and was located near the wartime capital of Chongqing. According to Hsieh Chiao-min, the Institute began with a staff of 50 and was composed of four departments: Human Geography, Physical Geography, Oceanography, and Geodetic Surveying.[10]

After consolidating power in 1949, the Communist government reorganized scholarly research activity. The Institute of Geography was included in the Academia Sinica and was moved to Beijing. The Chinese Academy of Sciences (CAS) was established in Beijing as the country's premier research center. One of its divisions, Earth Sciences, included geography. The discipline thus had a clear position within the Chinese scientific establishment largely because it emulated the Soviet model of geographic science. As Hsieh pointed out in 1959:

> China has taken the Soviet Union as its model in the transformation of an agricultural country into an industrialized one. It recognizes that geography can play an important role in national construction in surveying the physical environment and natural resources and in the planning of development projects concerned with irrigation, land use, soil erosion, transportation and hydrography.[11]

According to Wu Chuanjun, the leading economic geographer mentioned earlier, the Soviet approach to geography was based on fieldwork, sound technical knowledge of the physical environment, and the interrelation of physical and economic geography.[12] During the 1950's, the Soviet influence proved a significant counterpoint to the large number of Western-trained geographic specialists in China (who were much more concerned with historical and cultural geography), as can be seen by the numerous articles by Soviet geographers that appeared in China's main geographical journal, *Dili xuebao (Acta Geographica Sinica),* during that decade.

Geographic activity continued to expand in China, despite the Soviet withdrawal in 1960. In 1963 the third national assembly of the Geographical Association, China's main scholarly and professional association of geographers, was attended by 110 geographers. The 343 papers presented covered such subjects as agricultural and physical regionalization, land use, water conservation, soil and land resources, agroclimate regions, agricul-

[10] Hsieh, "Geography in Communist China."

[11] *Ibid.*

[12] Akademiia Nauk, SSSR Institut Geografii, *The Physical Geography of China,* 2 vols. (New York: Praeger, 1969); Ian Matley, "The Marxist Approach to the Geographical Environment," *Annals of the Association of American Geographers,* 56, 1 (Mar. 1966): 97–111; and Wu Chuanjun, "The Geographical Organization and New Trends of Development in Geography in the USSR," *Dili xuebao (Acta Geographica Sinica),* 24, 4 (Dec. 1958): 438–56.

tural mapping, and climatic cycles.[13] Chinese geography was then divided into seven subdisciplines, which were carried into the 1970's: economic geography, cartography, climatology, physical geography, hydrology, geomorphology, and historical geography.

## *THE STATUS OF GEOGRAPHY IN THE 1970's*

Before describing the content and status of these subdisciplines, I must emphasize that most geographical activity was suspended or disrupted during the Cultural Revolution and its aftermath (1966–76). Universities and research institutes were viewed as centers of privilege and elitism and became targets for Red Guards. For example, radicals hauled off new cartographic equipment from China's most important geographical research center, the CAS Institute of Geography in Beijing, and exhibited it as wasteful junk at a revolutionary display.[14] Many institutions closed and reopened only in the early 1970's, including the Institute of Geography and the then joint Geology and Geography Department at Beijing University. Geographers lost their jobs. Obviously it was impossible for geography—or indeed any scientific discipline—to develop or break new conceptual ground in such an atmosphere.

### *Economic Geography*

In China, this subject has been concerned mainly with agriculture. In the 1960's, economic geographers spent most of their time arguing about the definition of agricultural regions and the relation between economic and environmental factors. Activities at the 1963 meeting of the Geographical Association strongly suggested that this branch of Chinese geography was progressing at a modest pace, both in research and in practical application.[15]

After the Cultural Revolution (1966–69), land use and agricultural regionalization continued to dominate economic geography. The aim was to understand local environments in order to promote optimum farming practices. Economic geographers evaluated the impact of soil erosion and deposition in water conservancy studies associated with the Yellow, Wei, Hai, and Chang Jiang (Yangtze) rivers and studied the developmental potential of land resources in peripheral areas, such as Ningxia, Heilongjiang, and tropical areas in southern China. They also undertook many other kinds of studies designed to promote China's economic development. For example, they assisted in transportation studies for rail- and waterways, industrial location studies through surveys of regional resources, and studies of medium-sized cities with populations of 100,000 to 500,000 that concentrated

[13] "The Third National Assembly and Aid to Agriculture Comprehensive Scientific Annual Meeting of the Geographical Society of China," *Acta Geographica Sinica*, 30, 1 (Mar. 1964): 78–84.

[14] Ohio Academy of Science, "Oral Presentations."

[15] "Annual Meeting of Geographical Society of China."

on the rational location of resources between rural and urban areas and different regions in China. In 1978, Dr. Wu Chuanjun noted that China's economic geographers were responding to China's developmental needs through both agricultural and nonagricultural studies.[16]

### Cartography

Chinese cartography in 1963 was concerned with "county agricultural maps, base maps for programming provincial agriculture, the utilization of aerial photographs and map surveys as a means of improving production."[17] Whenever mapping at the local level (county and commune level) was undertaken, the main requirement was to map local agriculture in such a way that production was related to local environmental conditions. The goal was to assist commune-level production. The work did not appear to inspire theoretical or conceptual progress from academic or research cartographers.

The CAS-sponsored National Atlas of China project was a far more challenging and interesting job in which economic geographers participated. Initiated in 1954, with the late eminent meteorologist Zhu Kezhen (Koching Chu) as director, the National Atlas was to include general, physical, economic, and historical maps.[18] However, it came to a stop during the Cultural Revolution. In 1977 cartographers at the Institute of Geography said the project was under way again and would be completed soon. But as of 1979 none of the scheduled four volumes had appeared.[19]

Much cartographic work in the 1970's was done on map design and production techniques, as can be seen by the 1979 volume of *Acta Geographica Sinica* and a technical article by Wu Zhongxing on transforming map coordinates of points from one type of map projection to another projection.[20] Cartographers in Beijing were also conducting many experiments in scribing, three-dimensional relief mapping, and map production. Geographic cartography is, however, only a small part of China's total mapping effort; the State Bureau of Surveying and Mapping is the agency responsible for large-scale topographic mapping of China by aerial survey.

### Climatology

Climatology is treated as a physical science in China. It is studied for its relation to the environment and its impact on agricultural production.[21] In

[16] Wu Chuanjun, "Zhongguo jingji dilixue de fazhan" (The Development of Economic Geography in China), paper presented to the Ohio Academy of Science Delegation, Institute of Geography, Beijing, Aug. 12, 1977.

[17] "Annual Meeting of Geographical Society of China."

[18] Huang Bingwei, "Wo guo dili xuejie de biaoshuai, Zhu Kezhen tongzhi" (Comrade Zhu Kezhen: A Leader of China's Geography Circles), *Dili zhishi,* 1978, 2 (Feb.): 4–6; for an assessment in English of Zhu's work, see Hsieh Chiao-min, "Chu K'o-chen and China's Climatic Changes," *Geographical Journal,* 142 (1976): 248–56.

[19] Ohio Academy of Science, "Oral Presentations."

[20] Wu Zhongxing, "How to Transform Coordinates of Points from One Kind of Map Projection to Another," *Acta Geographica Sinica,* 34, 1 (Mar. 1979): 55–68.

[21] Kikolski, "Physical Geography"; and Wiens, "Geographic Science."

the 1960's climatologists were concerned mainly with identifying patterns of temperature and precipitation from which optimal crop patterns could be inferred and easily developed. Thus, climatology in the 1960's consisted mainly of classification. There was less concern with understanding weather processes. This contrasted with geomorphological work, which emphasized processual studies and attempted to improve the quality and precision of scientific measurement and inquiry.

Kikolski confirmed this evaluation in his critique of physical geography in China. He reported the limitations of locally oriented, applied studies in microclimatology and called for "fundamental research on the heat budget and the water balance."[22] Such work was under way in the 1970's. In particular, much work has been accomplished in long-range weather forecasting of China's monsoon climates. Examples include studies of the effects of tropical oceans on long-term fluctuations of the western Pacific subtropical high-pressure system and of heavy rainfall phenomena in South China.[23]

### *Physical Geography*

Physical geography in the 1960's focused on regionalization, specifically physical regionalization, a topic that merged readily into "physical regionalization serving agriculture."[24] In the 1970's, physical geography placed more emphasis on soils, vegetation, and plant growth (biogeography), and it blended with other natural sciences, such as geology and biology, particularly in natural resource surveys. For example, physical geographers participated in the 1973 expedition to Tibet and the 1977 survey of natural resources in Heilongjiang province.[25] In these surveys, data on geology, geomorphology, water resources, flora and fauna, and arable land were accumulated in order to increase scientific knowledge and understanding of China's resources and physical environments so that they could be exploited more rationally. The multidisciplinary approach based on extensive field observations was inspired by Soviet geography.

### *Hydrology*

Hydrology, another physical science in geography, was described in 1964 as "the balance of field water supply, hydrological regionalization and soil

[22] Kikolski, "Physical Geography"; and Cui Qiwu and Sun Yanjun, "On the Correlative Equation of the Heat and Water Balance," *Acta Geographica Sinica,* 34, 2 (June 1979): 167–78.

[23] Liang Biqi and Luo Huibang, "A Meso-analysis of Heavy Rainfall over South China," *Zhongshan daxue xuebao (Acta Scientarum Naturalium Universitatis Sunyatseni),* 53, 3 (Aug. 1978): 81–87; and Institute of Geography, Research Group on Long-range Weather Forecasting, "The Effects of the Tropical Ocean on Long-term Fluctuations of the Western Pacific High," *Kexue tongbao (Science Bulletin),* 22, 7 (July 1977): 313–17.

[24] "Annual Meeting of the Geographical Society of China."

[25] CAS, Qinghai-Tibet Plateau Scientific Expedition Team, "1973's Summary of the Tsinghai-Tibet Plateau Scientific Expedition of the Chinese Academy of Sciences," *Science Bulletin,* 19, 6 (June 1974), tr. Joint Publications Research Service (*JPRS*), 74: 287–88; and Zheng Du et al., "On the Natural Zonation in the Qinghai-Xizang Plateau," *Acta Geographica Sinica,* 34, 1 (Mar. 1979): 1–11.

moisture."[26] Studies of these phenomena are crucial to the development of arid lands, especially in China's northwest. Hydrologists have been concerned with techniques and equipment for measuring soil moisture. They have conducted research on the process of glacial melt and permeation and infiltration rates and capacities of streams and rivers. They have also tried to discover new information and the laws on the behavior of streams, groundwater, and soil moisture.

Work in hydrology and geomorphology merge and are difficult to separate. Work in these two areas provides some of the best examples of geographic scholarship in China. Recent work in hydrology based on satellite imagery, as reported in *Acta Geographica Sinica* in 1979, suggested that hydrology in China was up to current world standards in both content and methodology.[27]

### *Geomorphology*

Geomorphology has traditionally been one of China's main strengths in geography and has always enjoyed a good reputation and high prestige among professional geographers within China. There were probably about 1,000 practicing geomorphologists in China in the early 1960's.[28]

In the 1970's, Chinese geomorphologists worked on fluvial processes and slope developments. For example, they studied the behavior and discharge characteristics of China's major rivers, especially the Yellow River and the Chang Jiang, and loess erosional characteristics as a set of related phenomena and problems.[29] Quaternary studies were also important and were concerned mainly with Pleistocene and recent glaciation and related matters. Thus, a historical approach was taken to interpreting many surface geomorphic features, especially in northern and western China.

Yet another major topic in the 1970's was the karst landscape that covers as much as 600,000 km$^2$ in southwestern China, primarily in Yunnan, Guizhou, and Guangxi provinces. Work in Chinese karst has concentrated on practical matters of water supply and management, and little work has been completed on specific karst features.[30] Work has also been accomplished on slope development and mass wasting phenomena, and the Chengdu Institute of Geography has been conducting research on Tibetan mudflows.[31]

[26] "Annual Meeting of the Geographical Society of China."

[27] Jian Ning et al., "The Regulation of Flow and Sediment Based on the Principles of Fluvial Processes for the Improvements of the Lower Yellow River," *Acta Geographica Sinica*, 33, 1 (Sept. 1978): 13–26.

[28] D. R. Stoddart, "Geomorphology in China," *Progress in Physical Geography*, 2 (1978): 187–236; Hsieh, "Geography in Communist China"; Kikolski, "Physical Geography"; and Ma and Noble, "Chinese Geographic Research."

[29] Stoddart, "Geomorphology in China"; Jian et al., "Improvements of the Lower Yellow River"; Kikolski, "Physical Geography"; and M. M. Sweeting et al., "British Geographers in China 1977," *Geographical Journal*, 144, part 2 (1978): 187–207.

[30] Stoddart, "Geomorphology in China"; Ohio Academy of Science, "Oral Presentations"; and Sweeting et al., "British Geographers in China 1977."

[31] Joint United States–China Symposium, "Geographic Perspectives."

Coastal geomorphology received little attention, but there were signs in the late 1970's that the importance of coastal and estuarine phenomena was being recognized.[32] (For further discussion of this subject, see the chapter on fisheries, aquaculture, and oceanography in this volume.)

Geomorphologists seemed sure of the direction and value of their work in the 1970's. Their concern with methodology, technique, and precision indicated not only confidence in their activities, but also a healthy desire to improve the quality and precision of their work by keeping abreast of recent technological advances in equipment, technique, and experimentation.

*Historical Geography*

In contrast, historical geographers enjoyed modest status during the 1960's and 1970's, apparently due to their proclivity for scholarship that has little or no relation to production. In recent years some attempt has been made to accommodate studies in historical geography to practical matters. For example, recent work in historical geography has been concerned with the study of old gazetteers. Chinese geographers in 1977 said that they believed that useful geographical information could be derived from historical documents that would be relevant to solving contemporary problems. However, no specific details on such studies were presented to American geographers. Another question is the status of the eminent historical geographer Hou Renzhi of Beijing University. Although some Western geographers have met Hou and he was listed as dean of the Beijing University Department of Geography in 1979, he did not meet with the Ohio delegation in August 1977.[33]

*Preparing for the 1980's*

In December 1977, the popular geographical journal *Dili zhishi* (Geographical Knowledge) editorialized on the role of geography in the new modernization program that was being outlined by China's leaders. Geography, the editorial noted, was to continue to be closely associated with agriculture and was to support agriculture through better understanding and knowledge of the physical environment. Thus, the goal of geography was to study heat and moisture balance and man-environment relations, to evaluate agricultural potential based on environmental resources and parameters, and to predict better environmental change. It was to achieve these goals through improved scientific, analytical techniques, such as remote sensing, automated cartography, quantitative and computer analysis, and improved data collection.[34]

At the National Science Conference of March 1978, a change of direction in all scientific and technological areas in China was evident as the country's leaders undertook to accelerate the pace of modernization, beginning with

[32] Jian et al., "Improvements of the Lower Yellow River."
[33] Samuels, "Geography in China."
[34] "Xiang kexue jinshi."

an eight-year plan (1978–85).[35] The CAS Institute of Geography in Beijing was given the task of mapping the land resources and land use of China at a scale of 1 : 100,000, using remotely sensed data. (Such work had already been under way in some areas, e.g. Ningxia.) Geography was to play a role in improving industry, transportation, national defense, and medicine. Additional areas involving geographic applications were environmental monitoring and cartographic surveys based on automated techniques.

According to Huang Bingwei, director of the CAS Institute of Geography, the geographic portion of the eight-year plan was to concentrate on land and water resources. In connection with this, the Institute has been given responsibility for evaluating the environmental impact and coordinating some aspects of a massive, long-standing project to divert Chang Jiang water along the Grand Canal to the North China Plain for irrigation purposes.[36] Details available in 1979 indicated that geographers were to survey the water transfer route along the Grand Canal.[37]

The significance of this project, which involves China's largest and longest river and the fourth largest river in the world, is far-reaching. An enormous volume of water is discharged by the Chang Jiang (average discharge rate is 1,100,000 ft.$^3$/sec. at Nanjing). The project is intended to divert some of this water and pump it north via the Grand Canal to the water-deficient areas of the North China Plain and the environs of Beijing and Tianjin.[38]

Another major project outlined by Wu Chuanjun at the October 1978 Wingspread Conference was to regionalize China's landmass according to its agricultural quality and potential. There were signs in the late 1970's that geographers were increasingly in demand as city planners.[39] The geography departments at Beijing, Nanjing, Hangzhou, and Zhongshan universities had set up special sections concentrating on urban geography and city planning in response to a call from municipal construction bureaus for more urban planners with a geographic background. This represented a new direction for geography in China beyond its traditional environmental and agricultural orientation.

## *THE ORGANIZATION OF GEOGRAPHICAL RESEARCH*

According to the economic geographer Wu Chuanjun, there were approximately 6,000 professional geographers working and teaching in China

[35] *Ibid.;* and Huang Bingwei, "Statement," Joint United States–China Symposium, Wingspread Conference.

[36] Wu Chuanjun, "Statement," Joint United States–China Symposium, Wingspread Conference; Charles Greer, *Water Management in the Yellow River Basin of China* (Austin: University of Texas Press, 1979); and Gao Xia, "Yangtze Waters Diverted to North China," *Peking Review,* 1978, 38 (Sept. 22): 6–9.

[37] Huang, "Statement"; and Chen Shangkui, "Nan shui bei zhou gongcheng kancha ji" (Reporting on Investigating the Engineering Work for the Transfer of Southern Waters Northward), *Dili zhishi,* 1978, 11 (Nov.): 1–3.

[38] Greer, *Water Management;* Gao, "Yangtze Waters"; and Chen Shangkui, "Transfer of Southern Waters Northward."

[39] Wu, "Statement"; and Ohio Academy of Science, "Oral Presentations."

in 1978.[40] (Coincidentally the leading professional organization of geographers in the United States, the Association of American Geographers, also had approximately 6,000 members in 1978.) These geographers were employed in research centers or taught in university, normal school, and college departments (see the figure on page 178).

One obvious problem is the advanced age of China's geographers. The established scholars are 50 and older, and the advancing younger scholars are generally in their forties. The impact of the Cultural Revolution and the break in training is difficult to assess, but obviously there will be a shortage of trained younger specialists for some years. Whether the number of professional geographers in China is sufficient to meet the country's future needs is difficult to evaluate. If geographers begin to play an increased role in urban and regional planning and enlarge their position in such technical areas as automated cartography, remote sensing, and machine processing of information for a geographic data base, their numbers should probably increase.

*Institute of Geography, CAS, Beijing*

The Institute of Geography is the dominant geographical research institution in China. Established in 1950, it was directed by Huang Bingwei and had over 400 researchers in 1979. The organization of research groups largely paralleled the sevenfold division of research work evident at the 1963 assembly of the Geographical Association of China and in the 1970's. Historical geography was not included, however, and new research groups devoted to environmental protection and remote sensing had been established. (These additions indicated that the Chinese were well aware of developments in geography elsewhere in the world.)

A summary of the activities of the more important research groups is given below, based on briefings and papers presented to the Ohio Academy Delegation at the Institute of Geography in 1977, papers presented at the Wingspread Conference in 1978, and published works.

*The Comprehensive Physical Geography Group.* This group works on ground and land resources used to eliminate or modify unfavorable conditions, such as waterlogging and discontinuous permafrost in the virgin Great Xingan Mountain region. Some members have been involved in comprehensive studies of natural conditions and land attributes in Heilongjiang province. Research methodology includes the use of large-scale maps, field observations and site experiments, the study and prediction of soil moisture and water availability, and the examination of soil and biotic patterns and conditions. An essay published in 1979 on natural zonation on the Qinghai-Tibetan Plateau was probably written by this group.[41]

*The Environmental Protection Group.* This group is concerned with the rapid growth of industry and agriculture and the resulting stress placed on

[40]Huang, "Statement"; and Ohio Academy of Science, "Oral Presentations."
[41]Zheng et al., "Natural Zonation."

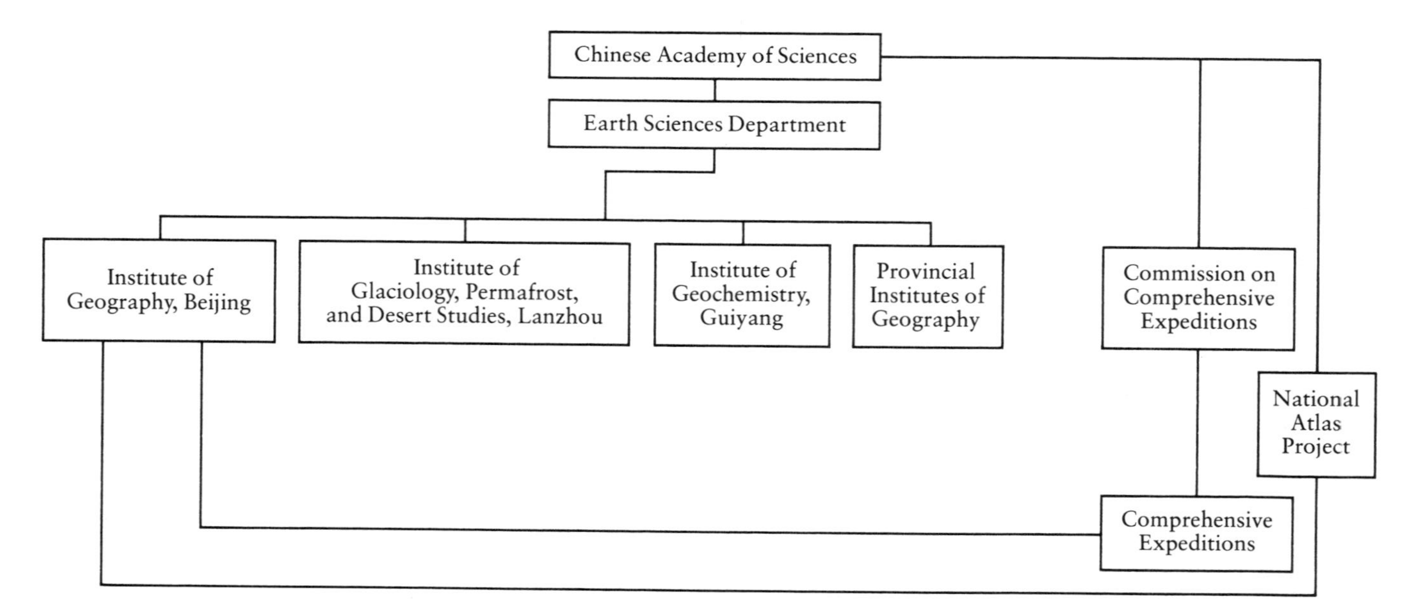


Position of Geography Within the Chinese Academy of Sciences

the natural environment. As of 1979, it planned to conduct chemical studies of soils, water, and plants and to relate them to the geography of various diseases. Researchers at the Institute of Geography candidly admitted that environmental research had been initiated only recently, and that they were inexperienced in dealing with the new subject matter. Some literature has been published in this field, however.[42]

*The Geomorphology Group.* These researchers do a great deal of work on fluvial studies, particularly the characteristics and physical behavior of channels. Recent activity has centered on the lower and middle reaches of China's largest river, the Chang Jiang. A large (600-m$^2$) hydrologic model of the lower Chang Jiang has been constructed in Beijing, and its channel characteristics were being examined through simulating variations in the stream flow. Work has also been completed on the silt load and channel behavior of the Wei He, an important tributary of the Yellow River. These studies were almost certainly aimed at predicting changes in channel behavior if the large northward diversion project were undertaken. A 1976 issue of *Dili jikan* (Geographical Collection), which is published by the Institute of Geography, showed that Chinese work in geomorphology and fluvial studies was highly technical in nature and conducted by joint teams of geographers, hydrological engineers, and hydrologists.[43] The articles in this volume again illustrated the Chinese penchant for the interdisciplinary, integrated surveys of natural resources noted above.

The Geomorphology Group also conducts research on karst topography in southwestern China. (This region, as noted earlier, has thousands of square kilometers of limestone deposits and some of the most unusual dome and tower karst landscapes in the world.) So far, research has focused on the availability and use of water resources under conditions of high rock porosity and the storage and utilization of the region's underground water resources.

*The Hydrology Group.* This group concentrates on small catchment basins (less than 100 km$^2$) and on measuring discharge rates. These are useful in bridge and culvert construction and in determining evaporation rates. To support this work, the Institute of Geography has constructed a storm-runoff analog model for the hydrologists. It simulates storm runoff and discharge and has been used in creating mathematical models for calculating peak storm runoff.

*The Cartography Group.* This group works on basic cartographic theory and techniques of map construction and design. Cartography is an important research area; not only has the Institute produced medium- and small-

[42] Zhang Shen, "Content of Trace Elements in Glacial Ice and Snow in the Mt. Jomolungma Region," *Acta Geographica Sinica*, 34, 1 (Mar. 1979): 12–17; and Tan Jianan, "Endemic Diseases and Chemical and Geographical Environments," *Dili zhishi*, 1973, 5 (May), tr. *JPRS*, 73: 29–31.

[43] "Dimao" (Geomorphology), *Dili jikan*, 10 (1976).

scale general thematic maps of China for specific research projects and atlases, but under the eight-year plan it will also map China's land resources and land use. The Institute, as noted earlier, does not produce general topographic maps, although some of its maps feature contour lines.

As of 1977, the Cartography Group had a modest role in the development of cartography nationally, but had mapped the natural conditions and resources of many areas on a 1 : 100,000 scale. For example, it had produced a set of maps of the land resources and capabilities of Ningxia Hui Autonomous Region. (The aim of this project was to promote optimal land use through cartography.)[44]

The Cartography Group also produced 1 : 25,000 scale maps of Mount Everest with 20-m contour intervals, following a survey of the Everest region. Another example of its work is a set of general physical/political maps of the entire country at a scale of 1 : 500,000 in which a 100-m contour interval was used.

*The Remote Sensing Group.* Little information is available here other than that the Institute of Geography has a special group for investigating this new methodology. The delegation of Chinese geographers that visited the United States in the fall of 1978 made cartography and remote-sensing technology one of its main areas of study. At least one paper based on Landsat multispectral imagery as a data source has been published.[45]

*The Regional Planning and Rational Allocation of Productive Forces Group* (Economic Geography). This group is engaged in applied work that encompasses what in the West would be called industrial location, transportation, urban geography, and regional planning. Water conservancy studies and environmental impact assessments are also included, and work in economic geography blends with other subfields. For example, the Chang Jiang diversion project will require some input from economic geographers. Specific topics include the size, growth, distribution, and economic functions of towns, cities, and industries and their rational location; the development of an integrated transportation network to serve the needs and demands of economic modernization; and means of minimizing differences between rural and urban areas.[46] Although it is difficult to identify specific projects, recent articles in *Acta Geographica Sinica* on highway regionalization and on mining towns were probably prepared by this group.[47]

[44] Ohio Academy of Science, "Oral Presentations"; and "Development in Surveying and Cartography," *Peking Review,* 1977, 45 (Nov. 4): 48.

[45] Pu Jingjuan and Wang Changyao, "Studies on Dynamics of Some River Deltas and Lakes Using Landsat-MSS Imagery," *Acta Geographica Sinica,* 34, 1 (Mar. 1979): 43–54.

[46] Ohio Academy of Science, "Oral Presentations."

[47] Geng Tading et al., "On the Highway Natural Regionalization of China," *Acta Geographica Sinica,* 33, 1 (Sept. 1978): 49–62; and Li Wanyan, "Industrial Development and City Planning in the Case of the Chinese Coal-Mining Cities," *Acta Geographica Sinica,* 33, 1 (Sept. 1978): 63–78.

*Committee on Comprehensive Expeditions, CAS*

The Committee on Comprehensive Expeditions was established in 1955 and has conducted expeditions and surveys in western China and the fringes of Inner Mongolia (1974), Tibet (1973–76), and the Northeast (1977), all little-known areas with only marginal human use of land and water resources. The goal of these expeditions has been to learn more about China's physical environment and resource potential through integrated field surveys of physical characteristics, natural resources, and development potential.[48]

These expeditions are typically composed of interdisciplinary teams of geographers, geologists, and biologists.[49] Geographers involved in the scientific exploration of the Qinghai-Tibetan Plateau have conducted studies of geomorphology, alpine landforms, climatology, and water resources and have identified natural environmental zones and assessed their potential for development and crop growing.[50] Geographers have also tried to identify water supplies and suitable trees and shrubs for arid lands, such as the Gobi, Ordos, A-la Shan, and Takla Makan (Tarim Basin) desert areas.[51] Officials at the Institute of Geography told the Ohio delegation in 1977 of an important survey expedition under way in the so-called new lands area of Heilongjiang province in which a number of geographers were involved, but disclosed no details about the survey's extent and duration.

*Provincial Institutes of Geography*

The Jiangsu Institute of Geography in Nanjing was set up in 1958 and is the leading geographical research center in southern China. As of 1977, it had 180 staff members divided into three departments: General Research, Limnology, and Cartography.[52] The General Research Department was engaged in studies of agricultural geography and environmental planning and was one of China's main research and training centers for work in urban geography. Little is known about the Limnology Department. Its research was apparently based on fieldwork on lakes in the lower and middle Chang Jiang basin. The Cartography Department was engaged in mapping and in training cartographers; few details were available about its specific activities.

[48] "Survey of Southern Ninghsia," *BBC Survey of World Broadcasts,* FE/W949/A/5 (Oct. 5, 1977); and "Scientific Exploration of Qinghai-Tibet Plateau Under Way," New China News Agency dispatch, May 18, 1974.

[49] Zheng et al., "Natural Zonation"; Ohio Academy of Science, "Oral Presentations"; and CAS, Qinghai-Tibet Scientific Expedition Team, "Tsinghai-Tibet Plateau Scientific Expedition."

[50] "Scientific Exploration of Qinghai-Tibet Plateau Under Way."

[51] *Ibid.;* "Survey of Southern Ninghsia"; Xia Xuncheng and Hu Wenkang, "Tulufan" (Turfan Depression), *Dili zhishi,* 1976, 9 (Aug.): 7–10; Lanzhou Institute of Glaciology, Permafrost, and Desert Studies, Theoretical Group, "Wenhua dageming tuidongle keyan shiyan de fazhan" (Advances in Scientific Research Motivated by the Cultural Revolution), *Dili zhishi,* 1976, 4 (Apr.): 1–3; Liu Zhaoqian, "Zonal Distribution of Soil in China," *Dili,* 1964, tr. *JPRS,* 32, 124 (Aug. 9, 1965): 106–10; and Ohio Academy of Science, "Oral Presentations."

[52] Institute of Geography, Jiangsu Branch, "Presentation to Ohio Academy of Science Geography Delegation," Nanjing, Aug. 22, 1977.

Less is known about China's remaining eight provincial institutes of geography. The Chengdu Institute in Sichuan province specializes in agricultural and physical geography and has analyzed mass wasting processes with emphasis on the investigation of debris flows. The Northeast Institute, located at Changchun, has focused its attention on marshland studies. The Institute of Glaciology, Permafrost, and Desert Studies is in Lanzhou. The Hebei Institute, located at Shijiazhuang, is involved in investigating underground water, as is the Henan Institute at Zhengzhou, which also specializes in agricultural geography. The Xinjiang Institute of Ürümqi does research on arid lands. The Guangdong Institute concentrates on agricultural studies and environmental matters, especially in tropical areas.[53]

As of 1979, according to Huang Bingwei, two other institutes were scheduled to open in the future: a center for economic geography in Changsha, Hunan province, and a center for tropical studies in Yunnan province. The provincial institutes of geography focus on local and regional needs, and the CAS has recommended that more be established. Since these regional institutes are funded at the provincial level, their creation and continued maintenance depend on their ability to convince local officials of their significance and contribution to local development.[54]

### *Geographical Association of China*

China's main scholarly association of geographers holds meetings of its 6,000 members every two or three years. In addition, nine specialized committees hold three or four individual symposia each year. As of 1979, the acting president was Huang Bingwei. The association sponsors China's main geographical periodical, *Acta Geographica Sinica,* which resumed publication in the fall of 1978. Some idea of its priorities may be gained from the titles of the main articles appearing in the first 1978 issue:

"Comrade Zhu Kezhen (1890–1974)"

"The Regulation of Flow and Sediment Based on the Principles of Fluvial Processes for the Improvements of the Lower Yellow River"

"Historical Variations in the Advance and Retreat of the Batura Glacier in the Karakoram Mountains"

"The Climatic Characteristics of Thunderstorms in China"

"On the Highway Natural Regionalization of China"

"Industrial Development and City Planning in the Case of Chinese Coal-Mining Cities."

## *EDUCATION*

A considerable amount of geography is taught in Chinese primary and secondary schools. The 1978 college entrance examination included papers on

[53] Huang, "Statement."
[54] *Ibid.*

earth sciences, China's regional geography, and world geography. The earth sciences section included basic knowledge about the solar system and earth-sun relations; geographic coordinate system and latitude and longitude; basic properties of maps; basic climatology; and the earth's interior structure. The regional geography section included knowledge of the physical and resource characteristics of every area and their relation to economic production, administrative units, patterns of physical geography (landform, climate, soils, main rivers, lakes, and water resources), and the distribution and nature of China's population. The world geography section emphasized physical and resource patterns, with some attention paid to political and economic geography, as seen from a Marxist perspective.[55]

Far more geography is taught in China at the precollege level than in the United States. Chinese schoolchildren almost certainly get a better training in earth sciences and physical geography than American children do. Chinese leaders clearly support geography as a means of linking better knowledge of the physical environment to improved agricultural production and better use of environmental resources.

Geographical education at the university level is less emphasized, however. In 1978 there were 31 departments of geography at the higher education and university level, compared with 41 in 1959 and 32 in 1961.[56] Of these, ten were major university departments. Among the most important geography departments are those at Beijing, Shanghai Normal, and Zhongshan universities.

*Beijing University.* This university reopened in 1970, and in August 1977 its Department of Geography had sections of geomorphology and quaternary studies, concentrating on studies of sediment and alluvium; general physical geography, focusing on environmental protection; and economic geography, specializing in urban planning, foreign-area studies, and historical geography.

The Department of Geography at Beijing University, like most of China's geography departments, is located within the natural sciences departmental structure. In 1977 there were about 500 majors in the natural sciences, of whom 150 were geographers.

In 1977 Beijing University offered only undergraduate degrees in geography. By 1979, however, a graduate program had resumed, although no degrees were granted. Before the Cultural Revolution, the program of study lasted five years and required 30 courses. The 400–500 graduates between 1953 and 1966 generally found employment in government research agencies and bureaus or city and regional planning agencies.[57]

[55] Clifton W. Pannell, "Geography," in Robert Barendsen, ed., *The 1978 National College Entrance Examination in the People's Republic of China* (Washington, D.C.: U.S. Department of Health, Education, and Welfare, Office of Education, OE Publication No. 79-19138, 1979), pp. 106–107.

[56] Huang, "Statement."

[57] Ohio Academy of Science, "Oral Presentations."

*Shanghai Normal University.* This institution had 12 teacher-training departments in 1977. Most students were destined to become middle-school teachers. The University as a whole had a faculty of 2,000 teaching a student body of 7,000; the Department of Geography had 230 students and a teaching staff of 55, with 80 research staff members and assistants. Several areas of specialization were offered: the geography of China and foreign areas (regional geography); the teaching of geography; physical geography; meteorology; and geomorphology and coastal studies. In 1977 the three-year program of study included mathematics, chemistry, meteorology, climatology, geology, surveying, mapping, soil studies, and in the third year foreign-area studies and six weeks of practice teaching.[58]

The Geography Department at Shanghai Normal University reported that it had a series of research projects under way in 1977 involving both staff and students. These included regional studies of China and foreign areas, coastal studies of the Chang Jiang delta and its mud flats, examination of the problems of saline soils, and work on medical geography. Reports presented by staff members in August 1977 included a description of the historical development of Shanghai with special attention to its industrial development and locational patterns and the agricultural geography and location of ports in the area around Shanghai.[59]

*Zhongshan University, Guangzhou.* This University had one of the first geography departments in China. Established in 1928, the Department specializes today in meteorology, hydrology, general physical geography, economic geography (agricultural and urban studies), and geology. It trains researchers and high-school teachers and conducts research on ways of improving agricultural production along the southeastern coast of China through better understanding of tropical weather and climate.

In 1978 the Department had a staff of 172, of whom 16 were professors and associate professors and 55 were lecturers. There were 308 students majoring in geography, 65 of them female. The library had 82,000 books and periodicals, 60 percent of which were in Chinese. There were also 32 laboratories, one meteorological station, one hydrological station, and a variety of other equipment, including satellite cloud-picture receivers and a radar for observing precipitation and weather activity.[60] In 1979 this Geography Department was divided into four separate sections: geography, geology, meteorology, and hydrology.[61]

[58] *Ibid.;* and Richard Kirkby, "Geography and Planning at Nanjing University," *China Geographer,* 7 (Winter 1977): 51–58.

[59] Ohio Academy of Science, "Oral Presentations"; and Shanghai Normal University, Department of Geography, "Briefings to the Ohio Academy of Science Delegation," Shanghai, Aug. 25, 1977.

[60] Shen Zanxin, "Statement," Joint United States–China Symposium, Wingspread Conference.

[61] Personal conversation with Dr. C. P. Lo, Department of Geography, University of Hong Kong, July 19, 1979. ed., University of Chicago, Department of Geography, Research Paper no. 138 (Chicago, 1971); and Leszczycki, "Geography in the People's Republic of China."

*PUBLICATIONS*

Leszczycki's full and detailed annotated listing of geographic periodicals published from 1949 to 1962 remains the most careful and thorough list compiled from standard sources, such as Harris and Fellman and the Union List of Serials.[62]

Although publications up to 1966 were considerable, the Cultural Revolution seriously disrupted publishing. Geographical periodicals simply ceased publication.[63] *Dili zhishi* reappeared in 1970 and the major geography periodical, *Acta Geographica Sinica,* in 1978. An issue of *Dili jikan,* an irregular compendium of research reports, was published in 1976 and was devoted to geomorphology. Despite these recent publications, scholarly activities, as indicated by major professional journals and periodicals, had not fully resumed by 1979. (For a complete list of Chinese geographic periodicals between 1949 and 1979, see the table on p. 186.)

Evaluating Chinese geography in the 1970's is difficult since conventional evaluations are based on the quality of published research and presentation of research findings at scholarly meetings. Because of the Cultural Revolution, few geographic research reports were published in China between 1966 and 1977 aside from the descriptive essays in the popular *Dili zhishi.* The main journal, *Acta Geographica Sinica,* resumed publication only in 1978, when Chinese geographers began attending a few international meetings. However, judging from available evidence, geographical research clearly is almost wholly applied, and its main emphasis falls on economic development. The strength of Chinese geography lies in physical studies: geomorphology, climatology, and water studies. Here the work is extensive and process oriented, in line with mainstream research in the West. The Chinese want to know more about the nature of physical processes that shape their environment in order to predict the results of any modifications they undertake.

Much less is known about Chinese work in cartography, remote sensing, and quantification. Although geographers are not primarily concerned with mapmaking in China, they have done a considerable amount of work in the field. The state of the art in color map production in the 1970's equalled that of the West perhaps 15 to 20 years earlier. As of 1977, there was no evidence of modern equipment, although Chinese geographers had toured modern cartographic and remote-sensing laboratories in the United States in the fall of 1978, including the headquarters of the U.S. Geological Survey in Reston, Virginia, and the Jet Propulsion Laboratory in Pasadena, California. The Chinese probably also lagged in computer graphics, automated car-

[62] Chauncy D. Harris and Jerome D. Fellman, *International List of Geographical Serials,* 2d ed., University of Chicago, Department of Geography, Research Paper no. 138 (Chicago, 1971); and Leszczycki, "Geography in the People's Republic of China."

[63] Chen Cheng-hsiang, "Ups and Downs of *Acta Geographica Sinica:* Some Personal Observations," *Geographical Review,* 57, 1 (Jan. 1967): 108–11.

*Chinese Geographic Periodicals, 1949–79*

| *Journal Name* | *Sponsor* | *Year of Publication* | *Language(s) employed for giving table of contents* |
|---|---|---|---|
| *Dili xuebao (Acta Geographica Sinica)* | CAS Institute of Geography | 1934–66, 1978– | Chinese, English, Russian |
| *Dili* (Geography) | Chinese Geographical Society and Institute of Geography | 1959– | Chinese |
| *Dili zhishi* (Geographical Knowledge) | Chinese Geographical Society and Institute of Geography | 1950– | Chinese |
| *Dili jikan* (Geographical Collection) | Institute of Geography, Beijing | 1957– irregular | Chinese, English, Russian |
| *Dilixue ziliao* (Memoirs of Geography) | Chinese Academy of Sciences | 1957–58 irregular | Chinese, English, Russian |
| Periodicals to which geographers have contributed include the following: | | | |
| *Scientia Sinica* | Chinese Academy of Sciences | 1957– | English, Russian |
| *Kexue tongbao (Science Bulletin)* | Editorial Committee | 1955– | Chinese, English |

tography, and remote sensing. Remote sensing was used for analyzing rock structure, evaluating vegetation and soil types, and examining hydrology and coastal estuaries, but budgetary limitations restricted the purchase of equipment necessary to analyze data from Landsat imagery or digital tapes.

One of the most difficult areas to evaluate is human geography. Work in agricultural geography has combined findings from physical geography and seems reasonably good. Its main thrust has been to assist in promoting rational and optimal land use. Efforts to regionalize China to assist proper land use seem particularly good. Urban and transportation geographers have resurfaced and are evidently much in demand, but as of 1979 the quality of the work was not high. The modest number of descriptive studies available by 1979 revealed little attention to theory or concern with understanding the basic social and economic processes at work.

In summary, geographical research in the 1970's was mostly applied, with some basic research in physical geography aimed at a better understanding of environmental resources and physical processes. In all cases the ultimate goals of research were to improve the environment and increase economic production, and these goals were linked directly to the designated role of geography as an environmental science within the context of the policy of the four modernizations.

*Edward C. T. Chao*

# *EARTH SCIENCES*

This chapter is based mostly on my three visits to the PRC in 1972, 1976–77, and 1978, a cumulative period of about five months. During these trips, which were aimed at promoting scientific exchanges, I visited 21 earth sciences institutes and their laboratories in 12 different cities. I have drawn additional information from the published literature, particularly *Scientia Geologica Sinica, Acta Geophysica Sinica, Geochimica, Acta Geologica Sinica, Vertebrata Palasiatica,* and *Scientia Sinica.* Trip reports, miscellaneous articles in the *Peking Review,* and news items from broadcasts, magazines, and newspapers have also proved useful.

In a field as wide and varied in discipline and subject matter as earth sciences, it is not possible to review every subject from an expert's point of view. Even if one could find an individual with the necessary expertise, the materials available for review would only be a minute fraction of those in existence. This chapter aims to be as informative as possible about geological activities in China during the 1970's, but is necessarily incomplete. The reader will be disappointed, for example, to find little information on China's extensive mineral deposits or estimates of mineral reserves. This is partly because such information is strictly classified in the PRC and partly because the Chinese have not finished exploring and assessing their reserves and do not know precisely what they consist of.

To put the development of earth sciences in China during the 1970's in proper perspective, it is necessary to appreciate the havoc created by the Cultural Revolution (1966–69) and the radical policies of the Gang of Four before their purge in 1976. It is also important to take China's subsequent

The style of the footnotes in this essay reflects the English-language table of contents in the works cited; thus, some citations are romanized according to the Wade-Giles system.

push to modernize science and technology and expand contacts with the outside world into account. The eight-year plan (1978–85) outlined by Vice-Premier Fang Yi at the National Science Conference held in Beijing in March 1978 had significant bearing on the future of earth sciences in China in the area of energy resources, materials, and space science and technology. With respect to energy resources, for example, Fang said:

> We should study the rules and characteristics of the genesis and distribution of the oil and gas in the principal sedimentary regions, develop the theories of petroleum geology and extend oil and gas exploration to wider areas; study new processes, techniques and equipment for exploration and exploitation and raise the standards of well drilling and the rate of oil and gas recovery; and actively develop crude oil processing techniques, use the resources rationally and contribute to the building [of] some ten more oilfields, each as big as Taching [Daqing].

As for coal, Fang called for: "Active research in basic theory, mining technology, technical equipment, and safety measures. At the same time research should be carried out in the gasification, liquefaction, and multipurpose utilization of coal, and new ways explored for the exploitation, transportation, and utilization of different kinds of coal." He also advocated paying "close attention to low-calorie fuels, such as bone coal, gangue, and oil shale, and marsh gas resources in the rural areas, and making full use of them where possible."

When he turned to materials development, Fang said:

> We should speed up research work on the paragenetic deposits at Panchihhua [Banzhihua], Paotow [Baotou], and Chinchuan [Jinzhuan], where many closely associated metals have been formed; solve the major technical problems in multipurpose utilization; intensify research on the exploitation of copper and aluminum resources; make China one of the biggest producers of titanium and vanadium in the world; and approach or reach advanced world levels in the techniques of refining copper, aluminum, nickel, cobalt, and rare-earth metals.

Regarding space science and technology, Fang said: "We should attach importance to the study of space science, remote sensing techniques, and the application of satellites; build modern centers for space research . . . and conduct extensive research in the basic theory of space science and the application of space technology."[1]

Fang Yi had many other things to say about the development of Chinese science and technology in general. For further information the reader is referred to Appendix A.

### *Organization of Earth Sciences Research*

Before the founding of the PRC in 1949, there were only about 200 geoscientists in China, including myself.[2] Geology assumed new significance

[1]Fang Yi, "Report to National Science Conference," New China News Agency, Beijing, Mar. 28, 1978.

[2]E. C. T. Chao, "Progress and Outlook of Geology," in Sidney H. Gould, ed., *Science in Communist China* (Washington, D.C.: American Association for the Advancement of Science, 1961), pp. 497–522; and idem, "Contacts with Earth Scientists in the People's Republic of China," *Science*, 179 (1973): 961–63.

under the Communist regime because the development of natural resources was crucial to China's social, economic, and political progress. Many geology institutes were established under the Chinese Academy of Sciences (CAS) between 1950 and 1954, including the Institute of Geology (Beijing), the Institute of Geophysics (Beijing), and the Institute of Geology and Paleontology (Nanjing). The Beijing Institute of Vertebrate Paleontology and Paleoanthropology was reorganized in 1953. The Ministry of Geology was established under the State Council around 1952 and later became the State Bureau of Geology. The State Seismology Bureau was established in 1971. In addition, many ministries, including those of Metallurgy, Fuels and Chemicals, Transportation, Agriculture, and Hydroelectric Power, acquired geology sections during the early 1950's. Since during this first decade China relied on the assistance of Soviet geologists and engineers, it is not surprising that the organization of earth sciences was patterned entirely on the Soviet model. As years passed, many reorganizations and modifications occurred. By 1978 an estimated 800,000 people were engaged in earth sciences activities in China, of whom about 50,000 were trained geologists and engineers. They were divided into four major research sectors, which are summarized in Table 1: the CAS, the State Bureau of Geology,* the ministries, and the universities.

The most important sector—from the point of view of basic research—was the CAS. Its nine earth sciences research institutes employed about 5,000 people. Each institute was divided into different research departments, many with their own laboratories. For example, in 1978 there were 11 departments in the Institute of Geology: Tectonics and Structural Geology, investigating a wide range of problems from neotectonics for earthquake forecasting to pre-Cambrian metamorphic rocks, application of remote sensing techniques, and structural modeling; Crustal Evolution, principally the historical tectonic development of northern China; Heat Flow Studies, with plans to resume high-pressure and -temperature studies; Modeling and Mathematical Geology; Stratigraphy; Mineralogy; Petrology; Engineering Geology; Sedimentology, including petroleum geology; Isotope Geology; and Mineral and Rock Analysis. The Institute staff numbered about 300. Similarly, the Institute of Geochemistry in Guiyang (Guizhou province) had ten research departments: Isotope Geology; Nuclear Geochemistry; Organic Geochemistry; Petrology and Mineral Deposits; Trace Elements Geochemistry; Sedimentology, Structure, and Mathematical Statistics; Environmental and Quaternary Geology; Mineral Physics; Chemical Analysis; and Instrumentation. This leading geochemical research center employed about 600 people in 1976–77.

The State Bureau of Geology, the largest earth sciences organization, is another major research sector in China. It is concerned with regional geologic mapping, geophysical prospecting and groundwater surveys, geologic mapping of mining districts, extensive sample analysis, drilling for ground-

* The State Bureau of Geology was once more renamed the Ministry of Geology in September 1979.

TABLE 1. *Principal Earth Sciences Institutions*

I. *Under the Chinese Academy of Sciences*

Institute of Geology (Beijing)
Institute of Geophysics (Beijing)
Institute of Vertebrate Paleontology and Paleoanthropology (Beijing)
Institute of Geochemistry (Guiyang)
Institute of Geology and Paleontology (Nanjing)
Institute of Engineering Mechanics (Haerbin)
Institute of Oceanology (Qingdao)
Institute of Oceanology of the South China Seas (Guangzhou)
Institute of Glaciology, Permafrost, and Deserts (Lanzhou)

II. *Under direct control of State Bureau of Geology*

Chinese Academy of Geological Sciences (with at least 12 research institutes)
Institute of Exploration Techniques and Design
Bureau of Hydrogeology and Engineering Geology
Bureau of Oceanography
Administration group
Foreign affairs group
Supplies group
Factories producing instruments (Beijing, Shanghai, Tianjin, Xian, Chengdu, and Chongqing)
Petroleum group
Geophysical prospecting group
Geological colleges (Changchun, Xuanhua, Wuhan, and Chengdu)
Aerial surveys brigade

III. *Each of the provinces and autonomous regions and the three cities of Beijing, Shanghai, and Tianjin has a bureau of geology under the joint control of the State Bureau of Geology and the individual provincial governments. Each of these bureaus has:*

1 research institute
1 central laboratory
1–2 regional mapping brigades
1 geophysical prospecting brigade
1 hydrogeology brigade
1 topographic surveying brigade
Many mine site and adjacent area geologic mapping and prospecting brigades

IV. *Ministries and Bureaus with Geologic Institutes*

Ministry of Metallurgy
Ministry of Petroleum
Ministry of Coal, with two research institutes
Ministry of Railways
Ministry of Hydroelectric Power
Ministry of Construction Materials
Ministry of Agriculture
Bureau of Oceanography
Bureau of Seismology

V. *Universities with Geology Departments*

Beijing University
Chinese People's University (Beijing)
Nanjing University
Northwest University (Xian)
Nankai University (Tianjin)
Chongqing University
Zhongshan University (Guangzhou)

water, and prospecting for mineral deposits. The Bureau runs its own system of research institutes and provincial geological bureaus that conduct applied research, mostly on mining and production problems.

However, the Bureau's principal research arm is the Beijing-based Chinese Academy of Geological Sciences (CAGS), an organization with a staff of about 2,500 people under the direction of Zou Jiayou. As of 1978, it had institutes and other organizational units in geology; mineral deposits; geophysical prospecting (Xian); geomechanics; ore deposits associated with volcanic rocks (Nanjing); ore deposits associated with Quaternary and pre-Cambrian sedimentary rocks (Tianjin); geology and mineral deposits (Xian); sedimentary ore deposits (Chengdu); oceanography; karst brigade; library; and a museum. As of 1978 two new units were being planned for oceanography, and mineralogy and petrology.

Many other bureaus, research groups, factories, and geological colleges are run by the State Bureau of Geology (see Table 1). The Bureau also shares responsibility with provincial governments for administering the provincial bureaus of geology in China's 21 provinces, five autonomous regions, and three municipalities. While the State Bureau of Geology supervises scientific and technical development, the provincial governments take care of manpower.

Each provincial bureau of geology has its own research institute and laboratory. Depending on the size of the province, it has either one or two regional mapping brigades, a geophysics brigade, a groundwater geology brigade, a topographic surveying brigade, and many mine site geologic mapping and prospecting brigades. There are 400 to 600 people in a small brigade and 1,400 to 1,700 in a large one. Similar administrative arrangements exist for the bureaus of geology in the three autonomous municipalities of Beijing, Shanghai, and Tianjin. About 5 to 10 percent of the State Bureau of Geology's 380,000 employees have had formal training in geology at universities or colleges; the remaining 90 to 95 percent consists of laborers, drilling crews, and other support personnel.

Geological research is also conducted by special units under ministries concerned with the exploitation of oil, gas, and other mineral deposits (see Table 1). For example, the Ministry of Petroleum conducts research on petroleum geology at its Institute of Scientific Research for Petroleum Exploration and Development in Beijing. Similarly, there are two institutes under the Ministry of Coal: the Fushun Institute of Coal Research, which works on coal mine safety in Fushun, and the Institute of Geological Exploration in Xian, which researches coal petrology and geology. The number of people in earth sciences working in the ministerial research sector in 1978 was estimated at 400,000.

A small amount of research is conducted at the universities. In 1978 there were 14 universities with geology departments, of which seven are listed in Table 1.

It should be emphasized that the strengths and weaknesses of research in the 1970's were directly related to the amount of sophisticated equipment involved. The PRC, as a developing country, could not manufacture all the elaborate and precision instruments it required. Thus, Chinese work in stratigraphy and paleontology, subdisciplines that do not depend on sophisticated equipment, is not far behind similar work in the United States, except the application of computers. On the other hand, the Chinese were lagging behind the West in isotope geochronology, some areas of geophysics, and descriptive and experimental mineralogy and petrology. These are subdisciplines in earth sciences that require precise and sophisticated instruments. I saw modern equipment for mineralogy and geophysics in a few Chinese laboratories, but they were exceptional; the equipment in most laboratories was of 1950's or 1960's vintage.

### *Geologic Maps, Journals, and Regional Geology*

The Chinese have compiled various important maps, including 1:4,000,000 geologic and tectonic maps of China and a 1:5,000,000 geologic map of Asia.[3] Unfortunately, the explanatory texts accompanying the geologic and tectonic maps are too brief. According to Li Tingdong, who was in charge of compiling these maps, a more complete 40,000-word explanatory text was due to be published in 1979 for the geologic map of Asia. Apparently, at the 1:200,000 scale geologic maps may have been completed for the whole of China.

Geologic, aeromagnetic, gravity, and radiometric maps of the 1:1,000,000 scale are for internal use only in China and are not generally made available to foreigners. Geologic maps of the scales 1:50,000 and 1:25,000 are available for most of the interior provinces of China, but are also restricted.

Scientific publications in geology ceased completely during the Cultural Revolution (1966–69), resuming only in early 1973. The quarterly journals *Scientia Geologica Sinica, Vertebrata Palasiatica, Acta Geophysica Sinica, Geochimica,* and the biannual *Acta Geologica Sinica* are the most important. Papers on geology also appear occasionally in such journals as *Scientia Sinica, Kexue tongbao (Science Bulletin),* and *Kexue shiyan (Scientific Experimentation).* Almost all papers are in Chinese, and only a very few journals publish English abstracts. These are usually very short (sometimes only a paragraph) and not very informative.

Most articles on regional geology published since the Cultural Revolution have described the geological characteristics of a region, for example, parts of Shandong province,[4] southern Hunan province, northern Guangdong

[3] CAGS, Compilation Group of the Geological Map of China, *An Outline of the Geology of China* (Beijing, 1976), 22 p. (explanatory text for the geologic map of China 1:4,000,000); CAGS, Institute of Geomechanics, *On Tectonic Systems* (Beijing, 1976), 17 p. (explanatory text for the tectonic map of China 1:4,000,000); and CAGS, *The Geologic Map of Asia, 1:5,000,000* (Map Publishing Co., 1975), 20 sheets.

These maps may be purchased from the Taching Import and Export Company, 11–27 44th Road, Long Island City, New York 11101.

[4] CAS, Institute of Geology, Shandong Seismogeologic Group, "Characteristics of the Block-Faulting Tectonics of Shandong Region and a Preliminary Division of Its Earthquake Belt," *Scientia Geologica Sinica,* 1974, 4: 315–28 (English abstract, p. 329).

province,[5] the Sungshan area of Henan province adjacent to the Qinling mountains,[6] and the Jilianshan region of Gansu province.[7] The articles generally revised earlier work and included some new data. The most outstanding regional geologic studies were on the geology of the north slopes of the Himalayan mountains near Chomolungma (Mount Everest).[8]

## *Stratigraphy, Paleontology, and Paleobotany*

A number of important descriptive contributions in paleontology and stratigraphy were published in the 1970's, mostly in *Acta Geologica Sinica.* These included articles on pre-Sinian,[9] Sinian,[10] Ordovician,[11] Sinograptidae,[12] Devonian,[13] and Carboniferous and Permian stratigraphy.[14] There

[5]Shui Dao, "Geological Development of the Chenxian-Huaiji Structural Belt with Special Reference to the Trending of the Cathaysian System," *Scientia Geologica Sinica,* 1975, 4: 327–42 (English abstract, p. 342).

[6]Ma Hsing-yüan et al., "Paleostructural Type of the Sinian System Sungshan Area, Honan Province," *Scientia Geologica Sinica,* 1975, 1: 12–30 (English abstract, p. 31).

[7]Wang Quan and Liu Xueya, "Paleo-oceanic Crust of the Chilienshan Region, Western China, and Its Tectonic Significance," *Scientia Geologica Sinica,* 1976, 1: 42–45 (English abstract, p. 55).

[8]CAS, Scientific Exploration Team of Tibet, *Scientific Investigation Report of the Region of Jolmo Lungma, 1966–68, Geology* (1974), 290 p.; Mu Enzhi et al., "Stratigraphy of the Mount Jolmo Lungma Region in Southern Tibet, China," *Scientia Geologica Sinica,* 1973, 1: 13–23 (English abstract, p. 25); and Chang Cheng-fa and Zeng Shilang, "Tectonic Features of the Mount Jolmo Lungma Region in Southern Tibet, China," *Scientia Geologica Sinica,* 1973, 1: 1–11 (English abstract, p. 12).

[9]Cheng Yu-chi, Chung Fu-tao, and Su Yün-jun, "The Pre-Sinian of Northern and Northeastern China," *Acta Geologica Sinica,* 1973, 1: 72–80.

[10]Lien Yu-tso and Tsao Jui-chi, *Biostratigraphic Significance of Stromatolites and Red Algae from the Sinian Subera of China* (Beijing, 1976), 15 p.; Research Group of Sinian Glaciation, *Sinian Glacial Deposits of China* (Beijing, 1976), 15 p.; and Zhu Shixing et al., "An Outline of the Studies on Stromatolites from the Stratotype Section of Sinian Suberathem in Chihsien County, North China," *Acta Geologica Sinica,* 1978, 3: 209–20 (English abstract, p. 220–21).

[11]Sheng Hsin-fu, "On the Classification and Correlation of Ordovician Formations of China," *Acta Geologica Sinica,* 1973, 2: 207–25; I Nung, "A Preliminary Study on the Stratigraphic Distribution and Zoogeographical Province of the Ordovician Corals of China," *Acta Geologica Sinica,* 1974, 1: 5–21 (English abstract, p. 22); and Xiao Chengxie, Xue Chunting, and Huang Xuecen, "Early Ordovician Graptolite Beds of Chongyi, Jiangxi," *Acta Geologica Sinica,* 1975, 2: 112–14 (English abstract, p. 125).

[12]Hsieh Chieh and Chao Yu-ting, "The Evolution and Systematics of the Family Sinograptidae," *Acta Geologica Sinica,* 1976, 2: 121–40 (English abstract, p. 140).

[13]Xiang Liwen et al., "Early Devonian Graptolite-Bearing Formation and Its Paleontological Features at Baoxing County, Sichuan," *Acta Geologica Sinica,* 1975, 2: 126–35 (English abstract, p. 135); Yang Jingzhi and Xia Fengsheng, "Devonian Bryozoans from Dushan and Wangyou near Huishui, Guizhou Province," *Acta Geologica Sinica,* 1976, 1: 57–72 (English abstract, p. 73); Qin Feng and Gan Yiyan, "The Paleozoic Stratigraphy of the Western Qinling Range," *Acta Geologica Sinica,* 1976, 1: 74–96 (English abstract, p. 97); Wang Xiaofeng, "The Discovery of the Latest Silurian and Early Devonian Monograptids from Qinzhou, Guangxi, and Its Significance," *Acta Geologica Sinica,* 1977, 2: 190–204 (English abstract, pp. 204–5); and Li Xingxue and Cai Chongyang, "A Type-Section of Lower Devonian Strata in Southwest China with Brief Notes on the Succession and Correlation of Its Plant Assemblages," *Acta Geologica Sinica,* 1978, 1: 1–11 (English abstract, p. 12).

[14]Ching Sung-chiao, "Subdivision and Correlation of the Lower Carboniferous of the Peishan Region, Kansu," *Acta Geologica Sinica,* 1974, 2: 159–74 (English abstract, p. 175); and Lin Baoyu, "A Preliminary Study on the Stratigraphical Distribution and Zoogeographical Provinces of the Permian Tabulate Corals of China," *Acta Geologica Sinica,* 1977, 2: 174–88 (English abstract, p. 189).

were also papers on Paleocene brackish-water ostracods and Eocene insects from amber.[15] Monographs have also been published on the freshwater Early Tertiary microfossils of the coastal region of the Gulf of Zhili (Bohai), which are important in petroleum geology.[16]

Most of the important papers in vertebrate paleontology and paleoanthropology have appeared in *Vertebrata Palasiatica.* Paleocene mammals from various red-bed basins, such as Nanxiong in Guangdong province, Zhaling in Hunan province, and Qianshan of Jiangsu province, have been discovered for the first time.[17] There have also been new finds of homonid fossils in Guizhou,[18] Yunnan,[19] and Guangxi provinces.[20] There have been very few contributions in paleobotany.

The CAS Institute of Geology and Paleontology in Nanjing and the CAS Institute of Vertebrate Paleontology and Paleoanthropology in Beijing led paleontological research in the PRC during the 1970's.

### *Mineralogy, Petrology, and Meteorites*

I visited a large number of laboratories concerned with mineralogy, petrology, analytical chemistry, and isotope geochronology and found that most of them identified minerals and rocks and solved analytical problems competently, despite their lack of modern petrographic microscopes, x-ray equipment, and electron microprobes. There was some sophisticated equipment, however, in a few laboratories. The Institute of Geology of the CAGS in Beijing had excellent x-ray and electron microprobe equipment. The CAS Institute of Geology, also in Beijing, had a modern mass spectrometer for rubidium-strontium isotope geochronological work. Chinese-made good-resolution gas chromatographs and British-made mass spectrometers were

[15] Wang Pin-hsien and Lin Ching-hsing, "Discovery of Paleocene Brackish-Water Foraminifers in a Certain Basin, Central China, and Its Significance," *Acta Geologica Sinica,* 1974, 2: 175–81 (English abstract, p. 183); and Hong Youchong et al., "Stratigraphy and Paleontology of Fushun Coal Field, Liaoning Province," *Acta Geologica Sinica,* 1974, 2: 113–43 (English abstract, pp. 145–50).

[16] Ministry of Petroleum and Chemical Industry, Institute of Planning for Petroleum Exploration and Development, and CAS, Institute of Geology and Paleontology, *Early Tertiary Ostracode Fauna from the Coastal Region of Bohai (Pohai)* (1978), 205 p. (English abstract, pp. 178–79); idem, *Early Tertiary Charophytes From Coastal Region of Bohai* (1978), 49 p. (English abstract, p. 48); and idem, *Early Tertiary Spores and Pollen Grains from the Coastal Region of Bohai* (1978), 163 p. (English abstract, p. 159).

[17] Chou Min-chen, "New Chapters in Asian Animal History: On New Discoveries of Paleocene Mammals in China," *Kexue shiyan (Scientific Experimentation),* 1973, 10: 26–27; Tong Yongsheng et al., "The Lower Tertiary of the Nanxiong and Qijiang Basins," *Vertebrata Palasiatica,* 14, 1 (1976): 16–25; Tang Yingjun and Yan Defa, "Notes on Some Mammalian Fossils from the Paleocene of Qianshan and Xuancheng, Anhui," *Vertebrata Palasiatica,* 14, 2 (1976): 91–99; and Wang Banyue, "Paleocene Mammals of Zhaling Basin, Hunan," *Vertebrata Palasiatica,* 13, 3 (1975): 154–62.

[18] Wu Maolin et al., "The Discovery of Fossil Human and Other Cultural Remains from Tung-tsu of Kweichou," *Vertebrata Palasiatica,* 13, 1 (1975): 14–23.

[19] Yunnan Museum, "Note on Lijiang Man's Skull from Yunnan," *Vertebrata Palasiatica,* 15, 2 (1977): 157–61.

[20] Zhang Yinyun, Wang Linghong, and Dong Xingren, "The Human Skulls from Zhenpiyen Neolithic Site at Guilin, Guanxi," *Vertebrata Palasiatica,* 15, 2 (1977): 4–12 (English abstract, p. 13).

also available at the CAS Institute of Geochemistry in Guiyang and the Petroleum College of Shengli Oil Field (Shandong province). Computers were available, but access to them for geological applications was limited. No high-voltage transmission electron microscopes were being used for geologic studies.

From 10 to 20 reports of new minerals were published between 1972 and 1979, including chengbolite[21] and other platinum group minerals,[22] taiyite (a new variety of aeschynite-priorite group),[23] furongite (an aluminum uranium hydrous phosphate),[24] nanlingnite (a new arsenite mineral),[25] laihunite (a new iron silicate mineral),[26] and changbaiite.[27] Most of these new minerals have been reviewed by Michael Fleischer in *American Mineralogist.*

The usual technique for obtaining data on abundance of minerals within rocks was by manual point count. But by 1978, a rash of image analysis systems had appeared on the market. Although the use of image analysis in modal analysis was just beginning in the United States at this time, it represented an important technical advance. This was one more area in which the Chinese have to catch up.

However, most other modern techniques for mineral and rock identification were lacking in the 1970's: the Gandolfi x-ray camera for obtaining a powder pattern from a single crystal fragment for quick identification was a novelty when I introduced it into Chinese laboratories, and research optical petrographic microscopes (imported) were available at only a few institutes. An important improvement in methodology in microscopic petrographic studies resulting from the study of lunar samples in the early 1970's was the preparation and use of polished thin sections without cover glass. I have no doubt that the Chinese will adopt this new methodology as soon as they have learned the basic techniques for making them. Mineralogists and petrologists had access to electron microprobes at only one or two institutes, and for this reason, there were no studies of the distribution of trace elements in coexisting rock-forming minerals.

[21] Sun Wei-chung et al., "Chengbolite: A Platinoid Mineral from Eclogite-Plagioclasolite of Pre-Sinian," *Acta Geologica Sinica,* 1973, 1: 89–93.

[22] Yu Tsu-hsiang et al., "A Preliminary Study of Some New Minerals of the Platinum Group and Another Associated New One in Platinum-Bearing Intrusion in a Region of China," *Acta Geologica Sinica,* 1974, 2: 202–17 (English abstract, p. 218).

[23] Ji Lingyi, "Taiyite, a New Variety of Aeschynite-Priorite Group," *Acta Geologica Sinica,* 1974, 1: 91–94 (English abstract, p. 94).

[24] Hunan 230 Institute, Hunan 305 Team, and Wuhan Geologic College, X-ray Laboratory, "A New Mineral Discovered in China: Furongite $Al_2[UO_2][PO_4]_2[OH]_2 \cdot 8H_2O$," *Acta Geologica Sinica,* 1976, 2: 203–4.

[25] Gu Xiongfei et al., "Nanlingnite, A New Arsenite Mineral from Southern China," *Geochimica,* 1976, 2: 107–12 (English abstract, p. 112).

[26] CAS, Guiyang Institute of Geochemistry, Research Group, and Liaoning Metallurgical and Geological Prospecting Company, 101 Geological Team, "Laihunite, a New Iron Silicate Mineral," *Geochimica,* 1976, 2: 95–103 (English abstract, p. 103).

[27] No. 8 Comprehensive Geological Brigade of Tonghua Region, and Jilin Institute of Geological Science, Laboratory of Petrology and Mineralogy, "Changbaiite ($PbNb_2O_6$), a New Mineral of Lead and Niobium from Eastern Kirin," *Acta Geologica Sinica,* 1978, 1: 53–61 (English abstract, p. 62).

In the field of petrology, there has been some work on the granites of southern China and on ultrabasic rocks, particularly in northwestern China. For example, Mo Zhusun has divided the Nanling granites of southern China into four main ages (pre-Cambrian, early Paleozoic, late Paleozoic and early Triassic, and middle and late Mesozoic) and described their characteristics in each.[28] Hou Shijun, who studied the general characteristics of the ultrabasic rocks in northwestern China, has found that they rarely occur in sheet form, but in veins and vertical elongate intrusive bodies along fractures and as intrusives in medium to basic volcanic series. He has shown all the places where they occur, dated them geologically, and classified them into three broad chemical types.[29]

Meteorite research is usually a good indication of the level of sophistication in mineralogical and petrologic investigations. The Chinese have done some high-quality work on the important meteorite shower (which is an H group olivine bronzite chondrite with a Pb-Pb age of 4,500 ± 70 m.y.) that occurred in Jilin province on March 8, 1976. (Total recovered weight was 1,700 kg.) Preliminary reports of the shower were followed by a systematic, competent study by members of the Institute of Geochemistry, which was published in *Geochimica.*[30] Tu Guangzhi, director of the CAS Institute of Geochemistry in Guiyang kindly gave me a 300-g sample of the Jilin meteorite in August 1978. I believe that this is the first time such extensive and systematic work on a valuable sample meteorite has been undertaken by the Chinese.

### *Geochemistry, Analytical Chemistry, and Isotope Geochronology*

The most important geochemical research is conducted at the CAS Institute of Geochemistry in Guiyang. In addition, many institutes engage in analytical work and isotope geochronology.

Besides the studies of trace elements in granites of southern China, there have been major efforts to find the cause of the fatal Keshan disease, which slowly weakens heart muscles.[31] This occurs along the Korean border of Jilin province, in the loess areas of Shaanxi province, and the Linxian area of Henan province. It may result from an imbalance of trace elements in the local diet or water supply. Another significant geochemical study in the 1970's involved the mineralogy and oxidation of sulfide minerals in various different climatic regions of China.[32]

[28] Mo Zhusun, "The Nanling Granites of Southern China, Ages and Origins," verbal communication, 1978.

[29] Hou Shijun, "The Ultrabasic Rocks of Northwest China, Origin and Geological Occurrences," verbal communication, 1978.

[30] Changchun College of Geology, Meteorite Research Group, and Jilin Institute of Geological Science, "Preliminary Study on the Mineral Composition, Structure, and Texture of the Meteorites," *Acta Geologica Sinica,* 1976, 2: 176–89 (English abstract, p. 190); *Geochimica,* 1: 1–76.

[31] Liu Dongsheng, "The Effect of Geology on Human Health and the Protection of the Environment," verbal communication, 1977.

[32] Li Xilin, "Characteristics of Oxidized Zones of Sulfides in China," verbal communication, 1977.

Most of the laboratories involved in mineral and rock analysis that I visited in China had moved away from gravimetric to instrumental rock analysis methods and were using Chinese-made atomic absorption single-channel machines or Japanese Shimazu x-ray generators. Shimazu optical spectrographs were used at the CAGS Institute of Geology and the Ministry of Coal's Institute of Coal Geology for Coal Ash Analysis in Xian. These instruments with digital output were clearly an improvement over optical spectrographic analysis using glass plates, which were employed in many Chinese laboratories. Use of isotope dilution and neutron activation began only in 1973. Isotope dilution methods were used in the laboratories of the Guiyang Institute of Geochemistry and the Beijing Institute of Geology.

Some idea of the activities of analytical chemistry on geological samples in China can be gained from the All-China Rock-Ore Analysis Symposium held in Guiyang in 1973.[33] Of the 424 reports presented at the meeting, 93 were on rare-element analysis, 83 on common-element analysis, 21 on mineral and rock analysis, 32 on polarographic analysis, 12 on ion electrode-specific methods, 41 on spectrographic analysis, and 22 on atomic absorption analysis.

In organic geochemistry and analysis, Chinese-made high-resolution gas chromatographs and British AEI mass spectrometers were being used at the Shengli oil field laboratory. A gas chromatograph was connected to a mass spectrometer for organic analysis at the Institute of Geochemistry's laboratory in Guiyang. This modern equipment reflected the importance placed on petroleum geology.

Isotope geochronology has not developed very far in the PRC. A national meeting on isotope geology was held in 1975, at which over 500 potassium-argon age determinations based on biotite or whole rock analyses were reported.[34] At the CAS Institute of Geology in Beijing, the geology departments at Beijing University, Changchun Geology College, and Guiyang Institute of Geochemistry, I saw only old-fashioned instruments for volumetric determination of argon on large samples. The mass spectrometers were of 1960 vintage, including those made in 1970. The $A^{39}$-$A^{40}$ release plateau method, which is widely employed in lunar sample studies, had not been explored in the PRC. Nor had computers been employed to automate mass spectrometers used in isotope geochronological determinations.

## *Geophysical Surveys, Earthquake Forecasts, and Tectonics*

Very little information about geophysical surveying and mapping done by air, land, or sea is available, although about 60 percent of China's land area has been covered by aeromagnetic surveys at a scale of 1:1,000,000 and additional coverage at scales of 1:100,000 and 1:50,000 is available for some

[33] "All-China Symposium on Rock-Ore Analysis," *Kexue tongbao (Science Bulletin)*, 19, 2 (1974): 95.

[34] "National Meeting Held in Exchanging Experience on Isotope Geology, Nov. 20 to Dec. 2, 1975 in Guiyang," *Science Bulletin*, 21, 2 (1976): 94–95.

areas. For example, aeromagnetic and gamma ray surveys for uranium, thorium, and potassium have been completed in Guangdong province at scales of 1:200,000 and 1:50,000. I have no information on gravity and seismic land surveys, although they must have covered the important oil and gas basins in the Songliao Basin of northeast China and the North China Plain. Fragmentary information suggests that magnetic and gravity profiles have been run offshore in the Gulf of Zhili, the Yellow Sea, and the East China Sea with irregular spacing of 18 to 20 km.

The instruments commonly used in areal surveys were magnetometers suspended from airplanes or mounted on their wing tips. These were manufactured in Beijing, Shanghai, Changchun (and perhaps Tianjin, Xian, Chengdu, and Chongqing as well) under the State Bureau of Geology's technical supervision. The magnetometers included the portable proton nuclear precession, cesium and helium optically pumped, and flux-gate airborne types. The sensitivity of proton magnetometers manufactured in Beijing was 1 gamma with a range of 0 to 100,000 gammas. The metastable helium and cesium magnetometers both had a sensitivity of 0.05 gamma.

The extensive use of aeromagnetic and radiometric surveys in prospecting, combined with large numbers of drilling teams, implies that Chinese geophysicists must be fairly successful in locating ore deposits. However, they do show signs of wanting to know more about modern sensing devices, computer handling and processing of geophysical data, and other modern methods of interpreting data.

In contrast to the relative paucity of information about regional geophysical surveying in China, a great deal is known about earthquakes. Many geophysicists and seismologists interested in earthquake prediction and seismology visited the PRC in the 1970's, including Robert S. Coe of the University of California at Santa Cruz in 1971, J. Tuzo Wilson of Canada in 1972, Bruce Bolt of the University of California at Berkeley in 1974, and three U.S. delegations in 1974, 1976, and 1978.[35]

The Chinese term "*dizhen yubao*" can be translated as earthquake forecasting. Considering the present stage of accomplishment in this field, this less precise term seems more appropriate than earthquake prediction, which is used in the United States and has a more scientifically rigorous sound.

China has many earthquakes. There were some 400 of magnitude six or greater on the Richter scale between 1900 and 1979, an average of five per year. Seven earthquakes of magnitude six or greater on the Richter scale occurred in China in 1976. The extensive Chinese earthquake study program is thus born out of necessity. Not surprisingly, information on earthquakes

[35] Robert S. Coe, "Earthquake Prediction Program," *EOS (Transactions of the AGU)*, 52, 12 (1971): 940–43; J. Tuzo Wilson, "Mao's Almanac, 3,000 Years of Killer Earthquakes," *Seismology Research*, Feb. 19, 1972, pp. 60–64; Bruce Bolt, "Earthquake Studies in the People's Republic of China," *EOS (Transactions of the AGU)*, 55, 3 (1974): 108–17; Frank Press et al., "Earthquake Research in China," *EOS (Transactions of the AGU)*, 56, 11 (1975): 838–81; and Barry Raleigh et al., "The Haicheng Earthquake Study Delegation, 1977: Prediction of the Haicheng Earthquake," *EOS (Transactions of the AGU)*, 58, 5 (1977): 236–72.

was far more abundant than on any other field of geoscience in the 1970's, and the majority of articles published in *Acta Geophysica Sinica* between 1973 and 1979 concerned them.

Severe earthquakes struck Xingtai in Hebei province on March 8 and 22, 1966, with a magnitude of 6.8 and 7.2, respectively. Premier Zhou Enlai visited the stricken area and called on geologists to deal with such natural disasters, along with the masses. Although the Cultural Revolution had just begun, earthquake studies did not suffer. They were considered a relevant field of applied science. (After all, earthquakes took human lives and disrupted the economy.)

However, there was no central direction for the study of earthquakes until the State Seismology Bureau was established in 1971, drawing scientific, technical, and engineering personnel from the CAS Institutes of Geology, Geophysics (both in Beijing), and Engineering Mechanics (in Haerbin). Within a few years about 10,000 geological workers were engaged in earthquake studies.

By 1974, there were 17 stations with short- and long-period seismographic systems in Beijing, Baotou, Chengdu, Changchun, Guiyang, Guangzhou, Kunming, Gushi, Lhasa, Lanzhou, Nanjing, Quanzhou, Shanghai, Taian, Wuhan, Ürümqi, and Xian.[36] There were eight geomagnetic observatories, some using very modern digital proton precession magnetometers to monitor variation in the magnetic field continuously, 60 mobile observation stations in seismically active areas of China, and about 250 regional stations and 5,000 observation points. In the densely populated and seismically active Zhangjiakou-Beijing-Tangshan-Tianjin area, there were 21 seismic stations alone.

The seismological instruments used at seismic stations included strong-motion instruments, among them the RDZ-1–12–66; short-period seismographic systems, such as the DD1, type 62 and 64; long-period seismographic systems, including the Kirnos and model SD-70; quartz bar extensometers and water tube tiltmeters; quartz pendulum tiltmeters; and laser Geodimeters, such as model JCY-2 for baseline surveys. Most of these instruments were made in China and were of good quality.[37]

In addition to the professional/scientific organization of earthquake studies, about 100,000 people are involved at the commune level. Using crude homemade instruments, peasants observe earth light, sound, water levels in wells, water temperature, variation in ground tilt, telluride currents, radon content in well water, and any unusual behavior in wild and domestic animals. This is the most outstanding characteristic of China's earthquake program.

[36] Press et al., "Earthquake Research in China"; and Robert M. Hamilton, "Earthquake Studies in China: A Massive Earthquake Prediction Effort Is Underway," *Earthquake Information Bulletin*, 7, 6 (1975): 3–8.

[37] Press et al., "Earthquake Research in China"; and K. Whitham et al., *Report on the Visit of the Canadian Seismological Mission to China, Oct. 20–Nov. 10, 1975* (Ottawa: Department of Energy, Mines, and Resources, Earth Physics Branch, 1975), 55 pp.

TABLE 2. *Successful Forecasts of Earthquakes*

| *Date* | *Magnitude on the Richter Scale* | *Location* | *Results* |
|---|---|---|---|
| Nov. 8, 1970 | 5.5 | Sichuan province | Evacuation |
| Feb. 3, 1972 | 5.8 | Yunnan province | Evacuation |
| July 17, 1972 | 4.8 | Yunnan province | ? |
| June 6, 1974 | 4.9 | Xingtai (Hebei province) | ? |
| June 15, 1974 | 5.0 | Aftershock of Yongshan earthquake, Yunnan province | Precautions |
| Feb. 4, 1975 | 7.3 | Haicheng, Liaoning province | Evacuation |
| Mar. 8, 1975 | 5.2 | NE Yunnan province | Evacuation |
| Sept. 4, 1975 | 5.0 | Yangbi county, Yunnan province | Evacuation |
| June 29, 1976 | 7.3 | Yongling, Yunnan province | ? |

SOURCE: K. Whitham et al., *Report on the Visit of the Canadian Seismological Mission to China, Oct. 20–Nov. 10, 1975* (Ottawa: Department of Energy, Mines, and Resources, Earth Physics Branch, 1975).

Earthquake forecasting goals in the PRC are established by the State Seismology Bureau; namely, to predict all earthquakes with magnitudes greater than five on the Richter scale, locate their epicenters to within 50 km, estimate their occurrence within two to three days, and gauge their magnitudes to within one unit on the Richter scale. Tectonic and regional stress analysis, geologic evidence of active faults, ground tilt or deformation, and periodicity of seismicity based on historical records are used in making long- to middle-term forecasts of one or more years. Foreshocks, variations of wave velocity ratios, water level and property changes, and abnormal animal behavior are among the premonitory indications used in making short-term forecasts. The Chinese tend to employ every forecast method that has ever been suggested. Although this does pay off in terms of results—nine earthquakes greater than magnitude five were forecast successfully between 1970 and 1976 (see Table 2)—it does reflect an inability to evaluate and sort out the validity of precursors on a scientific basis.

Chinese records on earthquakes, which cover some 3,000 years, are the longest and best in the world. They have been used to make historical earthquake maps, to calculate periods of seismic activity and quiescence in different regions, and to trace the geographic migration of earthquakes.[38] The historical records also contain a vast amount of information on earthquake damage, which may be useful for seismic engineering studies.

Another major subject of research in the PRC is the relation between seismicity and tectonics—seismotectonics. The relation between the two is clear

[38] Shih Chen-liang et al., "Some Characteristics of Seismic Activity in China," *Acta Geophysica Sinica,* 17, 1 (1974): 1–12 (English abstract, p. 13); and Li Yung-shan and Yüan Ting-hung, "The Chance of Occurrence of Future Great Earthquakes in the Light of Accumulation and Release of Elastic Potential in China," *Acta Geophysica Sinica,* 17, 4 (1974): 246–54 (English abstract, p. 254).

on maps that superimpose the distribution of the epicenters and magnitude of earthquakes on geological and tectonic characteristics, such as lines of faults.[39] There is a general consensus that a significant number of young faults, as well as rejuvenated movement along old faults (neotectonics), are related to regional stress patterns observable on the surface and can be mapped and analyzed. Thus, the Chinese are extremely interested in evidence of new or young faults observable from Landsat images. Correlating regional stress distribution with seismicity and seismic data is the principal basis for making long- and middle-term earthquake forecasts.

Tilt and crustal deformation were used to forecast the March 22, 1966, Xingtai earthquake in Hebei province and the Haicheng earthquake of magnitude 7.3 on February 4, 1975.[40] The Chinese have examined the validity of the hypothesis of dilation on seismic data for earthquakes in many parts of China, including the magnitude 7.1 Yongshan-Daguan earthquake of May 11, 1974.[41] Seismic velocity anomalies were not, however, a major factor in the successful forecasting of the Haicheng earthquake, although foreshocks were.[42] Unfortunately the problem of distinguishing foreshocks from earthquake swarms has not been resolved.[43]

[39] Shih et al., "Seismic Activity"; CAS, Institute of Geophysics, *Abbreviated Catalog of Strong Earthquakes of China (780 B.C.–1976)* (Map Publication Press, 1976), 29 pp.; and Teng Chi-tung et al., "Some Characteristics of Seismicity and Seismotectonics in China," *Science Bulletin,* 1978, 4: 193–99.

[40] Raleigh et al., "Haicheng Earthquake"; Whitham et al., *Canadian Seismological Mission;* National Seismological Bureau, Geodetic Survey Brigade for Earthquake Research, "Ground Survey Deformation of the Haicheng Earthquake of Magnitude 7.3," *Acta Geophysica Sinica,* 20, 4 (1977): 251–63 (English abstract, p. 263); Chen Yuntai et al., "The Focal Mechanism of the 1966 Xingtai Earthquake as Inferred from the Ground Deformation Observations," *Acta Geophysica Sinica,* 18, 3 (1975): 163–82; and Wang Guozheng et al., "Crustal Structure of the Central Part of the North China Plain and the Xingtai Earthquake, II," *Acta Geophysica Sinica,* 18, 3 (1975): 195–207 (English abstract, p. 207).

[41] Feng Deyi, Tan Aina, and Wang Kefen, "Velocity Anomalies of Seismic Waves from Near Earthquakes and Earthquake Prediction," *Acta Geophysica Sinica,* 17, 2 (1974): 83–89 (English abstract, p. 89); Feng Deyi, "Anomalous Variations of Seismic Velocity Ratio Before the Yongshan-Daguan Earthquake (M = 7.1) on May 11, 1974," *Acta Geophysica Sinica,* 18, 4 (1975): 235–39 (English abstract, p. 239); Feng Deyi et al., "Preliminary Study of the Velocity Anomalies of Seismic Waves Before and After Some Strong and Moderate Earthquakes in Western China," *Acta Geophysica Sinica,* 19, 3 (1976): 196–205, and 20, 2 (1977): 115–23 (English abstract, p. 124); and Feng Rui, "On the Variations of the Velocity Ratio Before and After the Xinfengjiang Reservoir Impounding Earthquake of M = 6.1," *Acta Geophysica Sinica,* 20, 3 (1977): 211–20 (English abstract, p. 221).

[42] Raleigh et al., "Haicheng Earthquake"; Whitham et al., *Canadian Seismological Mission;* and F. M. Zhu, "An Outline of Prediction and Forecast of Haicheng Earthquake of M = 7.3," *Journal of the Seismological Society of Japan,* 1976, pp. 15–26. (English tr., ed. P. Muller, *Procedures of the Lectures by the Seismological Delegation of the PRC,* special publication 43–32 [Pasadena: California Institute of Technology, Jet Propulsion Laboratory, 1976], pp. 11–19).

[43] Hsü Shao-hsüeh, "Characteristics of the Haicheng Earthquake of 1975," *Journal of the Seismological Society of Japan,* 1976, pp. 27–41 (English tr., P. Muller, ed., *Procedures of the Lectures by the Seismological Delegation of the PRC,* special publication 43–23 [Pasadena: California Institute of Technology, Jet Propulsion Laboratory, 1976], pp. 20–29); and Ke Longsheng et al., "A Study of Certain Features of a Foreshock Sequence and That of an Ordinary Earthquake Swarm," *Acta Geophysica Sinica,* 20, 2 (1977).

The validity of changes in groundwater conditions has not been precisely evaluated, because there have been no controlled experiments in active seismic areas. However, there is some understanding of their causes. Changes of water level in wells and changes in color and muddiness—observations made mostly by amateur earthquake workers—may well be related to the filling or draining of water from fractures and to porosity in rock masses as a result of local stress changes. Similarly, changes in temperature and radon content of well water may be earthquake related in that they are produced by radium decay. Radon solution in water could follow fractures in solid rock masses. The role of radon content changes has been discussed in connection with the Haicheng earthquake.[44]

As of 1978, the significance of geomagnetic and telluric current anomalies and their scientific basis as systematic earthquake precursors had not yet been understood. But geomagnetic anomalies were recorded for several strong earthquakes, including the 1974 earthquake near Xingtai of magnitude 4.9.[45] An anomalous drop of telluric current was also recorded as one of the precursors to the 4.8-magnitude earthquake in Yunnan province on July 17, 1972, and for the Yongshan-Daguan 7.1-magnitude earthquake near Zhaotong on May 11, 1974.[46] Telluric current changes were also described for the Haicheng earthquake.[47]

Abnormal animal behavior is a uniquely Chinese aspect of earthquake forecasting that is little understood. (It may be related to vibrations or tremors.) Reports of abnormal animal behavior have included snakes emerging from hibernation in February and being found frozen on the road, rats becoming so agitated that they acted dazed, geese flying, chickens refusing to enter their coops, pigs rooting at fences, cows breaking their halters and escaping, and goats and cows becoming unusually restless.[48]

There are several Chinese schools of interpretation of tectonics. The champion of the leading school was the late J. S. Lee, an engineer who founded the influential discipline of geomechanics and popularized it while minister of geology. Geomechanics is based on experiments on clay- and paraffin-structured modeling under rotational and tilting forces. The folds and fractures produced under various substrate conditions form patterns in the shape of Greek letters, such as epsilon ($\epsilon$) and nu ($\nu$). Lee emphasized the importance of deducing the vectors of the principal stress fields responsible for producing the linear fold and fault patterns that cover much of China. Lee's work on geomechanics has been well known in China since 1949. The discovery of oil in continental facies basins in China and the belief that earthquakes can be forecasted have both been attributed to his insights.[49]

[44] Raleigh et al., "Haicheng Earthquake."
[45] Press et al., "Earthquake Research in China."
[46] *Ibid.*
[47] Raleigh et al., "Haicheng Earthquake"; and F. M. Zhu, "Haicheng Earthquake."
[48] Raleigh et al., "Haicheng Earthquake"; Whitham et al., *Canadian Seismological Mission;* and F. M. Zhu, "Haicheng Earthquake."
[49] Lee, *Dizhilixue qailun* (Principles of Geomechanics) (Beijing: Kexue Chubanshe, 1973).

The most striking evidence of his influence, however, is the 1:4,000,000 tectonics map of China published in 1976, which portrays the north-south and east-west structural elements and the epsilon, nu, and vortex structures he proposed. (These are still unfamiliar to Western geologists.) The influence of Lee's school of geomechanics can also be seen in the successful applications of geomechanical stress analysis in exploring for mineral deposits by the CAGS Institute of Geomechanics.[50]

The 1970's saw the emergence of other schools of tectonics. Chen Guoda and his followers have emphasized depressions as major controls of tectonic evolution.[51] Huang Jijing's school has focused on the polycyclic evolution of folded belts.[52] Yet another school associated with Zhang Wenyou has proposed an evolutionary model of fracture and fault blocks.[53] These schools of tectonics were under the shadow of Lee's school of geomechanics until 1976. Since the purge of the Gang of Four and the revived emphasis on "letting a hundred schools of thought contend," other approaches have begun to gain ground.

The Chinese have considered the basic ideas of plate tectonics and continental drifts as applied to various regions of China.[54] But the application of plate tectonics to explain tectonics on the Chinese continent was still at a preliminary stage in 1978. Furthermore, the relation of seismic and geophysical data on earthquakes to plate tectonics had not been treated as thor-

[50]Hsiang Jung, "Geological Exploration: Achievements and Prospects," *Peking Review,* 20, 31 (1977): 16–19; Ti Li, "Follow China's Own Road of Developing Geological Sciences," *Scientia Sinica,* 19, 5 (1976): 616–25; and Shi Liqun, "The Application of Geomechanics to the Prospecting and Exploration of Oil in China," *Acta Geologica Sinica,* 1976, 2: 153–59 (English abstract, p. 159).

[51]Chen Guoda et al., "A Brief Review on the Geotectonics of China," *Scientia Geologica Sinica,* 1975, 3: 205–18 (English abstract, p. 219).

[52]Huang Chi-ching et al., "Some New Observations on the Geotectonic Characteristics of China," *Acta Geologica Sinica,* 1974, 1: 36–51 (English abstract, pp. 51–52); T. K. Huang et al., "An Outline of the Tectonic Characteristics of China," *Acta Geologica Sinica,* 1977, 2: 117–34 (English abstract, p. 135); and Huang Chi-ching et al., *An Outline of the Tectonic Characteristics of China* (Beijing: CAGS, Institute of Geology and Mineral Resources, Tectonics Group, 1977), 24 pp.

[53]Chang Wenyou and Zhong Jiayou, *On the Development of Fracture Systems in China* (Beijing: CAS, Institute of Geology, 1977), 17 pp.; Chang Wenyou, Ye Hong, and Zhong Jiayou, "Duankuai and bankuai" (Fault Blocks and Plates), *Zhongguo kexue (Scientia Sinica),* 1978, 2: 196–211; and CAS, Institute of Geology, Tectonic Map Compiling Group, "A Preliminary Note on the Basic Tectonic Features and the Developments in China," *Scientia Geologica Sinica,* 1974, 1: 1–16 (English abstract, pp. 16–17).

[54] Yin Tsan-hsün, "Emergence and Development of Plate Tectonics," *Scientia Geologica Sinica,* 1978, 2: 99–111 (English abstract, pp. 111–12); Wang Quan and Liu Xueya, "Paleo-oceanic Crust"; Li Ping and Wang Liangmou, "Exploration of the Seismogeological Features of the Yunnan–West Sichuan Region," *Scientia Geologica Sinica,* 1975, 4: 308–25 (English abstract, p. 326); Sichuan Geology Bureau, Geologic Team 106, "Pre-Sinian Geologic Features of Middle Section of Kangtian Axis and Its Relationship to Plate Tectonics," *Scientia Geologica Sinica,* 1975, 2: 107–12 (English abstract, p. 113); Li Chun-yu, "Tectonic Evolution of Some Mountain Ranges in China, as Tentatively Interpreted on the Concept of Plate Tectonics," *Acta Geophysica Sinica,* 18, 1 (1975): 52–73 (English abstract, p. 74); and idem, "Study of Earthquake Prediction on the Basis of Geotectonics," *Acta Geophysica Sinica,* 21, 1 (1978): 67–75 (English abstract, p. 75).

oughly and rigorously as plate tectonics applied to the margins of oceanic basins.

*Research on Petroleum Geology*

Among Chinese investigations of natural resources, petroleum, high-grade iron ore, and coal remain the most important. Information on mineral deposits and documented reserves is very difficult to obtain, however, because most of it is for internal use only and not available to foreigners. Data announced in the press and in radio broadcasts have been ably summarized by Wang.[55] It appears that the reserves of 17 mineral commodities are very large by world standards. These include coal, petroleum, iron ore, and salt, as well as tungsten, antimony, molybdenum, magnesite, limestone, fluorspar, and phosphates.

Searches for high-grade iron ore and coal for consumption in southern China are widely quoted national goals. A great deal of recent work has been done on the latter by the Institute of Coal Geology and Exploration of the Ministry of Coal in Xian and by geological units working in various individual provinces.[56] Some bituminous and a large amount of low-grade anthracitic Paleozoic coal have been found in southern China.[57] However, because I was not briefed by geologists at the Ministry of Metallurgy, I am not aware of the progress or lack of progress in their search efforts for high-grade iron ores.

There have been many estimates of the oil and gas reserves of the PRC, but all of them are highly speculative, bordering on the meaningless. In 1975 the CIA estimated that PRC oil reserves on land were about 40 billion barrels and that there was a similar reserve for the offshore areas. This approached a round figure of about 100 billion barrels.[58] However, in 1978, Wang quoted an indirect Japanese source as saying that Vice-Premier Deng Xiaoping had estimated China's oil reserves at about 400 billion barrels, a fourfold increase over the U.S. estimates.

Based on the current steady production of Taching, Shengli, Dagang, and Renqiu oil fields, the vast potential oil and gas basins of Tsaidam and Tarim, and other small but deep sedimentary basins scattered in northern and southern China, the PRC is clearly an important oil-producing country in the Far East. In the late 1970's China's annual crude oil production was about 100 million metric tons. (For further information about China's petroleum industry and production statistics, see Vaclav Smil's chapter on energy.)

[55] K. P. Wang, *Mineral Industries of the People's Republic of China* (Washington, D.C.: Department of Interior, Bureau of Mines, 1978), 11 pp.

[56] Lin Ji, *Early Paleozoic Anthracite of Southern Shaanxi and Southern China* (Xian: Institute for Coal Geology and Exploration, 1978), verbal communication.

[57] Ministry of Coal Industry, Institute of Geology and Exploration, tr., *Paleogeographic Environment and Coal Distribution of the Tse-hsui Coal Series of the Fu-jung Region of South China,* 4 pp.; and idem, tr., *Mesozoic Coal-Bearing Series of Hunan and Kiangsi Province,* 4 pp.

[58] U.S., Central Intelligence Agency, *China: Energy Balance Projection* (Washington, D.C.: Library of Congress, 1975), 33 pp.

Research on petroleum geology is conducted at the Institute of Scientific Research for Petroleum Exploration and Development in Beijing under the Ministry of Petroleum. In 1978 the Institute had a staff of about 1,600 and was about to establish laboratories in organic geochemistry, stratigraphy and paleontology, sedimentology and mineralogy, and physical properties, such as porosity and permeability.

There are also research centers in important oil fields such as Daqing and Shengli, which I visited, and also at Dagang. The equipment of the Shengli Oil Field Petroleum Institute was probably among the best in China. It had a small computer to carry out experiments on boundary conditions between oil and water. It was also equipped with Chinese-made gas chromatographs and British AEI mass spectrometers for studying problems related to the origin and migration of oil. Imported, portable, truck-mounted, computer-processed well logging equipment with 8,000 m of cable from Dresser-Atlas Industries in Houston, Texas, was being installed at the Shengli oil field. The Chinese were developing their own down-hole logging instruments and computerized logging output devices.

I was very impressed with the detailed and systematic core data kept in the oil fields by production brigades. These consisted mostly of young people with high-school backgrounds, trained on the job, who were responsible for monitoring and maintaining the production of oil wells and water injection wells. Stratigraphic logs based on recovered cores were also well kept. A wealth of information had been published on the micropaleontology and general stratigraphy of the Daqing, Dagang, and Shengli oil fields.

The correlation of channel sands and lacustrine beds offset by closely spaced faults has made the exploration and development of the Dagang and Shengli oil fields a geological nightmare. Computer processing of seismic profiles and their interpretation, as well as correlation of well logs by modern logging equipment, were unresolved problems in the late 1970's.

According to one of the geologists whom I met at the Institute of Scientific Research for Petroleum Exploration and Development, there are two major categories of structural sedimentary basins producing oil and gas in China: compressional basins elongated in an east-west direction in western China and tensional or extensional basins elongated in a northeast-southwest direction in eastern China. The compressional basins include the northern basin on the Tibetan plateau, the huge Tarim basin of southern Xinjiang, the Zungarian basin north of Tianshan in northern Xinjiang, and the Tsaidam basin of Qinghai province. They are intermontane or platform basins formed and preserved as the result of the northward advance of the Indian plate against the main Asian plate. The tensional basins were formed essentially parallel to the western boundary of the Pacific plate. The eastward movement of the Asian plate was stopped by the westward subduction of the Pacific plate. For this reason, the northeast-southwest Sichuan and Ordos basins are broader on the west, where they are further from the Pacific suture, whereas the Songliao and the basins of northeastern China and the

North China Plain are intermediate in their widths. Those basins offshore the Yellow and East China Seas are narrow and more elongated because they are closer to the plate boundary. In spite of the different geologic setting and structural and depositional evolution of these two types of basins, oil and gas, mostly of continental origin, have been found in each of these basins.[59]

The characteristics of source beds of continental origin in China have been briefly discussed by Jin.[60] Information on the petroleum geology of the Tarim and the Tsaidam basins was not made available to me. However, if oil and gas have been found there, China's potential reserves must be quite remarkable.

The oil and gas from the giant Daqing oil field are being produced from a broad anticlinal structure in continental Cretaceous formations with minor undulations covering an area of up to 30 km in width and 50 to 60 km in length, trending north northeast.[61] This structure is located in the northern part of the Songliao basin, about 160 km northwest of Haerbin in Heilongjiang province. All of the current source and reservoir rocks and producing horizons are from Cretaceous formations. Except for scattered and sparse outcrops, most of the Cretaceous rocks are not exposed. The stratigraphic information is based almost entirely on the study of drill cores.

The lower Cretaceous formation rests unconformably on Jurassic strata consisting of volcanic rocks of intermediate composition interbedded with green clay shales and thin coal seams. The Institute for Petroleum Development at the Daqing oil field has divided the Cretaceous section into seven formations, in ascending order:

1. Denglouku formation, which is 1,500 m thick, consisting of coarse clastic sandstone, gravelly sandstone, unfossiliferous, flood-type, fluvial deposits.

2. Zhuantou formation, which consists of brownish red mudstone and shale, and fine-grained sandy mudstone interbedded with grayish green and grayish white sandstone and is divisible into four units on the basis of lithology and grain size.

3. Qingshankou formation, which consists of black mudstone, shale and grayish green mudstone and thin sandstones, grading into red mudstone, sandy gravels, and brownish red mudstone and grayish green sandstone at marginal areas of the basin.

4. Yaojia formation, which is 15 to 200 m thick. The lower part consists of grayish white and grayish green sandstone interbedded with grayish green

[59] Chiu Chung-chien, *Introduction to the Activities of the Institute and the General Petroleum Geology of China* (Beijing: Ministry of Petroleum, Institute of Scientific Research for Petroleum Exploration, 1978), verbal communication.

[60] Jin Yan, "Some Basic Geological Characteristics of the Continental Oil Source Rocks in China and Their Originating Conditions," *Acta Geologica Sinica*, 1977, 1: 19–27 (English abstract, p. 28).

[61] Wang Heng-chien, "Petroleum Geology of Taching Oil Field," verbal communication, 1978.

and purplish red mudstone with sandstone and conglomerate with pebbles of mudstone and volcanic rocks. The upper part consists of grayish green and grayish black mudstone interbedded with grayish green sandstone with some brownish red mudstone and contains ostracods, conchostrea, and some plant and charophyte fossils. To the southeast and northwest, scattered ostracods, pelecypods, and gastropods are present in the formation.

5. Fulongzhuan (or Nenjiang) formation, which is the best-developed formation of the Cretaceous sequence and is widespread within the basin. It consists mainly of grayish black mudstones, shale and oil shale, grayish green mudstone with intercalated gray and grayish white fine sandstone, and some red mudstone in the upper part. It is richly fossiliferous.

6. Sifangtai formation, which is up to 410 m thick and consists mainly of interbedded grayish green sandstone and brownish red mudstone. This formation contains small snails and large clam fossils and scattered ostracods.

7. Mingshui formation, which consists of brownish red, grayish green fine sandstone, sandy mudstone, and black shale.

The important petroleum source beds are black mudstones and oil shales in the first unit of the Qingshankou formation, and the first and second units of grayish black mudstones of the Fulongzhuan formation. Most of the fine- to medium-grained sandstones are good reservoir rocks. There are seven principal producing sands, but in places there are as many as forty. Most of the several thousand currently producing wells are drilled to a depth of about 1,200 m. The water-oil boundary is usually located at a depth of about 1,050 m. Work has also been done on the stratigraphic sections from which the oil and gas from the Shengli and Dagang fields are produced.[62]

The Tertiary sedimentary rock of the Gulf of Zhili region where the two fields are located consists mainly of gray and variegated colored mudstones and sandstones with minor amounts of carbonate rocks, oil shale, and gypsum. It is divided into three formations and ten stratigraphic units. The three formations in ascending order are: (1) the Kongdian formation of red and gray clastic sediments, 1,100 to 1,600 m thick; (2) the Shahejie formation, more than 2,000 m thick, consisting mainly of gray and dark gray mudstones, with some sandstones, oil shales, and two limestone members; and (3) the Dongying formation, 1,000 to 1,500 m thick, of gray mudstones interbedded with carbonaceous shales and sandstone, and grayish green mudstones with interbedded sandstones. The main source beds are the gray and dark gray mudstones and oil shales of the Shahejie formation and the 300–500 m thick middle part of the Dongying formation. Most of the sandstone and carbonate are good reservoir rocks. Oil is produced from many horizons, however, due to the closely spaced faults and the complex basement rises and depressions, and the elevations of the oil and gas boun-

[62] *Ibid.;* Ministry of Petroleum and Chemical Development, and CAS, Institute of Geology and Paleontology, *Early Tertiary Ostracode Fauna;* and Li Shaoguang, *Petroleum of the Dagang Oil Field* (Beijing: Ministry of Petroleum, 1978).

daries vary greatly, causing great difficulty in exploration and development. Wells are mostly drilled to depths of a little more than 3,000 m.

Only fragmented information is available about the small, newly discovered Renqiu oil field about 120 km south of Beijing, which has nine or ten wells producing about 1,000 metric tons of crude oil per day.[63] My sources have indicated, however, that the oil is produced from pre-Cambrian Sinian cavernous limestones. The source beds are the Tertiary lacustrine beds that overlap unconformably and abut the cavernous limestones or are in fault contact with them.

### *Remote Sensing Applications*

The Chinese showed considerable interest in and enthusiasm for remote sensing applications during my 1976–77 and 1978 visits. They were aware of the valuable contributions Landsat data could make in mapping little-known parts of Tibet and Qinghai and western Sichuan provinces. Landsat images could save time and money in providing planimetric base maps, assisting in geologic mapping, and providing detailed lineament analysis to aid mineral exploration, structural and tectonic analyses, and earthquake forecast studies. If Landsat images or computer-compatible tapes were supplemented by medium-scale (1:500,000 and 1:1,000,000) and large-scale (1:50,000) geologic maps and aerial photography and by aeromagnetic, radiometric, and gravity data, vast areas of China could be quickly surveyed for mineral resources.

By 1973 the State Bureau of Geology had obtained black and white, colored infrared, and thermal infrared images using scanning techniques and good-quality equipment. Use of Landsat band 4 to 7 images began in 1975. The Chinese have been assisted in the revision of provincial geological maps at scales of 1:1,000,000 and 1:500,000 and filled in blank spaces in such areas as Tibet. Chinese geologists have used Landsat's 1:4,000,000 mosaic of China to revise the 1:4,000,000 geologic map of China. (Landsat revealed previously unknown buried structures and large faults that extend over several hundred kilometers.) Chinese geologists have also successfully used Landsat images in making a geological analysis of sites selected for large dams, such as the Yandan water reservoir in Guangxi province, and hydro-electric stations, such as Datong in Shanxi province. The Chinese had just begun to apply remote sensing techniques and satellite images to geology and mineral exploration in the late 1970's. The importance of remote sensing technology and equipment was stressed by Fang Yi at the National Science Conference in Beijing of March 1978.[64]

### *Earth Sciences Personnel and Their Training*

There were fewer than 200 geologists in China in 1949, some of whom had received advanced degrees and training in Europe and the United States.

[63] Wang Chieh, "Petroleum Geology of Shengli Oil Field," verbal communication, 1978.
[64] Fang Yi, "Report to National Science Conference."

By 1979 most of these geologists and geophysicists were still active, although in their late sixties, seventies, or even eighties. (When I went to China in 1978, I met ten geoscientists over 70 years old.)

Another important group of Chinese geologists was educated in the Soviet Union in the 1950's. The rest were trained in China for periods of three to five years.

The best geology departments are at Beijing and Nanjing Universities. An example of the type of education provided by these universities can be seen in the Geology Department at Beijing University. As of 1978, it had four main areas of specialization: geochemistry, geomechanics, earthquake geology, and paleontology and stratigraphy. There was also a group conducting geothermal investigations. The basic courses included general geology, historical geology, paleontology, mineralogy, petrology, structural geology, field geology, and mineral deposits. In addition there were elective courses in mathematics, physics, chemistry, and foreign languages. Majors in paleontology took courses in biology, zoology, and botany as well. The department had 16 graduate students—its first since the Cultural Revolution. The four geological colleges of the State Bureau of Geology in Changchun, Beijing (now moved to Wuhan), Xuanhua (Hebei province), and Chengdu also train geologists. The number of technical high schools in the provinces that train geologists is not known, but is probably quite large. It is safe to assume that professional training at institutions of higher education was quite limited in the 1960's and 1970's and that on-the-job training was far more important.

Judging from my visits to the PRC in 1972, 1976–77, and 1978 and the eight-year plan announced at the National Science Conference in 1978, two broad areas in earth sciences will be emphasized in the 1980's and 1990's: the development of natural resources, with an emphasis on petroleum geology and energy resources, and the application of remote sensing in space science. A third area of emphasis that is clearly important, although it was not singled out at the National Science Conference, is the earthquake study program.

In this chapter I have attempted to show that Chinese geologists in the late 1970's were not far behind their Western counterparts in the less equipment-oriented disciplines in earth sciences, such as paleontology, stratigraphy, and geologic mapping. But in the more equipment-oriented disciplines, they lagged behind. If all the policies announced by Fang Yi in 1978 are faithfully carried out, it would take Chinese geologists from 5 to 15 years to catch up. The most important of these policies are the advanced training of experienced geologists in the 40–50 year age group at universities and research institutes in Europe, Australia, Canada, and the United States; the modernization of laboratories; favorable treatment of intellectuals and scientists; and a renewed emphasis on basic and theoretical research.

As in the other sciences, the PRC has shown great interest both in the acquisition of modern equipment and instruments for the modernization of earth sciences laboratories and in increasing contacts between Chinese geologists and their foreign colleagues in universities, private institutes, and government laboratories. Such contacts have served as an important communications bridge between the two research communities.

The first stage of scientific contacts represented by exchange visits of small delegations is now progressing to more substantive collaboration in which scholars and scientists are visiting institutes and laboratories for longer periods of time to compare data and jointly explore topics of mutual interest. There is no doubt that closer collaboration between Chinese and American scientists will continue to play an important role in promoting understanding and friendship between the two countries.

*Richard J. Reed*

# *METEOROLOGY*

Chinese interest in meteorology has a long history. Examples of weather lore and wisdom can be found in songs and poetry dating back to the early Zhou period (1122 B.C. to 221 B.C.). The recording of wind and snow information as well as of natural events that provide indirect evidence of weather conditions, such as the time of peach tree flowering, began during the same era. In later centuries the records increased in volume and scope, and the range of meteorological interests broadened to include attempts to explain and predict weather events and to construct simple measuring devices. The first wind vane was invented in China in A.D. 132, long before its appearance in Europe. The rain gauge, too, seems to have been first used in China; a calculation of the capacity of a gauge is contained in a mathematical treatise of the Song period (A.D. 960–1279).[1] In the centuries preceding the modern scientific era in Europe, knowledge of meteorology in China equaled and in some respects surpassed that of the Western world.

European influence on Chinese meteorology began about 1670 when Verbiest, a Jesuit missionary and astronomer, introduced the thermometer. In the following century, two missionaries, Goby and Albro, made meteorological observations at Beijing. During the nineteenth century, meteorological observatories were established under foreign auspices, one in Beijing by the Russian Academy of Sciences in 1849 and another, the well-known Zikawei (Xujiahui) Observatory, in Shanghai by French Jesuit missionaries in 1873. The scientific efforts of these observatories were, however, conducted with little or no Chinese participation.

[1] Zhu Kezhen (Chu K'o-chen), "Some Chinese Contributions to Meteorology," *Geographical Review,* 5 (1918): 136–39; and idem, "From the History of Chinese Meteorology," tr. Ya. M. Berger, *Voprosy geografícii,* 35 (1954): 278–84 (English translation available from library of British Meteorological Office).

Because of limited contact with modern science and the resistance of the Chinese bureaucracy to Western thought and culture, China did not begin to enter the mainstream of meteorological science until after the Revolution of 1911. Progress was slow at first, but by the late 1930's meteorology in China was practiced much as in other countries, though on a more modest scale. A meteorological society that published scientific papers on a quarterly basis was founded; a meteorological institute was established at Nanjing; and college-level training courses were inaugurated at the institute, at Nanjing University, and at Qinghua (Tsinghua) University in Beijing. The network of weather observing stations, begun by Europeans at the turn of the century, was expanded, and weather reporting and forecasting services were initiated.

The reasons China has supported the development of meteorology differ little from those that have fostered its growth worldwide: concern for the effects of weather on agricultural, industrial, aeronautical, and shipping operations and the need to provide public warnings of severe weather. The government of the PRC has been particularly aware of the importance of weather to economic development and has accordingly stressed the consolidation and expansion of weather services and a wide-ranging development of meteorology emphasizing practical applications. The result has been accelerated growth in the field since the Liberation, with most of the advances taking place between 1949 and 1966. In some areas, particularly basic research, growth was arrested by the Cultural Revolution, and a period of decline set in. Only lately has this shown signs of ending.

This chapter reviews the development and status of Chinese meteorology in three main subject areas: meteorological services and weather forecasting; research; and education. There has been relatively little meteorological research since the Cultural Revolution began in 1966; hence the bulk of the discussion concerns development in the 1950's and early 1960's.

The information in this chapter is based partially on the trip report of a delegation of the American Meteorological Society (AMS) in 1974, of which I was a member.[2] For example, the section on education is drawn mainly from this report. Other sources have also been consulted, including an article by Rigby on meteorology, hydrology, and oceanography in China from 1949 to 1960[3] and a survey of dynamic meteorology and numerical weather prediction during the same period by Blumen and Washington.[4] Both of these articles contain extensive bibliographies of meteorological articles by Chinese authors.

[2] W. W. Kellogg et al., "Visit to the People's Republic of China: A Report from the A.M.S. Delegation," *Bulletin of the American Meteorological Society*, 55 (1974): 1,292–350.

[3] Malcolm Rigby, "Meteorology, Hydrology and Oceanography, 1949–1960," in Sidney H. Gould, ed., *Sciences in Communist China* (Washington, D.C.: American Association for the Advancement of Science, 1961), pp. 523–614.

[4] W. Blumen and W. M. Washington, "Atmospheric Dynamics and Numerical Weather Prediction in the People's Republic of China, 1949–1966," *Bulletin of the American Meteorological Society*, 54 (1973): 502–18.

Historical material has been taken mainly from the writings of Zhu Kezhen,[5] the preface to a collection of Academia Sinica papers,[6] and Needham.[7] The few items that appear on developments in Chinese meteorology since 1974 come from a report by the Australian physicist K. D. Cole, who visited Beijing in March 1978;[8] information provided by Professor C. C. Chang of Catholic University, Washington, D.C., who made extended visits to China in 1973 and 1975 and has since maintained contact with Chinese meteorologists; information provided by the U.S. National Academy of Sciences; and recent meteorological articles in Chinese journals.

## METEOROLOGICAL SERVICES AND WEATHER FORECASTING

### Historical Background

In the late 1800's and early 1900's the British-controlled Chinese Maritime Customs established 30 meteorological stations at lighthouses along the coast and the Chang Jiang (Yangtze) River. Reports from these stations, collected at the Zikawei Observatory in Shanghai, provided the basis for the first weather warning and forecasting services in China. The Observatory was particularly noted for its work on typhoon forecasting.

Following the overthrow of the Ch'ing dynasty in 1911, the Central Meteorological Observatory was established in Beijing. This appears to have been the first institution operated by native Chinese that attempted to provide meteorological services. However, it was hampered by its inability to secure weather reports from the Maritime Customs stations and by a lack of funds for establishing many branch stations of its own. In 1928 the Institute of Meteorology was founded in Nanjing as part of the newly formed National Research Institute (the predecessor of the Academia Sinica and Chinese Academy of Sciences [CAS]). Shortly thereafter the Institute began analyzing weather maps and issuing forecasts for a major part of China. It also made strenuous efforts, although with limited success, to expand the network of observing stations, which in 1928 consisted only of about 50 stations, including those located in Japan, Formosa, and the Philippines.[9]

Some further expansion of meteorological facilities and services occurred during the ensuing years as stations were added by provincial bureaus of

[5] Zhu, "Some Chinese Contributions to Meteorology" and "From the History of Chinese Meteorology."

[6] Chinese Academy of Sciences, *Collected Scientific Papers: Meteorology, 1919–1949* (English-language ed.) (Beijing: CAS, 1954), 625 pp.

[7] Joseph Needham, "Meteorology," in *Science and Civilisation in China, Vol. 3* (Cambridge, Eng.: Cambridge University Press, 1959), pp. 462–94.

[8] K. D. Cole, "A Glimpse of Atmospheric, Space and Earth Physics in the People's Republic of China," *EOS: Transactions of the American Geophysical Union,* 60 (1979): 17.

[9] National Research Institute of Meteorology, *The Institute of Meteorology: Its Organization and Work* (Nanjing, 1929), 7 pp.; idem, *The Institute of Meteorology: Its Organization and Work* (Nanjing, 1931), 12 pp.; and idem, *The Institute of Meteorology: Its Organization and Work* (Nanjing, 1935), 18 pp.

industry, agricultural experimental stations, and water conservation and aeronautical organizations. However, by 1949 the network of weather stations was still inadequate and consisted of about 70 unevenly distributed stations for surface observations and a handful of stations for upper-air observations. Moreover, because of the multiplicity of sponsoring agencies, uniform standards were lacking in instrumentation and observation.

Recognition of the importance of meteorology to the national economy, especially to agricultural production, resulted in a reorganization and expansion of weather services in the 1950's, first under the Military Commission Meteorological Bureau established in 1949 and later, after 1953, under the Central Meteorological Bureau (CMB). Under the leadership of these organizations, there was a dramatic improvement in the synoptic observing network. By 1960 the number of surface stations had grown to about 400 and the number of upper-air stations to 60 or 70. The station network has continued to expand, and today China has one of the finest synoptic networks in the world. There has also been a remarkable expansion in the number of stations making climatological observations. Moreover, the quality of the observations has improved as a result of emphasis in the 1950's on developing uniform standards of instrumentation and observation. During this decade the manufacture of meteorological instruments began in China; previously they had been obtained almost exclusively from abroad. Since then China has more or less kept pace with the development of specialized, modern instrumentation, such as electronic computers, weather radars, and satellite receivers.

One noteworthy aspect of the development of weather services under the Communist regime has been the heavy emphasis placed on satisfying the meteorological needs of all segments of the population. This has led to greatly expanded services at the local level, aided in part by the training of subprofessional help in agricultural communes. Similar services exist elsewhere in the world, but the degree to which the Chinese have utilized subprofessional help seems unique.

### *Organization and Operation of the Meteorological Service*

The civilian meteorological service is organized at four levels: central, provincial, district, and county or local. The armed services have a separate weather service, and some other organizations, such as the salt industry, the forest service, and civil aviation, have their own specialized services. These groups, however, receive technical guidance from the CMB. An important adjunct to the government's civilian service are the weather posts found at many communes. These are operated by commune members, often on a spare-time basis, and are not part of the official service.

The CMB, with headquarters in Beijing and a staff of over 1,000, serves as the overall coordinating and guiding body for the civilian service. It is composed of the National Weather Center, which collects and analyzes current weather data and issues nationwide forecasts; a research institute, which in 1974 specialized in dynamic and synoptic meteorology, climatology, cloud physics, weather modification, and instrument development; and a data office for collecting, storing, and publishing data. Administratively, the CMB reports to the State Council. The civil meteorological service as a whole employs over 20,000 full-time workers and about 7,000 part-time workers. (These figures do not include workers in other agencies with specialized weather services.)

Offices in the 29 autonomous regions, municipalities, and provinces also prepare weather maps and issue public forecasts, as do some of the 200 district or subprovincial offices. These provincial and district offices are administratively and financially independent; the role of the CMB is confined primarily to providing technical guidance. The stations in the main synoptic network, for instance, are managed at the provincial level under standards set by the CMB; the training of technical personnel is also the responsibility of provincial governments.

Stations at the county level may be either climatological stations or synoptic stations that report regularly to the network. Both types of stations supply weather forecasts as well as observations. These stations issue "supplementary forecasts" based on information received from the provincial weather centers, generally by telephone, and on the forecaster's knowledge of local meteorological peculiarities. To gain this specialized knowledge, forecasters are encouraged to do research and to assimilate local weather lore, often expressed in the form of proverbs, from peasants and fishermen. Supplementary forecasts are also made at commune weather posts despite the limited meteorological backgrounds of the personnel.

*Weather Forecasting—Methods and Practice*

The AMS delegation toured a number of forecast centers and stations during its visit in 1974 and obtained a detailed picture of their facilities and operations. These included the National Weather Center in Beijing, the provincial center in Shanghai, the Jiading county weather station near Shanghai, and the weather post of Bingzhou Commune near Guangzhou. It is unlikely that routine weather operations have changed much since then.

At the National Weather Center data were collected from the entire northern hemisphere, and twice-daily hemispheric charts were analyzed for surface and 500-mb levels. Other standard-level charts were analyzed for the eastern half of the hemisphere only, covering the area from the Greenwich meridian eastward to the date line. The analyses were of high quality and maintained the excellence of the Beijing daily map series described by Rigby.[10]

The prognostic and forecast charts were prepared both by conventional subjective methods and by numerical prediction. Analyses and prognoses were transmitted in code to regional forecast centers by teletypewriter and

[10] Rigby, "Meteorology, Hydrology and Oceanography," p. 529.

radio. Facsimile transmission, the usual method for disseminating such information in more advanced nations, was not yet in use, although some facsimile equipment was available for special purposes. It was stated at the time that more general application was planned for the future and according to recent information, facsimile transmission went into full operation in 1978.

The method used in operational weather prediction was a combined dynamical-statistical method that employed, as input, observations over a period of time rather than at a single instant (the usual procedure). The method was devised and described many years ago by Gu Zhenzhao.[11] A more recent description of the method can be found in an article by Zheng and Du in the English-language journal *Scientia Sinica*.[12] The numerical model used in operations was formulated jointly by scientists from the Institute of Atmospheric Physics of the CAS, Beijing University, and the CMB.

In 1974 only 500-mb barotropic forecasts were made routinely. The forecasts were carried out for the eastern half of the hemisphere and were for periods as long as 96 hours, but only the forecasts for the first 48 hours were distributed to field stations. Normally the prediction model employed data extracted from a sequence of three maps. Verification statistics reveal that the level of performance was only slightly below that of comparable forecasts in the United States. The data for the machine computations were extracted manually from plotted, hand-analyzed charts, and the output data were in turn manually plotted and analyzed. Data input and output were by paper tape. The procedure was reminiscent of American methods in the mid-1950's, when numerical prediction was just becoming operational.

The computer employed in making the forecasts, the DJS-6 manufactured in Beijing, was similar to American computers of the 1950's. Deficiencies in computing power, rather than the meteorologists' knowledge and capabilities, accounted for the primitive state of numerical prediction. The DJS-6 computer performs 50,000–70,000 operations per second and has a 32,000-word core memory plus drum storage (16,000 words per drum) and a 48-bit word length. It requires two hours to perform a 96-hour forecast. The delegation was told that considerably faster computers had been built, but that there was no immediate prospect of one becoming available for meteorological purposes.

However, a vast improvement in computing power followed the installation of two M-160 and one M-170 computers purchased from Japan in March 1978. These are equivalent in performance to the IBM 158–168 series (over one million operations per second, 0.5–0.6 megabites of memory, and 32-bit word length).

[11]Gu Zhenzhao (C. C. Koo), "On the Equivalency of Formulations of Weather Forecasting as an Initial Value Problem and as an 'Evolution' Problem," in *The Atmosphere and the Sea in Motion* (New York: Rockefeller Institution Press and Oxford University Press, 1959), 505–9.

[12]C. L. Zheng and H. Y. Du, "A New Numerical Weather Prediction Model Utilizing Multiple-Instant Observation Data," *Scientia Sinica*, 16 (1973): 396–406.

The National Weather Center also contains medium- and long-range forecast sections that prepare five-day, ten-day, and monthly forecasts of mean conditions. The Chinese said that the results of these forecasts were superior to predictions based on climatological data alone, but admitted that it was difficult to define an acceptable standard of verification. As is common elsewhere, circulation indices, trend charts, analogs, and statistical methods of regression and spectral analysis form the basis of long-range forecasts.

At Shanghai, the American delegation saw a provincial weather center. It was located in the building that formerly housed the Zikawei Observatory, and was responsible for Shanghai, its vicinity, and adjacent seas. The forecasts were disseminated by the methods commonly employed in metropolitan areas of developed nations. Radio broadcasts were made four times daily over Shanghai radio and twice daily from coastal stations for foreign and domestic shipping. Local newspapers published two-day forecasts. Both automatic and direct answering services were available for receiving forecasts by telephone, and extended-range predictions were provided in written form to various government agencies.

The provincial weather center at Shanghai also provided "mobile weather posts." Teams of forecasters and observers were temporarily detached during the fishing season to provide on-the-spot weather advice to the fleet. Judging from public information bulletins, the mobile weather posts, which also serve agriculture, are a source of special pride to the Chinese authorities, perhaps because they show the extreme efforts that are made to identify and serve the meteorological needs of all segments of the population.

As part of the typical operation of a provincial center, weather maps were drawn and forecasts prepared. But there were some surprises: a radiofacsimile that received analyses and prognostic charts from Tokyo and a new DJS-6 computer that was soon to be used for numerical prediction. It appears from recent information that electronic computers have also been installed at provincial centers in Hunan, Hubei, Xinjiang, and Heilongjiang. Their use is primarily for research on numerical prediction.

In evidence at the Shanghai center were cloud pictures obtained from U.S. satellites via the automatic picture transmission (APT) system. There are at least 40 receiving sets in provincial weather centers, research centers, universities, and other meteorological facilities throughout China. Those seen by the American delegation were manufactured by a toy factory in Shanghai. They appeared to be well made and solid state throughout. The display device, an adaptation of a newspaper wirephoto receiver, was also of Chinese manufacture and delivered a good-quality picture. A CMB official said that China intended to launch a meteorological satellite of its own and added that it was not possible to give a definite date, but that "when it is in space you will know." A PRC delegation that visited the United States in November 1978 stated that plans to launch a meteorological satellite were proceeding, but that it was not quite ready.

At Jiading county, northwest of Shanghai, the American group observed the operations of a county weather station. It was well equipped to make synoptic and climatological observations and, as is typical of county stations, devoted its main effort to supporting local agricultural activities. As part of this support, it gave advice on the proper times for planting seeds and using herbicides. Temperatures were monitored to warn of extremes that could prove harmful at different stages of rice development. Warnings of late frosts, typhoons, rainstorms, and cold waves were also issued. Forecasts were made three to five times daily and more often during periods of severe weather. Warnings could be broadcast directly to the fields and peasants' living quarters via loudspeaker systems.

Consistent with Chinese practice, the forecasters were engaged in research projects bearing on local forecasting problems. One project aimed at developing a formula for predicting the probability of rainfall. The multiple regression technique employed resembled that used in many parts of the world. More peculiarly Chinese was research on peasant weather proverbs. These proverbs were tested for their reliability and refined in accordance with results. For example, one proverb stated that if the clouds look like castles, showers are sure to come. Testing of the proverb revealed, not surprisingly, that it was generally reliable; if the cloud tops grew rapidly and dissolved, the showers would be heavy; if the castle-like clouds occurred in conjunction with lenticular clouds, there would be widespread rain.

The weather post at the Bingzhou Commune outside Guangzhou was a revelation to the AMS group because it demonstrated that useful weather information could be gathered and disseminated at the grass-roots level by subprofessional help. The post was operated by a bright, enthusiastic young woman who had the equivalent of a junior high school education and had received some meteorological training at the nearby county weather station. On the walls of her small office were displayed hand-painted posters illustrating peasant weather proverbs, and on a table there was a tank containing two fish of a species that she said became active when a change of weather was imminent. She used her knowledge of proverbs and plant and animal signs to predict the weather. For example, she said that when dragonflies hovered low in concentration, moderate rainfall would follow. And when crabs walked on the road, there would be heavy rainfall within a couple of days. The core of her forecasts, however, had a more substantial base; she received twice-daily briefings from the nearby county weather station and monitored the weather broadcasts of the provincial center in Guangzhou. In the event of approaching severe weather, she had the responsibility of telephoning warnings to the commune headquarters and to the various production brigades. A well-equipped weather station used to collect climatological data was located in a rice field not far from her weather office. The data were used, in part, to prepare long-range outlooks for assisting production brigades in their planning, but the methods used to prepare the forecasts were not clear.

The Zikawei Observatory was the only place visited at which a weather radar was in evidence, but the delegation was told that a national radar network exists. At the time it consisted of about 30 3-cm Chinese-made radars (presumably the model 711 whose characteristics are described in the AMS report) and a few 10-cm coastal radars for typhoon detection (also Chinese made), 5-cm weather radars imported from Japan, and older 3-cm Marconi radars. In addition about 70 Rawinsonde systems, operating at 400 MHz with balloon-borne transponders, were used to measure winds aloft as part of the synoptic observing system.

*Assessment of Meteorological Practice*

What are some of the strengths and weaknesses of China's weather services compared with current facilities and practices in scientifically advanced nations? A primary strength is the synoptic observing network. In both the density of stations and the quality of observations, it ranks with the best in the world. The growth of the network after 1949 was truly phenomenal. Equally remarkable was the expansion in the number of climatological stations. A second strength is the demonstrated ability to manufacture and utilize the high-technology items that are essential to modern weather practice: electronic computers, radars, satellite receivers, and facsimile equipment, although most of these are not equal to modern facilities elsewhere. Still another strength is the great stress on identifying and satisfying users' requirements. Forecasters visit harbors, factories, and communes to learn the needs of various user groups and are encouraged to do research that will improve their services.

At least two weaknesses can be singled out. First is the shortage of some of the high-technology items mentioned above, particularly electronic computers. There is every indication that Chinese meteorologists are capable of constructing sophisticated computer models that would put operational numerical weather prediction more on a par with that in the advanced nations, but they lack, or at least until recently have lacked, the computing power to do this. The lack of modern graphic devices and of sufficient facsimile equipment to disseminate weather maps is another handicap that must be fully overcome. Although more radars could be usefully employed, they do not appear to be in short supply. Nor is there a notable deficiency in the satellite area, since excellent equipment is available at many places for receiving signals for satellites placed in orbit by other nations.

A second weakness, or potential weakness, is the emphasis on using weather lore and proverbs in formulating local or supplementary forecasts. Although proverbs may have some small value in extremely short-range, local predictions, they cannot be used to tell what lies beyond the horizon, as can radar, satellites, and synoptic mapping. In most parts of the world the use of weather proverbs ended a century ago with the advent of scientific forecasting. Their use in China today seems a sign of backwardness and could perhaps, if encouraged further, retard the development of sounder methods of weather prediction.

Many Chinese meteorologists are convinced that weather proverbs are valuable for long-range or seasonal prediction and believe that farmers through centuries of careful observation have discerned relationships between current and future states of the weather. Meteorologists in general, however, would be reluctant to accept this view unless it is substantiated by rigorous testing.

The use of weather lore and proverbs originated in the period following the rectification campaign of 1957–58 and was thus at least partly politically and ideologically inspired. In summarizing the development of meteorological services during the 1950's, Lu and Wang stated that at the end of that decade "service agencies first recognized that their work must rely upon the Party and the masses," and that "central and local administrations should act in harmony and take the course of consolidation between science and the experience of the masses, between science organizations and mass movements."[13] The use of weather proverbs was a step toward implementing this directive.

What are the future prospects for meteorological services in China? Clearly the potential for advancement exists, but the rate of progress is difficult to predict since it will depend largely on political decisions. If meteorological practice is to achieve the level of performance that prevails in scientifically advanced nations, access to more powerful computers is needed urgently. Without such access, operational numerical prediction cannot progress. Fortunately, electronic computers were designated by the National Science Conference (held in early 1978) as one of eight scientific and technical spheres to be given prominence in the 1978–85 period. However, even if computer technology progresses rapidly and more powerful computers are purchased from abroad (as was happening in the late 1970's), meteorology will not benefit quickly unless it is favored more than it has been in the past. Sounder and more extensive scientific training for some of the personnel providing local forecast services is also needed. Such training may be available soon, now that the Gang of Four has been deposed and a new regime more sympathetic to science is in command.

## *METEOROLOGICAL RESEARCH*

### *Historical Background*

Some research on meteorology and climatology was conducted in China in the late nineteenth century by H. Fritsche, the director of the Observatory of Terrestrial Magnetism and Meteorology in Beijing, and by the French Jesuit fathers at the Zikawei Observatory near Shanghai. The observatory in Beijing was established by the Russian Central Academy of Sciences in 1849 and ceased to function soon after Fritsche returned to Russia in 1883. The

[13] H. Lu and P. F. Wang, *Development of Meteorological Services in the Past Decade,* tr. JPRS from *Acta Meteorologica Sinica,* 30 (Oct. 1959): 197–201, JPRS-2452 (Washington, D.C.: JPRS, 1960).

Zikawei Observatory was operated by the Jesuit fathers until 1950 when the CAS and the CMB took over joint control. It was not until the formation of the Chinese Meteorological Society in 1924 and the establishment of the Institute of Meteorology of the National Research Institute in 1928 that Chinese meteorologists began to engage in research activities. In 1925 the Meteorological Society started to publish proceedings. These consisted largely of translations of foreign papers and reports of observations. By 1935, however, 27 original contributions had appeared.

The Institute of Meteorology at Nanjing carried on a variety of endeavors, including the collection and publication of meteorological observations, the issue of weather forecasts, and the training of meteorological personnel. It also published the *Memoirs of the National Research Institute of Meteorology,* an occasional series of research papers, the first of which appeared in 1929. By 1949 at least 16 papers had been published. The author of the first paper and the director of the Institute of Meteorology, was Zhu Kezhen, a remarkable figure in Chinese meteorology who at the time of his death in 1974 was a vice-president of the CAS. Zhu received his Ph.D. in meteorology from Harvard University in 1918, where he studied under Robert DeCourcy Ward. He had previously studied astronomy and meteorology at the University of Illinois. He was a man of wide-ranging interests and accomplishments who will be best remembered for his work on climatic change in China.[14]

By 1949 about 400 research papers on meteorological subjects had been published by Chinese scientists. In the early 1950's, 25 of them were selected by the CAS for publication in the volume discussed above, *Collected Scientific Papers: Meteorology, 1919–1949*. The contents reveal that the interests of Chinese meteorologists in the 1930's and 1940's were similar to those of meteorologists elsewhere. Ten papers on synoptic meteorology dealt with such subjects as cyclones, air masses, fronts, wind circulation aloft, and typhoons. Among the topics treated in eight papers on general circulation and dynamical meteorology were the monsoon, the trade winds, macroscopic turbulence, the formation of semipermanent centers of action, and energy dispersion in the atmosphere. There were also seven papers on China's climate. Much of the work was descriptive and relatively poor, but several papers were theoretically oriented and of high quality.

A noteworthy feature of the volume is the number of articles originally published in foreign journals (American, English, and German) and the number of authors temporarily affiliated with Western institutions. Foreign contacts played a significant part in the early development of Chinese meteorology. A number of meteorologists were educated abroad. In addition to Zhu Kezhen's stay at Harvard, Li Xianzi obtained a Ph.D. at the University of Berlin in 1934, as did Zhao Jiuzhang, who in 1937 became director of the Institute of Geophysics and Meteorology. Zhao also visited universi-

[14]For a detailed account of Zhu's career, see C. M. Hsieh, "Chu K'o-chen and China's Climatic Changes," *Geophysical Journal*, 142 (1976): 248–56.

ties in the United States in 1946. Du Zhangwang, at one time secretary of the CAS, studied at the University of London. Ye Duzheng and Xie Yibing earned Ph.D.'s under Rossby at the University of Chicago in the late 1940's, and Gu Zhenzhao was associated with Rossby in 1947–48, while the latter was at the Institute of Meteorology, University of Stockholm. These scientists, notably Zhu, Ye, Xie, and Gu, were major contributors to the development of meteorological research after 1949. Although the names and numbers are not well documented, Chinese students are also known to have received graduate education at meteorological institutes in the U.S.S.R. during the 1950's.

*Liberation to Cultural Revolution (1949–66)*

The importance of meteorology to China's economy, particularly agricultural production, was recognized early in the PRC. As described above, this recognition resulted in a vast expansion of the weather observing network and weather services during the 1950's. It also resulted in a substantial increase in research, as can be seen by the large number of research papers that appeared and the wide range of topics treated. Some idea of the volume and diversity may be gained by examining the bibliography appended to Rigby's review of meteorology in China during the period 1949–60.[15] All together 750 papers are listed, the majority of them from two journals: *Tianqi yuekan (Weather Monthly)*, essentially a journal of applied meteorology, and *Qixiang xuebao (Acta Meteorologica Sinica)*, a more substantial journal carrying reports of original research. The majority of articles listed date from the 1957–60 period. The range of subjects treated and the numbers of papers published in the various categories can be seen from the compilation in Table 1.

Some differences from Western research at this time are worth noting. Agricultural meteorology and meteorological instruments received unusually heavy emphasis, for example. This coincided with the campaign in the late 1950's to make support of agricultural production the main focus of meteorological services and with the general effort to improve the meteorological observing system. Nearly all papers on these subjects were published in *Weather Monthly.* Another difference was the absence of research in cloud physics, an expanding field elsewhere at the time. This was remarkable, since weather modification, an applied branch of cloud physics, was already being practiced in China. Development of weather radar, an important tool in cloud physics research, also appeared to be lagging.

A series of papers by staff members of the CAS Institute of Geophysics and Meteorology published in the Swedish journal *Tellus* attracted some attention in the West during the 1950's, but on the whole Chinese meteorological research had little impact on the subject's development in the rest of the world. From the limited information available, it is difficult to judge the

[15] Rigby, "Meteorology, Hydrology, and Oceanography," pp. 558–601.

TABLE 1. *Meteorological Publications in China, 1949–60*

| *Subject* | *Number*[a] |
|---|---|
| Dynamic meteorology | |
| General | 13 |
| Boundary layer turbulence | 10 |
| Numerical weather prediction | 28 |
| Synoptic meteorology | |
| Weather forecasting (short-, medium-, and long-range) | 89 |
| Large-scale circulations, cyclones, and anticyclones | 62 |
| Typhoons | 20 |
| Weather elements and weather phenomena | 69 |
| Weather modification | 14 |
| Climatology | |
| General | 19 |
| Climatic change | 4 |
| Microclimatology | 4 |
| Agricultural meteorology | 86 |
| Meteorological instruments | 51 |
| TOTAL | 469 |

SOURCE: Malcolm Rigby, "Meteorology, Hydrology, and Oceanography," in Sidney H. Gould, ed., *Sciences in Communist China* (Washington, D.C.: American Association for the Advancement of Science, 1961), pp. 523–614.
[a] Papers by foreign, chiefly Soviet, scientists are omitted.

originality of the research. It is possible that many articles essentially repeated work already done elsewhere. Such papers could have served a useful purpose, however, by educating the rapidly expanding research force.

A survey of authors' institutional affiliations, based on the higher-quality journals (*Acta Meteorologica Sinica, Scientia Sinica, Scientia, Science Record,* and the *Peking University Journal*), reveals that nearly all articles emanated from four research centers: the Institute of Geophysics and Meteorology of the CAS (60 percent), Beijing University (20 percent), the Institute of Meteorology of the CMB (10 percent), and Nanjing University (10 percent). These percentages remained essentially unchanged in the late 1970's. An abbreviated and updated version of Rigby's table listing all institutions engaged in meteorological research or operations appears in Table 2. Table 3 provides an updated version of Rigby's list of periodicals publishing meteorological articles.

A final characteristic of the research effort of the 1950's was the collaboration between Chinese and Soviet scientists. For example, a paper by Zeng[16] was based on a thesis completed in 1960 under Professor I. A. Kibel of the Institute of Geophysics of the Soviet Academy of Sciences, and global predic-

[16] C. T. Zeng, "Adaptive Processes and Development in the Atmosphere," *Acta Meteorologica Sinica,* 33 (1963): 163–74, 281–89.

TABLE 2. *Primary Meteorological Institutions*

| *Institution* | *Location* | *Year established* |
|---|---|---|
| Central Meteorological Bureau | Beijing | 1953[a] |
| Nanjing Institute of Meteorology | Nanjing | 1960 |
| Chengdu Institute of Meteorology | Chengdu | 1978 |
| Chinese Academy of Sciences | | |
| Institute of Atmospheric Physics | Beijing | 1966[b] |
| Institute of Plateau Meteorology | Lanzhou | 1966? |
| Chinese Meteorological Society | Beijing (11 branches) | 1924 |
| Beijing University (Geophysics Department) | Beijing | 1959[c] |
| Nanjing University (Department of Meteorology) | Nanjing | 1930 |

[a] Possibly a date of reorganization, since there is evidence that it or a similar organization existed earlier.
[b] Preceded by the Institute of Geophysics and Meteorology established in Beijing in 1950 and by the Institute of Meteorology founded in Nanjing in 1928.
[c] The meteorology program in this department stems from the meteorology division established at Qinghua University in 1935 and transferred to the Physics Department of Beijing University in 1952.

tion models of Kibel and Blinova formed the basis of work accomplished by Zhu and Zeng[17] while visiting the Institute of Geophysics of the Soviet Academy of Sciences. The Soviet meteorologist E. M. Dobrischman, performed research on medium- and long-range prediction during a visit to Beijing in 1958. As still another example, a paper on dynamic meteorology listed in Rigby's bibliography is attributed to C. M. Proskov, Russian meteorological adviser, CMB, Beijing. Prior to 1960, the Chinese freely acknowledged the extensive assistance received from the U.S.S.R. in the post-Liberation period.

A later survey of meteorological research in China, based entirely on English-language translations of articles and abstracts, was published by Blumen and Washington in 1973.[18] Their study was more restricted in scope than Rigby's and treated mainly dynamic meteorology and numerical weather prediction. However, it covers a longer time period (1949–66) and is particularly valuable because it affords an opportunity to discern research trends in the 1960's before the Cultural Revolution. The topics treated in their article and the numbers of papers in each topic are given in Table 4. (Because of differences in selection procedures and in criteria for categorizing articles, Tables 1 and 4 are not comparable.)

Blumen and Washington's analysis and data show that the research of the 1960–66 period was broadly based. They commented that although the Chinese "frequently noted that theoretical studies were undertaken with a view to application in terms of the country's economic development, the scope of research activities during the early 1960's was relatively indistin-

[17] Y. T. Zhu, "On the Calculation of the Dynamic Influence of the Mountain Massifs in the Nonlinear Problems of Long-Range Forecasting of Meteorological Elements," *Bulletin of the Academy of Sciences, U.S.S.R.: Physics and Solid Earth*, 1959, pp. 1,266–73; and C. T. Zeng and Y. T. Zhu, "The Nonlinear Problem of Atmospheric Motion on a Planetary Scale," *Bulletin of the Academy of Sciences, U.S.S.R.: Physics and Solid Earth*, 1961, pp. 98–99.
[18] Blumen and Washington, "Atmospheric Dynamics."

TABLE 3. *Meteorological Journals and Series*

| *Publication* | *Sponsor* | *Date* |
|---|---|---|
| *Qixiang xuebao (Acta Meteorologica Sinica* or *Chinese Journal of Meteorology)* | Chinese Meteorological Society | 1925–66 |
| *Tianqi yuekan (Weather Monthly)* | Chinese Meteorological Society | 1950–66 |
| *Qixiang (Weather)* | Central Meteorological Bureau | 1975– |
| *Scientia Sinica* [an English-language *publication*] | Chinese Academy of Sciences | 1952–66; 1973– |
| *Kexue tongbao (Scientia* or *Science Bulletin)* | Chinese Academy of Sciences | 1950–66; 1973– |
| *Science Record* | Chinese Academy of Sciences | 1957–60 |
| *Diqiu wuli xuebao (Acta Geophysica Sinica, formerly Journal of the Chinese Geophysical Society)* | Chinese Geophysical Society | 1950–66; 1973– |
| *Wuli xuebao (Acta Physica Sinica)* | — | 1953–66; 1974– |
| *Peking University Journal of Science (Acta Scientiarum Naturalium Universitatis Pekinensis)* | Beijing University | 1955–66 |
| *Memoirs of the National Research Institute of Meteorology* | Academia Sinica, Nanjing | 1929–48 |
| *Meteorological Memoirs of the Research Institute of Meteorology* | Central Meteorological Bureau | — |
| *Dilixue ziliao (Geophysical Memoirs)* | Chinese Academy of Sciences | 1957–? |
| *Scientia Atmospherica Sinica* | Institute of Atmospheric Physics | 1976– |
| *Collected Papers of Institute of Atmospheric Physics* (formerly *Geophysics and Meteorology)* | Institute of Atmospheric Physics | 1957– |
| *Collected Papers of Geography* | Institute of Geography | 1957– |
| *Acta Geographica Sinica* | Institute of Geography | 1934– |

guishable from similar activities taking place in all countries with vigorous research centers."[19] Thus, the work on theoretical analysis of large-scale motions begun in the 1950's continued to increase in scale. Work on developing numerical prediction models also continued, although model development was slow by U.S. standards. The development of primitive equation models, which require more powerful computers, tended to lag behind. However, the effort was impressive when compared with that in many advanced nations.

In addition to these expanded research activities, work began in other, previously neglected areas. Theoretical papers appeared on the formation of typhoons and weaker tropical disturbances. The deficiency of research in cloud physics began to be remedied; there were eight contributions on the subject in a 1963 monograph.[20] Other papers involving field measurements

[19] *Ibid.*, p. 503.

[20] Gu Zhenzhao (C. C. Koo) et al., *Some Theoretical Problems Concerning the Microphysical Processes of Precipitation, Cloud and Fog,* tr. U.S. Air Force, Cambridge Research Laboratories, AFCRL-68-0665, Tr. no. 39 (Cambridge Research Laboratories, 1968), 59 pp.

TABLE 4. *Publications on Dynamic Meteorology and Numerical Weather Prediction, 1949–66*

| *Subject* | *Number* |
|---|---|
| Dynamic Meteorology | 107 |
| Synoptic and planetary-scale motion | 60 |
| Typhoons | 9 |
| Mesoscale systems: mountain waves, gravity waves | 4 |
| Cumulus dynamics, including microphysical processes | 23 |
| Boundary layer turbulence | 11 |
| Numerical weather prediction | 17 |
| TOTAL | 124 |

SOURCE: W. Blumen and W. M. Washington, "Atmospheric Dynamics and Numerical Weather Prediction in the People's Republic of China, 1949–1966," *Bulletin of the American Meteorological Society*, 54 (1973): 502–18.

of ice nuclei and observations of ice and snow appeared at the same time. The development in the early 1960's of a vigorous research effort on cumulus dynamics under the leadership of Zhao Ribing is noteworthy.[21]

A significant characteristic of research in the 1958–66 period was the rise in the papers' quality and originality. For example, in turbulence research, Su proposed the wind-profile equation that later became known as the KEPYS formulation. (Although the letters in the acronym denote the names of turbulence researchers who contributed to the formulation, the *S* stands for the name of an investigator other than Su.)[22] In research on large-scale dynamics, Wu derived a necessary condition for the instability of a zonal current with both vertical and lateral shear that was essentially equivalent to the result established concurrently by Pedlosky.[23] Yang made a contribution to the problem of the development of waves in the tropical easterlies that was in the forefront of contemporary theoretical work.[24] In mesoscale research, Zhao, Zhang, and Yan produced a paper dealing with the formation of a pressure jump in the lee of a mountain that has become known and referenced in the Western literature.[25] These are only a few examples of the important advances that were made in meteorological research in China im-

[21] For a summary of this work, see R. B. Zhao and S. P. Zhou, *Cumulus Dynamics* (Beijing: Source Press, 1964), 131 pp.

[22] T. S. Su, "On the Basic Law of Turbulent Exchange in the Surface Boundary Layer of a Stratified Atmosphere," *Acta Meteorologica Sinica*, 30 (1959): 114–18; and H. A. Panofsky, "Determination of Stress from Wind and Temperature Measurements," *Quarterly Journal of the Royal Meteorological Society*, 89 (1963): 85–94.

[23] J. S. Wu, "The Influence of Large-Scale Topography on the Instability of Disturbances," *Acta Meteorologica Sinica*, 34 (1964): 11–19; and J. Pedlosky, "The Stability of Currents in the Atmosphere and Ocean, Part 1," *Journal of the Atmospheric Sciences*, 21 (1964): 201–19.

[24] T. C. Yang, "Dynamic Instability of Easterly Disturbances," *Acta Meteorologica Sinica*, 35 (1965): 189–99.

[25] J. P. Zhao, K. K. Zhang, and S. M. Yan, "A Preliminary Investigation on the Formation of Pressure Jump Produced by Mountains in a Two-Layer Model," *Scientia Sinica*, 15 (1966): 723–29.

mediately before the Cultural Revolution. Had the momentum of the early 1960's been maintained, China today undoubtedly would have become a major contributor to the worldwide growth of meteorological knowledge.

No comprehensive study of meteorological research in the 1960's similar to Blumen and Washington's study of research in dynamic meteorology exists for other areas of meteorology. However, from a survey of titles of articles appearing in *Acta Meteorologica Sinica* and *Scientia Sinica*, it is evident that meteorological research progressed on several fronts. Three papers were published on atmospheric radiation, three on atmospheric electricity, and one each on radar meteorology and measurement of atmospheric ozone. Several papers also appeared on stratospheric meteorology, one of them on sudden warmings, a subject that was at that time beginning to attract considerable attention among Western meteorologists.

*Post–Cultural Revolution (1966–present)*

With the advent of the Cultural Revolution in 1966 and the subsequent discontinuance of research journals, meteorological activities in China disappeared from sight for eight years. Basic research suffered a severe setback, although its full extent cannot be ascertained from the report of the 1974 AMS delegation. It is quite possible that the level of research activity was greater than appeared from the delegation's unavoidably cursory inspection. Certainly Chinese scientists seemed well informed on recent Western literature. (Apparently, they circulated research papers informally during the period when journals were suspended.)

The AMS group visited three organizations that were especially prominent in research during earlier years: the Institute of Atmospheric Physics of the CAS, an offshoot of the earlier Institute of Geophysics and Meteorology; the Department of Geophysics, Beijing University; and the CMB, which has as one of its components the Institute of Meteorology. At the Institute of Atmospheric Physics the delegation was shown experimental equipment for the reception and display of direct broadcast signals from the very high resolution radiometer (VHRR) on the national oceanic and atmospheric polar-orbiting satellites. Completed in October 1973, the Institute's VHRR receiving station was one of the first to operate in the world. The pictures received were of high quality and were apparently used cooperatively with the CMB in practical application to weather analysis and forecasting. No mention was made of their use in basic research.

At the Institute of Atmospheric Physics the group also observed two hydrodynamical simulation experiments. One was a rotating annulus experiment using water, and was aimed at studying the effect of the summer air flow over Asia and the movement of atmospheric vortices that cause floods along the Chang Jiang. The other was a rotating table using heated smoke in air to simulate the formation of typhoons and the effects of topography on their structure and movement. The results of the experiments were reported in *Scientia Sinica* following its resumption.[26] The attempt in these experiments to simulate directly situations that are important in practical weather prediction seems unique. In similar modeling done elsewhere, it is customary to treat idealized states that lend themselves more easily to theoretical interpretation. Professor C. C. Chang of Catholic University participated in the research during a five-month visit to Peking in 1973 and a four-month visit in 1975.

Cloud physics, weather modification, and radar meteorology were areas singled out for special inquiry by the AMS representatives. At Beijing University they saw a laboratory experiment in the freezing of supercooled drops and learned of other research that was under way in cloud physics, but the visit was too brief to determine the extent of the effort. Cloud physics and weather modification were listed as one of five major research areas of the Institute of Atmospheric Physics. Again there was little opportunity to evaluate the effort. Conversations were held with two scientists, one in charge of weather modification and the other in charge of the cloud-seeding rockets, and some aspects of the research were discussed. It emerged that the Institute had been conducting modest hail suppression experiments in Xiyang county in Shanxi province for two years. Rockets bearing silver iodide (one of which was seen at the Institute) were fired into hailstorms from two or three launchers. A radar was used in the experiments, but not quantitatively, and there was no organized rain-gauge or hailpad network for evaluation. Because the program was in progress, results were not available.

The group was told that China has a long history of hail suppression efforts conducted by peasants; some communes still fire blank cannon shells, rockets, and artillery shells at hailstorms. Recent news releases carried many accounts of these efforts. It is claimed that successful hail suppression has been carried out in 500 of the 800 Chinese counties susceptible to hail damage. However, since the explosion method has long been discredited in other parts of the world, it is hard to believe that the results have been evaluated by normal scientific procedures.

A somewhat larger program of weather modification was being conducted in Hunan province of southern China by the Institute of Meteorology of the CMB. The program involved seeding of warm cumulus clouds using pulverized salt released by aircraft near the cloud base and was aimed at increasing rainfall during arid summers. The work had been in progress for about five years at the time of the delegation's visit. A Chinese spokesman noted that the CMB could not organize a properly controlled scientific experiment because seeding had to be carried out on all occasions when conditions were favorable to rainfall enhancement rather than on a random basis.

[26] T. C. Ye and C. C. Chang, "A Preliminary Experimental Simulation on the Heating Effect of the Tibetan Plateau on the General Circulation over Eastern Asia in Summer," *Scientia Sinica*, 17 (1974): 397–419; and C. C. Chang, T. W. Wei, and F. H. He, "An Experimental Simulation on the Structure and Topographical Influence of Typhoon," *Scientia Sinica*, 18 (1975): 384–95.

The only research effort in radar meteorology the AMS group had an opportunity to view was a small program at the CMB's Institute. In this program, observations from the network of operational radars were being used in applied research on short-range weather forecasting. In addition to microwave radars, the Chinese also had experimental laser radar (lidar) cloud base indicators. Two different models were on display at the Guangzhou Commodities Fair, which the group visited. It should also be mentioned that spherics equipment was in use at the Institute of Atmospheric Physics for research on hail detection.

There are two reliable sources for assessing the progress of meteorological research in China since the AMS visit in 1974: an account by the Australian physicist K. D. Cole of his visit to Beijing in March 1978[27] and the titles or contents of articles published in Chinese journals since 1973. Cole reported only on the meteorological activities of the Institute of Atmospheric Physics. His account of the Institute's work on dynamic meteorology showed that the dynamics of mesoscale systems was a topic of interest there, as everywhere. The Institute's scientists also shared the growing worldwide interest in physical processes important in understanding climate and climate change and in developing methods of long-range prediction. But it appears from Cole's account that the program in dynamic meteorology had progressed little since 1974. There was no evidence of a further advance in numerical weather prediction, for example.

In cloud physics, Cole was told that the emphasis was on the formation of precipitation in warm clouds and that the effects of silver iodide seeding from airplanes were being studied. Again, there was no evidence of new initiatives. However, it does appear that new programs have been started in remote sensing. The equipment developed for this purpose included a microwave radiometer for measuring atmospheric temperature and precipitation, a lidar for measuring atmospheric opacity, and devices for acoustic sounding of the lower atmosphere. Some new initiatives were also evident in the area of boundary layer studies; a tower 320 m high had been built to make special measurements. The emphasis in this area had recently shifted to urban problems, presumably because of growing concern about air pollution.

The first meteorological articles to appear since publication of journals resumed are contained in *Scientia Sinica, Scientia* (or *Science Bulletin),* and *Qixiang (Meteorology)*. Some meteorological articles, primarily of a review nature, have also appeared in *Kexue shiyan (Scientific Experimentation),* a journal that was not referenced before the Cultural Revolution. Of the 11 papers published in *Scientia Sinica,* four describe the laboratory experiments performed at the Institute of Atmospheric Physics described earlier, two are theoretical papers on turbulence, two are on numerical weather prediction, and single papers deal with climate change, long-range weather prediction, and the theory of squall line formation.

[27] Cole, "Atmospheric, Space and Earth Physics."

Only a few titles of articles published in *Scientia* are known. The subjects treated are statistical weather prediction, climate fluctuations in China, worldwide circulation changes in relation to sea surface temperatures, and numerical weather prediction. The contributions in *Meteorology* range from short informational items to practical studies relating to weather observing and forecasting. One contribution on the short-range prediction of rainfall was submitted by the meteorological workers of a commune weather post.

The development and state of meteorological research in the PRC is characterized first by 15 years of remarkable achievement that brought meteorological research in China almost abreast of that in the most advanced nations by the time of the Cultural Revolution. There was considerable emphasis on applied meteorology, and basic research flourished. Scientists acknowledged practical goals, but their research priorities differed little from those prevailing in the world at large. Research activities were guided mainly by the urge for fundamental knowledge and by the state of the art, not by pressures to satisfy immediate practical needs.

The Cultural Revolution brought an end to this situation. Basic research was not entirely eliminated, but it had to take a back seat. The researcher was advised to remold his outlook and to concentrate his efforts on more useful goals. Thus, research in radar and satellite meteorology was devoted almost entirely to applications in practical weather prediction. Laboratory modeling experiments were conducted to duplicate on a miniature scale flow patterns that were of special interest to Chinese meteorology, rather than as a means of gaining fundamental knowledge of fluid dynamics. Practical necessity prevented scientific experiments in weather modification from being properly evaluated. Peasants engaged in hail prevention based on unsound methods. The idea that premature use of a science might result in a waste of time, labor, and resources was lost on policymakers, or else they did not care so long as alleged scientific activity served political and ideological goals.

Just as basic research suffered a sharp decline in the late 1960's as a result of political events, it now seems destined to move forward again, as a result of the new policy of making science and technology the crux of modernization plans. But whether progress in meteorological research can be as fast as in the 1950's and the goal of reaching advanced world levels by the year 2000 thereby achieved is a matter of question. In some ways, the resources are not as favorable now for accelerating the pace of meteorological research as they were a quarter century ago. The technological gap between China and the most advanced nations is larger than it was in the 1950's. The use of computer technology in meteorology is an example of a vital area in which the gap has steadily widened.

Moreover, the human resources needed for carrying out a rapid expansion may be lacking. In the 1950's there existed a nucleus of capable young leaders, many of them trained abroad, to spearhead the effort. Today these leaders are no longer young and in many cases have disappeared. It is not known whether there are young meteorologists of equal stature and similar vision to replace them. Conceivably there are none, since over a decade has elapsed since advanced instruction was given at the university level. Surely a penalty will have to be paid for this neglect. A more favorable scientific climate, if maintained, will undoubtedly accelerate progress in meteorological research. But how rapidly meteorology in China will catch up with that in the advanced nations is difficult to predict in view of the many uncertainties involved.

In 1972 China joined the World Meteorological Organization (WMO), the arm of the United Nations that coordinates and supports international activities in weather forecasting and other areas of applied meteorology. This action followed the vote of the WMO membership to recognize mainland China rather than Taiwan as the rightful representative of the Chinese people. Because of the unwillingness of the International Council of Scientific Unions (ICSU) to act similarly, China has not yet become officially involved in international programs of meteorological research. It withdrew from the International Geophysical Year (IGY) after taking part in the initial planning, in protest at Taiwan's inclusion. However, it proceeded unilaterally with its observing program and did not withhold its observations.

In 1979 a global weather experiment of unparalled magnitude, known as the First GARP Global Experiment (FGGE), was conducted under the joint sponsorship of the WMO and ICSU. China provided two ships for making observations in the tropical Pacific.

## METEOROLOGICAL EDUCATION IN CHINA

The first meteorological education or training program in China began in 1929 at the Institute of Meteorology in Nanjing. Initially, the training program was quite limited, and the first class of 14 students received only a six-week course of instruction in elementary meteorology and observing. By 1931 the number of students had increased to 40, and the course had been lengthened to six months and expanded to include mathematics and physics.

In 1930 a second university-level program of meteorological instruction was initiated when the Central University of Nanjing (now Nanjing University) created a meteorology division within its Department of Geology and Geography. During a nationwide reorganization of education in 1952, divisions of meteorology that then existed within the Geography Department of Zhejiang University and the Astronomy Department of Qilu University were moved to Nanjing and combined with the department there. At the same time, two-year training classes were started at Nanjing University to accelerate the training of meteorologists, but these were discontinued in 1956 when emphasis shifted to improving the quality of the personnel. It is claimed that from 1949 to 1958 the number of technical personnel trained by the Department of Meteorology of Nanjing University was eight times the number trained in the 20 years prior to the Liberation.[28]

In 1935 Qinghua University in Beijing established a meteorological division. Earlier, in 1929, a meteorological observatory had been started there as a laboratory and research center for students in the Department of Geography. In the reorganization of 1952, the meteorology division was transferred to the Physics Department of Beijing University, and in 1959 it was combined with the geophysics division of the same department to form a separate Geophysics Department.

The rapid growth of operational or applied meteorology after 1949 required greatly expanded facilities for training of technical personnel. To meet this need, intermediate-level meteorological schools were established in Beijing and in Sichuan and Guangdong provinces. The Beijing facility, the Beijing Institute of Meteorology, at one time had a faculty of nearly one hundred and could enroll 1,700 students.

The AMS delegation was told in 1974 that there were three institutions offering complete instruction in meteorology at the university level: Beijing and Nanjing Universities and the Institute of Meteorology of Nanjing. (The latter is a branch of the CMB, not a continuation of the earlier Institute of Meteorology established by the Academy of Sciences.) In addition, five universities offer some meteorology courses in departments of geography, geophysics, and physics; nothing is known about them, except their locations: Kunming, Lanzhou, Hangzhou, Changchun (Jilin University), and Guangzhou (Zhongshan University). It is surprising, in view of the substantial growth that has taken place in meteorology since 1949, that the number of universities with programs in meteorology remains the same as 40 years ago.

The AMS group was told that the Beijing Institute of Meteorology had been discontinued and that training of subprofessional personnel was handled by provincial bureaus. The part-time meteorologists who operate weather posts at agricultural communes receive their training in seminars conducted by county weather stations. Another factor of possible importance in meteorological education is the provision of training at research institutes.

The AMS delegation gained a limited picture of university-level training in meteorology on the basis of a half-day visit to Beijing University's Geophysics Department. Thirty students and 24 teachers in the atmospheric physics course specialized in cloud physics, air pollution, and satellite meteorology, all fashionable areas. They had built their own equipment for receiving visible and infrared pictures transmitted from U.S. satellites. In January 1974 a class of 45 students completed the course in atmospheric physics,

[28] K. T. Xie, I. P. Xie, and K. H. Ye, *Meteorological Education in the Last Decade,* tr. JPRS from *Acta Meteorologica Sinica,* 30 (Oct. 1959): 202–5, JPRS-2505 (Washington, D.C.: JPRS, 1960), 11 pp.

becoming the first to graduate since 1966. The meteorology course specialized in synoptic meteorology, dynamic meteorology, and numerical weather prediction. Training in the latter included some use of the high-speed computer located in the CMB a few miles away.

At the time of the AMS visit, the students were preoccupied with criticizing Lin Biao and Confucius; hence it was not possible to observe classes in session. The group, however, did visit university workshops in which young technical workers, not students, were fabricating electronic devices and growing silicon crystals for commercial use. Evidently the students also worked in the shops and in nearby factories when not occupied with studies or political activities.

Clearly meteorological education is ill equipped to assist China in achieving its modernization goals. The number of educational institutions and students is relatively limited. With a population of over 900 million, China trains only about a quarter as many students in meteorology as the United States. And the Cultural Revolution left a huge gap in the ranks of professionally trained workers. Eight years elapsed without a class graduating from Beijing University at the baccalaureate level, and there was no graduate education for 12 years. The deficiencies produced by these gaps cannot be overcome quickly.

Moreover, the students of the post–Cultural Revolution era have been trained to fulfill immediate practical needs. However, despite a meager educational base, meteorological science surged forward rapidly in the 1950's and by the mid-1960's was on the verge of catching up with the scientifically advanced nations. It is therefore not beyond the realm of possibility that a new surge will occur now that graduate education has been renewed, the value of theoretical research openly recognized, and substantial numbers of students sent abroad for advanced training. The present leadership's goal of catching up in science and technology by the turn of the century may not be totally unrealistic in meteorology, provided political stability is achieved and the present science and education policies are maintained on a steady course.

# *FISHERIES, AQUACULTURE, AND OCEANOGRAPHY*

Fishing and aquaculture have a long and impressive history in China.[1] The earliest known mention of fishing dates to 1122 B.C., and nets, silk lines, and tapered bamboo rods were probably among the first examples of Chinese fishing technology to be transferred and adopted throughout the world. The Chinese were also among the first peoples to develop pisciculture, the art of breeding and rearing domestic fish, and to devise a system of strict fishing laws.

When the PRC was founded in 1949, China's system of freshwater aquaculture was probably the most extensive and sophisticated in the world. The marine fishing industry, however, was comparatively primitive. Traditional junks were limited to within 10 to 12 miles of the coast, and there was no deep-sea fishing. The interdisciplinary subject of oceanography was a recent Western import and only tenuously established in the country's universities and research institutes.

By the 1970's, China's coastal and inland fisheries were generally acknowledged to be among the most productive in the world; the PRC, along with Peru and Japan, ranked among the world's top three fishing nations. Somewhat paradoxically, however, there were no precise, comprehensive, or

[1] This chapter is based on a trip report by the U.S. Oceanography Delegation sponsored by the Committee on Scholarly Communication with the People's Republic of China (CSCPRC) that visited China in September and October 1978. Chaired by Dr. Walter Munk, the Delegation members included Daniel Cohen, John Edmund, Dirk Frankenberg, Douglas Inman, Peng Tsung-hung, John Ryther, Manik Talwani, David Wallace, Carl Wunsch, and Tu Wei-ming. The trip report, edited by Caroline Davidson, who also adapted the report for this volume, was due to be published by the National Academy of Sciences in 1980 under the title *Oceanography in the People's Republic of China.*

reliable statistics of the country's aquatic production until June 1979, when the Chinese State Statistical Bureau reported total 1978 output of aquatic products at 4.66 million metric tons (MT)—40,000 MT below the 1977 output.

The Chinese clearly wanted to quantify their aquatic resources and to exploit them more fully. At the National Science Conference of March 1978 Fang Yi, minister in charge of the State Science and Technology Commission, called for a comprehensive, quantitative survey of fishery resources, a rational, ecologically sound strategy for exploiting them, and a scientific basis for increasing food production. He said that it was necessary to increase freshwater production, marine fishing, fish breeding, and fish processing to improve the diet of the country's 950 million people. Fisheries were identified as one of 108 key areas for nationwide scientific and technological research.

Exactly one year later, a Beijing New China News Agency domestic service broadcast announced that China's annual harvest of marine aquatic products had fallen to just over 3 million MT in recent years. Moreover, the average Chinese consumed only about 7 catties (3.5 kg) of marine aquatic products and 3 catties (1.5 kg) of freshwater aquatic products a year, "less than one-third of the average world consumption of aquatic products." The broadcast concluded that "to raise the living standards of the Chinese people and speed up the four modernizations, we must try 1,001 ways to obtain more protein from the vast seas." The two methods specifically cited in the broadcast were development of offshore and deep-sea fisheries and increased fish breeding in seawater.

The main organizations concerned with implementing China's plans to exploit its aquatic resources to their maximum potential are:

1. The Chinese Academy of Sciences (CAS), which runs the Institute of Oceanography at Qingdao, the South China Sea Institute of Oceanography at Guangzhou, the Institute of Hydrobiology at Wuhan, and the Institute of Geology at Beijing.

2. The State Aquatic Products Bureau, whose three regional fishery institutes for the South China, East China, and Yellow seas are based in Guangzhou, Shanghai, and Qingdao, respectively.

3. The State Oceanography Bureau, which runs the Institutes of Oceanography at Hangzhou, Xiamen, and Guangzhou and the Institute of Marine Instrumentation at Tianjin.

4. Colleges and universities, such as Shandong College of Oceanography at Qingdao, Xiamen University, and Jiaotong University in Shanghai.

As of 1978, a National Research Center for Oceanography was to be built at Qingdao, and a second center was also under consideration for the south, perhaps at Xiamen or Hangzhou.

*Marine Fisheries*

Marine fisheries are located throughout China's 14,000-km coastline.[2] As of the late 1970's, most of the fishing fleet stayed within 200 miles of shore.

[2] Some of the information for this section was kindly provided by Dr. G. Ian Pritchard of the Canadian Department of Fisheries and Oceans.

Within this coastal area, however, lies one of the largest and most productive fishing shelves in the world, extending from the Bohai Gulf in the north through the Yellow and East China seas into the South China Sea. The shelf has a great variety of fish, crustaceans, and seaweeds. There are more than 80 commercially important species, including herring, mackerel, sardine, cutlass fish, yellow croaker, butterfish, shark, eel, flounder, prawn, sea cucumber, oyster, jellyfish, and seaweed.

The State Aquatic Products Bureaus in Guangzhou, Shanghai, and Qingdao collect regional statistics on marine fisheries, but these are by no means comprehensive. The catch for 1977 in the South China Sea and the East China Sea was estimated at 1.0 million MT each, followed by 800,000 MT in the Yellow Sea, and 300,000 MT in the Bohai Gulf, a total of 3.1 million MT. (These estimated landings did not include catches by other countries.) Catches in the Yellow Sea and the Bohai Gulf by all countries, including China, were estimated at 2 million MT. (Japan and Korea conduct substantial fishing in the Yellow Sea, but no foreign fishing takes place in the Bohai Gulf.)

More than a million families engage in the small-scale collective and private fishing enterprises that account for 80 percent of production. The remainder comes from fishing corporations run by provinces, municipalities, cities, counties, and regions. In the 1970's the largest enterprises were at Lüda (Lüshun and Dalian) in Liaoning province in the north, at Yantai and Qingdao on the Shandong Peninsula, and at Shanghai. Smaller state corporations were located on Hainan Island, on the Leizhou Peninsula in western Guangdong province, at Fuzhou in Fujian province, and at Wenzhou in Zhejiang province.

As of 1978, the second largest fishing corporation was the Oceanic Fishery Company of Dalian. It employed 7,000 persons, operated about 150 ships, and had its own shipyard, freezer–cold storage, 400-MT per day ice plant, net-manufacturing plant, and processing, packing, and sales facilities.

The company divided its operations among five brigades, which fished mostly in the Yellow and East China seas and south to Taiwan. Fishing was primarily pair trawling. Each boat was about 100 tons with a 600-HP motor. Larger vessels up to 200 tons went further south to purse seine for mackerel, using seines about 1,500 m long and 20 m deep. The company also engaged in some night-light fishing for mackerel and pompano-like species in the East China and Yellow seas and a small amount of whaling. The company's total annual production varied between 70,000 and 80,000 MT.

As of 1978, little information was available about trends for key species. Personnel at the East China Sea Fisheries Institute in Shanghai claimed, however, that most major species were declining in the East China Sea. Without specifying the years involved, they said that annual catches of the big yellow croaker had dropped from 200,000 to 80,000 MT. Catches of the little yellow croaker had declined even more sharply, from 150,000 to 20,000 MT a year. Annual cutlass fish catches fluctuated between 540,000 and 400,000 MT, and catches of squid between 50,000 and 80,000 MT. Filefish catches, on the other hand, had risen rapidly since the establishment of a new fishery—from 45,000 to 200,000 MT between 1974 and 1977.

*Freshwater Aquaculture*

Recent materials published in the Chinese press suggest that freshwater production from ponds and lakes has declined in absolute terms since the 1950's.[3] For example, in January 1979, a deputy head of the State Aquatic Products Bureau, Cong Ziming, told the Beijing *Guangming ribao (Guangming Daily)* that the average yearly catch from rivers and lakes in the six water-rich provinces along the Chang Jiang (Yangtze) had dropped 29 percent between 1953 and 1973. Cong explained that the main reason for the decline lay in the conversion of fishing ponds, rivers, and lakes into cultivated fields for growing grain, and said that in Hubei, "the province of a thousand lakes," the total water area had declined about 75 percent between 1949 and 1979 as a result of reclamation. Similarly pessimistic views were expressed in a March 1979 Beijing Radio broadcast that said that between 1959 and 1978 the proportion of freshwater fish output had declined from 40 percent to 23 percent of China's total output of aquatic products.

This pessimism may be unwarranted, however. China's total freshwater area is usually estimated at about 20 million ha, of which approximately half can be used for fish culture. About 5 million ha consists of large natural lakes or man-made reservoirs, with annual yields of 50 to 5,000 kg/ha. The other 5 million ha consists of small, intensively farmed ponds, managed by agricultural and fish-farming communes, with annual yields varying from 1,000 to 10,000 kg/ha. Production is concentrated around the big lakes and river systems, the Chang Jiang and Pearl River basins being the largest.

If the yield of the less intensively managed larger lakes and reservoirs is estimated at a conservative 500 kg/ha and that of the smaller fish farms at 3,000 kg/ha, the annual production of the entire 10 million ha of freshwater fish farms could potentially amount to about 17 million MT.

The Chinese, however, take a much less optimistic view of their potential for aquaculture production. In 1974 the New China News Agency said that only 4.6 million ha of water surface could be used for breeding fish, and that just 70 percent, or 3.2 million ha, was actually in use. In October 1978, however, the *People's Daily* said that "fish breeding currently takes place on less than 1.3 million ha of revervoir area (about two-thirds of the total). However, average production is only about 82.5 kg/ha."

Whatever the actual and potential productivity of Chinese freshwater aquaculture, its importance is easily explained. In addition to the obvious requirements of the population for protein, much of the marine catch consists of less desirable sizes and species of fish than those grown in culture, and China has logistical problems with distributing marine landings due to its

[3]Useful information for this section was kindly provided by Mr. Jaydee R. Hanson of the University of Hawaii.

relatively small number of fishing ports, its poorly developed transportation system, and its minimal refrigeration and processing facilities.

Chinese aquaculture is successful for three reasons. First is the policy of making multiple use of aquatic resources. Reservoirs and smaller farm ponds may be constructed primarily for irrigation or domestic water supply, but they are simultaneously used for fish production. Second, fish pond culture in China has always been considered an integral part of agriculture. The agricultural and aquatic farms supplement each other in several important ways that increase yields of each component part. Third, mixed species cultivation, or "polyculture," is universal, with particular emphasis on species low in the food chain.

There are about 20 species of freshwater fish of economic importance in China. They are virtually all members of the cyprinidae (the carp and minnow family). The four most important are the so-called major Chinese carps, or "family fishes": the grass carp (*Ctenopharyngodon idella*), which feeds on grasses and other aquatic plants; the silver carp (*Hypophthalmichthys molitrix*), a phytoplankton feeder; the bighead carp (*Aristichthys nobilis*), a zooplankton feeder; and the black, or snail, carp (*Mylopharyngodon piceus*), which feeds on mollusks and snails. Smaller numbers of mud carp (*Cirrhinus molitorella*), common or mirror carp (*Cyprinus carpio*), crucian carp (*Carassius auratus*), and the cichlid *Tilapia* spp. are also cultivated.

The grass carp is a herbivore that normally eats aquatic macrophytes (both rooted and unattached and both submerged and floating aquatic weed species), but it also feeds voraciously on terrestrial plant wastes, such as grass clippings and vegetable tops. Although fast growing (5–10 kg or more a year), the grass carp is highly inefficient in its food utilization. Fortunately it produces large quantities of organic wastes. This material settles to the bottom and supports a community of benthic invertebrates that, in turn, serve as food for the black, common, mud, and crucian carps. Their organic wastes also decompose, liberating nutrients that support communities of phytoplankton and zooplankton to feed the silver and bighead carps, respectively. The planktonic populations are further increased by periodic fertilization of the ponds with fermented pig manure or other organic wastes, such as those of the ducks that are often raised in the same ponds with the fish. The combination of polyculture, which takes advantage of every feeding niche in the pond ecosystem, and the use of agricultural wastes explains the low costs and high yields of Chinese pond culture.

The major carps normally live in large river systems and are unable to spawn naturally in the stagnant farm ponds. Chinese aquaculture once depended on the annual collection of fingerlings from their natural environment, but in 1958 the major carps were successfully induced to spawn artificially by injecting them with pituitary extract of human chorionic gonadotrophin.

By 1978, most counties in China's fish-farming regions had a commune specializing in hatchery production and rearing of fry for distribution to grow-out ponds located on other communes in the county. The personnel of these fishing communes were surprisingly accomplished in such fields as controlled induced spawning, selective breeding, larval rearing, disease prevention and treatment, and nutrition—areas thought in the West to require highly specialized training. Workers at the Jingbu Fish Farm outside Shanghai, for example, said that 90–95 percent of their fingerlings normally survived to become marketable adults, and that they could usually diagnose, treat, and cure a disease in three days.

Research on freshwater aquaculture is conducted at national and provincial institutes. For example, the CAS Institute of Biochemistry in Shanghai and the CAS Institute of Zoology in Beijing successfully synthesized, tested, and evaluated an ovulating agent for inducing spawning by farm fishes during the early 1970's. By 1978 this substance (an analog of the nonapeptide LH-RH) was available to fish-farming communes, although the extent to which it had replaced the use of carp pituitary is not known.

The Institute of Hydrobiology at Wuhan conducts basic and applied research on fish breeding and genetics and applied research on fish breeding and genetics, disease control, and the basic productivity of lakes and reservoirs. It has successfully cultured a cyprinid herbivore, the Chinese bream *Megalobrama amblycephala,* to replace the grass carp as the essential first-stage herbivore in the Chinese polyculture system. (The grass carp is difficult to rear in the early stages of its life cycle and is subject to severe disease problems; the Chinese bream is hardier and as good a table fish.) By 1978 the Chinese bream was cultured in 20 provinces.

Scientists at the Institute of Hydrobiology have also discovered that the nace, *Plagiognathops microlepis,* another cyprinid, is a detritus feeder. (One of the primary needs in pond polyculture is to have a detritus or humus feeder. For a long time, no species was available.) Researchers have developed techniques to propagate the species artificially in ponds in order to ensure adequate stocking fry and found that the nace does not compete with other fish for food, is very adaptable to climatic variations, can be bred artificially, and grows rapidly. By 1978 the species was widely used in China.

The Institute of Hydrobiology has also successfully crossbred several closely related cyprinids. The goal of this kind of breeding is to utilize the hybrid vigor of the first generation. The male mirror carp (*Cyprinus carpio*), an inbred strain of the common carp, was mated with the female red carp (*Cyprinus carpio haematopterus*). The growth rate of the resultant cross exceeds that of the parents by 60 percent. The fish is omnivorous and tasty. By 1978 the cross had been used in 23 provinces. Another successful cross mates male red carp *Cyprinus carpio haematopterus* with the female crucian carp, *Carassius auratus gibelio.* The resulting cross grows to 250 g in five months and is flavorful and nutritious. It was successfully bred in 1977, but had not spread beyond the province of Hubei.

Scientists at the Institute have also attempted to hybridize species that are not closely related. They crossed grass carp and bream (*Megalobrama am-*

*blycephala*), but found that the hybrid was sterile. They tried to overcome this problem by artificial stimulation. They also tried a cold shock five minutes after fertilization by reducing the water temperature to 0–2°C for 20 minutes, with limited success. However, they obtained better results by waiting 44 minutes after fertilization and maintaining the cold shock for 25 to 30 minutes. Another technique was to use cortisone treatment in a solution of 50 parts per million. By 1978 only preliminary results had been obtained, and no viable progeny had been developed.

The Institute's ecological unit studies the productivity of China's larger lakes and reservoirs to determine their maximum carrying capacities and the proper stocking ratios of different species in order to utilize all feeding niches fully. The unit carries out basic studies of chemical nutrient cycling, primary productivity, and the abundance and distribution of phytoplankton, zooplankton, and benthos. They hope eventually to predict potential fish yields.

*Marine Aquaculture*

In the 1970's Chinese mariculture was less highly developed than freshwater aquaculture. According to a Beijing New China News Agency domestic service broadcast in March 1979:

> Methods of production and management in the field of seawater fish breeding remain backward in most parts of China. . . . Strengthening of fish breeding in seawater is of great significance in increasing fishery output, improving production quality, satisfying people's needs for high-quality protein, and providing funds for national construction. For this reason, we must regard seawater fish breeding as an important task and pay full attention to it.

The report referred to a National Fishery Working Conference held by the State Aquatic Products Bureau that had decided to incite the commune- and brigade-level breeding farms to greater efforts, consolidate the country's 77 state-run seawater breeding farms by 1981 (their annual output apparently accounted for less than 7 percent of the nation's total mariculture production), and set up base areas for marketing fish, shrimp, and other aquatic products. All coastal counties were to build one or two state breeding farms by 1985.

In the mid-1970's, the bulk of Chinese mariculture production consisted of the brown seaweed *Laminaria japonica,* a cold-water species of kelp introduced from Hokkaido, Japan. Some 10,000 dry MT were produced each year in China's northern coastal waters. A typical kelp hatchery near Qingdao consisted of two large (5,200-m$^2$) greenhouses fitted with shallow tanks through which fertilized, refrigerated (7–9°C) seawater is circulated. Kelp spores attach themselves to strings mounted on frames in the hatchery tanks in the spring and develop into sporelings 2–4 cm long by late summer. When the coastal seawater temperature drops below 20°C, the strings with sporelings attached are moved outdoors and attached to buoyed ropes.

Over the following six to eight months the plants are manually thinned, transferred to larger ropes, individually brushed to remove sediments and epiphytes, and fertilized daily. They grow to mature sporophytes ranging in length from 3 m around Qingdao to over 5 m near Dalian, where the water cools more quickly in the fall and the growing season is longer. In 1978 annual production of kelp in the two areas was respectively about 30 and 50 dry MT/ha.

A small species of red seaweed, *Porphyra yezoensis,* imported from Japan, was also grown in northern China in the 1970's. This highly prized food is reared in hatcheries for part of its life cycle. The spores are attached to string nets in the spring and are put out on buoyed ropes in October; the mature plants are harvested in March and April. Annual yields of *Porphyra* are about 0.5 MT/ha.

Varieties of *Laminaria* have been successfully bred through 5 to 15 generations at the Institute of Oceanography at Qingdao. Selective breeding and x-ray–induced mutation have improved growth characteristics and led to higher iodine content and higher temperature tolerance, the latter permitting culture to be extended south. Fundamental research on the genetics of *Laminaria* at Shandong College of Oceanography has resulted in the production of parthenogenetic clones of female gametophytes, an important step in the ultimate genetic control and establishment of pure breeding lines of this important seaweed.

Marine fish and invertebrate culture in China was less well developed than that of marine plants or freshwater fishes in the 1970's. The large penaeid shrimp *Penaeus orientalis* was, however, apparently grown with considerable success in the East China and Yellow seas. According to Liu Jiuyu, crustacean specialist and head of the Zoology Department of the Qingdao Institute of Oceanography, *Penaeus orientalis* matures sexually in captivity in holding ponds. (This contrasts with other penaeid shrimp species cultured elsewhere, where gravid females are usually taken from commercial fisheries to obtain the young.) Larval stages are hatchery reared, and post-larvae grow over 15 cm (25 g) in less than five months, making it possible to grow the animals to marketable size in one brief growing season. Yields of *P. orientalis* in China are not impressive, however, mainly because their culture depends on natural food in the ponds, rather than on heavy artificial feeding as in most other parts of the world.

The communes that culture seaweeds around Qingdao and Dalian also engage in several kinds of invertebrate culture. Mussels (*Mytilus edulis*) are grown on buoyed ropes, using the same techniques developed in Spain and adopted in many other countries around the world, including the United States. A period of about 18 months is required for the mussels to reach marketable size, and two crops a year, spring and fall, totaling about 480 MT/ha (shells included), are produced. (This compares with some 600 MT/ha per year in Spain for the same species.)

Smaller numbers of scallops (*Chlamys farreri*) are also grown, the juveniles suspended from ropes in layered lantern baskets. Sea cucumbers (*Stichopus japonicus*) are released on the bottom, where they live on detrital material falling from the plants and animals above. The culture of mussels, scallops, sea cucumbers, and several other invertebrate species such as abalone, clams, and oysters was still at the experimental stage in the late 1970's.

Production units did most of the empirical experimentation in mariculture, including spawning and larval rearing in hatcheries. They were assisted, however, by research organizations. In 1978, for example, the Yellow Sea Fisheries Institute in Qingdao maintained several species of phytoplankton and provided the hatcheries with seed cultures for larval rearing. The Qingdao Institute of Oceanography also carried out studies aimed at increasing production and survival of mussel spat.

### *Biological Oceanography*

As of 1978, biological oceanographic research in China was almost exclusively descriptive and concerned with the distribution and quantification of organisms in a given oceanic region. It was oriented toward two clear objectives: increasing production of marine organisms and assessing the impact of pollution. The 1978 volume of *Oceanologia et Limnologia Sinica,* for example, contained articles on the quantitative distribution of *Sticholonche zanclea* in the western part of the East China Sea, the geographical distribution of cephalopods in Chinese waters, and the geographical distribution and evolution of pelagic polychaetes from the South China Sea islands. Coral reefs were being studied at the South China Sea Institute of Oceanography in Guangzhou and the Qingdao Institute of Oceanography. Studies of East China Sea plankton and benthic communities were reportedly under way at Xiamen University and the State Oceanography Bureau's Institutes of Oceanography at Xiamen and Hangzhou. The East China Sea Fisheries Institute in Shanghai was studying the Chang Jiang estuary. The Qingdao Institute of Oceanography was investigating the communities of the East China and Yellow seas, and the Northeast Oceanographic Station in Dalian was conducting research on the plankton of the Bohai Gulf.

Many research organizations had new programs, research units, or departments devoted to pollution studies in 1978. They were mostly preoccupied with developing analytical techniques and paid relatively little attention to the related activity of measuring the effects of pollutants on the environment.

The potential accumulation of heavy metals and pesticides in fish to the extent that they would become unsafe for human consumption was of some concern. Fish flesh was monitored in some areas for that reason. According to the Institute of Hydrobiology in Wuhan, fish from certain local ponds had been designated unsafe by public health officials.

A few organizations undertook ecological surveys to establish baseline conditions in areas unaffected by pollution and to compare present conditions in polluted areas with those observed earlier, when pollution was absent or less severe. The Shandong College of Oceanography has conducted a series of oil pollution surveys in the Yellow Sea's coastal waters, for example. The Northeast Oceanographic Station at Dalian has compared the abundance and distribution of plankton in the Bohai Gulf in May and August of 1976 with that observed in May and August 1959. But both these studies suffered from a serious lack of modern equipment and methodology that precluded the possibility of drawing valid conclusions from the results.

Most of the work in systematics in the 1970's consisted of isolated species descriptions, reviews of small groups restricted to China, and catalogues. An exception to this was the two-volume monograph on Chinese cyprinoid fishes by Professor Wu Shenwen and his colleagues at the Institute of Hydrobiology published in 1964 and 1977, which was comprehensive and paid some attention to modern theory.

As of 1978, most Chinese systematists were engaged in producing the 60-volume *Fauna Sinica.* (A companion *Flora Sinica* was being prepared as well.) The fish section of the *Fauna Sinica* was to comprise 17 volumes and had been assigned to four institutes for preparation. The East China Sea Fisheries Institute in Shanghai, which has a collection of about 1,100 species, was responsible for the sections on sharks and rays, tetraodontoids, gobioids, and scorpaeniforms. The Qingdao Institute of Oceanography, which has a collection of 40,000 specimens of 1,200 species, was covering perciforms. The Institute of Zoology in Beijing, which has the largest collection in China (60,000 fishes), was dealing with clupeiforms, eels, myctophoids, mugillids, flatfish, catfish, and other miscellaneous groups. The Institute of Hydrobiology, which maintains an important collection containing 750 of the approximately 800 known species of Chinese freshwater fishes and many specimens from Asia, parts of Africa, and North America, was preparing the section on cyprinoid fishes.

The centers of research on crustacean systematics were Xiamen University, where Professor Zhong Zheng was working on the distribution and taxonomy of copepods, and Qingdao Institute of Oceanography, where the head of the Department of Zoology, Dr. Liu Jiuyu, specialized in stomatopod and penaeid taxonomy.

Further work in systematics was needed. Researchers on benthos were handicapped by having to use outdated, extraterritorial Soviet manuals for identifying samples. In some cases fishery biologists lacked information about the taxonomic identity and numbers of species on which they kept data. Studies of ichthyoplankton were in a preliminary state partly because of the absence of good information about adult fish fauna. To catch up with the rest of the world, Chinese systematists needed to expand their explorations into deep-sea and distant water areas, study material from non-Chinese regions, and relax restrictions on sending specimens out of China on loan or exchange.

Little other biological oceanographic research was under way in the

1970's. For example, there were almost no biofouling studies or basic research on primary production, trophic efficiency of marine food chains, nutrient uptake kinetics, remineralization rates, or other topics designed to elucidate functional features of marine ecosystems.

*Chemical Oceanography*

Marine geochemistry, in common with the other areas of oceanography, embraces many diverse areas of study, some of which are not directly connected to work in the deep sea. In the United States, interest in chemical processes in estuaries, continental freshwater systems, the atmosphere, and sediments is increasing. Classical descriptive chemical oceanography, originally an adjunct to marine biology and physical oceanography, is rapidly losing the centrality it once occupied in the field.

As of 1978, Chinese research in marine geochemistry was resuming after a long hiatus during the Cultural Revolution and its aftermath (1966–76). Many of the scientists working in the field were chemists with no oceanographic background. Few data were available.

Groups doing systematic water column sampling for oxygen, nutrients, and the carbon system were active at the Institute of Oceanography at Hangzhou, the CAS Institute of Oceanography and Shandong College of Oceanography at Qingdao, and the Institute of Oceanography and Xiamen University at Xiamen. The Institute of Oceanography in Hangzhou was developing and standardizing various analytical techniques. At the Institute of Marine Instrumentation, development work was well advanced on a self-recording salinity temperature depth instrument, with sensor design and configuration rather similar to the Bisset-Berman model widely used in the United States until about 1973, and a salinometer closely resembling the Hytech model currently used in the United States.

Studies of the mineralogy and geochemistry of shelf sediments were being pursued at the South China Sea Institute of Oceanography, at the Institutes of Oceanography at Hangzhou and Qingdao, and at Shandong College of Oceanography. Geochemists at the last institution were also working on oil reservoir studies using drill core samples. Their techniques included x-ray diffraction, x-ray emission spectrography, and differential thermal analysis. Induced thermoluminescence studies were under way at the Qingdao Institute using a clinical x-ray source. The instruments at these different institutions were of widely varying vintages and origins. The x-ray diffraction systems were mostly Chinese. The emission spectrometers ranged from an old German system to Soviet machines and a new Zeiss (Jena) PQ22 purchased in 1977. The differential thermal analyzers were either Hartman-Braun (Frankfurt) or Shimazu (Japan) equipment.

The South China Sea Institute of Oceanography was analyzing local shelf sediments and organisms (principally seaweeds). The Hangzhou Institute was conducting a systematic survey of the East China shelf and Okinawa

Trough, and studying recent foraminifera, ostracoda, diatoms, and pollen. Samples were taken with surface grabs, gravity cores (up to 4 m), and surface-powered vibrating cores (restricted to water depths of 100–150 m).

At the Institute of Oceanography at Qingdao, similar work was being done on the East China and Bohai shelves, but there were no studies of their sediments' organic content and composition. Sedimentological studies were conducted independently of the geophysical work, including profiling, under way in the same regions.

Many oceanographic institutions had pollution chemistry laboratories. Their analytical caliber, however, was unimpressive. Laboratory supplies for sample containment and preparation were in short supply, and laboratory fixtures were often dilapidated, a sure source of contamination. The level of instrumentation varied widely. Most uniform were Chinese-made gas chromatographs for pesticide and hydrocarbon analyses. (Most laboratories had three or four models.) Heavy metal analyses were performed by direct polarography, anodic stripping, or atomic absorption (flame or flameless) on preconcentrated samples. The polarographs were generally primitive by U.S. standards. The atomic absorption units included Hungarian-made machines (either flame or with a Chinese-made graphite furnace); well-designed, compact Chinese flame instruments; and modern American machines, such as the Perkin Elmer 503. A Chinese-made cold-vapor atomic absorption analyzer for mercury was perhaps the most common instrument.

### *Geophysical Oceanography*

The Marine Geology Division of the State Geology Bureau was probably the most active organization in offshore geophysical work in China in the 1970's. Its personnel had prospected for hydrocarbons on the South China Sea's continental shelf and slope and had discovered a large sedimentary basin lying slightly below the shelf edge. They had also conducted drilling operations and had some oil shows in this basin and in the East China Sea.

Scientists at the South China Sea Fisheries Institute were also active in geophysical work. They had carried out work on magnetic anomalies in the South China Sea basin. However, their method of interpreting gravity measurements on the shelf and its islands used isostatic anomalies calculated by conventional methods and was considerably out of date. Scientists at the Institute of Oceanography at Qingdao had produced a good map of magnetic anomalies in the East China Sea that traced their continuation to Taiwan and Japan. They had also conducted shallow seismic reflection work with sparkers as a sound source, mainly to look for bedrock and thickness of sediments in harbors. The scientists used some of the shelf data quite imaginatively, especially in the area where the Chang Jiang carries its sediments into the ocean.

The conduct of marine geophysics by scientists at colleges and universities was limited, since they had no ships and their equipment was poor. They were, however, familiar with work carried out by others in areas such as the

South China Sea, concepts of plate tectonics, and geological implications of geophysical work done at sea.

As of 1978, there were about 30 vessels of more than 800 tons engaged in oceanographic activities and many small vessels engaged at the local level. Marine geophysical work was carried out in ships belonging to three main categories of institution:

1. The Marine Division of the State Geology Bureau. One of its ships for oil and mineral prospecting, *Haiyang No. 2*, was built in 1972 at a shipyard in Shanghai. Its displacement was 3,000 tons, and its length, beam, and draft were 106 m, 15 m, and 5.5 m, respectively. Although it could cruise at a speed of 18 knots, it usually employed a speed of 6 knots for its main work, seismic reflection profiling. It carried 25 scientists and a crew of 40.

2. The South China Sea Institute of Oceanography. Its ship, the *Shiyan*, carried gravity and magnetic equipment, but no seismic equipment.

3. The State Oceanography Bureau's Institutes. These Institutes share a number of ships. *Xiang Yang Hong Nos. 5* and *11* exceed 10,000-tons displacement and have been used for long-range cruises in the Pacific.

The Chinese were engaged in a vigorous program of modernizing their research ships. In early 1979 they acquired two ships from the United States with up-to-date seismic and navigational equipment for hydrocarbon exploration on the South China Sea shelf and slope and for geophysical research. A large modern research ship equipped for carrying out deep-sea research in various oceanographic disciplines was being built in Japan for delivery in 1981. In addition, about half a dozen 800-ton vessels were to be built in China by 1985.

Equipment for gathering marine geophysical data was limited in quantity and not up to the best U.S. standards. The Chinese were able to make marine magnetic measurements in almost the same way as American institutions, but lagged somewhat behind in marine gravity and substantially behind in marine seismology studies. A proton precession magnetometer manufactured by the Geological Instruments Factory in Beijing was widely used. It was a direct gamma-reading instrument and used kerosene as the source of protons, with a reported accuracy of 0.25 gamma.

As of 1978, the Geological Instruments Factory had apparently manufactured ten surface-ship gravimeters. One being tested when the American Oceanography Delegation visited the factory had a single, flat, vibrating string 30 mm long made of beryllium-bronze, platinum-iridium, and phosphor-bronze. Its frequency was about 2,000 hertz, and the meter had to be kept at a constant temperature of +0.1°C. A drift figure of 2 milligals/month was indicated. Because the gravimeter's response was nonlinear to vertical acceleration, the filtering out of these accelerations was no trivial matter. An on-line digital computer (the DJ57, of Chinese manufacture) was used for this purpose. The computer printed out a value every 3.5 minutes, which was presumably the length of time needed to carry out filtering. No on-line digital to analog conversions were performed, although such conversion and automatic continuous plotting of the analog values have proved of great value in monitoring gravimeter performance elsewhere. The Chinese were not fully satisfied with the accuracy of their gravimeters—to judge by the number of inquiries about the quality of those made in the West. As of 1978, they were ordering German Graf-Askaina gravimeters.

Seismic reflection equipment was manufactured at a factory in Chongqing in Sichuan province. A 225-kjoule sparker provided the sound source. Four air guns provided a total capacity of 16 liters and used air at a pressure of 150 kg/cm². However, it appeared that the sparker was preferred as a sound source. On *Haiyang No.* 2 the streamer had 24 sections, and its active portion was 1,200 m long. It was towed behind an inactive section 500 m long. The seismic signals were recorded on magnetic tape as analog records. By means of an analog to digital converter manufactured by the Geological Instruments Factory in Beijing, the analog records were digitized and later processed at a computer installation in Beijing. Using this equipment, the Chinese were able to penetrate 3,000–4,000 m of sediments under the South China Sea continental shelf. They wanted to import up-to-date equipment from the United States, but as of 1978, they had been unable to do so because of American export restrictions.

### *Physical Oceanography*

The main areas of interest in physical oceanography in the late 1970's tended to have fairly obvious, immediate applicability: tides, storm surges, waves and wave forcing, the circulation and water masses of the adjacent seas, and the oceans' influence on typhoons. These and related topics were nearly universally emphasized at all institutes.

In 1978 the South China Sea Institute of Oceanography was reportedly studying the structure of seawater masses, currents, and wave spectra and developing marine instrumentation, including oceanographic and meteorological buoys. The Institute of Oceanography in Hangzhou was working on storm surges, using regression methods involving weather maps provided by the Central Meteorological Bureau and tide records obtained from tide gauges along the coast. As of 1978, there had been no attempt to make predictions. The Hangzhou Institute had also studied the Kuroshio (the Pacific counterpart of the Gulf Stream), with the necessary data obtained from Japan's Cooperative Studies of the Kuroshio. This work was shorebound and did not involve deep-water observations. The Institute sent its data and programs to be computed at the Shanghai Institute of Computer Technology. Finite element techniques using diagnostic methods were applied to a study of the Kuroshio's flow. Results had not been tested, however, against direct observations.

The Institute of Oceanography at Qingdao had departments of marine hydrography and physics (physical oceanography) and marine instrumentation. Scientists were studying the hydrography of the East China Sea and nearby waters, in particular the summer circulation of cold water in the Yellow Sea, the circulation near the mouth of the Chang Jiang, and the Kuroshio system. A wave group focused on the engineering of offshore structures and was developing instruments for measuring waves at sea. An equal-resistance wave recorder and a digital wave staff had been completed. Fundamental work was under way on the theory of surface and internal waves. A tide group was working on tidal predictions and shallow-water ports, using a quasi-harmonic method, and on numerical solutions to the equations of motion to study the effects of friction on the tidal bore and tidal dissipation in the Yellow Sea.

Shandong College of Oceanography was conducting research on storm surges using dynamical and regression models and was apparently beginning work on topographic shelfwaves. The staff was studying ocean currents and circulation using modifications of the dynamic method, the prediction of sea surface temperatures by statistical methods for fisheries purposes, and the dynamics of tides in shallow water and their prediction, including numerical models based on Hansen's method.

The Tianjin Institute of Marine Instrumentation develops and constructs instruments for physical oceanography and is responsible for standardizing and calibrating all instruments in China. Among the instruments being developed in 1978 were an airborne infrared thermometer (which had been used at sea and compared with shipboard data), an in-situ salinity temperature profiler device with a self-contained magnetic tape recording system, an acoustic velocity profiling system, a laboratory induction salinometer, and instruments for intercalibration. Most of this equipment was comparable to that available in the United States around 1963.

However, the actual equipment on oceanographic vessels tended to be of the Nansen bottle era (1910–50). The ships were used to carry out classical physical oceanography: lowering current meters to make velocity profiles in shallow seas, doing Nansen cast work, measuring wavefields with crude but qualitatively correct Tucker wave gauges, and measuring water surface temperatures. Most ships were equipped with winches, although the cables usually could go only to about 2,500 m even if the drums could take longer ones. But there were no lifting cranes or A-frames of the types normally found in the West and no shipboard computers.

Computers are in fact a special problem for the development of Chinese oceanography. As of 1978, the small number of special-purpose, Chinese-made computers available at oceanographic laboratories were slow by current standards and lacked peripheral equipment, particularly plotters. Some computing was done away from laboratories in regional computing centers, and many institutions were improving their facilities by purchases from abroad, but there was a lot of ground to make up.

### *Coastal Oceanography (Shore Processes)*

Study and understanding of the complex interactions among waves, tides, currents, and sediments on beaches and in adjacent waters require an interdisciplinary approach involving physical oceanographers, geologists, chemists, biologists, coastal engineers, and marine archaeologists. Since the early 1960's, Western scientists from two or more of these disciplines have combined forces to make concentrated studies of nearshore phenomena and have contributed to their understanding. This interdisciplinary approach, referred to as coastal oceanography or shore processes, became common in most large oceanographic institutions in the United States during the 1970's.

As of 1978, however, there were no comparable interdisciplinary groups in China. The Chinese were active in all the constituent disciplines, but had not advanced to the point of making concentrated interdisciplinary efforts to understand and solve coastal problems. They worked within the constraints of traditional disciplines, and consequently, their understanding of shoreline problems lagged behind the West. This was unfortunate, given China's 8,000-km-long, diverse coastline, which makes the study of shore processes particularly attractive. In addition, the practical need for research in coastal oceanography is clear because of China's growing demand for harbors and coastal structures to handle shipping and oil drilling. (China's major rivers carry large quantities of sediment, and the country's all-important river mouth harbors require extensive dredging. Tianjin-Tanggu New Harbor, for example, requires maintenance dredging of $3.5 \times 10^6$ cubic meters of mud each year.)

Chinese coastal engineers needed input on sedimentation and transport from physical oceanographers and marine geologists, the benefit of nearshore survey teams, and instruments for measuring environmental factors.

### *Education*

The two main centers of education in fisheries, aquaculture, and oceanography in 1978 were Shandong College of Oceanography at Qingdao and Xiamen University. Both of these were among China's top ("key") 89 colleges and universities. Shandong College of Oceanography had six departments: Physical Oceanography and Meteorology, Marine Physics, Marine Chemistry, Marine Biology, Marine Geology and Geophysical Exploration, and Marine Fisheries and Aquaculture. The curriculum in the Department of Physical Oceanography and Meteorology consisted of mathematics, physics, statistics, hydrodynamics, general oceanography, and general meteorology. The curriculum in the Department of Marine Biology, which had over 100 majors and five undergraduate students, was as follows:

*First year:* Botany (morphology and taxonomy with a marine emphasis), zoology, mathematics (calculus), chemistry, politics, and physical training.

*Second year:* Zoology, plant physiology (metabolism, biochemistry, and photosynthesis), organic chemistry, phycology (life cycles), politics, and physical training.

*Third year:* Embryology and histology, biochemistry, general oceanography, oceanographic surveying (including two two-week training cruises), plankton (plant and animal morphology and taxonomy), microbiology (bacteria), ecology (natural history, biogeography, physical factors of the ocean), and politics.

*Fourth year:* Genetics (Mendelian, molecular, population, and statistics), pollution, evolution, politics, and thesis and research.

As of 1978, Xiamen University's Oceanography Department had 200 majors, with the biologists among them taking basic courses in general biology, vertebrate and invertebrate zoology, and specialized courses in ichthyology, marine physiology, and marine ecology. Ocean-oriented biology majors took similar general courses, but went on to take specialized courses in genetics and embryology.

Both institutions planned to expand their enrollments. Shandong College of Oceanography aimed to increase its undergraduate enrollment from 1,000 in 1978 to 4,000 by 1985. During the same period, Xiamen University expected to increase its undergraduate intake from 1,000 to 3,500. Both institutions also planned to increase their intake of graduate students.

The years lost to the Cultural Revolution and its aftermath had clearly created major problems in reeducating and retooling thousands of displaced college students as well as in educating those ready to enter college. The need for updated textbooks, new laboratory equipment, library facilities, and information channels was urgent.

The important policy changes of 1978 favoring basic research and improved education in fisheries, aquaculture, and oceanography, along with China's increasing contacts with the West, meant that the prospects for all three activities were good. Chinese scientists had much to teach their colleagues abroad about aquaculture, just as they had much to learn about the more quantitative, technical, and interdisciplinary aspects of oceanography. There were many areas of potentially fruitful collaboration among individual scientists, oceanographic institutions, and laboratories, including geophysical studies of the China Sea's shelves (particularly the dynamics of shallow water circulation on the shelves), quantitative resource evaluation, research on marine ecology, and investigations of the Western Pacific's circulation, mean flow, and eddy field. Also significant were opportunities for research on marine algae and for geochemical work on China's large rivers and estuaries in conjunction with studies of beach and nearshore processes. Collaborative studies of fish productivity in reservoirs and their interpretation in terms of energy flow and food chain efficiency theory were of particular interest to American scientists, and many developing nations wanted information on the multispecies culture of Chinese carp.

The main obstruction to such joint efforts was the Chinese denial of free access to data, samples, and field sites. As of 1978, guidelines for data and sample exchange had not been clearly and comprehensively spelled out, and no mechanism for working out problems on a case-by-case basis existed. There were, however, promising signs that such guidelines would be produced, and that the Chinese would become more open in the near future.

*H. M. Temin*

# *BASIC BIOMEDICAL RESEARCH*

In 1966 *Science* acclaimed the synthesis of insulin in China as a major scientific achievement and concluded that it might be an important indication that the Chinese scientific effort was about to achieve "quality in a growing number of fields."[1] Then came the Great Proletarian Cultural Revolution, Lin Biao, and the Gang of Four. During this period, which lasted from 1966 to about 1978, almost all scientific research was severely disrupted. The insulin work now seems less of a harbinger of quality in basic biomedical research and more an isolated accomplishment without much effect. Indeed, until 1978 basic biomedical research was unimportant to decision makers in the PRC. Major infectious and parasitic diseases were under control, and agricultural and energy problems were much more important. The small amount of basic biomedical research in China capitalized on special talents or opportunities and did not require major resources.

China's National Plan for the Development of Science and Technology, announced in March 1978, seemed to mark a change of direction. Genetic engineering was singled out as one of eight priority areas, and biomedical research was recognized as having national importance. Vice-Premier Fang Yi, the administrator of China's research programs, said he hoped that the country would have strengthened its organization and coordination and started basic studies in genetic engineering by 1981. He predicted that studies of molecular biology, molecular genetics, and cell biology would make "fairly big progress" by 1985. (See Appendix A.)

This chapter discusses the major subject areas of basic biomedical research in the PRC as they appeared in 1978 and attempts to evaluate the base underlying the ambitious eight-year plan (1978–85), outlined by Fang

[1] V. K. McElheny, "Total Synthesis of Insulin in Red China," *Science*, 153 (1966): 281–83.

Yi. Persons accustomed to the depth and breadth of U.S. basic biomedical research will be amazed to find many areas not even represented in China. They will also find that distinctions between basic and clinical biomedical research are less sharp than they are in the United States. (For example, China's foremost institute of biochemistry runs a large primary screening program for liver cancer in rural areas.) The chapter also describes the main centers of basic biomedical research. These are mostly research institutes in Beijing and Shanghai within the Chinese Academy of Sciences (CAS) or the Chinese Academy of Medical Sciences (CAMS).

The information in this chapter, which mostly dates from before 1978, comes from trip reports,[2] my experiences, and study of the relevant literature, which is extremely sparse. There is a great paucity of information and much repetition in the experiences of Western visitors and in the published reports. Either much of basic biomedical science has not been seen and reported, or more likely, it does not exist.

## MAJOR SUBJECT AREAS

*Protein synthesis: insulin.* The synthesis of biologically active insulin in 1965 is one of the PRC's most significant scientific accomplishments.[3] It was performed primarily at the CAS Institutes of Biochemistry and of Organic Chemistry in Shanghai and was a major impetus in the development of biochemical factories. It led to research on x-ray crystallography of insulin (at the Beijing Institutes of Biophysics and Physics, and at Beijing University), study of structure-activity of insulin analogs (at the Beijing Institute of Zoology), and synthesis of other polypeptide hormones.[4]

As of 1977, work involved further study of various insulin analogs, including synthesis, biochemical characterization, x-ray crystallographic study, and measurements of biological activity and receptor binding. The work paralleled and, in some cases, supplemented or contradicted work in Western laboratories.[5]

[2] Royal Danish Academy of Sciences and Letters, "Natural Sciences in China, Oct.–Nov. 1975"; Royal Society, London, *Scientific Visits to China Under Arrangements Between the Academia Sinica and the Royal Society, Nov. 10–25, 1973* (London: Royal Society, 1974); Rockefeller University, "Report on a Visit to the People's Republic of China, May 5–22, 1977" (Sept. 1977); University of Arizona, China Delegation, *China, 1976* (Tucson: University of Arizona, 1976); "Biomedical Science in China, Report of Trip by H. M. Temin, April 20 to May 6, 1977"; and other reports.

[3] McElheny, "Insulin."

[4] Royal Danish Academy of Sciences and Letters, "Natural Sciences in China," pp. 76–77; Beijing Insulin Structure Research Group, "Studies on the Insulin Crystal Structure: The Molecule at 1.8 A Resolution," *Zhongguo kexue (Scientia Sinica)*, 17, 6 (Dec. 1974): 752–78; and CAS, Beijing Institute of Zoology, Division of Endocrinology, Insulin Research Group, and CAS, Shanghai Institute of Biochemistry, Insulin Research Group, "Studies on the Mechanisms of Insulin Action: I. Interactions of Insulin and Its Analogs with Insulin Receptor," *Scientia Sinica*, 17, 6 (Dec. 1974): 779–92.

[5] P. DeMeyts, E. Van Obberghen, and J. Roth, "Mapping of the Residues Responsible for the Negative Cooperativity of the Receptor-Binding Region of Insulin," *Nature*, 273 (1978): 504–9.

The insulin work was sophisticated and up to date in methodology and as such appeared to be almost unique in Chinese biomedical science. It also represented a large portion of China's total basic biomedical research effort. But between 1976 and 1978 it led to only five papers with English summaries and was cited only occasionally in the non-Chinese literature.[6]

*Proteins and enzymology.* A small amount of basic protein and enzyme biochemistry has been published. There is also practical interest in enzymes for use in the chemical, pharmaceutical, and agricultural industries. The basic biochemistry has come mostly from the Institute of Biochemistry, with some papers from Beijing Medical College, the Institute of Microbiology, Baotao Special Medical School, and Zhongshan, Beijing, and Nankai Universities. In addition, visitors have reported work on characterization of oxidative enzymes at the Institute of Biochemistry and of rabbit muscle glyceraldehyde-3-phosphatase at the Institute of Biophysics.

The only published work of general interest to enzymologists and protein chemists concerned the purification of alpha fetal protein by immunoabsorption-affinity chromatography and the solid-phase synthesis of peptides.[7]

*Nucleic acids.* Primary interest here has been agricultural; nucleic acids and their hydrolysates are used as growth factors for rice and other crops. There has been almost no published work on nucleic acids, except that involved in cell biological experiments and studies of plant virus genomes.

Visitors have reported some ribonucleic acid (RNA) sequencing at the Institute of Biophysics (although it was not clear what work was done in China and what performed on fellowship in England), deoxyribonucleic acid (DNA) synthesis at the Institute of Cell Biology, transfer RNA characterization at the Institute of Biochemistry, and cell-free synthesis of tobacco mosaic virus (TMV) proteins at the Institute of Microbiology. This unpublished work was relatively sophisticated, but still years behind the respective fields in the West.

*Genetics.* A number of review articles have appeared especially on the philosophical implications of genetics. However, most experimental work has related to plant and animal breeding. The lack of genetics contrasts strongly with Western biomedical research, where genetics is a central field. This is a result not only of the Lysenko doctrine, as it was in the Soviet Union, but also of the lack of senior geneticists and the need to use geneticists in practical work.

*Microbial genetics.* Interest in microbial genetics was related mainly to industrial microbiology and infectious diseases. There was also some theo-

[6] *Science Citation Index,* 1977, 1978.

[7] Royal Society, London, *Scientific Visits to China;* and H. Y. Zhang et al., "Purification of Antigens and Antibodies by the Immunoabsorption-Affinity Chromatography Method: The Purification of AFP," *Shengwu huaxue yu shengwu wuli jinzhan (Advances in Biochemistry and Biophysics),* 1974, 2 (June): 36–39.

retical interest, as shown by review articles, especially in mechanisms of mutation. Most work was done at the Institute of Microbiology, with some at the Institute of Genetics. Visitors have also reported some work on nitrogen fixation and on infectious diseases at the Institute of Plant Physiology. At Fudan University in Shanghai, a group was studying microbial, radiation, and population genetics, but all fundamental work there was halted by the Cultural Revolution.

Research work in microbial genetics was generally competent, but equivalent to Western work of 1960 to 1970.

*Animal virology.* The development and characterization of vaccines was actively pursued at many different institutes in different cities in China, and there was much interest in the relation of viruses to cancer.

Vaccines were prepared using attenuated live measles virus, inactivated type-B encephalitis virus, attenuated live influenza virus, absorbed epidemic cerebrospinal meningitis agents (bacterial?), asthma (unclear what agents were involved), and attenuated live poliomyelitis virus, as well as against smallpox. An inactivated vaccine was prepared against type-B Japanese encephalitis virus grown in cell culture and used with good results. Attempts to develop a vaccine against hepatitis B virus and to improve the Sabin polio vaccine were under way. However, sensitive radioimmuoassays were not used to evaluate these vaccines.

Diagnostic virology was performed using cell culture, and simple comparisons of different viruses in cell culture were performed. Fluorescent and electron microscopy were used. An article on antiviral chemotherapy against influenza was published, and visitors have reported the use of polynucleotides as interferon inducers.[8] In sum, the virology was limited in quantity and similar to Western work of about 1960.

Rickettsia, especially the trachoma agent that was isolated in the PRC in 1957, were also studied.[9] There was also some plant and insect virology at the Institutes of Microbiology and Biochemistry.

*Cell culture.* PRC research institutes had facilities for cell culture, although some of the media were imported. Cell culture was used in virology and cancer research. The establishment of a human diploid cell strain, growth of human esophageal cancer epithelial cells, and study of liver cancer cells *in vitro* have been reported.[10] These appear to be valid results

[8] Rockefeller University, "Report on a Visit."

[9] F. F. Tang et al., "Studies on the Etiology of Trachoma with Special Reference to Isolation of the Virus in Chick Embryo," *Zhonghua yixue zazhi (Chinese Medical Journal)*, 75, 6 (June 1957): 429–46.

[10] Kunming Institute of Medical Biology, Prevention and Treatment Department, Substitutes Research Group, "Establishment of a Human Diploid Cell Strain (KMB-13) and Preliminary Observation of the Biological Characteristics," *Yichuan xuebao (Acta Genetica Sinica)*, 1974, 2 (Dec.): 147–55; CAMS, Institute of Tumor Prevention and Treatment, Cytobiology Group, "Establishment of a Strain of Human Esophageal Cancer Epithelial Cells," *Chinese Medical Journal*, 1974, 7 (July): 412–15; and J. M. Chen et al., "Preliminary Report on the Establishment and Characteristics of a Strain of Human Liver Cancer Cells (BEL-7402) *in Vitro*," *Kexue tongbao (Science Bulletin)*, 20, 9 (Sept. 1975): 434–36.

and could be of general use. As of 1978, there had been little exploitation of these strains, in the PRC or elsewhere.

*Cytogenetics.* Chromosome-banding techniques have been applied at Hunan College of Medicine, the Institute of Oncology, Beijing Medical College, and the Institute of Genetics. The techniques were being applied to clinical material and used in experimental oncology. The work was of little general interest.

Work on radiation cytogenetics was under way at Suzhou Medical College, repeating basic results obtained elsewhere. The Institute of Experimental Biology was doing relatively unsophisticated work on chromosome structure.

*Developmental biology.* Work of general interest has been carried out in this field in the PRC. Since the mid-1960's, Dong Dizhou and his colleagues at the Institute of Zoology have been studying transplantation of nuclei from somatic cells into eggs, between different species, and even between subfamilies of fish, especially goldfish and bitterlings. Their results have extended earlier Western work and are generally known in China, although they have unfortunately received little notice or reference abroad.[11] Dong and an American collaborator, M. C. Niu of Temple University, have reported the effects of nucleic acid injection on goldfish development, including the heritable effects of RNA injections.[12] As of 1977, an attempt to provide biochemical evidence for these observations by studying certain isoenzymes was under way. Preliminary observations of a reverse transcriptase in goldfish eggs, which could provide a mechanism for the genetic effects of RNA, have been made. The biochemical work was less convincing than the nuclear transplantation.

*Cell biology.* Interest in this area was mainly theoretical, with review articles on cell fusion and cell membranes. Some work on reconstituting membrane adenosine triphosphatase (ATPase) in cancer cells was carried out at the Institute of Zoology.

*Cancer research.* There was an abundance of cancer research in China. The major efforts were clinical and epidemiological in nature, with relatively little laboratory or experimental work. Cytogenetics, cell fusion, and cell culture have been used. Visitors have reported attempts to promote immunogenicity by cell fusion, nuclear transplantation, and nucleic acid injection, as well as interest in studying possible virus involvement in certain human cancers.[13]

As of 1977, chemical carcinogenesis was little studied, although the possi-

[11] Dong Dizhou et al., "Transplantation of Nuclei Between Two Subfamilies of Teleosts (Goldfish–Domesticated *Carassius auratus* and Chinese Bitterling–*Rodeus sinensis*)," *Dongwu xuebao (Acta Zoologica Sinica)*, 19, 3 (Sept. 1973): 201–12.

[12] Dong Dizhou and M. D. Niu, "Nucleic Acid–Induced Transformation Goldfish," *Scientia Sinica*, 16, 3 (Aug. 1973): 377–84.

[13] Henry S. Kaplan and Patricia Jones Tsuchitani, eds., *Cancer in China* (New York: Alan R. Liss, 1978).

ble importance of chemicals in the etiology of human cancer was recognized. In some cases attempts had been made to isolate or remove chemical carcinogens from the home and work place.

Attempts to isolate new anticancer compounds were being made, especially at the Shanghai Institute of Materia Medica. Microbiological systems were often used in initial screening. Some new drugs had been isolated, but I am not sure whether these were truly novel or attempts to improve existing ones.

*Radiobiology.* Dogs and mice were used to study radiation injury and protective effects at the Institutes of Biochemistry and Biophysics and at Lanzhou University. The observations were straightforward and not of general interest.

*Acupuncture.* This practice has aroused considerable interest in the United States. It is used in the PRC both in the traditional way for treatment and more recently for anesthesia.

There has been much basic research on the mechanisms of these acupuncture effects, especially at the Institutes of Physiology and of Psychology and at Zhongshan Medical School. Most of the work has focused on the mechanism of acupuncture anesthesia. Western visitors have been impressed with this research and have accepted the Chinese results. These reject traditional yin and yang explanations, indicate a central humoral effect of the peripheral needling, and show the requirement of afferent transmission by sensory neurons of Groups I and II leading to inhibition of spinal transmission and of certain central discharges. They also point to additional involvement of psychic factors (in humans at least) and possible involvement of the "still reaction."[14]

Dr. M. Vogt, a member of the Royal Society Delegation, summarized the Institute of Physiology's work:

> In experiments on themselves, the physiologists obtained evidence that the pain threshold (to electrical stimulation of the skin), is, in fact, modified by the insertion of acupuncture needles; the effect is produced by the needles stimulating deep structures, not the skin, since the skin can be treated with a local anesthetic without interfering with the rise in pain threshold. In contrast, procainization of the deep tissues around the needle tip (which includes nerve branches going to the muscle) abolished the analgesic effect. Furthermore, occlusion of the circulation in the arm did not interfere with the effectiveness of a needle inserted into the hand. Thus the humoral transmission of some substance liberated at the site of the needle was ruled out as the cause of analgesia.
>
> The subjective feeling by the anesthetist of "suction" when the needle is in a "good" position, and of "numbness or heaviness" experienced under such conditions by the subject, can be related to muscular contractions at the needle tip and have thus a rational explanation.
>
> The effect of a needle inserted on one side of the body affects the pain threshold bilaterally in symmetrical regions. The most effective points for the insertion of

[14] See trip reports cited in note 3.

needles appear to be those which have the same segmental innervation from the cord as the site where analgesia is desired.[15]

These, and other conclusions, derive mainly from personal conversations; not all of them or the data supporting them have been published. However, this work is of general interest and at the forefront of pain and anesthesia studies.

*Other fields.* (1) Endocrinology. There is interest in reproductive physiology. A few papers on prostaglandins have appeared. (2) Parasitology. According to the Rockefeller University trip report, the limited research efforts were firmly practical, akin to those of state agricultural experiment stations in the United States.[16] (3) Keshan disease. This endemic myocardial disease of unknown etiology was studied at Haerbin and Jilin Medical Schools. (4) Pharmacology. The search for new drugs, especially from traditional medicines, was carried out at the Institutes of Materia Medica in Beijing and Shanghai. Modern techniques were used in this work. An interesting abortifacient and a "morning-after" contraceptive pill have been reported.[17]

## THE ORGANIZATION OF BASIC BIOMEDICAL RESEARCH

As in the Soviet Union, basic biomedical research is generally carried out at institutes unaffiliated with universities. Since the early 1960's these research institutes have gone through three phases of organization. In the early 1960's they were primarily affiliated with the CAS or the CAMS. After the Cultural Revolution, most were affiliated—administratively and financially—with provincial or municipal governments, but still kept some academy ties. (It is not clear if this was as true in Beijing as in Shanghai.) Since 1978 affiliations have shifted back to the original academies.

There is also some research at universities and at smaller institutes affiliated with them, although this was interrupted by the Cultural Revolution. Since the educational reforms introduced at the end of 1978, graduates who finish their university course work can carry out laboratory work at academy institutes. I was told in 1977 that institute and university research staffs were chosen from those graduates "most fit and best in political approach, as well as in knowledge." Institutes appeared to have no choice in hiring new staff in 1977. Graduating students could express their preferences on a form, but the Ministry of Education assigned them to jobs as needed. It will be interesting to see if this system is modified.

According to visitors' reports and judging from the volume of publications, the Shanghai Institutes of Biochemistry, Cell Biology, and Physiology,

[15] Royal Society, London, *Scientific Visits to China,* p. 1.

[16] Rockefeller University, "Report on a Visit," p. 19.

[17] Shanghai Institute of Experimental Biology, Second Laboratory, "Studies on the Mechanisms of Abortion Induction by Trichosanthin," *Scientia Sinica,* 19, 6 (Nov./Dec. 1974): 811–30; and Frederick S. Jaffe, "China Trip Notes, 1977."

and the Beijing Institute of Biophysics are paramount in basic biomedical research. These grew out of an original Institute of Physiology founded in 1950 with a staff of 20. The new CAS Institute of Molecular Biology in Beijing, which is headed by Zuo Zhenglu of the Institute of Biophysics, has the same lineage.

Institutes are separated into divisions or departments that work in different subject areas. These were determined, it was described in 1977, in two ways. Either all members of a laboratory decide on areas and approaches (given the Chinese respect for age and experience this usually means accepting the senior scientist's ideas), or the academies, in response to exterior political pressures, assign problems to institutes.

There is no information on how staff is transferred within an institute, how progress is monitored, or how budgets are allocated between groups within an institute or between institutes.

## *MAJOR INSTITUTES VISITED BY WESTERN SCIENTISTS*

*Institute of Biochemistry, CAS, Shanghai.* The Institute of Biochemistry was the lead institute in the synthesis of biologically active insulin mentioned earlier. This work led to a significant capability in synthesis of peptides, a capability used to synthesize modified insulins for structure-activity studies and several other polypeptide hormones as well, including oxytocin, vasopressin, angiotensin II, and glucagon, which are used in medical research.

The Institute of Biochemistry was founded in 1950 and by 1977 had between 400 and 500 personnel, including 200 research workers and 90 staff members in a nearby biochemistry factory. Areas of research interest included protein synthesis (methods, modified insulins, polypeptide hormones), nucleic acid synthesis, structure of nucleic acids (transfer RNAs), enzymology (immobilized and oxidative enzymes), liver cancer (diagnosis, alpha fetal protein, hepatitis B virus, therapy), plant viruses, and acupuncture anesthesia. There was also an instrument division. In general the equipment was less plentiful and less sophisticated than that in U.S. laboratories.

Several projects involved applied research. The Biochemical Factory and Instrument Division prepared reagents and equipment; immobilized enzymes were used industrially and to prepare nucleic acid fragments for research; and nucleic acids and digests were used to promote plant growth in communes.

*Institute of Biophysics, CAS, Beijing.* This Institute has been involved in x-ray diffraction studies of native and modified insulins. (Some visitors report that the Physics Department of Beijing University has led this work; all agree that the two cooperated.) The Institute was founded in 1958 as an offshoot of the older Institute of Physiology and Biochemistry in Shanghai. It employed 220 to 230 personnel in 1972 and 400 in 1975. The Institute does not have one central laboratory building, but is scattered among various other CAS institutes. Research is conducted in five departments, as well as at workshops and an animal farm. As of 1977, there were five departments:

Radiation Biology, which studied the effects of long-term low-level external radiation on rats and monkeys, internal radiation from internal radionuclides, radiation dosimetry especially by thermoluminescence, and environmental pollution.

Molecular Biology, which studied cyclic adenosine monophosphate phosphorylase and glyceraldehyde-3-phosphatase, the structure and function of RNA (including sequencing messenger RNA and the use of poly [IC], an interferon inducer, in clinical trials), and x-ray crystallographic analysis, especially of modified insulins.

Sensory Receptors, which studied mechanical receptors in pigeon legs, the structure and function of the retina, and mechanisms of acupuncture anesthesia.

Submicroscopic Structure of Cells, which concentrated on the structure and function of yeast mitochondria and oocytes of *Chisocephalus* (a crustacean that undergoes sex reversal).

Technical Projects and Instrumentation, which manufactures equipment, especially for radiation studies, including electron spin resonance and fluorescence spectrophotometers.

Visitors in the mid-1970's reported an increased number of projects, which might explain the small number of publications from any of these departments.

*Institute of Cell Biology, CAS, Shanghai.* This Institute shares a building with the Institute of Biochemistry. Its staff increased from 200 to 340 between 1974 and 1977. There were four departments, including:

Cancer Research, which studied environmental carcinogenesis in industrial situations, the relationships of Epstein-Barr virus (EBV) to nasopharyngeal carcinoma (NPC) and of hepatitis B virus to liver cancer, the role of nucleic acids in cell cultures and as a messenger, and the immunology of hepatoma cell membranes. Most of this work was of low quality.

Physiology of Reproduction and Chinese Medicine, which concentrated on an abortifacient isolated and purified from root tubers. It has been characterized biochemically and physiologically with interesting results.

DNA and Chromosomes, which studied calf thymus chromatin, the effects of histone and nonhistone proteins on transcription by *Escherichia coli* RNA polymerase, and the chemical synthesis of an oligonucleotide. These efforts were many years behind related Western work.

The laboratories were similar to those in the Institute of Biochemistry. Both Chinese and Western equipment were used.

*Institute of Zoology, CAS, Beijing.* This Institute housed the laboratory of Dong Dizhou, whose nuclear transplantation work is of particular interest. The Institute has its own relatively large building, constructed in 1956. The staff numbered over 500, with 350 research and technical workers; 50 of these were "experienced workers" (equivalent to associate or full professor), and the rest were "younger experimental science members." This ratio of experienced to young researchers was unusually high.

The Endocrinology Department studied reproductive endocrinology, the mechanisms of insulin action (with the Institute of Biochemistry), and the mechanism of acupuncture anesthesia. The insulin work involved receptor isolation and was current with Western work. The reproductive endocrinology involved testing hormones and analogs synthesized elsewhere for field application. The Royal Danish Academy was told in 1975 that luteinizing releasing hormone was "first found and synthesized in China."[18]

The Cell Biology Department studied nuclear cytoplasmic relationships (injection of nuclei from differentiated tissues into enucleated eggs of the same species, nuclear transplantation between species and subfamilies, injection of nucleic acids into fertilized eggs), immunology (fusion of Walker sarcoma cells with chicken red blood cells to test antigens for tumor therapy and inhibition of leukocyte attachment for lung cancer diagnosis), and membrane ATPase dissociation and reconstitution. The nuclear transplantation and ATPase studies were of good quality, and some of the nuclear transplantations were unique. But the nucleic acid and immunological studies aroused skepticism in Western visitors.

*Institute of Cancer Research, CAMS, Beijing.* This Institute is concerned with human cancer primarily from the viewpoints of treatment and epidemiology. However, a preclinical group was doing some laboratory studies. The epidemiology, primarily of esophageal cancer, was of high quality and great interest. The laboratory studies involved testing suspected foodstuffs for carcinogenicity and chicken esophageal tumors for viruses (electron microscopy). There was also work with NPC involving cell culture and screening for EBV and with liver cancer using a radioimmunoassay to help purify alpha fetal protein. The laboratory work was relatively limited and unsophisticated by Western standards.

*Institute of Physiology, CAS, Shanghai.* The staff here numbered 300, with 170 scientific and technical workers and 50 in the workshop. There were five groups: acupuncture anesthesia, muscle biophysics and nerve-muscle trophic relationships, sense organs, hypoxia, and reproduction. The work on acupuncture anesthesia (summarized above) was considered to be of high quality, as was that of the muscle group. (Both were headed by Western-trained scientists.)

*Institute of Genetics, CAS, Beijing.* This Institute was mainly concerned with plant breeding, transplantation of fertilized sheep eggs, and microbial genetics. Established in 1951 as an office, it was enlarged to an Institute in

[18] Royal Danish Academy of Sciences and Letters, "Natural Sciences in China," p. 19.

1959 and moved to its present building in 1964. There were about 350 members in 1977, including 200 scientists and technicians.

Most of the work was directed toward practical applications. Exceptions were transformation by DNA of factors for sporulation in *Bacillus subtilis* and synthesis of certain enzymes for industrial purposes. Visitors have not been impressed with the quality of this work.

*Institute of Microbiology, CAS, Beijing.* Established in 1958, this Institute had a staff of 400 in 1977, including 300 scientists and technicians. Research areas included taxonomy of microorganisms, industrial microbiology, microbial genetics, and biochemistry. In the latter area there were basic studies on TMV RNA and protein synthesis *in vitro* and on plasmids of *Staphlococcus aureus* and *Escherichia coli.* The work was competent, but years behind related Western work. There was also a factory for ribonucleoside triphosphates, an offshoot of the TMV work.

*Institute of Materia Medica, CAMS, Shanghai.* This Institute was founded in 1958 and had a staff of 550 in 1976, with about 300 scientific workers in its five departments of Synthetic Chemistry, Plant Chemistry, Pharmacology, Analytic Chemistry, and Antibiotics. A contraceptive drug, Anordin, has been developed and tested for effectiveness and toxicity. It has been used postcoitally since 1969 in clinical trials. Screening was also being done for drugs for cancer, coronary disease, chronic bronchitis, and nervous diseases. Efforts were made to isolate, characterize, and prepare synthetically the active components of traditional Chinese medicines. The work was competent.

*Institute of Pharmacology, CAMS, Beijing.* There were 300 workers here in 1976, doing research on influenza, the common cold, bronchitis, cancer, cardiovascular diseases, contraceptives, and indigenous herbs. The chemistry and pharmacology were of high quality despite the relatively unsophisticated equipment.

*Beijing Medical College.* The College conducted work on bronchitis, stomach cancer, and acupuncture anesthesia (methods of locating acupuncture points and evidence of the involvement of central humoral factors).

*Tumor Research Institute, Zhongshan Medical College, Guangzhou.* This Institute had four departments—Anticancer Drugs, Etiology and Pathology, Immunology, and Epidemiology—as well as clinics. In etiology, unsophisticated studies of chromosomes in EBV and NPC and of the role of nitrosamines and aflatoxins in NPC were under way.

There are numerous other institutes at a provincial and municipal level, which have not been reported on and are primarily known from one or two research citations or mention to visitors. The work in these institutes is almost certainly of lesser quality than that in the national institutes. Much of the equipment used in biomedical research in China was Western-made or copied from Western equipment.

*PUBLICATIONS*

The majority of biomedical papers in *Kexue tongbao (Science Bulletin)* are neurophysiological in nature and deal especially with mechanisms of acupuncture anesthesia. *Zhongguo kexue (Scientia Sinica)* publishes work on insulin and developmental biology. Applied and industrial microbiology is primarily published in *Weishengwu xuebao (Acta Microbiologica Sinica)*, with some reports of isolations and classification. Traditional zoology is mostly published in *Dongwu xuebao (Acta Zoologica Sinica)*, as is research on physiology, panda life, rat control, parasitology, and animal cancer. *Shengwu huaxue yu shengwu wuli jinzhan (Progress in Biochemistry and Biophysics)* publishes review articles on biochemistry and biophysics, as well as descriptions of equipment and techniques. *Shengwu huaxue yu shengwu wuli xuebao (Acta Biochimica et Biophysica Sinica)* also publishes articles on a variety of topics in this area.

Other journals publishing related work include *Dongwuxue zazhi (Journal of Zoology)*, *Kexue shiyan (Scientific Experimentation)*, *Zhonghua neike zazhi (Chinese Journal of Internal Medicine)*, *Zhonghua yixue zazhi (Chinese Medical Journal)*, and *Yichuan xuebao (Acta Genetica Sinica)*. Most of these journals are available in an English edition; others have English summaries.

As of 1977, basic biomedical research in the PRC was very thin. Except for work on the structure and function of insulin (Institutes of Biochemistry, Zoology, Biophysics, Organic Chemistry, and Physics),[19] nucleo-cytoplasmic relationships (Institute of Zoology),[20] an abortifacient and a possible contraceptive (Institutes of Experimental Biology and Materia Medica),[21] and mechanisms of acupuncture anesthesia (Institute of Physiology and perhaps Beijing Medical College),[22] research was unsophisticated and 10 to 20 years behind the West. Much of the first-rate work was by groups headed by older, Western-trained scientists.

An unsympathetic observer might conclude that there was almost no basic biomedical research in the PRC. Certainly Western research rarely referred to Chinese work and was independent of it. (Acupuncture anesthesia and the introduction of certain new drugs may provide minor exceptions to this generalization.) Furthermore, clinical research and practice in the PRC

[19] Institute of Zoology, Insulin Research Group, and Institute of Biochemistry, Insulin Research Group, "Mechanisms of Insulin Action."

[20] Dong et al., "Transplantation of Nuclei"; and Dong and Niu, "Nucleic Acid–Induced Transformation Goldfish."

[21] Shanghai Institute of Experimental Biology, Second Laboratory, "Mechanisms of Abortion Induction."

[22] C. Y. Jiang, K. S. Xu, and C. H. Jiang, "The Intraspinal Afferent Pathway and the Analgesic Effect of Acupuncture," *Chinese Medical Journal*, 1976, 11 (Nov.): 701–4; and J. Y. Wei et al., "Observations on Activity of Some Deep Receptors in Cat Hind Limb During Acupuncture," *Science Bulletin*, 18, 4 (Oct. 1973): 184–86.

was based on Western biomedical research and traditional Chinese practices, with little obvious input from modern, indigenous basic research. For example, visitors have reported that research on acupuncture anesthesia has not affected its use in China.

In addition, the Chinese research effort was very inefficient in terms of the total number of published papers. Only a relatively small number of papers were published and available in the West in relation to the number of scientists reported to be engaged in research.

These gaps and inefficiencies can be explained by the shortage of trained personnel, the need to make all reagents and thus the unavailability of many of them, isolation from Western scientists, and lack of interest and money from the government combined with political disruption. However, one would be hard put to justify the assertion that China has suffered in any way as a result of this.

What of the future, especially now that genetic engineering is a national priority? I see little chance of the rapid progress anticipated by Vice-Premier Fang Yi. Whole fields will have to be created; supplies of sophisticated reagents and equipment, especially radionuclides, established; and a cadre of scientists trained. All this has to be accomplished with a few experienced middle-aged scientists. Meanwhile, the pace of scientific advance in the rest of the world is accelerating.

However, the planned establishment of the Beijing Institute of Molecular Biology and the increased contact permitted between Western and Chinese biomedical scientists indicate that the Chinese are aware of at least some of the problems and anxious to solve them. The next few years will provide an exacting test to the proposition that a community rich in basic scientists is needed as a basis for successful applied science.

*Myron E. Wegman*

# *BIOMEDICAL RESEARCH: CLINICAL AND PUBLIC HEALTH ASPECTS*

Studies in curative and preventive medicine are influenced not only by the individual researcher's natural urge to follow promising leads, but also by the recognition that new knowledge in these fields may well affect the well-being and even the lives of millions of people. In China, the high premium public policy puts on individual productivity and on the need for every ablebodied person, the "common man," to function effectively inevitably affects the stimulus and the priorities for research in clinical medicine and public health.

China's basic concept regarding medicine and health sciences was reiterated in January 1978 by Huang Jiasi, president of the Chinese Academy of Medical Sciences (CAMS): "The problems of prevention and treatment of diseases most frequently seen always top the list of medical services and medical research."[1] This concept has held true despite the state of flux in organizational relationships that began in 1949 and is emphasized to all visitors as still present. Even with shifts in political leadership and changes in contemporary priorities, the emphasis has remained on practical health priorities designed to improve the lot of the common people.

This chapter attempts to review, against a brief historical background, some general characteristics and illustrative examples of China's current medical research. The information comes from a variety of sources, chiefly

[1]Huang Jiasi, "Resolute Struggle for the Realization of Chairman Mao's Behest in the Creation of a New Medicine Combining the Essence of Both Traditional Chinese and Western Medicines," *Chinese Medical Journal* [English ed.], n.s., 4, 1 (Jan. 1978): 14.

reports in Chinese scientific journals, printed and oral reports of United States medical delegations and individual visitors, including my own visits, to the PRC, and the analytic books and papers of sinologists who have made health studies a special interest.[2] Hitherto, most visits have been brief and cannot provide the detail that comes from extended observation or actual work in a laboratory. Some recent delegations, however, have obtained valuable information and insights by concentrating on single subjects, such as cancer or herbal pharmacology.

Interpreting medical information is seriously complicated by the absence of epidemiological data on the incidence and prevalence of major diseases in China. Also lacking is a demographic base from which key health indices, such as infant mortality, can be calculated or estimated. Birthrates and death rates are available only for localized areas, without the underlying data, making it difficult to evaluate the impact of preventive and therapeutic measures. Although establishing a comprehensive reporting system may not be a priority, the routine for reporting all births and deaths to local authorities is already in place, and it would not seem an overwhelming problem to obtain mortality rates—important data for health practices and research. Similarly, it should be possible to establish a few representative areas for regularly gathering demographic and epidemiologic information.

## *HISTORICAL PERSPECTIVE*

Despite the lack of specific data, it is generally recognized that in 1949 China suffered from many acute and chronic infections and parasitic diseases, a high incidence of severe and debilitating malnutrition, and an infant mortality estimated at 200 per 1,000 live births, due chiefly to diarrhea and other environmentally related diseases.[3]

Although the new Communist government launched campaigns to reduce mortality and to control diseases like cholera, malaria, tuberculosis, and kala-azar, the Ministry of Public Health and China's 41,000 Western-trained physicians continued to put considerable stress on high-quality research and medical education that would compete with that in the West.[4] When Lampton analyzed the contents of *Zhonghua yixue zazhi (Chinese Medical Journal)* from 1955 to mid-1960, he found a distinct preponderance of articles dealing with uncommon diseases requiring elaborate and costly treatment.[5]

[2] Pi-chao Chen, *Population and Health Policy in the People's Republic of China* (Washington, D.C.: Smithsonian Institution, 1976); and D. M. Lampton, *The Politics of Medicine in China* (Boulder, Colo.: Westview Press, 1977).

[3] Zhu Futang, "Accomplishments in Child Health Since Liberation," *Zhonghua yixue zazhi (Chinese Medical Journal)*, 79 (Nov. 1959): 385.

[4] Chu-yuan Cheng, "Health Manpower," in M. E. Wegman, Tsung-yi Lin, and E. F. Purcell, eds., *Public Health in the People's Republic of China* (New York: Josiah Macy, Jr., Foundation, 1973).

[5] Lampton, *Politics of Medicine.*

Theoretical, sophisticated research continued to gain popularity, and the "pro-research" bias extended to heart and brain surgery. As infections came under control, research interest started turning to chronic diseases—a trend reflected in the 1958 founding of the Beijing Institute of Cancer Research. At the same time, the Ministry of Public Health paid more attention to improving urban health standards and strengthening county hospitals than to commune health services, which served the majority of the population. At first, this bias toward professionalism had general support. But as economic conditions improved and aspirations grew, discontent grew over the inequality of services for the peasants.[6]

In the view of some persons, including Chairman Mao, the whole medical establishment was becoming increasingly isolated from the people and their true needs. Mao took the devotion of so much effort to advanced research as evidence of lack of concern for the "little" man. On June 26, 1965, he attacked the "Ministry of Urban Gentleman's Health," and doctors who had read too many books:

> They work divorced from the masses, using a great deal of manpower and materials in the study of rare, profound and difficult diseases at the so-called pinnacle of science, yet they either ignore or make little effort to study how to prevent and improve the treatment of commonly seen, frequently occurring and widespread diseases. I am not saying that we should ignore the advanced problems, but only a small quantity of manpower and material should be expended on them, while a great deal of manpower and material should be spent on the problems to which the masses most need solutions.[7]

The ensuing Cultural Revolution (1966–69), with its attack on excessive intellectualism, seriously affected the medical research establishment. The Ministry of Public Health received virulent and cutting criticism for its elitist and overly professional attitudes. Medical research lost its priority status and suffered from lack of support, and opportunities for intellectual exchange were interrupted. Nevertheless, studies and collection of data apparently continued, but at a far slower pace, and the need to learn from observation and experiment was reiterated.

When medical schools began to reopen in 1970, their curricula had been drastically shortened to produce "fully" trained physicians in three rather than five or six years. Whatever the effect this restriction on education may have had on medical practice, it was a serious impediment in the development of trained research workers. Despite this, as China returned to relative stability, research was emphasized once more, particularly in the new contacts made with the West.

A new research dimension, consistent with Mao Zedong's injunction to combine Western and traditional medicine, was investigation by Western-trained doctors of all aspects of Chinese medicine, especially acupuncture, moxibustion, and the use of herbs.

[6] *Ibid.*

[7] Mao Zedong, "Directive on Public Health Work," in Stuart Schram, ed., *Chairman Mao Talks to the People* (New York: Pantheon, 1974).

It is worth noting that although Chairman Hua Guofeng and Vice-Premier Deng Xiaoping did not refer specifically to the medical field at the National Science Conference of March 1978, documents from the conference noted biomedical research as a key area. Quoting Dr. Huang further: "We must emphasize the study and research of basic medical sciences, without which clinical and preventive medicine cannot make a new leap forward."[8]

By 1979 the renewed emphasis on both basic and applied research was reflected in the requirement of at least five years of middle school prior to medical studies, extension of the medical curriculum to a minimum of five years, and the institution of a national medical admissions test, on which the academic grade became the key factor in admission. In 1979, 100,000 applicants were said to have sought the some 30,000 places in the 95 Western and 26 traditional medical schools. Academic performance while in medical school has been given far greater prominence.[9] Furthermore, the medical school attached to the Capital Hospital, a unit of CAMS, is aiming to produce teachers and researchers and will therefore require three years of general university work after middle school and before entering the medical program.

## *CHARACTERISTICS OF CURRENT MEDICAL RESEARCH*

Any collection and presentation of information that adds to human knowledge is, in broad terms, research and may have implications for immediate clinical use and future investigational studies. In many fields of science, observation and the recording and analysis of observations constitute almost the sole form of research. In medicine the definition is often narrower, with a tendency to think of research as primarily experimental innovation in prevention, diagnosis, or treatment.

Chinese medical literature suggests a broader concept of what medical research may include. For example, great emphasis is laid on integration of Chinese and Western medicine in education, in practice, and in research. Chinese medical papers continually refer to traditional theories of energy (*qi*), organ (*zangxiang*), meridian (*jingluo*), yin-yang, and the Five Elements—concepts that are not easy for the Western-trained medical scientist to comprehend. Yet, if the value of traditional Chinese medicine is to be assessed objectively, then the observed phenomena will have to be explained, not only to satisfy the laws of physics, chemistry, and biology, but also in terms of mental and emotional factors still little understood in the West.

It may surprise some that considerable space in Chinese medical journals is devoted to aspects of the environment that in the Western world would be thought of as the province of the engineer. One would hardly expect to find in the *Journal of the American Medical Association (JAMA)* such articles as

[8] Huang, "Resolute Struggle," p. 14.
[9] Huang Jiasi, personal communication, June 24, 1979.

recent ones in the *Chinese Medical Journal* on such subjects as high-temperature composting to remove infectious organisms from night soil, the larvicidal action of pisciculture, the modification of battery production to eliminate mercury pollution, and an air-water jet dust suppressor for coal mines. Not topics one would see in *JAMA* but clearly consonant with Chairman Mao's dictum that one should put prevention first.

Chinese medicine also differs from its Western counterpart through its philosophic orientation to the thought of Marx, Lenin, and Mao. Medical journals publish constant reminders to research workers to "serve the people," and research priorities are decided by political directives rather than by relative availability of research grant funds.

An interesting example of the constant interplay between politics and science in China may be seen in the highly successful mid-1960's campaign against syphilis.[10] In line with Chairman Mao's dictum of "putting politics in command," the campaign was based on a universally administered questionnaire, supported by vigorous propaganda, that discovered suspects in numbers that could be tested serologically; such testing would simply not have been feasible for the whole population. All this was combined with an active treatment program and a political attack on prostitution based on rehabilitation and job training rather than on imprisonment.

A classic weakness, in Western eyes, of most Chinese research on therapy is the absence of true controls, essential if one is to draw secure conclusions about the efficacy of a particular treatment. The Chinese belief that no remedy should be consciously withheld was explained by the Herbal Pharmacology Delegation:

> In revolutionary Chinese medicine nothing is ideally more sacred than the welfare of the patients, who are supposed to be kept totally informed about the medical measures being taken to cure them. It is thus politically and professionally immoral for a physician to do anything other than provide the best possible medical treatment with the full knowledge of the patient. Double-blind experiments are ruled out, and placebos may not be administered.[11]

The Steroid Chemistry Delegation noted that the lack of an epidemiological approach and the absence of controlled studies made it difficult to evaluate not only the effectiveness of a therapeutic regimen but also its association with adverse side effects.[12]

Instead of true controls, comparisons are often related to historical or regional experience. Such comparisons are difficult enough to interpret under

[10] Ma Haide, "With Mao Tse-tung Thought as the Compass for Action in the Control of Venereal Diseases in China," *China's Medicine*, 1 (Oct. 1966): 52–68.

[11] National Academy of Sciences, *Herbal Pharmacology in the People's Republic of China: A Trip Report of the American Herbal Pharmacology Delegation* (Washington, D.C.: Committee on Scholarly Communication with the People's Republic of China, 1976).

[12] Josef Fried, Kenneth J. Ryan, and Patricia Jones Tsuchitani, eds., *Oral Contraceptives and Steroid Chemistry in the People's Republic of China: A Trip Report of the American Steroid Chemistry and Biochemistry Delegation* (Washington, D.C.: Committee on Scholarly Communication with the People's Republic of China, 1977).

the best of conditions, but even more so when one does not know the research workers and few reports contain enough detail on the characteristics of the groups studied for one to be sure that sampling variations do not vitiate the conclusions drawn.

It seems evident that another reason for not doing properly controlled studies is lack of experience. To be sure, medical research everywhere has long suffered from inadequacy of controls. Only recently, in fact, with the development of "scientific medicine" in this century have medical students been taught not to accept "empirical" conclusions. What is sometimes forgotten, however, is that despite a lack of convincing evidence, empirical conclusions or interpretations based on judgment may in fact be quite correct. In the Chinese situation, where there is extensive dependence on primary-level personnel trained to do careful observation, there may be even more reason to utilize research that is not as thoroughly supported by strict evidence. Thus, despite shortcomings in research design and the inadequacy of data, there may well be nuggets of important findings in many of the reported studies.

One compensation is that reports on the effectiveness of different types of therapy may be made in conformity with criteria established at a national medical conference. Unfortunately, the criteria are not generally available in the published literature, and they cannot be compared easily with similar statements from other countries.

A corollary problem concerns the use of statistical methods in medical research. During the Cultural Revolution, "statistics" came in for a particularly hard time. Visitors have been told that wall newspapers put up during the Cultural Revolution denounced excess use of analytic and statistical techniques and decried the emphasis on their study and utilization. As a consequence, research design suffered severely, and only in the last few years have detailed quantitative tables and analyses begun to appear once again.

## *RESEARCH FACILITIES AND RESOURCES*

Every medical delegation has had contact with China's major base for medical research, the Chinese Academy of Medical Sciences (CAMS), and the Chinese Academy of Traditional Medicine (CATM), which run numerous research institutes covering everything from antibiotics to tuberculosis. Other institutions, however, are also concerned with medical problems. The Herbal Pharmacology Delegation found, for example, that the biochemistry and analytic chemistry departments of the Institute of Organic Chemistry of the Chinese Academy of Sciences (CAS) was studying the structure and function of chemicals used as medicinal drugs.[13]

In addition to the research institutes organized at the national level, many provinces have their own establishments, working sometimes on local prob-

[13] NAS, *Herbal Pharmacology.*

lems. Research is also carried out by military hospitals, brigade stations, people's hospitals, medical schools, and other organizations. This diversity has advantages, but can lead to unnecessary repetition and duplication.

Although the Chinese treat organizational charts rather casually, with much de facto decentralization, the way some institutions are organized may give an idea of what are considered major subject areas for research. The Institute of Materia Medica in Beijing, visited by several delegations, has a staff of 300, organized in seven major departments: synthetic organic chemistry, phytochemistry, pharmacognosy, analytical chemistry, pharmacology, antibiotics, and cultivation of medicinal plants. The Institute of Chinese Materia Medica of the CATM has four research areas: identification of herbs, drug forms and processing, chemistry, and pharmacology. The Shanghai Institute of Experimental Biology, originally a laboratory of developmental physiology with a staff of 20, had in 1977 over 300 research workers in four laboratories: cancer etiology and prevention, with three subgroups—tumor immunology, nucleic acid research, and etiology; reproductive physiology, in particular steroid contraceptives and the effects of Chinese herbs and medicines on the reproductive tract; DNA structure and function, chromosome structure and function, and organic synthesis of nucleic acid components; and new techniques such as radio isotopes, ultrastructure, radioimmunoassay, and development of new types of equipment.

While the brevity and superficiality of contacts with Chinese medical research make any real assessment of quality impossible, the Herbal Pharmacology, Cancer, and Rockefeller University Delegations were quite impressed with many of the scientists they met and disappointed with others. Some laboratories were poorly equipped, with little evidence of productive activity, but several institutions had excellent equipment, on a par with facilities in the United States. A few laboratories were "near worldwide standards."[14] Some laboratories had up-to-date Japanese or other foreign equipment, and others had excellent Chinese-made apparatus. Shortage of advanced equipment, especially computer facilities, led to much sharing among laboratories. The Rockefeller University Delegation, which had a broad sampling of different kinds of research, reported a generally favorable impression of the institutions they visited, despite a number of low spots.

In recognition of the fact that medical research often involves much more than a single institution, a series of research coordinating groups have been established in such key areas as cardiovascular diseases, traditional anesthesia, child development, and cancer.[15] These groups are responsible for coordinating and improving research and control work done throughout the PRC. Given China's communication problems, it is not clear how effective the coordination is, but the goal is not only praiseworthy but potentially of

[14] Rockefeller University, *Report of a Visit to the People's Republic of China* (New York: Rockefeller University, 1977).

[15] Henry S. Kaplan and Patricia Jones Tsuchitani, eds., *Cancer in China* (New York: Alan R. Liss, 1978).

great significance as an example for other countries. Institutionalization of the process, so long as it does not discourage individual initiative, may reduce unnecessary duplication and encourage cooperative activity.

## MEDICAL PUBLICATIONS

In 1965, there were at least six journals in specialty fields of clinical medical practice besides the *Chinese Medical Journal,* which was started in 1887. There were also journals in closely related fields such as physiology and pharmacology and general scientific journals in zoology, genetics, and other basic sciences. Suspended during the Cultural Revolution, these publications have reappeared slowly. The *Chinese Medical Journal,* for example, resumed publication in 1973, as a Chinese monthly with English abstracts; a bimonthly English version started in January 1975. The *Zhonghua neike zazhi (Chinese Journal of Internal Medicine)* was reestablished in 1975 and the *Zhonghua waike zazhi (Chinese Journal of Surgery)* in 1977, both bimonthly Chinese publications. The *American Journal of Chinese Medicine* was started in the United States in 1973; it carries occasional papers from China.

Readers accustomed to Western research journals must do some adjusting to understand the Chinese scientific press. Rhetorical statements about revolutionary goals, Marxist dialectics, and Mao Zedong thought are frequent, but, as in Ma's paper on syphilis, can be an essential part of a scientific presentation. In general, papers describe reasonably well the broad goals of the investigation, the materials studied, and the tools used. There is considerable dissimilarity in the presentation of details, statistical or otherwise, and little uniformity in the way material is described and analyzed. Brevity is a common characteristic—often frustratingly so when essential details are omitted.

There is also unevenness in the way statistics are presented and used. Gross numbers and the total scope of study are given, but essential information is often provided only in averages or percentages. Conclusions are frequently stated in unassessable generalities. Even when probabilities are discussed, it is unusual to find raw data that a reader can analyze or interpret for himself. Percentages and ratios are given to two decimal places, implying exaggerated precision. Data on group comparability, when available, are usually inadequate. For example, most papers contain basic information on the subjects covered and state that they are from a particular population. Too often, however, descriptive data on the populations are insufficient to assess representativeness. This paucity of specific information is reflected in the format of tables and numerical material. An occasional article is backed up by a master table, but in general only partial information is given, and cross-checking and comparison are difficult. References are generally given carefully, and there are a fair number of citations of foreign literature.

Delegations and individuals who visited China in the 1970's have reported that important Western journals were available in research institutions. Although there is disagreement about how familiar Chinese medical scientists are with current world medical literature, many delegations met scientists who knew a good deal about the foreign literature in their particular field. For example, the Cancer Delegation commented on their hosts' frequent references to significant current publications on basic cancer research, and the Herbal Pharmacology Delegation found several of the groups they visited well versed in their field.[16] A study by the University of Lund of the opinions of 144 scientists who visited the PRC between 1971 and 1974, 44 of whom were in the medical field, found that the majority of respondents thought Chinese knowledge of world science and technology was as good as that of comparable workers in their own countries.[17] It is evident that some of the familiarity is selective, as in the West—certain institutions have well-read scientists and other institutions have staff knowledgeable only in a specific area.

A rather significant difference between Chinese and Western medical research articles is that the former frequently combine reports of research results with a discussion of clinical implications and recommendations on patient management. The lack of such recommendations is a frequent complaint by American practitioners about Western academic journals. Some fear this practice may give unwarranted and premature significance to the research. On the other hand, it may well lead more practitioners to read and criticize the research itself.

## *SCOPE OF RESEARCH ACTIVITY*

Any attempt to analyze the scope of the PRC's medical research effort is complicated by the scarcity and unevenness of information, mentioned earlier. Attempts such as Kao's[18] to do a systematic analysis face the problems of uncertainty in categorization and difficulty in deciding what is research. This chapter accordingly attempts only to appraise a few key areas and even then not in real depth. Individual experiences are cited not to give numbers and details but to illustrate how data may be presented and used by various research institutions.

### *Infectious and Parasitic Diseases*

As noted earlier, infectious and parasitic diseases were a major concern immediately after 1949. The Chinese directed their efforts to improving the environment and performing immunizations, giving attention chiefly to the characteristics of severe infectious diseases and to the use of Chinese drugs in therapy. One important area of concentration was on acute fulminating

[16] *Ibid.;* and NAS, *Herbal Pharmacology.*

[17] University of Lund, *Lund Studies of China's Science and Technology* (Lund, Sweden: University of Lund, Research Policy Program, 1975).

[18] F. F. Kao, *Respiratory Research in the People's Republic of China* (Bethesda, Md.: John E. Fogarty International Center, 1975).

diseases characterized by shock and circulatory collapse. Chinese research looked at microcirculatory changes as revealed through microscopic studies of the nail fold.[19] The drug anisodamine, extracted from *Datura,* appears to have favorable results in diseases like fulminant meningitis and toxic dysentery showing these changes.[20]

Many infectious diseases, however, were still prominent in the 1970's. *Hepatitis* continues to excite great interest in China, as elsewhere; many questions are directed at visitors on epidemiology and treatment, and the subject is frequent in the literature. For instance, the March 1977 issue of the *Chinese Journal of Internal Medicine* published (1) a call for an intensified research attack on the disease; (2) an army hospital report on a blood donation project used to make a detailed antigen survey of 14 families in a rural village[21] (the variations observed in the ratios between cases and asymptomatic carriers suggested that the simple anal-oral route of transmission could not explain all findings or answer the questions they raised); and (3) an article from an infectious disease hospital which concluded that the Plant Hemagglutinin Agent skin test was a valuable tool for diagnosis and selection of cases for various therapies.[22]

Many papers in the 1970's dealt with various forms of therapy for hepatitis, most frequently a combination of Western and traditional medicines. A fairly vigorous search for an effective hepatitis vaccine was also under way. An HB vaccine from the sera of antigen carriers was tested on guinea pigs and rabbits.[23] Inoculation of human volunteers produced low immunity, with no toxic effects. A Beijing army hospital carried out studies on epidemiology, transfer of hepatitis B in the family, and prevention, using gamma globulin and vaccine. It used traditional Chinese drugs for improvement of nonspecific immunity and/or elimination of HBcAG, but an ideal drug or combination of drugs had not been identified.[24]

*Influenza and the common cold.* Both are prevalent in China, as elsewhere. Several visitors have commented on the relatively undeveloped state

[19] CAMS, First Research Laboratory of an unnamed institute; Beijing Friendship Hospital, Department of Pediatrics; and CAMS, Institute of Materia Medica (Beijing), Department of Pharmacology, "Changes in Nail-Fold Microcirculation and Amines in Diseases Manifesting Acute Microcirculatory Disturbances," *Chinese Medical Journal* [English ed.], n.s., 1, 3 (May 1975): 216–24.

[20] Beijing Friendship Hospital, Department of Pediatrics, "Anisodamine in Treatment of Some Diseases with Manifestations of Acute Microcirculatory Insufficiency," *Chinese Medical Journal* [English ed.], n.s., 1, 2 (Mar. 1975): 127–32.

[21] People's Liberation Army, 269th Hospital, Hepatitis Prevention and Treatment Group, "B Type Hepatitis Antigen Survey of 14 Families in a Rural Village," *Zhonghua neike zazhi (Chinese Journal of Internal Medicine),* 1977, 2 (Mar.): 75.

[22] Su Sheng and Cui Jenyou, "Value of Clinical Application of Plant Hemagglutinin Agent (PHA) Skin Test in Viral Hepatitis," *Chinese Journal of Internal Medicine,* 1977, 2 (Mar.): 82–86.

[23] Tao Jimin et al., "A Preliminary Study on Hepatitis B Vaccine," *Chinese Medical Journal* [English ed.], n.s., 4, 2 (Mar. 1978): 101–10.

[24] Jia Keming, "Outline of Progress in B Type Viral Hepatitis Research in Recent Years," *Chinese Journal of Internal Medicine,* 1977, 2 (Mar.): 112–15.

of virology in China, and the papers on the subject vary widely from the sophisticated to the simple, from how to establish observation posts to the proper technique for fluorescent antibody studies. A virologic study of common colds reported in 1974 used techniques prevalent in the United States in the 1960's.[25] Primary human embryo kidney cells, a resource no longer available to American scientists, were used for inoculation. The study was straightforward, the electron microscope pictures very good, and the findings of some 90 different strains of respiratory viruses, with serologic confirmation of the general prevalence of neutralizing antibodies to coronavirus, useful.

Some attention has been paid to influenza immunization, but there is more interest in using sprays to prevent the disease. In one county where influenza A was prevalent from April to June 1976, rooms occupied by students and workers were sprayed twice a week with either ganitheria mist or ilex oil; a third group was observed as control. Incidence of influenza was 11.5 percent for ganitheria, 15.3 percent for ilex, and 46.8 percent for the control.[26] Since similar experiments in the West have not been successful, one wonders about the comparability of the controls.

*Enteric infections.* Although enteric infections are a decreasing problem in morbidity and mortality, they continue to arouse interest. Case reports of typhoid fever describe therapy based on antibiotics and Chinese herbs. Bacillary and amebic dysentery receive more attention, particularly on the management of shock.

Papers on parasitic diseases include ancylostomiasis, filariasis, amebiasis, clonorchiasis, paragonimiasis, acanthocephalosis, and sparganosis. They deal chiefly with prevention, usually through environmental sanitation, and treatment with traditional medicines. A problem developed with the latter—25 cases of sparganosis were reported to have spread through the common use of raw frog's flesh as a poultice for sore eyes.[27]

*Schistosomiasis.* This disease continues to be of great concern to the Chinese, as it has been for many decades. The major focus of prevention—breaking the chain of transmission—is environmental and is discussed below.

The Herbal Pharmacology Delegation, which visited China in 1974, reported that an extensive screening program had identified some antimonial compounds (antimony ammonium gluconate, antimony sodium dimercap-

[25] People's Liberation Army, Hou No. 59170 Unit, Common Cold Prophylactic and Therapeutic Group, "Virologic Study of Common Colds in Peking, 1971–1973," *Chinese Medical Journal* [English ed.], n.s., 2, 4 (July 1976): 279–84.

[26] Tian Qiwei et al., "Prevention of Cold and Influenza with Ganitheria Mist and Ilex Oil Mist," *Chinese Journal of Internal Medicine*, 1977, 2 (Mar.): 126.

[27] Hainan College of Specialized Medicine: Ophthalmology Teaching and Research Group and Parasitology Teaching and Research Group, "Clinical Observations and Investigation of Epidemiologic Factors of 25 Cases of Sparganosis," *Chinese Medical Journal*, 1977, 6 (June): 343.

tosuccinate [Sb-58], and thiouracil-antimony-1) and some medicinal plants (*Cucurbita pepo* [pumpkin seed], *Hemerocallis thunbergii, Lobelia radicans, Euphorbia pekinensis,* and *Veratrum nigrum*) with some potential.[28]

The Schistosomiasis Delegation, which visited China in April 1975, came away with a negative impression of research progress.[29] The Delegation did not see any laboratories in operation. They had the impression that reviews of the effectiveness of Chinese drugs had mostly stopped during the Cultural Revolution, and that attention had turned to imported drugs.

The literature, however, continues to show substantial interest in schistosomiasis and the use of traditional drugs to complement antimony compounds. Studies on the biology of schistosomiasis range from those on the life cycle of the organism to the serological status of infected human beings. A paper on *in vitro* egg cultivation gives an idea of the kind of research, the tools used, and results reported:

> Under adequate conditions, the eggs of *Schistosoma japonicum* can be cultured *in vitro* without undergoing the host-tissue phase . . . the entire course of development ranging from 12 to 13 days. Serum and RBC are indispensable in the culture medium. . . .
>
> Systematic study of cell division and egg maturation was made by phase microscopy . . . [through] four stages: (1) single-cell; (2) cell-cleavage; (3) embryonic development; and (4) maturation and miracidium formation. . . .
>
> The present study shows that any mature female worm obtained by autopsy from an experimental host is able to lay eggs *in vitro,* whether it is still embraced by a male worm or already separated from the latter after copulation. The number of eggs laid *in vitro* is influenced by the type of culture medium used and the viability of the female worm.[30]

Studies on mice and rabbit tissue compared *S. japonicum* eggs dying naturally with those dying as a result of drugs, using fluorescence staining to distinguish dead and living eggs in host tissues.[31] A human study, noting that serum gamma globulin and IgG levels were highest in late-stage schistosomiasis, led to the inference that Chinese drugs capable of improving cell-immune mechanisms may be important in comprehensive treatment.[32] This theme, that Chinese drugs often help through supporting homeostatic mechanisms, recurs in connection with many diseases. One gathers that schistosomiasis is still a research priority.

[28] NAS, *Herbal Pharmacology.*

[29] "Report of the American Schistosomiasis Delegation to the People's Republic of China," *American Journal of Tropical Medicine and Hygiene,* 26 (1977): 427–62.

[30] Xu Shie, "Culture of *Schistosoma japonicum* in Vitro, with Special Reference to Production and Development of Eggs," *Dongwu xuebao (Acta Zoologica Sinica),* 20, 3 (1974): 231–39.

[31] Yan Zizhu, Jiu Lizhu, and Xue Haizhou, "The Development and Changes of the Eggs of *Schistosoma japonicum* in the Tissues of Animals," *Acta Zoologica Sinica,* 20, 3 (1974): 263–71.

[32] Suzhou College of Medicine, First Hospital, Department of Infectious Diseases, "Preliminary Observation of the Immune Mechanism of Victims of Schistosomiasis," *Chinese Journal of Internal Medicine,* 1977, 5 (Sept.): 278–81.

*Heart, Circulation, and Related Systems*

These subjects have occupied more space in the *Chinese Journal of Internal Medicine* than any other subject except infectious diseases.

Disorders of the conductive system and improved techniques of electrocardiography, including vector cardiography, were discussed frequently throughout the 1970's, and interest in lipids was high. Animal experimental work included a controlled study purporting to show that feeding high-cholesterol diets to domestic rabbits can produce coronary atherosclerosis.[33] Human studies in the same field were reported frequently, like one from Sichuan: "In a study of 543 normal adults, 21 patients with hypercholesteremia, and 35 patients with coronary diseases, hypercholesteremia occurred among those whose fat intake was 8.4–10.6 percent of total calories. A portion of serum triglycerides, cholesterol, and free fatty acid must originate as a conversion of sugar metabolism."[34]

A survey of nomadic herdsmen in Inner Mongolia showed higher levels of cholesterol and lower levels of trigylceride than elsewhere. The differences were attributed to the local diet, which is high in protein and fat and low in carbohydrates.[35]

A number of reports dealt with open heart surgery as well as other surgical approaches to the regulatory and circulatory system.

Concern with preventive techniques was particularly manifest in organized programs for reducing hypertension in factories. It should be noted that the work place is a convenient locus for health research and care in China, since health personnel at factories provide general health services to employees and are not limited to treating work-connected conditions. In two factories, for example, a controlled study, although not double-blind, was possible because an antihypertension program had been started in one factory three and one-half years previously. Through a detailed analysis of the subjects' age, sex, occupational status, and prevalence of essential hypertension or coronary heart disease, the relative occurrence of complications was compared in the two factories, with the conclusion that the prevention program had been successful.[36]

[33] Zhejiang Medical College: Pathology Teaching and Research Section, Coronary Heart Disease Unit, and Cardiovascular Disease Research Laboratory, "Characteristics of Experimental Coronary Atherosclerosis in Domestic Rabbits," *Chinese Medical Journal*, 1977, 5 (May): 292–93.

[34] Sichuan College of Medicine: Biochemistry Teaching and Research Group and Department of Internal Medicine, Cardiovascular Group, "A Study on Sugar and Fat Metabolism in Hypercholesteremia and Coronary Diseases. I: Sugar Tolerance Test and Fat Tolerance Test," *Chinese Journal of Internal Medicine*, 1977, 5 (Sept.): 294–97.

[35] Xia Huiming et al., "A Survey of Herdmen's Serum Lipids in Inner Mongolia," *Chinese Medical Journal* [English ed.], n.s., 3, 5 (Sept. 1977): 343–46.

[36] Chaoyang Hospital, Beijing, Department of Medicine, "Prevention and Treatment of Essential Hypertension and Coronary Heart Disease in Factories," *Chinese Medical Journal* [English ed.], n.s., 3, 3 (May 1977): 157–62.

*Cancer*

A cancer delegation visited China in September and October 1977. Its careful and detailed report covering clinical practice, public health procedures, and research is my main source of information.[37]

The epidemiologists on the Cancer Delegation found a special excitement in the regional epidemics of cancer in China. Nasopharyngeal and hepatocellular carcinoma occur excessively in the South and esophageal cancer in certain areas of the North. For example, in Linxian county, Henan province, esophageal cancer has an annual incidence as high as 140 per 100,000, which is about 50 times greater than the normal world rate. The environmental origin of these cancers is suggested by their regionality, their lower incidence among Chinese who have migrated to nonepidemic areas, and the parallels between human and animal cancers observed in affected regions.

Since animals have narrower environments than human beings and shorter latent periods in developing cancers, they provide an excellent means of identifying potential carcinogens. This was dramatically illustrated when 50,000 Linxian residents moved to an area 360 miles away. Although they took only one chicken with them, 12 of their 5,484 chickens in the new area developed gullet cancer, in five instances associated with esophageal cancer in owners or neighbors. No cancer occurred among chickens owned by natives of the area. The leading etiological suspect was identified as a pickled vegetable mix, a Linxian specialty, which was fed to the chickens along with other food scraps.

In seeking specific environmental chemicals that might be involved in induction of esophageal cancer, Chinese investigators have considered trace elements, nitrosamines, and fungal contaminants, as well as personal eating habits. The Beijing Institute for Cancer Research, in studying high-incidence areas, has implicated low molybdenum content in the environment as shown by serum, urine, and hair samples. Other studies have suggested that excess production of nitrosamines in foods and water may be related to the low molybdenum content. Samples of grains and pickled vegetables have been compared for nitrosamine content by thin layer chromatography. Of 124 samples from Linxian county, 23 percent were "positive" for nitrosamines, whereas only one out of 86 samples (1.2 percent) from Fanxian county, which has little esophageal cancer, were positive. It is suspected that both nitrosamines and unidentified fungal products are the carcinogenic factors.[38]

Many studies have been carried out on fungi as etiologic agents, but the Cancer Delegation was not impressed with any of the conclusions. Studies on Epstein-Barr virus (EBV) in nasopharyngeal cancer confirmed results reported in the West.

There have not been vigorous studies of the association of hepatitis virus and cancer. Nor have potentially important etiologic and preventive studies

[37] Kaplan and Tsuchitani, *Cancer in China.*
[38] *Ibid.*

on smoking, radiation, and occupational hazards been conducted. Although a high proportion of lung cancer patients are smokers (mortality doubled between 1964 and 1975), no studies have been published on this subject or on the association of lung cancer with workers in coke ovens who are exposed to high concentrations of polycyclic aromatic hydrocarbons; 82 percent of coke oven workers with lung cancer are smokers, suggesting a possible combined effect. The association of mesotheliomas with asbestos has not been studied either. Furthermore, extensive use has been made of x-ray and fluoroscopy in screening programs in factories and rural areas, apparently without considering carcinogenic effects.

The Cancer Delegation found unevenness in cancer research. For example, work on nasopharyngeal cancer was often incomplete and was less systematic than that on esophageal cancer. Similarly, the numerous approaches to carcinogenesis were not well integrated. Although interesting data on leukemia were available—notably that chronic lymphatic leukemia, which constitutes 30 percent of adult leukemia in the United States, constitutes about 1 percent in China—epidemiological studies of leukemia were purely descriptive. On the plus side, the Chinese have been systematically preparing maps of site-specific cancer mortality; as of 1977, six had been completed for Jiangsu province, for lung, liver, stomach, and cervical cancers, leukemia, and all cancers combined. The work parallels similar projects in the United States and should provide some extremely interesting comparisons.

The Chinese believe that cancer has multifactorial interacting causes, susceptible to a combined approach without necessarily identifying specific etiologies. The concept has merit but may well divert attention from promising etiologic leads.

Mass surveys and systematic screening may be key techniques in the control and prevention of many forms of cancer. Although the Chinese have paid little attention to measuring the relative utility of surveys, screening, and other methods of case finding, they have improved mass examination methods and adapted them to unskilled hands. In southern China, for example, 600 barefoot doctors screened 436,786 persons by nasopharyngoscopy, increasing the proportion of stage one and two cancers found from 24 percent (in 1970) to 87 percent (in 1975), as well as identifying some patients with nasopharyngeal hyperplasia.[39] Similarly, barefoot doctors in Linxian county have screened for esophageal cancer by using a mesh-covered balloon attached to a single or double lumen tube, which is passed into the stomach. The proportion of carcinoma found in those surveyed was 1.2 percent, of which 71 percent were in early stages and amenable to therapy.[40]

[39] *Ibid.;* and Henan Province and CAMS: Coordinating Group for the Research of Esophageal Carcinoma, "Studies on Relationship Between Epithelial Dysplasia and Carcinoma of the Esophagus," *Chinese Medical Journal* [English ed.], n.s., 1, 2 (Mar. 1975): 110–16.

[40] Henan Province and CAMS: Coordinating Group for the Research of Esophageal Carcinoma, "The Early Detection of Carcinoma of the Esophagus," *American Journal of Chinese Medicine,* 2, 4 (Oct. 1974): 367–74.

Mass screening has been particularly successful in attacking cancer of the uterine cervix. In 1958, 66.1 cases were found per 100,000 examinations, but by 1976 the figure had dropped to less than 17. Late stages had also become less common; when 79,348 women were screened in Shanghai in 1975–76, none were found.

The usual forms of cancer therapy—surgery, radiation, and chemotherapy—are widely and effectively used in China. In addition, extensive use is made of Chinese herbs and traditional medicines, usually to complement Western drugs, either for general support or, in a few instances, for supposed antineoplastic effect. Little evidence is available to document the effects of combining Chinese and Western drugs. Some of the reported results seem unusually good, particularly the long-term survival rates after surgical removal of esophageal carcinoma, both in early cases *in situ* and in those with advanced lesions that have penetrated the esophageal wall. An interesting new chemotherapeutic agent isolated in China is the harringtonine group of plant alkaloids from *Cephalotaxus* species. On the basis of pharmacologic analysis and animal trials, it is said to be effective against brain tumor and leukemia.[41]

Cancer, because of its chronicity and complexity, is one area where the problem of evaluating therapy, whether it be surgical or medical, becomes particularly difficult. Group comparability is a vital component of meaningful research, but as noted earlier, the Chinese have not appreciated the importance of controlled, random clinical trials. Abundant data are often available, but the fact that groups are not comparable from a prognostic standpoint makes true evaluation virtually impossible. The Cancer Delegation thought the climate was right for a change in attitude on this important aspect of therapeutic research.

### *Respiratory Disease*

The field is of considerable interest because of the prevalence of chronic bronchitis in China. The editor of the *American Journal of Chinese Medicine* categorized the research under way in 1975 under a number of headings: pulmonary, anatomical, and functional geometrics; kinematics of the lung system; lung dynamics; gas exchange and gas diffusion in the pulmonary system; regulation of ventilation; pulmonary function testing; instrumentation; smoking and air pollution; altitude and hyperbaric physiology; acupuncture and the respiratory system; and clinical and public health concepts, including respiratory drugs.[42]

China is in a unique position for studies of high-altitude physiology. More than 25 percent of the country lies at an altitude of over 3,000 m and more than 10 percent over 5,000 m, the altitude of Mont Blanc. Most of the at-

[41] CAMS, Institute of Materia Medica, Department of Pharmacology, "The Antitumor Effects and Pharmacologic Actions of Harringtonine," *Chinese Medical Journal* [English ed.], n.s., 3, 2 (Mar. 1977): 131–36.

[42] Kao, *Respiratory Research*.

tention, from the medical standpoint, has been given to pulmonary edema and cor pulmonale. Papers on these subjects and on chronic bronchitis continued to appear regularly in the Chinese literature of the late 1970's. One study found that the rheumatoid factor was present in the serum of chronic bronchitis patients in significantly higher proportion than in that of normal people.[43]

Several papers deal with specific causative agents of lung disease, such as tuberculosis and fungi. There has been much interest in using ultrasound as a diagnostic tool. Therapeutically, most papers discuss Chinese drugs, although hyperbaric oxygen, not well thought of in the West, has been used for a variety of patients.

An enormous and almost bewildering variety of drugs for respiratory disease have been tested. Some 150 Chinese herbs for bronchitis are said to have yielded special promise, including Manshanhong, Duzhuanha, and Aidizha, although, as usual, lack of controls makes their evaluation difficult. One drug under study at the Shanghai Institute of Materia Medica is Hancai (*Roripa montana*). Two related compounds, rorifone and rorifamide, were isolated before the Cultural Revolution. They have been synthesized and used with an "81% effective rate," although no controls are mentioned.[44]

### Surgery

From Chinese publications and visitors' reports, it seems that surgical practice covers the whole gamut of what goes on in the West. The literature covers the more common surgical procedures, ranging from individual case studies to analyses of group experience. Considerable space is given to rare and difficult operations, such as open heart surgery, complex brain operations, and limb reimplantation. Chinese surgeons have pioneered in reattaching limbs; almost all visitors have seen actual cases or films of major reimplantations. In the reports, much is made of the social importance of such operations in restoring workers to the labor force. Basic research has concentrated on developing techniques for reuniting severed tendons, nerves, and blood vessels.

The experience of the Sixth People's Hospital in Shanghai gives some idea of the magnitude of China's efforts in reattaching limbs.[45] From January 1963 to June 1976, the hospital had 438 cases of reimplanted limbs or digits, an average of about three a month. In 301 cases severance was complete. The overall survival of reimplanted limbs was 83.2 percent. After the introduction of microsurgical techniques in 1973, the survival rate of digits improved to 92.3 percent.

[43] Shanghai Chronic Bronchitis Etiology, Prevention, and Treatment Coordinating Group, "Preliminary Study on the Relation Between Allergy and Chronic Bronchitis," *Chinese Medical Journal* [English ed.], n.s., 2, 1 (Jan. 1976): 63–68.

[44] Kao, *Respiratory Research*.

[45] Shanghai Sixth People's Hospital, "Extremity Replantation," *Chinese Medical Journal* [English ed.], n.s., 4, 1 (Jan. 1978): 5–10.

Plastic surgery, particularly for severe burns, is also prominent in China, and cryosurgical tools for use in a variety of operations have been produced. Novel approaches include using human skin as a filler for the perforated area in surgery of detached retina and a garlic slice to close a perforated eardrum.[46]

Several papers make a case for combined Western and Chinese medical therapy instead of surgery for acute appendicitis and other abdominal diseases, claiming that end results are not appreciably different from those achieved by early surgery.[47] As usual, insufficient data make objective evaluation impossible.

The *Chinese Journal of Surgery* resumed publication in the fall of 1977, with papers on such subjects as lymphocyte reaction and prognosis in breast cancer, and the relation of the cutaneous nerves at the surgical site to the success of acupuncture anesthesia.

### *Orthopedics*

The treatment of fractures is often cited as an important area for combining Western and traditional medicine; ancient bonesetting techniques are now skillfully combined with diagnostic x-rays and fluoroscopy. Once a fracture has been reduced, the limb is bound in herb-soaked bandages and the fractured bone held in position with a relatively flexible splint. Although visitors have been told that there is slippage and nonunion in a small number of cases, the overall results are considered worthwhile in terms of less disuse atrophy and more rapid return to work. The papers describing the techniques used do not attempt to quantify successes and failures, and there have been no studies comparing results with those of the rigid splinting and immobilization of joints practiced in the West.

There is considerable interest in diagnosis and therapy of protruding and ruptured intervertebral disks. Many papers on the subject describe the various techniques for manipulation and traction, including ancient techniques of hand pressure on the spine. The theoretical base for the particular treatment recommended is often analyzed. One paper cites a Qing dynasty source as the basis for a modern physical formula.

There is also great interest in arthritis and concern with various forms of physiotherapy and the use of spas. Postpoliomyelitic paralysis has been treated by burying catgut in acupuncture points; before and after pictures were presented as evidence of effectiveness. It is interesting that a report on

[46] Wuhan College of Medicine, First Hospital, Department of Ophthalmology, "Application of Human Skin in the Surgery for Detachment of Retina," *Chinese Medical Journal*, 1977, 9 (Sept.): 567–68; and Xu Weizheng, "Garlic Slice in Repairing Eardrum Perforation," *Chinese Medical Journal* [English ed.], n.s., 3, 3 (May 1977): 204–5.

[47] Nankai Hospital, Institute for Acute Abdominal Diseases, and Zunyi Medical College, Acute Abdominal Condition Research Group, "Treatment of Acute Abdominal Diseases by Combined Traditional Chinese and Western Medicine," *Chinese Medical Journal* [English ed.], n.s., 4, 1 (Jan. 1978): 11–16.

this technique for periarthritic inflammation of the shoulder was by a barefoot doctor.[48]

### *Acupuncture*

Acupuncture is the medical subject that has perhaps aroused greatest interest in Western research workers, and extensive studies on this fascinating subject are going on in many countries. For detail that cannot be incorporated into this brief review, the interested reader is referred to two readily available studies: the Report of the Acupuncture Anesthesia Study Group and that of the Fogarty Center.[49] The following is an overview of Chinese acupuncture research reported from within China in the 1970's.

Long before the PRC was established in 1949, the Western world, scientific and lay, was intrigued by reports of the use of acupuncture in relieving pain, especially in musculoskeletal disorders and arthritis, and in treating a broad variety of ills. Despite thousands of years of use, however, the evidence about specific effects was too scanty to persuade Western physicians to treat acupuncture other than as a curiosity.

Acupuncture consists of the insertion of fine needles (0.4–0.28 mm in diameter and 1–10 cm in length) into one or more of the 600 or so acupuncture points on the body's various meridians. The needle is advanced until "*deji*" (tingling, distension, heaviness, numbing) is experienced by the patient. The needle point is stimulated by manual rotation or electrical impulse. Acupuncture points may also be stimulated by moxibustion, burning a wick of floss, *Artemisia vulgaris,* on the point or by the ancient technique of cupping, which until recently was widely practiced in the West as a local circulatory stimulant.

In the mid-1950's medical workers in Xian, Shanghai, and Hebei province, stimulated by success in controlling the pain of sore throat and toothache with acupuncture, initiated the use of acupuncture as a pain deadener for related surgical procedures. Apparently, the first operation with acupuncture was a tonsillectomy. The use of acupuncture as an anesthetic, or more precisely, an analgesic or hypalgesic, spread rapidly. It was used increasingly in both simple and complicated procedures. Reputable eyewitnesses attested to the validity of the claims.

Four major theories attempt to explain acupuncture.[50] The meridian theory is based on the belief that there is a free flow of *qi* (vital energy) through the meridians of a healthy body. This flow is governed by the interplay of two opposing forces, the yin (negative) and the yang (positive). Disease or

[48] Gao Shuhua (Barefoot Doctor), "Burying Catgut in Acupuncture Points for the Treatment of Periarthritic Inflammation of the Shoulder," *Chinese Medical Journal,* 1977, 4 (Apr.): 216.

[49] NAS, *Acupuncture Anesthesia in the People's Republic of China: A Trip Report of the American Acupuncture Anesthesia Study Group* (Washington, D.C.: Committee on Scholarly Communication with the People's Republic of China, 1976); and J. Y. P. Chen, *Acupuncture Anesthesia in the People's Republic of China* (Bethesda, Md.: John E. Fogarty International Center, 1973).

[50] NAS, *Acupuncture Anesthesia.*

pain results from their imbalance. Needling appropriate points on the meridians may restore balance and cure the ailment or relieve the pain.

A second, neurophysiologic theory is built around the concept of "gate-control" of pain: acupuncture effectively is thought to prevent pain sensations from reaching the brain. This has now been modified to a "multiple-gate" theory, which suggests that stimulation of other areas, e.g. the thalamus, the lower brain stem, or mid-brain, might be similarly inhibitory. The effect is lessened by such factors as excitement or increase in blood pressure. Working on this theory, the director of Shanghai's Institute of Physiology, Zhang Xiangtong, showed that neurons in certain areas of the thalamus discharged characteristic patterns of impulses in response to "painful" stimuli.[51] These responses could be abolished by morphine and were diminished significantly by electrical needling of certain acupuncture points.

The humoral theory holds that certain chemical agents of an analgesic nature are produced by acupuncture. The pain threshold of rabbits has been increased by injection of subarachnoid fluid or blood from rabbits treated with acupuncture; Western research on endorphin production may be relevant.

Finally, the theory that pain relief is due to hypnosis and conditioning is not under investigation in China.

The early numbers of the *Chinese Medical Journal* were full of papers about the physiology and techniques of acupuncture. By the 1970's the number of papers had dropped off, but a good deal of research was still under way. Two coordinating groups were active, and the Shanghai Institute of Physiology was trying to answer such questions as:

Which receptors does needling stimulate?
Where do resulting afferent impulses go in the thalamus?
Does direct electrical stimulation inhibit the pain responses in nearby areas of the thalamus?
How do general levels of excitation of the brain affect acupuncture analgesia?
Why is acupuncture more effective on pain impulses entering the same segment of the spinal cord than on those entering adjacent or distant segments?
How does one reconcile the short time course of the neurophysiological events with the long time course of the analgesia?[52]

Several papers in the 1970's reviewed acupuncture results on the clinical side. Some 5,400 operations, of 10 types, under acupuncture anesthesia were studied in several hospitals around Shanghai and anesthesia graded as excellent, good, moderate, poor, or failure. In 96.5 percent of pulmonary

[51] Rockefeller University, *Report;* and Chen, *Acupuncture Anesthesia.*
[52] Rockefeller University, *Report;* and Chen, *Acupuncture Anesthesia.*

resections, at least moderate anesthesia was achieved, but operations for detachment of the retina had a rate of 80.7 percent.[53]

A similar study of 1,474 operations under acupuncture done at the Beijing Children's Hospital between 1966 and 1973 showed roughly the same level of satisfactory results.[54]

Research on acupuncture therapy is less impressive. Absence of controls is particularly serious because acupuncture is often said to be effective in diseases from which recovery is usually spontaneous and acupuncture is often used in conjunction with drugs, chemotherapy, and other Western methods of therapy. An example is a report on the treatment of pediatric pneumonia from the People's Liberation Army 202d Hospital.[55] Of 4,060 cases admitted between 1969 and 1975, 35 percent were under one year old and 40 percent were aged between one and two. Eighty percent were mild and 20 percent were severe; they all received acupuncture, and 47 percent received other therapy as well. Of the 4,060 cases, 93.5 percent were cured, 5.9 percent improved, and 0.6 percent died. It is obviously impossible to measure the benefit of acupuncture from these data, particularly since pediatric pneumonia had a high cure rate even in pre-antibiotic days.

There continues to be a great distance between the consistently observed effects of acupuncture and their satisfactory explanation. Acupuncture may well have utility in clinical application and may help explain pain mechanisms and how they influence recovery. Every medical delegation to China has been told that Chinese scientists are interested in collaborative research on acupuncture. It is to be hoped that this will soon begin.

### *Pharmacology and Therapy*

Interest in drugs from Chinese herbs and other natural sources is very strong in both lay and scientific literature on traditional medicine. There has recently been great expansion in the study of various Chinese herbs and systems of treatment, particularly in combination with Western medicine. The research is widely distributed among the institutes of the CAMS and the CATM, as well as hospitals and traditional medical schools.

Three major difficulties quickly come to light in analyzing research on the usefulness of Chinese drugs. First, as emphasized repeatedly, the lack of experimental controls in human studies makes them exceedingly difficult to evaluate. Second, there is the conviction of Chinese physicians that every symptom must be treated with a specific drug. Little concern is given to the

[53] Shanghai Acupuncture Anesthesia Coordinating Group, "Acupuncture Anesthesia: An Anesthetic Method Combining Traditional Chinese and Western Medicine," *Chinese Medical Journal* [English ed.], n.s., 1, 1 (Jan. 1975): 13–27.

[54] Beijing Children's Hospital, "A Clinical Analysis of 1,474 Operations Under Acupuncture Anesthesia Among Children," *Chinese Medical Journal* [English ed.], n.s., 1, 5 (Sept. 1975): 369–74.

[55] People's Liberation Army, 202d Hospital, "Acupuncture Point Injection in Pediatric Pneumonia: Report of 4,060 Cases," *Chinese Medical Journal* [English ed.], n.s., 4, 1 (Jan. 1978): 51–54.

fact that a great number of diseases are self-limited, and that homeostasis may be aided better by supportive measures than by drugs of doubtful effectiveness. Finally, the "placebo effect" is a significant factor in any study of drug usage. Critical analysis of Chinese pharmacology is thus an even more parlous task than evaluating drug usage in the West.

One member of the Herbal Pharmacology Delegation had some interesting observations on the difference between the Western and the Chinese approach to drug therapy:

> The Westerners were interested in direct observation of *cause,* i.e., active principle, and kept directing their questions in an effort to isolate the drugs used: what drug is given for what *disease*? The Chinese answer, repeated many times, was directed toward the *cure* and therefore denied isolation: drugs are given together in varying dosages according to the time of day and the needs of the individual patient. They insisted, therefore, that they could not provide a list of prescriptions for each malady. Instead prescription depended upon the skill of the observing doctor, in tandem with the physique and needs of the patient.[56]

The Herbal Pharmacology Delegation saw little evidence of basic pharmacologic research but met a number of investigators they characterized as excellent. They did see drugs of plant and animal origins being screened for pharmacologic activity and attempts being made to isolate their active principles. The Delegation noted that

> in general, the Chinese seem to be focussing primarily on clinical studies as a first step in pursuing leads supplied by traditional medicine. One cannot fault the Chinese for studying traditional drugs in man before they study them in animals. Before wasting time on laboratory attempts to explain an action or to isolate an active principle, it is imperative to know that the medicine *has* useful activity. And since these traditional medicines have been used on so many people for so long with apparent safety, ethical and medical fears should be minimal.[57]

To list all the traditional Chinese drugs under study is beyond the scope of this chapter, but it is worth noting a few that may have potential for worldwide use, including anisodamine (for treatment of diarrhea, meningitis, Ménière's syndrome); pueraria lobata (hypertension); sarcolysine (cancer); and securinine (poliomyelitis, neurasthenia). Anisodamine, which the Fogarty Center Report called as important a discovery as ephedrine, is an alkaloid from the Chinese solanacea plant *Anisodus tanguticus* that proved very effective in improving microcirculatory blood perfusion and in treating shock.

The Rockefeller University Delegation saw considerable promise in trichosanthin, which has been studied since 1973 as an abortifacient in the second trimester of pregnancy.[58] Recorded in the Ming dynasty (1500's) as an agent for inducing menses or eliminating retained placenta, it is said to

[56] NAS, *Herbal Pharmacology.*
[57] *Ibid.*
[58] Rockefeller University, *Report.*

induce abortion in 99 percent of women even in the middle of the second trimester.

The drug also seems to be effective in treating choriocarcinoma. The active principle is a protein of molecular weight 18,000, isoelectric point 9.4. Following injection, it localizes in the placental trophoblast and the kidneys. It damages the syncytiotrophoblast but not the cytotrophoblast.

Another potentially useful drug comes from *Ilex chinensis,* a ligneous plant known for centuries as helpful for treating burns. A medical school, teaching hospital, pharmaceutical plant, and college of pharmacology combined in a study that showed that the plant reduced the death rate for burns by 12.9 percent and shortened hospital stay by 14.5 days.[59]

The Herbal Pharmacology Delegation concluded that a rational basis might exist for about half the prescriptions given in a guide for barefoot doctors, and that many of the 248 plant and animal drugs used in their preparation were excellent candidates for pharmacologic and phytochemical investigations.

With all the interest in acupuncture anesthesia, work on herbal anesthesia is sometimes overlooked. Most of this deals with drugs of the atropine variety. A combination of anisodine (from *Anisodus tanguticus*) and scopolamine (from *Flos daturae*) has been successfully used as an anesthetic for patients who were in shock but required surgery. The combination helped to raise blood pressure and combat shock.[60]

### *Family Planning*

Support of family planning is a matter of Chinese policy; thus, there is considerable research interest in new contraceptives, particularly oral preparations from herbs and estrogens.

Anordrine, an A-Nor-steroid with a published structural formula, has been used at the Institute of Materia Medica in Shanghai to prevent implantation.[61] Administered orally to mice, rats, rabbits, hamsters, and dogs in single- and multiple-dose experiments at various intervals after mating, a relatively small dosage level completely prevented implantation; all control animals became pregnant after mating. Acute and chronic toxicity studies, although with relatively small numbers of animals, were essentially negative, as were tests for teratogenicity. In a clinical trial, 109 wives of couples who lived apart used anordrine during the *tanqing* (family visit) period.[62]

[59] Nantong Medical College Pharmacology Group, "Treating Burns with Herbal Medicine," *China Reconstructs,* 26 (May 1977): 33–35.

[60] Ningbo District, Zhejiang, Coordinating Group for Research on Traditional Chinese Medicine Anesthesia, "Traditional Chinese Medicine Balanced Anesthesia on Patients in Shock," *Chinese Medical Journal* [English ed.], n.s., 2, 1 (Jan. 1976): 33–41.

[61] Gu Zhibing et al., "Pharmacological Studies of a Contraceptive Drug Anordin," *Chinese Medical Journal* [English ed.], n.s., 2, 3 (May 1976): 177–84; and Fried, Ryan, and Tsuchitani, *Oral Contraceptives.*

[62] Shanghai Municipal 53-Tanqing Anti-Pregnancy Tablet Clinology Cooperative Group, "Clinical Observation of 53-T'an-ch'ing Anti-Pregnancy Tablets," *Chinese Medical Journal,* 1977, 10 (Oct.): 618–20.

The drug was reported to be 99.5 percent effective in preventing pregnancy. If its effectiveness is confirmed and toxicity and teratogenicity are indeed insignificant, this drug has great promise.

Another contraceptive for couples who live apart most of the year, a combination of two progestational agents, quinegestanol acetate and megestrol acetate, was completely effective on rabbits and, when tried on 2,000 women, showed 98 percent effectiveness if at least eight pills were taken each month.[63] Preliminary studies in rats and monkeys of a male contraceptive, gossypol, a cottonseed oil pigment, showed a four- to five-week effect, followed by restoration of fertility. Electron microscope studies indicated that the sperm head was affected. Thought to be reliable, safe, and low-cost, the drug is now being examined for long-term effects.[64]

Abortion is generally available in China, and the search for new and safer methods to carry out the procedure continues, as elsewhere. Various forms of the suction method were developed some time ago; attention in the 1970's focused on the use of Chinese herbs and their derivatives, notably trichosanthin.

Although comprehensive and detailed report forms on family planning are in use,[65] there have been no reports on large-scale field trials in the general population.

In this field, as in others related to health, subjects like motivation, how well advice is carried out, and the efficacy of health education all cry out for investigation. Perhaps with the renewed national interest in social sciences, this important research area will be neglected no longer.

*Environment and Health*

Roughly 10 percent of the papers published in ten numbers of the *Chinese Medical Journal* in 1977 dealt with health aspects of environmental protection. Proper treatment of human wastes has justified priority in a country accustomed to using night soil as fertilizer. Several papers dealt with designs of tanks and duration of holding periods and their effect on reducing parasites and bacteria.[66] Many papers were on the practical side—how to improve general environmental sanitation in a community and how to carry out and use surveys of rural villages to develop worker interest in health.

Interest in protecting workers was abundantly evident. Toxic elements in the worker's environment, including lead, mercury, benzene, and rodenticides, received a great deal of attention. One study of the effect on workers of a chlorinated hydrocarbon pesticide sprayed with an ultra-low-volume

[63] Fried, Ryan, and Tsuchitani, *Oral Contraceptives*.

[64] National Organization of Male Contraceptive Drugs, "A New Male Contraceptive Drug: Cotton Phenol (Gossypol)," *Chinese Medical Journal*, 1978, 8 (Aug.): 455–58.

[65] F. S. Jaffe and D. Oakley, "Observations on Birth Planning in China, 1977," *Family Planning Perspectives*, 10, 2 (Mar./Apr. 1978): 101–8.

[66] People's Liberation Army, Research Institute of Military Medical Sciences, "Night Soil Digestion for Bacterial Destruction," *Chinese Medical Journal* [English ed.], n.s., 3, 6 (Nov. 1977): 361–63.

atomizer showed that they were not affected if the spraying took place for less than 22 minutes, but that if it continued longer, protective measures were needed.[67] Another occupational health paper described an air-water dust-settling device to combat injurious gas spread by explosions in coal mining.[68] Ejecting a mixture of air and water under high pressure in a conical mist caused the dust and poison gas to settle down. The paper reported data on dust concentrations at the face, with and without the suppressor, and on samples of air at various distances, as well as describing the construction of the device and its technical properties.

Environmental measures are also used to combat malaria and schistosomiasis. For example, areas are seeded with fish to control *Anopheles;* planting a grass carp in ditches can greatly reduce larval density after 15 months. Similarly, in controlling the vector of schistosomiasis, peasants have made mass attempts to kill snails by filling in ditches. A comprehensive water snail extermination program in an autonomous region achieved a 90 percent reduction of contaminated areas.[69]

### *Public Health, Epidemiology, and Nutrition*

Little has been published in China on health administration, organization of services, or epidemiologic methods. To some extent this reflects the fact that administration has not been taught as a separate subject in public health schools, but is included in the study of individual diseases. Similarly, epidemiology is basically disease oriented, and preventive aspects are prominent in papers on each disease or disease group. There is, regretfully, no evidence of a total community studied by epidemiological methods. This approach—quite feasible given China's extensive local organization—could provide models for collection of data needed for public health programs and advance the study of disease as a community phenomenon. As noted earlier, the Chinese pride themselves on treating the person rather than the disease, and the same principle could be extended to the community.

Consistent with China's decentralization, there is great variation in the way health services are paid for, ranging from employer prepaid service in many factories and government offices, to voluntary health insurance in the communes, to fee-for-service. In general, however, the Chinese themselves do little writing about organizational and economic aspects of health services, and very little research and analysis is apparently done on such subjects.

Maternal and child health care has been a priority area for three decades. Studies on the progressive decline in infant mortality are beginning to ap-

[67] Shanghai Municipal Labor Health and Occupational Diseases Prevention and Treatment Center, "A Survey of Effects on the Human Body of Drug Application with an Ultra-low Volume Atomizer in the Rice Paddies," *Chinese Medical Journal,* 1977, 10 (Oct.): 594–96.

[68] Xuzhou Health and Anti-Epidemic Station; Xuzhou Coal Mining Administration, Zhuandai Coal Mine; and Xuzhou Medical College, "New Air-Water Jet Dust Suppression for Coal Mines," *Chinese Medical Journal* [English ed.], n.s., 4, 1 (Jan. 1978): 47–50.

[69] Bao Zhengying, "Characteristics of Watersnail Distribution in Guangxi and Measures for Its Extermination," *Chinese Medical Journal,* 1977, 5 (May): 259–62.

pear, usually as either part of the description of a general maternal and child health program or an analysis of the results in a single hospital.[70] The extraordinarily low mortality figures reported may be related, in part, to demographic factors.[71] Studies of growth and development are of basic importance in preventive pediatrics, and such studies are in progress.[72]

Papers on nutrition, per se, are conspicuous by their absence. This may reflect the well-attested general improvement in nutrition, as well as more concern with providing adequate food and nutrients than with studying nutrition as a medical phenomenon. There are, however, papers on specific problems like favism and prevention of goiter by iodine administration.

One comes away from preparing a chapter like this with an overwhelming impression of unevenness in both the clinical and the public health aspects of China's biomedical research.

There are bright, able, and imaginative investigators in the PRC, and there are those who are pedestrian, to say the least. There is a substantial amount of medical and public health research going on, ranging from the very good to the mediocre. The breadth of subjects covered is impressive; the depth is not. Individual persons and groups are working in most of the medical and biological areas being studied elsewhere, even some relatively fundamental ones, and are exploring some new areas, such as traditional medicine, with approaches and theories little understood in the West.

Methods, research design, and facilities, while sometimes good, are often sadly lacking and leave much to be desired. Reports of research results can be frustratingly inadequate. Nevertheless, substantial progress has been made and, despite the defects, some significant findings are coming out—although not always in areas of prime interest outside China—and more may be expected.

Given the present Chinese leaders' support of science, the October 1978 signing of a collaborative research agreement between the World Health Organization and the PRC, and China's deep interest in preventive and curative medicine, the outlook for biomedical research seems bright.

[70] Guangdong Provincial Institute of Maternal and Child Health Care, "Rural Child Health Care in Kwangtung Province," *Chinese Medical Journal* [English ed.], n.s., 4, 2 (Mar. 1978): 85–88; and Shanghai Bureau of Textile Industry, Second Hospital, Department of Obstetrics and Gynecology, "Survey and Analysis of Deaths of Less Than One Year Old Infants Among Textile Workers' Families in 22 Years," *Chinese Medical Journal*, 1977, 9 (Sept.): 563–66.

[71] M. E. Wegman, "Annual Summary of Vital Statistics, 1972, with Some Observations on China," *Pediatrics*, 52, 6 (Dec. 1973): 873–82.

[72] Coordinating Study Group on the Physical Development of Children and Adolescents, and CAMS, Institute of Pediatrics, "Studies on the Physical Development of Children and Adolescents in New China," *Chinese Medical Journal* [English ed.], n.s., 3, 6 (Nov. 1977): 364–72.

*Jack R. Harlan*

# *PLANT BREEDING AND GENETICS*

Regardless of ideology, genetics is genetics and plant breeding is plant breeding. The experience with Lysenkoism in the U.S.S.R. has shown clearly that it is impossible to repeal fundamental laws of biology. However, national priorities and ideologies, organizational structures, educational levels, recent history, and current attitudes do profoundly influence the subject matter, quality, and effectiveness of research. Plant breeding and genetics in the PRC have uniquely Chinese characteristics.

This chapter briefly describes the research milieu, pertinent organizational and educational structures, and principal subjects under investigation and assesses the present state of the art. This account is inevitably biased. One month in the PRC is not enough to become well grounded in these subjects. China is changing rapidly. Some trends visible in 1974 when the U.S. Plant Studies Delegation visited China have flourished; others have changed direction.[1] The available literature is also biased. Although several good journals report research in genetics and plant breeding—often with summaries in English—they do not consider that the ordinary is worth publishing. Good, solid, old-fashioned plant-breeding research is seldom mentioned. Journals contain many papers on such subjects as tissue culture, mutation breeding, wide crosses, and protoplast fusion, but few on breeding methodology, theory, or basic genetic research. Consulting with colleagues who have also visited China and reading their reports does not solve these problems because they had largely the same experience that I had. With some exceptions, we went to the same places, talked with the same people, and saw the same work.

[1] National Academy of Sciences, *Plant Studies in the People's Republic of China* (Washington, D.C.: NAS, 1975).

Recent history suggests that China intends to strengthen itself by using its own resources. This resolve requires not only a strong military posture and a disciplined people, but a strong economy and agricultural self-sufficiency. Consequently, agriculture is given a very high priority in national goals; it is China's number one industry and is treated accordingly. This approach is in striking contrast to that of most developing nations, even where the economy is based on agriculture.

In addition, agriculture has a central importance in China's Marxist ideology. Agriculture is where the masses are in China. It is central to the whole concept of self-reliance and self-sufficiency. It is an industry in which hard work and enthusiasm pay off even with relatively low capital investment. It is also an industry that requires the participation of the masses, where one can, in fact, "learn from the people" and where elitism can be controlled, if not eliminated.

Following Hua Guofeng's call for modernization of science and technology, agricultural scientists may do less field extension work and more basic research in the future. But the basic environment for genetic and plant-breeding research is likely to prevail for some time: peasants' talents and resources will be used wherever possible, and most of the work in genetics will continue to be applied. Patriotism, revolutionary fervor, pride in Chinese accomplishments, and peer pressures for hard work and self-sacrifice will maintain the momentum of agricultural development, regardless of policy shifts.

## *ORGANIZATION*

The organization of genetics and plant-breeding research in the PRC is diffuse and decentralized in keeping with the principle of local and regional self-sufficiency. Some of the academies and institutions were separated from the control of the Chinese Academy of Sciences (CAS) during the Cultural Revolution, and some are relatively new organizations established at the provincial, municipal, and county levels.

Research is organized on seven levels: national, provincial, district, county, commune, brigade, and team. The national academies and institutes have branches or counterparts at the provincial, municipal, and autonomous region levels. Colleges and universities have been established to serve educational needs at provincial and municipal levels as well. Specialists from institutes, colleges, and universities participate in joint research projects involving the masses at scientific experiment stations of districts, counties, communes, production brigades, and production teams.

### *The Chinese Academy of Sciences* (CAS)

The Chinese Academy of Sciences conducts a wide range of scientific investigations and is directly responsible to the Science and Technology Commission of the State Council. Four of its institutes are important to genetics and plant breeding:

1. The Beijing Institute of Genetics has some 200 professional researchers working in four divisions: molecular genetics, developmental genetics, microbiology, and plant genetics. The last division has laboratories of cell fusion, heterosis and male sterility, and mutation breeding. Facilities include a rice research station on Hainan island and laboratories for the study of the genetics of rice, wheat, sorghum, maize, potatoes, animals, and bacteria.

2. The Beijing Institute of Botany is administered jointly by the CAS and the Beijing municipal government and has a staff of some 300. It is divided into sections researching taxonomy, geography, paleontology, morphology, cytology, somatic hybridization, fat and oils, and haploid breeding. Facilities include a botanical garden, the largest herbarium in the PRC, and laboratories for the study of protoplast fusion, metabolism, hormones, photosynthesis, nitrogen fixation, electron microscopy, cytology, morphology, paleobotany, physiology, and biochemistry.

3. The Guangdong Institute of Botany is administered jointly by the CAS and the Guangdong provincial government. It includes the South Seas Rice Paddy Research Station, an anther culture group, a genetics laboratory, and a botanical garden (which does research largely on medicinal plants).

4. The Shanghai Institute of Plant Physiology is directly under the CAS and has some 275 researchers working in divisions of cell physiology, photosynthesis, nitrogen fixation, phytohormones, and microbiology. There is also a phytotron, a laboratory with facilities for growing plants under various combinations of strictly controlled conditions. Genetic research includes cell fusion, irradiation, and microbial genetics for production of antibiotics.

### *The Academy of Agricultural and Forestry Sciences*

The Academy of Agricultural and Forestry Sciences (Beijing) is under the direction of the Ministry of Agriculture and Forestry and operates the following institutes: Agriculture, Atomic Energy Utilization in Agricultural Research, Citrus, Cotton, Crop Breeding and Cultivation, Farm Field Irrigation, Mutation Genetics, Olericulture, Plant Protection, Sericulture, Soil and Fertilizer, Tea, and Tobacco. Some of these are located in Beijing; others are in the provinces.

In addition, the Academy has provincial branch Academies of Agricultural Sciences in Zhejiang, Heilongjiang, Henan, Jilin, Guangdong, Guizhou, Liaoning, Shanghai, Shandong, Shaanxi, Xinjiang, and Sichuan. The organization of two of these can serve as examples. The Jilin Academy of Agricultural Science operates five institutes: Crop Breeding; Animal Husbandry Research; Fruit Research; Plant Protection; and Soils, Fertilizer, and Cultivation Research. The Guangdong Academy operates eight institutes: Grain Crop, Fruit Breeding, Economic Crop Research, Livestock and Veterinary Science Research, Plant Protection, Sericulture, Soil and Fertilizer, and Tea Research.

### *Other Academies and Institutes*

Academies and institutes have proliferated all over the country and are supported at the provincial, county, or municipal level. For example, the Fujian Agricultural Science Research Institute is under the jurisdiction of the Fujian provincial government. Other provincial academies and institutes are found in Anhui, Hebei, Hunan, Jiangsu, Guangxi, and Yunnan. Municipal academies of agricultural sciences exist in Beijing, Shanghai, Kunming, and Tianjin. Most of the autonomous regions, which all have provincial status, have their own agricultural academies.

The minor academies and institutes seem to come in swarms. Reports appear in the literature from such organizations as Yifeng County Institute of Agricultural Science (Jiangxi), Shijiazhuang District Institute of Agricultural Sciences (Hebei), Huiming Regional Bureau of Agricultural Science, and Hunan Institute of Flax. Their number is legion.

### *Agricultural Colleges and Universities*

The list of universities and agricultural colleges is much shorter, but over 20 can be identified from genetics and plant-breeding literature. Most are relatively new institutions, but a few have been in operation 40 years or more. The Northwest College of Agriculture at Wugong (Shaanxi) was established, for example, in 1934 and may be considered typical of the older institutions. Some, like Jiangxi Communist University, have a revolutionary ring to their names; others are more conventional—Beijing, Xiamen, Inner Mongolia, Lanzhou, and Sichuan Universities. Some appear to be small and provincial (e.g. Qianyang Agricultural School); others are designated as teachers colleges (e.g. Beijing Normal University, Hebei Normal University, Guangxi Normal College); and still others are of national significance (e.g. Chinese University of Science and Technology in Beijing).

All conduct research in genetics and plant breeding. In the agricultural institutions this takes place in such standard departments as agronomy, horticulture, and animal husbandry; in the universities and normal schools in such departments as biology or the equivalent. Joint research projects with counties and communes are common.

### *People's Groups*

To enable the masses to participate in the scientific improvement of agriculture, scientific experiment stations have been established at the county, commune, and production brigade levels. Even the production teams may get into the act by organizing "production team scientific experiment small groups." Literally millions of peasant technicians have been trained for specific technical tasks, including identifying and monitoring diseases and pests, devising integrated control measures, and noting differences in cultivars with respect to resistance or tolerance, and they may be involved in testing and selecting trial material or material treated by mutagens.

These peasant organizations proliferated as a result of the Cultural Revolution in the late 1960's. Academies from the institutes, colleges, and universities were "sent down" to the communes to help solve the farmers' problems. There were not enough specialists to cover the whole country, so "basic contact points" were set up at selected production brigades, small experiment stations established, and field extension work organized. Research personnel were expected to devote a large portion of their time "squatting on the points" in order to share their expertise directly with the farmers. The proportion of time devoted to extension activities of this kind seems to have been flexible, but in recent years varied from one-third to two-thirds. Party Chairman Hua Guofeng has called for a reduction in order to get more research done, and as more people are trained, there should be less need for this procedure in the future.

### *Miscellaneous Organizations*

There are a number of other organizations involved in genetic research. Apparently, the Suzhou Monosodium Glutamate Fermentation Plant conducts applied work in microbial genetics, and it is likely that other industrial organizations are doing similar work. The Guangzhou Scientific Research Institute of Tropical Crops has sections on agrotechnology and oil palm culture, plant protection and latex-processing laboratories, and three field stations, two of them on Hainan island.

Finally, but most importantly, the winter nursery facilities on Hainan island should be mentioned. Their organization is not clear, but

> more than 18,000 workers, poor and middle farmers, leaders and technicians of research agencies, and teachers and students of colleges and special schools are coming from all 27 provinces [includes Taiwan—ed.] and municipalities every year to devote to crop breeding in the over 120,000 mou (8,000 ha) of crop breeding fields of the island. Most of the acreage is used for breeding paddy rice, with emphasis on heterosis utilization; the remaining is used for breeding such upland grains as wheat, corn, and kaoliang and such economic crops as cotton.[2]

The Institute of Genetics of the CAS also has a rice research station on the island, and other rice breeders throughout China have access to Hainan for winter generation progenies. This is an invaluable way of speeding up plant-breeding programs nationwide and presumably helps plant breeders and geneticists to exchange information.

## *EDUCATION AND TRAINING*

A relatively small number of research personnel received training in the United States and Europe before 1949 or in the U.S.S.R. in the 1950's, but most plant-breeding researchers are young and were educated within the

[2]Guangdong Provincial Academy of Agricultural Sciences, Survey Group, "Progress in China's Crop Breeding with Reference to Propagating and Breeding in the South," *Yichuan yu yuzhong (Genetics and Breeding)*, 1977, 1: 2–3.

PRC. Members of the older generation often occupy positions of responsibility, and foreign experience is no handicap. They are often extremely capable and competent scientists. The main work force in the national plant-breeding effort, however, is young and is enthusiastic and dedicated. Until 1978 education and training in the PRC produced many enthusiastic, well-trained technicians, but few scientists with a thorough grounding in basic biology, theoretical and quantitative genetics, statistics, and experimental design. On the other hand, there were many young people ready and willing to learn.

Until the educational reforms of 1978, the formal education of this corps of research workers varied. Most had finished primary school (five years), and some had completed middle school (usually four years). A small number had attended college or university (usually three and one-half years), but until the late 1970's no degrees were awarded. A few students did more work, but left when it was mutually agreed that they would gain little by further studies. Courses were applied and field oriented. Many young people were trained at commune or brigade scientific experiment stations by specialists from academies, institutes, and universities.

There was great enthusiasm for learning among the youth. Classes, seminars, or courses were given by scientists at the "basic points" selected brigades and communes. There were short courses and night classes, and technicians learned from each other. There was also strong peer pressure to "serve the people" and a commitment at all levels to assimilate the experience, expertise, and wisdom of the peasant. Young people were full of ideas, and every commune and production brigade became something of an experiment station. There appeared to be a great willingness to try new things and incorporate workable ideas into current operations.

The Chinese came to realize, however, that since the older generation of foreign-trained scientists was beginning to disappear, the bright eagerness of young researchers might be inadequate. The new push for modernization once again changed China's approach to education, which reverted to a more traditional approach to scientific education.

The scientists in responsible positions appear to be well informed and fairly well up on the literature. The major publications of the Western world are available to most of them, often reproduced in China. Chinese publications virtually ceased during the Cultural Revolution, but the more important ones resumed in the 1970's, and some new ones were started. *Yichuan yu yuzhong (Genetics and Breeding)*, published by the Institute of Genetics, is typical and is devoted to reporting the activities, results, and experiences of popular breeding and genetics research; introducing breeding techniques and experimental methods; promoting common knowledge concerned with botany, zoology, microbiology, and medical genetics; and discussing problems in breeding and genetics research.

Other important journals are *Yichuan xuebao (Acta Genetica Sinica)*, *Yichuanxue tongxun (Bulletin of Genetics)*, *Zhiwu xuebao (Acta Botanica*

*Sinica), Zhiwu zazhi (Journal of Botany),* and *Zhiwu fenlei xuebao (Acta Phytotaxonomica Sinica).* These are all published in Beijing. There are also some provincial reporting services, such as *Guangdong nongye kexue (Guangdong's Agricultural Science).*

These journals are not an adequate means of exchanging information. Academies and institutes appear to be isolated one from the other, and laboratories in the same city are frequently unaware of research in neighboring institutions, which results in much duplication.

Most laboratories visited by American delegations were modestly equipped and engaged in pedestrian work. Sophisticated equipment was being manufactured in China, but had not yet been distributed to the provinces. Electron microscopes were produced in Shanghai, for example, and the Shanghai Institute of Plant Physiology had one, but few other laboratories were so equipped. The Chinese were aware of these deficiencies, and the call to modernize science suggested that corrective measures would be taken.

## RESEARCH

### Haploid Breeding

Haploid breeding is perhaps the best example of a plant-breeding procedure that has developed a unique character in China. Haploid plantlets from anther cultures were first reported in 1966, and haploid breeding has become the focus of much research. But the Chinese were the first to use the procedure on wheat, and the only people to popularize the methodology on a wide scale and for an array of crops. The general procedure is as follows: (1) pollen grains at an early stage of development are cultured on a sterile medium; (2) the calli so formed are induced to form plantlets; and (3) the chromosome number of the haploid plantlets usually is doubled spontaneously, but sometimes artificially (colchicine). Theoretically, this produces completely homozygous diploids. Studies of variability of the lines so produced and of chromosome counts confirm that this is usually what happens, although aneuploids or higher ploidy levels are sometimes obtained.

In self-fertilizing species such as wheat, rice, barley, and soybeans, a common plant-breeding strategy is to cross carefully selected parents, allow the $F_1$ to self-fertilize, and then advance generations to the $F_3$–$F_6$ (or more) before selections are made. This allows time for linkages to be broken up, new genetic combinations to be established, and plants to approach homozygosity so that selected lines are reasonably uniform and have some identity. But in haploid breeding, homozygosity and uniformity are achieved immediately. The Chinese believe that this process bypasses three to six years of selfing and thus speeds up plant breeding. They claim that a "degenerated" variety can be rejuvenated by this technique. Workers at the Lichuan County Institute of Agriculture (Jiangxi) even claimed that heterosis in hybrid rice can be stabilized by haploid breeding.

In cross-fertilizing species, such as maize, where commercial hybrids are desired, inbreeding following hybridization is routinely practiced to provide lines with uniformity and identity. The Chinese argue that they can produce uniform lines immediately by haploid-breeding techniques and that this saves several generations of inbreeding.

Another advantage of haploid breeding is the quick identification of specified genotypes. The genotype AABBCC, for example, can be spotted immediately because there are no heterozygotes in the population to cause confusion. The technique is valuable wherever homozygotes are desired.

Research on haploid breeding in the PRC got under way in 1970. Tobacco, as usual, proved to be easy. It produces calli readily, and the calli are rather easily induced to form plantlets. Some measure of success was obtained in rice and wheat by 1971. In September 1972, the Haploid Breeding Section of Beijing Botanical Institute and Heilongjiang Agricultural Testing Laboratory published a report indicating that tobacco haploids could be mass-produced in the PRC but, it added, more research was needed on rice, wheat, maize, sorghum, cotton, and rape. Apparently, there was already at that time a considerable commitment to this essentially untried procedure.

There have been steady improvements in the technique since then, especially with wheat. Culture media were improved; chilling the wheat spikes before removing the pollen helped the formation of calli; and the rate of induction was improved with experience and refinements of technique. (One of the difficulties with wheat was that in late spring and early summer when pollen is produced, the temperatures are too high for good callus growth. Adequate temperature controls for laboratories were not available, but the problem was solved by storing the calli in the dark at 5°C until fall. They greened up readily when taken out of storage, and the induction rate was good.) In some materials, under certain conditions, embryoids are formed instead of calli. These are more likely to produce plantlets than calli, and Chinese scientists found ways of increasing embryoid formation.

A 1974 report by the Wheat Group of Kunming Municipal Institute of Agricultural Science (Yunnan) described a new variety of wheat, Huapei No. 1, that was being released for production. It was derived from a doubled haploid produced in 1971 by the Institute of Genetics. The breeding time of three years was indeed short. Since then, a number of varieties of both rice and wheat have been developed and distributed for production.

Maize was apparently more difficult, but by 1975 the secrets were unlocked by the 401 Research Group of the Institute of Genetics. Successful anther culture has been reported for rape, red pepper, eggplant, rubber, poplar, and wheat × *Agropyron* hybrids as well as for tobacco, rice, wheat, and triticale. An interesting sidelight is a report of induction of grape endosperm callus (triploid) in order to produce seedless grapes.[3]

[3] Wu Xijin et al., "Induction of Callus in *Vitis* Endosperm Cultured in Vitro," *Zhiwu xuebao (Acta Botanica Sinica)*, 19 (1977): 93–94.

Haploid breeding was popularized. For example, *Acta Botanica Sinica* published an article entitled "The Great Proletarian Cultural Revolution Made 'the Lowly' More Intelligent: How We Undertook Haploid Breeding of Eggplants."[4] The authors were from the Agricultural Experiment Station, Dongsheng Commune, Beijing. Another article in the same volume was entitled "We 'Bumpkins' Can Undertake Haploid Breeding Too."[5] A partial answer to the question of how country "bumpkins" became so sophisticated was provided a few pages later in another article by a Scientific Research Station in Henan province and the Department of Biology, Beijing University.[6] Apparently teams from the communes and brigades were trained at the universities and institutes and received instruction by scientists "squatting on the points." This kind of interchange is basic to the Chinese system, and peasants can indeed do haploid breeding or any other kind for which they have received training.

The technical excellence is unquestionable, the enthusiasm and involvement of the masses is remarkable, and the system is unique to the PRC, but the genetic reasoning behind the operation is somewhat obscure. It has not been demonstrated either in the PRC, or in the West that a completely homozygous line is the most desirable and useful line. Indeed, a number of studies by Allard and others indicate that a high degree of homozygosity is unstable even in self-fertilizing species, and that there is a selective advantage for some heterozygous allelic combinations. Moreover, there are theoretical reasons for believing that a too rapid approach to homozygosity does not provide for adequate recombination and breakage of linkage. Immediate fixation can cause loss of highly desirable combinations.

On the other hand, if haploid breeding is practiced on a sufficiently large scale, there is a chance of producing and fixing highly superior genotypes. Good results may be achieved simply by the volume of the project, and this is probably the reasoning behind the popularization of haploid breeding. Outsiders may speculate, however, that better results might have been obtained if the immense resources of the peasant-technician population had been devoted to making more crosses rather than to haploid breeding.

But the involvement of the masses in a national plant-breeding effort is perhaps more critical to success than are the procedures used. Given the number of crosses being made and millions of technicians searching for superior genotypes, many will be found. Even if the testing lacks statistical refinement and experimental designs are elementary, excellent lines will be

[4]Beijing, Dongsheng Commune, Agricultural Experiment Station, "The Great Proletarian Cultural Revolution Made 'the Lowly' More Intelligent: How We Undertook Haploid Breeding of Eggplants," *Acta Botanica Sinica,* 18 (1976): 220–24.

[5]Henan, Yanshi County, Dagou Commune, Xiaozun Production Brigade, Scientific Research Station, "We 'Bumpkins' Can Undertake Haploid Breeding Too," *Acta Botanica Sinica,* 18 (1976): 216–19.

[6]Henan, Yanshi County, Dagou Commune, Xiaozun Production Brigade, Scientific Research Station, and Beijing University, Department of Biology, Class of 1973, Practice Group, "Studies on Photorespiration in Several Different Varieties of Wheat by Biochemical Methods," *Acta Botanica Sinica,* 18 (1976): 299–304.

identified, propagated, and used. Chinese agriculture is rational and pragmatic. Whatever works is used. The theory and genetics of the situation are, for the time being, irrelevant. It seems safe to predict, however, that haploid breeding will be integrated into conventional systems, and that the emphasis of the 1970's was only a phase in the national plant-breeding effort.

### *Mutation Breeding*

The literature on mutation breeding in the PRC resembles the literature in most countries. Fewer data may be presented, but the experiments are similar to those conducted in other countries—except that irradiated seeds are occasionally widely distributed so that selections can be made by the masses. This not only counteracts scientific elitism, but provides thousands of observers to look for useful mutations.

Mutagenic sources have included x-rays, gamma rays from Co-60, fast neutrons, lasers, and chemicals. Species treated have included tobacco, rice, wheat, soybean, cotton, maize, sorghum, millet, rape, cabbage, silkworms, yeast, and bacteria. Tissue cultures, seeds, flowers, and whole plants have also been treated. Diversity has been increased, and some of the mutants appear to be useful. Mutation breeding has sometimes been combined with haploid breeding with good results, since haploids are more sensitive to radiation and the results more conspicuous. Some studies have reported using methods of quantitative genetics to analyze induced mutations.

Radioactive tracers have been used much as they are elsewhere to study the physiology of growth and nutrition. Elements supplied in fertilizers are traced from uptake by the roots through translocation within the plant to final deposition, loss, or recovery.

The Chinese have released several new varieties of more important crops, claiming that they were developed by mutation breeding. They also claim that the method accelerates the breeding process over conventional crossing and pedigree systems since selections can be made in $M_4$ or even earlier. The approach may be appreciated more in China than in other countries because mutational defects often result in a shortening of the life cycle, i.e. earliness, and this is an overriding Chinese goal for most crops.

### *Wide Crosses*

A considerable effort has been made in China in the area of wide crosses. These may be roughly divided into two categories: legitimate and illegitimate. The legitimate wide-cross studies have clearly defined goals and involve species that are likely to be crossable. Diploid and tetraploid cottons have been crossed, for example, in order to improve upland cotton in specified ways. Work has continued for a number of years with material derived from upland and sea island cottons (*Gossypium hirsutum* × *G. barbadense*). Hexaploid and tetraploid wheats have been crossed to transfer disease resistance, and seedless triploid watermelons have been produced much as Kihara produced them decades ago. Different species of *Brassica* have been hybridized, and much effort has been devoted to *Triticale.* Such studies are being conducted worldwide and are standard components of most plant-breeding programs.

The illegitimate hybrids are perhaps reported for purposes of sensationalism or propaganda. These include rice × sorghum, rice × bamboo, rice × reed (*Phragmites*), and millet × sorghum. The following report by a peasant experimenter illustrates the tenor of this activity.

> I am a peasant. In order to increase the yields of cotton and grain and breed new varieties, I began to work on distant crossing of cotton in 1957. The sexual hybridization was conducted by using the cotton and castor bean plants possessing the characters such as strong adaptability, short branches of fructification, better fructifying ability, resistance to insect pest and drought and cold injury. In 1962, I obtained the new variety "Mine-Nung No. 14" which had the characteristics of both cotton and castor bean plants. Later the petwheat was also obtained. Now the two new varieties of cotton and wheat have been put on production, and well received by the poor and lower-middle peasants.
>
> In the 20 years engaged in research on distant crossing, I was constantly opposed by class enemies and the bourgeoisie in the party. The bourgeois specialists and "authorities" also said that "I was a 'bumpkin,' who did not know science," "the castor-cotton was false," etc. But in accordance with the need of production, I did away with all fetishes and superstitions, emancipated the mind, followed Chairman Mao's teaching of the primary standpoint of practice, upheld the philosophy of struggle, finally obtained certain results. I consider that all genuine knowledge originates in direct experience, and both the theory and practice of distant crossing have wide prospects for future development.[7]

Chinese activities in the area of protoplast fusion, however, are valid. For example, fusions of wheat × wheat and maize × maize protoplasts have been reported in *Acta Genetica Sinica.*[8] The fusion of protoplasts from different species of *Nicotiana* has also been reported. Less legitimate combinations, such as tobacco × turnip, wheat × lima bean, wheat × saffron, and broad bean × saffron, are not as likely to develop into plants. The combinations, however, are real enough in cell culture.

The work has been considered sufficiently important that conferences and meetings have been held in the Shanghai Institute of Plant Physiology with delegates from throughout China. The field is relatively new, and no one yet knows its potential. Laboratories to study protoplast fusion have been established in a number of countries, and the technique is being explored vigorously. The Chinese do not intend to be left behind.

### *Conventional Breeding*

Sensational results are published more frequently than is justifiable; as in other nations, the bulk of conventional plant-breeding work in the PRC is straightforward, time consuming, and pedestrian.

[7] Zhang Sizhou, "Take Class Struggle as the Key Link and Persist in the Work on Distant Crossing," *Acta Botanica Sinica,* 18 (1976): 333.

[8] "Isolation and Induced Fusion of Plant Protoplasts," *Yichuan xuebao (Acta Genetica Sinica),* 1974, 1: 59–68.

Conventional plant-breeding programs are conducted for all major crops, and at least some form of mass selection is practiced for all minor ones. It is apparent from the literature that considerable effort has been devoted to the development of male sterile and restorer systems. Hybrid maize and sorghum are routine, but the Chinese are trying to extend the procedure to other major crops. Typically, they have coined a term—"three lines" (male sterile line, maintainer line, restorer line)—for the system and are popularizing it. Students, peasant technicians, and field workers are searching for male sterile plants of rice, wheat, and other crops. A successful three-lines system was claimed for rice in 1977.[9] But although the lines were found as early as 1973, work continues on developing an appropriate gametocide for rice.[10] Gametocides for hybrid wheat are also being tested.

Male sterile and restorer systems for wheat have been available in the United States for many years, but seed production has been troublesome. Apparently it is difficult to convert a self-fertilizing plant into an outcrossing one. One would expect the same problems with rice. On the other hand, production practices that are uneconomic in America might be viable in the PRC. At any rate, serious attempts have been made to produce both hybrid wheat and hybrid rice by the three-lines system or alternatively by use of gametocides.

According to Li Jingxiong, a maize breeder at the Academy of Agricultural and Forestry Sciences, the T-type (Texas) male sterile system was used for maize in the 1950's.[11] In 1961 it was discovered that most sterile lines were susceptible to northern corn leaf blight (*Helminthosporium turcicum*), and the T system has been little used since. This association between susceptibility to northern corn leaf blight and T cytoplasm has not been confirmed elsewhere. It was not until 1970 that a serious epidemic of southern corn blight (*H. maydis*) swept the United States because T cytoplasm hybrids were susceptible to race T of the pathogen. Both Chinese and American scientists have been researching new sources of cytoplasmic sterility. In 1976 Chinese scientists claimed to have found one that is satisfactory.[12] Li Jingxiong, however used the C cytoplasm.[13] The male sterile system used in sorghum is the one developed in Texas some decades ago. The Chinese obtained their lines from Africa, where the male sterile system is in use.

Early maturity is an almost universal objective of Chinese breeding because of the evolution of complex and precise cropping patterns, including double cropping, triple cropping, intercropping, and relay cropping. Even at the latitude of Beijing (ca. 40°N), wheat followed by rice is a common crop-

[9] Li Bihu, "How We Studied Hybrid Rice," *Acta Botanica Sinica,* 19 (1977): 7–10.

[10] "No. 73010: Rice Chemical Male Sterilizer," *Genetics and Breeding,* 1977, 3: 15.

[11] Li Jingxiong, "Maize Culture and Seed Increase" (Paper delivered at Centro Internacional de Mejoramiento de Maíz y Trigo, El Batan, Mexico, Aug. 26, 1977), mimeographed transcript.

[12] "Preliminary Report of Study on Male Sterile System of Corn," *Genetics and Breeding,* 1976, 5: 18–19.

[13] Li Jingxiong, "Maize Culture."

ping system. Among the several systems described to our Delegation at Nanking was the following. In the cool months of the year, a crop of barley is grown (barley is preferred to wheat because it matures earlier). About a month before the barley is ready for harvest, workers part the barley and transplant cotton by hand in rows. The barley is harvested when it is ready, and the cotton matures during the summer. But the cotton must be picked and the plants turned under in time to plant a crop of rice before it is time to plant barley again. The system is labor intensive and exacting. Everything must be done on time. Furthermore, the barley, the cotton, and the rice must all mature early.

Transplanting helps to speed up a cropping system and is much used in China. Cotton, rice, maize, sorghum, and even wheat are transplanted to accommodate demanding cropping patterns. Shorter life cycles are thus an essential goal of Chinese plant-breeding efforts.

### *Rice*

Rice is China's major cereal. Some 117 million metric tons of paddy were reported for 1976.[14] Understandably, rice breeding is emphasized. The area devoted to rice is being expanded to the limits of adaptation, far into the northeast provinces, higher into the western mountains, and on new lands wherever irrigation can be developed.

Earliness, yield, and resistance to diseases and pests are the main objectives. The Chinese have their own sources of semidwarf growth habit and developed short-statured, high-yielding varieties before these were popularized by the International Rice Research Institute (IRRI). The first Chinese high-yielding indica variety was released in 1960 in Guangdong province; the first release by the IRRI was IR-8 in 1966. IR-8 was tested in China in 1967, and all available IRRI releases have since been tested. Since the visit of the Plant Studies Delegation in 1974, interchange, including some exchange of germplasm, between Chinese and IRRI rice specialists has become more active.

In general, the IRRI varieties are not suited to China's demanding multiple-cropping systems. The IRRI varieties mature too slowly and the high-tillering habit of most of them is not considered efficient by the Chinese, who achieve the same effect by transplanting more seedlings in each hill. The IRRI germplasm has, however, provided Chinese scientists with excellent sources for yield and disease and pest resistance.

Breeding for resistance has a high priority. Among the most serious diseases are bacterial leaf blight (*Xanthomonas oryzae*), blast (*Pyricularia oryzae*), and a sheath blight caused by *Corticium sasakii.* Major insect pests are those found in many rice-growing countries, namely, brown planthopper (*Nilaparvata lugens*), green leafhopper (*Nephotettix virescens*), stem borer (*Chilo suppressalis*), and whorl maggot (*Hydrellia* sp). Screening for resis-

[14] U.N., Food and Agriculture Organization, *Production Yearbook* (Rome: FAO, 1976).

tance and incorporation into high-yielding, short-season types are routine procedures at the major rice-breeding stations. Although genetic resistance is important, effective disease and pest control is provided by combinations of cultural practices, sprays, predators, vector control, careful rotation, and other integrated measures. Peasant technicians monitor buildups and devise protection strategies that appear to be highly effective.

### *Wheat*

Wheat ranks second among China's cereal crops and is grown in every province of the country. Forty-three million metric tons were reported for 1976.[15] About 85 percent is winter wheat, although the use of spring wheats is probably increasing as they are integrated into the more complex cropping patterns. In addition to earliness, which has a special value in the farming system, Chinese wheat breeders are working to develop the semidwarf high-yielding types that have swept the world in the past decade or so. The Chinese have their own sources of dwarfing genes, but nevertheless became interested in materials developed in Mexico as early as 1962. These were tested widely in the 1960's, and in the early 1970's China purchased some 15,000 tons of stock seed of about 15 Mexican-developed varieties. These met with rather limited success since they tended to be susceptible to scab, *Helminthosporium,* and stripe rust, and in the south the seeds tended to sprout in the head. Chinese scientists then began to cross the Mexican materials with local varieties. A large percentage of the wheat now grown in China evolved from Chinese-Mexican germplasm.

Breeding methodologies are similar to those in other wheat-producing countries except for the emphasis on haploid breeding described earlier. Some breeders are able to advance three generations a year. For example, crosses can be made in March in the Beijing area and harvested in June. These hybrids are then planted in July in the cool mountainous areas of Guizhou. By November, the seeds are harvested and planted in the winter nurseries on Hainan. These plants in turn mature in time for spring planting in Beijing. Most breeders, however, get only two generations a year.

Three-way and four-way crosses are apparently common (i.e. $F_1$'s may be crossed to a third variety, or $F_1$'s with different parents may be intercrossed). These populations may either be advanced several generations before selection or be made homozygous by doubling haploids. Spring × winter crosses are also common; the goal is to develop a high-yielding intermediate type relatively insensitive to day length.

Disease resistance has high priority in breeding programs, although recent success in disease control can probably be attributed more to cultural and cropping practices than to genetic resistance. Head blight caused by *Gibberella zeae,* with its associated scab of kernels, is probably the most important disease of wheat and the only one not under control.[16] It is especially trou-

[15] *Ibid.*

[16] National Academy of Sciences, *Wheat in the People's Republic of China* (Washington, D.C.: NAS, 1977).

blesome because it is one of the few diseases actually encouraged by multiple-cropping systems. The sexual *Gibberella* stage is formed on stalks of both maize and rice, and this is the major source of inoculum. Intensive screening under artificial infection at the Jiangsu Academy of Agricultural Sciences (Nanjing) has identified wheats with considerable resistance. The development of resistant, high-yielding types is under way.

The races of stripe rust (*Puccinia striiformis*) and stem rust (*P. graminis*) are monitored, the former at the Shaanxi Academy of Agricultural and Forestry Sciences at Wugong, and the latter by the Jilin Academy of Agricultural Sciences at Gongzhuling. The races are identified by inoculation of a standard set of varieties, and the corresponding resistance genes in wheat are known. The races of stem rust appear to be remarkably stable in China, suggesting that the sexual stage does not occur. Resistant varieties have been grown for many years without loss of resistance. Stripe rust, on the other hand, does show shifts in race patterns and poses some epidemic threat. Leaf rust (*P. recondita*) does not seem to be a serious problem, and no breeding work was observed by the U.S. Wheat Studies Delegation in 1976.[17]

There is no research on wheat quality; all wheats are considered adequate for human consumption. The overriding priorities are yield and earliness.

### *Maize*

Maize ranks third among cereals produced in China. Thirty-four million metric tons were reported for 1976.[18] It is increasing in importance at the expense of sorghum and millet as more land is brought under irrigation. This native American crop was introduced in China early in the sixteenth century. The first written record of maize in China shows that it was being grown in Fujian province by 1511. Within 200 years the crop was established throughout the country. This history is important; the number of introductions was probably small and the variability limited.

Some early inbreds from the United States were introduced during the 1940's, and some of them were still in use when the Plant Studies Delegation visited the PRC in 1974. More modern American inbreds have since been introduced and are probably being modified for hybrid seed production in China. Meanwhile, the Chinese maize breeders developed inbreds from local sources and produced hybrids from materials at hand. Double crosses were used at first, but the program suffered a severe setback in 1966, when an epidemic of northern corn leaf blight swept through North China.

This occurred during the Cultural Revolution, when the entire system of plant breeding was reorganized and the masses became involved. Resistant plants were identified, mostly from local open-pollinated varieties, often by peasant farmers. Winter nurseries on Hainan were provided to speed up the work, and there was a shift toward single-cross hybrids. By 1977 Li Jing-

[17] *Ibid.*
[18] FAO, *Production Yearbook.*

xiong reported that in the main maize belt of North China, 55 percent of the area in hybrids was in single crosses, 40 percent in double crosses, and 5 percent in three-way or top-cross materials.[19] About 30 different hybrids were involved.

Hybrid seed production is performed by commune production brigades. Often the responsibility for maintaining and producing an individual inbred is delegated to a production team. Different teams deliver their inbred seed to a brigade, which then establishes crossing fields. The operation seems to run very smoothly; isolation of inbreds is excellent and seed production adequate. The system is coordinated by a county seed farm that conducts performance trials and demonstrations and also provides inspection, distribution, and contracts for seed production with brigades. There are seed departments at prefectural and provincial levels that coordinate seed production activities on a broader scale.

As with other crops, disease resistance is a prime concern. In addition to the leaf blights, virus diseases are becoming troublesome. Maize chlorotic virus and maize stunt are spreading in the north. Most American inbreds are susceptible, but resistance has been found in European sources. The Chinese have recently become aware of the narrowness of the genetic base of the maize crop. As in the United States, most of the production is derived from a small number of inbred lines.[20] There is little work under way on quality.

### *Soybean*

The soybean is a crop of Chinese origin, but it does not receive much attention in plant breeding. It is given third-rank status, and production is low. The state of Illinois alone produces about as much as the whole of China. The reasons for this are not entirely clear, especially since an increase in animal production is a stated national goal.

## *ASSESSMENT*

The human resources devoted to genetics and plant breeding in the PRC are impressive. There are scientists of first-rate competence and capability. There may not be as many as the size of the country and its needs would suggest, but the best are as good as can be found anywhere. There are many bright, enthusiastic young researchers who are highly motivated and eager to learn. There is also an army of peasant technicians who are keen, interested, and motivated. The vast population of peasant farmers know their crops and have accumulated experience and wisdom. The three-in-one research system appears to function well because the masses are involved and intrigued by experimentation.

Agriculture and the food supply have been given top priority by the government. In the eight-point modernization program announced in 1978,

[19] Li Jingxiong, "Maize Culture."
[20] *Ibid.*

agricultural science and technology was listed first. This is a favorable indicator of continued support. Too many developing countries have tried to modernize at the expense of agriculture, with discouraging results.

Another point in favor of the plant-breeding effort is the high percentage of arable land under irrigation. China has more land under irrigation than any other country in the world. With appropriate inputs of fertilizers and good soil-management practices, the genetic makeup of the crop varieties is more likely to become a limiting factor than the water supply. Plant breeding can improve this component. If the water supply limits production, improvement through plant breeding is lessened.

The organization of agricultural sciences has some strengths. Sending scientists to the countryside produced some good results: a vast network of experiment stations for field trials and seed production, and a corps of technicians. The winter nursery facilities are also a national asset of great importance.

The need to improve testing programs still exists. Uniform national nurseries were closed during the Cultural Revolution and were never really reopened. Statistical treatment of data and experimental designs could also be improved; with the number of tests being conducted on the communes, vast amounts of data are generated that could be very useful.

Although the genetic resource holdings available to plant breeders are not known in detail, my impression is that scientists are working with a rather narrow base in many crops. Maize has probably the worst base because of its rather late and limited introduction; rice and soybeans probably the best because of their long histories in China. But Chinese plant breeders do not seem to be collection-minded. Old landrace populations of wheat and rice are rapidly being replaced by the short-statured, high-yielding types, but breeders seem unconcerned about genetic erosion. Some of the most valuable genetic resources in the world may well disappear, with no attempt made to salvage them. It is difficult to find genuine, old-fashioned kaoliang in the sorghum belt of China; it has been replaced by modified Texas hybrids. Soybean germplasm appears to be more secure because there are not many plant-breeding programs devoted to it.

The figures in the 1976 FAO *Production Yearbook* show that China has done well compared with some countries, but still falls short of the best standards. For example, the national average yield of rice (paddy) in China was 3,294 kg/ha; the figure for the United States was 5,244; for Japan 5,503; South Korea 5,967; and Spain 6,071. For wheat, the yield in China was 1,387, in Japan 2,496, and in Europe as a whole 3,187 kg/ha. For maize, the yield in China was 1,954 compared with 5,489 kg/ha in America. For barley, China produced 1,556, Japan, 2,619, and America 2,411 kg/ha. Yields of 5,437 kg/ha of wheat and 4,280 kg/ha of barley in the Netherlands indicate the potential of these crops.

Some of China's modest yields are due to the multiple-cropping systems and the need for early varieties. Crops are often in competition with each other for part of their life cycle, and soil management is more difficult in a tight cropping sequence. Nevertheless, there is plenty of room for improvement through plant breeding, and higher yields can confidently be expected in the future.

Characteristics that seem uniquely Chinese include popularization of such techniques as haploid breeding and the involvement of the masses in screening desirable mutants, looking for resistant genotypes or male sterile lines, and monitoring diseases and pests. The three-in-one and open-door research efforts are endemic Chinese approaches. Other components of genetics and plant breeding in the PRC resemble applied genetics and plant breeding elsewhere.

*Robert L. Metcalf and Arthur Kelman*

# *PLANT PROTECTION*

China's 5,000 years of continuous, recorded civilization provide an unequaled perspective on pests and pest control. During this time, the Chinese have experienced almost every pest problem incident to agriculture and have practiced an astonishing array of pest control methods. The Western world has remained largely unaware of some of these approaches because of cultural and linguistic barriers, as well as our missionary emphasis on teaching in lieu of learning. Today the West recognizes that satisfactory pest control is not a simple matter of routine application of pesticides, but rather a far more complex ecological effort. This makes it important to understand the experience of China, where some of the principles of integrated pest management (IPM) were enumerated over a thousand years ago.

The earliest organized efforts in Chinese insect control came with the depredations of the migratory locust, *Locusta migratoria manilensis*. Recorded locust plagues date back to 707 B.C.[1] Since then, more than 800 outbreaks have occurred in various parts of China, chiefly in the flood plains of the Yellow, Huai, and Chang Jiang (Yangtze) rivers and in the coastal regions to the south.[2] Successive outbreaks of locusts in a given locality have occurred every three to 15 years and lasted from 12 to 30 months. The severity of the problem can be judged by the 1929 outbreak, which devastated more than 11 million acres of cropland, causing illness, starvation, death, and wholesale migration from the affected area.[3] Another disastrous locust outbreak occurred in 1938 when Nationalist soldiers destroyed the dikes along the Yellow River and locusts breeding in the flood plain caused damage in about 140 counties in ten provinces.

[1] Tsien-hsi Cheng, "Insect Control in Mainland China," *Science*, 140 (1963): 269–77.
[2] Ma Shijun, "The Population Dynamics of the Oriental Migratory Locust," *Acta Entomologica Sinica*, 8 (1958): 1–40.
[3] Cheng, "Insect Control."

The earliest record of locust control dates back 3,000 years to a Shang dynasty (1523–1027 B.C.) poem, which describes the use of flames to destroy locusts. About 2,000 years ago in the Han dynasty (202 B.C.–A.D. 220), Wang Zhong (Wang Chung) described the use of open ditches to trap immature locusts. In A.D. 2, government officials offered bounties to peasants for collecting locusts. During the Tang dynasty (618–906), burning and trapping were successfully combined. During the Song dynasty (960–1279), the digging of locust eggs from the soil was described, and in 1182 a law required citizens to collect and destroy locusts. This is very likely the first example in the world of a legal requirement for pest control.

By the Ming dynasty (1360–1644), the locust's behavioral traits, such as migrating upwind and early in the morning during heavy dew, its reluctance to fly at dusk or in the afternoon after mating, and its tendency to fly low after rain and toward flames, were described in simple texts and diagrams. Around 1600 Shen Guangjing (Shen Kuang-ching) described the high prevalence of locust invasions in Hebei, Shandong, and Henan provinces and said that individuals acting alone could not control locusts. He called for advance preparations at a national level. Other innovations of this period, described in the *Jin Book,* include the planting of legumes, which locusts do not eat, and the herding of ducklings in fields to act as biological control agents.[4]

In ancient times, Chinese peasants were virtually helpless in combating the recurring locust plagues. Confucius (551–479 B.C.) regarded locust outbreaks as natural calamities and part of heaven's will. Despite advances in knowledge in the subsequent two millennia, effective locust controls only became available after the Revolution in 1949 when China began to mobilize rural manpower on a vast scale to combat the locust menace. During the early 1950's manual destruction was practiced extensively. Locust eggs were dug up in winter and destroyed; trenches 20–30 km long were dug and young locusts driven into them and buried. Thousands of peasants encircled infested areas and killed locusts with bamboo brooms and locust swatters made from shoe soles nailed to wooden poles.[5]

Dusting with benzene hexachloride (BHC) began in 1951, and the use of aircraft for dispersal increased from 13.7 percent of locust-infested areas in 1958 to about 54 percent in 1960 and about 70 percent in 1963.[6] By this time the Chinese realized that they had relied too heavily on aircraft application of insecticide, which was not only costly, but also had deleterious effects on beneficial insects and environmental quality. Moreover, the use of insecticides had weakened locust control organizations and delayed the accumulation of sound ecological information about locust control. As a result, the use of aerial spraying was curtailed and by 1975 was virtually

[4] "The Influence of the Struggles Between Confucians and Legalists on the Development of Entomology in Ancient China," *Acta Entomologica Sinica,* 18 (1975): 121–27.

[5] Cheng, "Insect Control."

[6] *Ibid.*

abandoned.[7] Yang Xiandong, vice-minister of agriculture, said in 1963 that the density of locusts had been reduced but the area of infestation had not decreased.[8]

More than 120 million persons participated in pest management programs for migratory locusts following a major outbreak in 1951. Primitive methods of surrounding, whipping, and burning locusts were supplanted gradually by more ecologically oriented efforts directed by stations surveying locust-breeding grounds and forecasting outbreaks on the basis of meteorological conditions. From 1958 on, enormous efforts were made to convert breeding areas in reedy, riverside flood plains into rice paddies, forest belts, and pastures. More than one million ha of mountainous farmland in the Yellow River valley were converted into terraced fields. Four million check dams and 27,000 small reservoirs were constructed on small tributaries to control flood surges, and trees and grasses were sown on four million ha.[9] This campaign represents the most extensive pest management program of all time.

Today, rather than relying primarily on insecticides for locust control, China has developed an imaginative and practical approach for the environmental control of locusts that knits together biology, ecology, meteorology, water management, and cultural practices, and relies heavily on forecasting.[10]

## *IMPORTANCE OF PLANT PROTECTION IN CHINA*

China is still an agrarian society. Four-fifths of its population reside in rural areas where they engage primarily in producing food and fiber for themselves and urban dwellers.[11] Thus, agriculture is China's most important activity, and its farms currently produce enough food to feed about one billion people, or more than one-fifth of mankind. Scientists and agriculturists who have visited China since 1972 have commented, almost without exception, that the Chinese people appear exceptionally healthy and well nourished.

China has only about 107 million ha of arable land (compared with 156 million ha in the 48 contiguous United States), yet China is essentially self-sufficient in food production.[12] The success of plant protection and improved agronomic practices can be gauged by the increase in grain produc-

[7] National Academy of Sciences, *Insect Control in the People's Republic of China,* Report no. 2 (Washington, D.C.: Committee on Scholarly Communication with the People's Republic of China, 1977), pp. 144–46.

[8] L. T. C. Kuo, *The Technical Transformation of Agriculture in Communist China,* Praeger Special Studies in International Economics and Development (New York: Praeger, 1968), p. 173.

[9] Cheng, "Insect Control."

[10] "Elimination of Locust Infestation by Combined Efforts of Reconstruction and Chemical Control in Weishan Lake District," *Acta Entomologica Sinica,* 17 (1974): 247–58.

[11] NAS, *Insect Control,* pp. 6–8.

[12] S. Wortman, "Agriculture in China," *Scientific American,* 232 (1975): 13–21.

tion from about 113 million tons in 1949 to 175 million tons in 1974.[13] In comparison, the United States produced about 194 million tons in 1973. In 1976 China imported about four million tons of wheat, but this was partially offset by the export of about one million tons of rice.[14] These figures, as well as the observations of visiting scientists, indicate a highly successful and expanding agriculture.

China has attained virtual self-sufficiency in agriculture largely because of the enforcement of Mao Zedong's dictum: "Take agriculture as the foundation." Thus, hundreds of millions of man years of effort are concentrated in rural areas, and large monetary resources are channeled into food production. The financial commitment of China's leaders to agriculture has been estimated at U.S. $5 billion annually.[15] Agricultural production has been bolstered by extensive double and triple cropping in Central and South China, vast terracing and irrigation schemes in the North China Plain, and the development of area fertilizer programs to ensure optimum crop fertility.[16] It appears that crop yields and production have now been pushed close to their natural limits.[17] Therefore, the great challenge for the future will be to cope with a natural population increase now estimated at slightly over 1.5 percent each year. This population growth will require about 4.0 million additional tons of grain annually and ensures that agriculture will remain an area of paramount concern.

Protecting plants from pest attack is an obvious way to increase the available food supply without creating new agricultural land, improving irrigation, or increasing fertilizer production. Indeed, China's success in maintaining self-sufficiency in food production will be a measure of its achievements in controlling plant diseases and insect pests.

Detailed recent data are not available on crop losses from pest attack in China. However, in 1937 it was estimated that insects and plant diseases caused a 10–20 percent reduction of Chinese grain crops.[18] Another, more recent report in 1962 by the Bureau of Plant Protection of the Ministry of Agriculture suggested that in seriously affected areas grain losses were about 10 percent, cotton 20 percent, and fruits 30 percent of total production.[19] In 1967, in a thorough compilation of pest losses in agriculture, Cramer presented data for crop losses in the PRC based largely on comparable data from Southeast Asia and the U.S.S.R.[20] These data suggest that the percent-

[13] D. H. Perkins, "Estimating China's Gross Domestic Product," *U.S.–China Business Review,* Sept./Oct. 1976, pp. 13–17.

[14] "Wheat Exporter's Notes," *U.S.–China Business Review,* Mar./Apr. 1976, p. 51; and A. L. Erisman, in U.S., Congress, Joint Economic Committee, ed., *China: A Reassessment of the Economy* (Washington, D.C.: Government Printing Office, 1975), pp. 324–49.

[15] Perkins, "Estimating China's Gross Domestic Product."

[16] NAS, *Plant Studies in the People's Republic of China,* Report no. 1 (Washington, D.C.: Committee on Scholarly Communication with the People's Republic of China, 1975).

[17] NAS, *Insect Control,* pp. 7–8.

[18] John Lossing Buck, *Land Utilization in China* (Nanjing: Nanjing University, 1937), p. 4.

[19] Kuo, *Technical Transformation,* p. 171.

[20] H. H. Cramer, *Plant Protection and World Crop Production,* tr. J. H. Edwards, Pflanzenschutz-Nachrichten no. 20 (Leverkusen, Germany: Farbenfabriken Bayer, 1967).

age losses to insects and plant diseases respectively are: rice—15 and 8 percent; corn—12.4 and 9.4 percent; wheat—5 and 9 percent; cotton—16.1 and 12 percent; sugarcane—20.1 and 19.2 percent; apples—13.2 and 13.7 percent. The accuracy of these loss estimates may be subject to question because they were 10 to 20 years out of date and did not allow for the immense progress made in plant protection in China under communism. Moreover, the possibility that crop losses to pests in China are significantly higher than in the United States is not substantiated by the observations of American scientists who have visited China recently.[21]

Although the precise magnitude of pest damage to Chinese agriculture is unknown, it is clear that losses are significant enough to warrant major governmental efforts to curb them. In 1958 the Ministry of Agriculture's Bureau of Plant Protection planned for thorough control of plant diseases and insect pests on 12.03 million ha of cropland in 18 provinces and municipalities.[22] Included were 3.6 million ha of rice, 1.6 million ha of wheat, 4.1 million ha of miscellaneous grains, 2.4 million ha of cotton, and 0.22 million ha of fruits and other economic crops. It was estimated that this control program could save 1.25 million metric tons of grain and 90,000 metric tons of cotton.

Such strenuous efforts have had substantial effects, and in April 1960 Vice-Premier Tan Zhenlin reported to the Second National People's Congress that substantial blocks of farmland had been freed from such pests and diseases as aphids, red spider mites, locusts, armyworms, rice borers, and wheat smut.[23] In 1962 Vice-Minister of Agriculture Yang Xiandong said that pests on 27 million ha were controlled by pesticides and that more than 2.67 million tons of grain had been harvested as a result.[24]

## *ORGANIZATION OF RESEARCH*

Before 1949, China's research in entomology and plant pathology was not organized on a regional or provincial basis.[25] A relatively small number of Chinese entomologists and plant pathologists who had been trained in Europe or in American land grant universities dominated the teaching at universities and colleges. A few carried out research in systematics and descriptive biology at important educational centers, such as Nanjing, Lingnan, Beijing, and Zhongshan Universities. The National Agricultural Research Bureau was the governmental center for applied plant protection research. The cataclysmic events of World War II largely destroyed these meager scientific foundations.[26]

[21] A. Kelman and R. J. Cook, "Plant Pathology in the People's Republic of China," *Annual Review of Phytopathology*, 15 (1977): 409–29; NAS, *Insect Control.*
[22] Kuo, *Technical Transformation*, p. 171.
[23] *Ibid.*, p. 172.
[24] *Ibid.;* and Yang Xiandong, "Strive for Greater Success on the Plant Protection Front," *Chinese Agricultural Bulletin*, 1963, 6: 1–4.
[25] Wortman, "Agriculture in China."
[26] Cheng, "Insect Control."

Although a few of these scientists emigrated to Taiwan in 1949, a large number hold key administrative positions in universities and important agricultural research centers, such as the Institute of Zoology in Beijing, the Institute of Entomology in Shanghai, and the Institute of Entomology in Guangdong. Other Western-trained scientists are influential in applied research at regional centers such as the Northwest Institute of Agriculture in Wugong in Shaanxi, the Jilin Academy of Agricultural Sciences in Gongzhuling, and the Hunan Institute of Plant Protection in Changsha.

The period immediately following 1949 was one of turmoil for many of these foreign-trained scientists. Some were accused of individualism and bourgeois leanings and were forced to denounce themselves publicly. The Communists' basic policy was to emphasize practical applied research for the immediate improvement of public health and increased food production. Therefore, a number of prominent scientists found it expedient to change their research emphasis from systematics to applied ecology and plant protection.[27] The First Five-Year Plan (1953–58) marked the beginning of organized plant protection research in China. This was carried out under the Chinese Academy of Sciences (CAS), which organized a group of 21 research institutes and experiment stations where modern research facilities were concentrated. Most of the country's trained entomologists and plant pathologists were brought together in these centers. The Academy began publishing a series of high-quality scientific journals, including *Acta Entomologica Sinica.* Their publication was discontinued from 1967 to 1973, but resumed in 1974.

The number of universities and agricultural institutes increased from about 200 in 1949 to 400 in 1963, and student enrollments rose from 117,000 to 819,000.[28]

With the Great Leap Forward in 1958, a new era in plant protection began. A number of new research institutes were established: the Guangdong Institute of Entomology (1958), the Shanghai Institute of Entomology (1959), and the Shanghai Academy of Agricultural Sciences (1960). Other institutes, including the Northwest Institute of Agriculture and the Hunan Institute of Plant Protection, were strengthened and tied more closely to provincial governments. The Twelve-Year Plan for Scientific Development (1956–67) standardized Chinese biological nomenclature and simplified the characters used in scientific words. Entomology and plant pathology were combined into a single applied discipline—plant protection. Under the Twelve-Year Plan, most agricultural scientists directed their research toward solving practical and applied problems. Plant protection specialists devoted their efforts to keeping plants healthy, rather than to developing basic knowledge about plant diseases and insect enemies. Plant protection in China today is therefore a combination of extension and mission-oriented research sup-

[27] Wortman, "Agriculture in China."
[28] Zhou Peiyuan, *China Reconstructs,* Feb. 1963.

ported by a minimum of basic research.[29] However, administrators have made extensive efforts to keep abreast of basic research elsewhere by reprinting and circulating Western and Japanese journals. For example, 1,400 copies of the U.S. *Journal of Economic Entomology* are reprinted and supplied to local libraries and plant protection specialists. As a result, Chinese research workers are usually aware of scientific progress in their fields and refer to foreign research in their publications.

Plant protectionists work at all levels of government. At the commune and brigade levels, they diagnose plant health problems, forecast outbreaks of pests, apply control measures, and organize demonstration plots. Sometimes a plant protectionist, together with a veterinarian and a paramedic ("barefoot doctor"), is assigned to a production brigade. More commonly, plant protectionists and agronomists function as a special research group in a commune. In Shaanxi province, for example, 50 percent of the communes are reported to have such teams, as well as simple laboratories designed specifically for crop protection work. These are equipped with microscopes, manuals for the identification and control of plant diseases and insect pests, reference collections of common insects and diseased plant specimens, and beautifully detailed posters illustrating the life cycles of important pests. Facilities for culturing organisms used in biological control of insects—for example, *Bacillus thuringiensis* and *Beauveria bassiana*—are sometimes available.

The plant protectionists are usually enthusiastic, dedicated, and well acquainted with the applied aspects of their disciplines. They are almost unique in the cooperation and rapport they have with peasants, who regard the control of plant diseases and insect pests as one of their primary responsibilities to the state and their communes.

The extent of decentralization can be gauged by the number of cooperative research projects that research institutes and colleges have with various communes. For example, in 1976 the Jiangsu Institute of Agricultural Sciences in Nanjing had 300 such projects, and the Agricultural Research Institute in Gongzhuling had 500.[30] Thus, China's total agricultural extension effort in plant protection is greater than that of any other country.

## *INSECT PEST CONTROL: RESEARCH AND TECHNOLOGICAL DEVELOPMENTS*

The most successful aspects of plant protection research in China have been the large-scale development and implementation of programs for practical integrated pest management. IPM is the evaluation and consolidation of all available techniques into a unified program to manage pest populations so that economic damage is avoided and adverse side effects on the environment are minimized.[31] It involves determining means of modifying

[29] Kelman and Cook, "Plant Pathology."
[30] *Ibid.*
[31] NAS, *Insect Pest Management and Control,* Publication 1,965 (Washington, D.C.: NAS, 1969).

the life system of a pest to reduce its numbers to tolerable levels, applying biological knowledge and current technology to achieve the desired modification, and devising procedures for pest control suited to current technology and compatible with economic and environmental quality aspects.[32]

In China, as elsewhere throughout the world, the need to develop IPM programs for major agricultural pests is urgent. The strategy of single-factor pest control interventions using insecticides that seemed so promising in the post–World War II era has become increasingly less attractive because of the development of resistance to insecticides by hundreds of species of important insect pests, disturbances of agro-ecosystems that have resulted in the appearance of devastating secondary pests, widespread environmental pollution by persistent and toxic chemicals, and rapidly increasing costs of pesticides.

IPM has a higher social priority in China than in any other part of the world, and it has brought steady and sometimes spectacular success to agricultural production and crop protection.[33] The following sections review China's major pest problems and methods of control.

*Migratory Locust Program*

The locust program at Weishan Lake, Shandong province is typical of the efforts that have been made.[34] The physiography of the area was analyzed, and in the 1960's, with 40 million man-days of labor and 550 million m$^3$ of concrete, the landscape was transformed by building 423 irrigation systems with 21,552 ditches having a total length of 15,492 km. This system prevented flooding in 39,000 ha and resulted in a good harvest despite 100 rainless days. In 1973, 58,890 ha were planted with rice, and 8,190 ha around the rice fields were planted with shade, fruit, and medicinal trees. This program largely eliminated locust infestations in the region by destroying breeding areas. It also reclaimed thousands of hectares of arable land and improved the local environment.

A similar program in Fengnan county, Hebei province, was very successful. Starting in 1959, six reservoirs and several hundred irrigation ditches were built to drain water from 385 km$^2$, and three million trees were planted. This program reclaimed 11,700 ha, and as a result the area needing insecticide treatment was reduced to 2 percent of the pre-1959 area.

Flight behavior in the locust's solitary and migratory phases has been carefully analyzed. Most migratory locusts fly at 15–20 m height, and some cover 25 miles or more in a single flight. The female's flying speed is greater than that of the male. At 19°C the averages were 4.7 m/sec for males and 3.7 m/sec for females; at 29°C the figures were 7.4 m/sec for males and 5.0 m/sec for females. Females were found to outnumber males by 10–37 percent, and

[32] P. W. Geier, "Management of Insect Pests," *Annual Review of Entomology*, 11 (1966): 471–90.

[33] NAS, *Insect Control*.

[34] "Elimination of Locust Infestation," pp. 247–57.

it is believed that females play the leading role in setting the course of the migration. Chinese entomologists believe that the reflection of moonlight from lakes and rivers is the positive phototropic response that guides locusts to the adjacent breeding grounds. They also believe that the migratory flights are triggered by the arousal of the mating instinct as a result of moonlight and other appropriate environmental conditions.[35]

As an indication of the paucity of information in the Western world about Chinese insect control, Uvarov in his authoritative *Grasshoppers and Locusts* cites only eight Chinese references about *Locusta migratoria manilensis.*[36]

*Biological Control of Insects*

Biological control, a basic component of IPM programs, consists of introducing and encouraging natural enemies to regulate insect pest populations.[37] The natural enemies involved may be insect predators and parasitoids, as well as a variety of insect pathogens: bacteria, viruses, fungi, and nematodes. Biological control has been widely and effectively used in China since the Cultural Revolution and has, in its various manifestations, become an integral and widely practiced component of IPM programs in the culture of cotton, corn, rice, fruit and vegetable crops, and forests. Biological control in China has not followed the traditional pattern of exploration for suitable parasitoids and predators abroad and their importation, culture, and release, which has been the most successful procedure in North America, Western Europe, and Australia. But this deficiency coincides with several years of political isolation after the Cultural Revolution and may be corrected in the future.

The vigor of Chinese biological control efforts is apparent, however, in the widely developed and effective programs for the culture and mass release of the hymenopterous parasitoids *Trichogramma* and *Anastatus,* in

[35] Yu Jijing, Chen Yonglin, and Ma Shijun, *Acta Entomologica Sinica,* 4 (1954): 1; Chen Ningsheng, "The Olfactory Response of *Locusta migratoria manilensis* Meyen and the Function of Its Antennae," *Acta Biologiae Experimentalis Sinica,* 9 (1964): 27–37.

[36] B. P. Uvarov, *Grasshoppers and Locusts,* 2d ed. (London: Center for Overseas Research, 1977); Chen Yongguang and Jin Zhunde, "A Preliminary Investigation of Wing Frequency of the Oriental Migratory Locust," *Acta Entomologica Sinica,* 10 (1961): 436–38; Chen Ningsheng, "Olfactory Response"; Jin Zhunde, Guo Fu, and Zheng Zhuyun, "Food Specialization and Food Utilization of the Oriental Migratory Locust and the Influence of Different Food Plants on Its Growth and Fecundity," *Acta Entomologica Sinica,* 7 (1957): 143–66; Huang Guanhei and Ma Shijun, "Fat Consumption and Water Loss of the Oriental Migratory Locust During Flight and Their Relation to Temperature and Humidity," *Acta Zoologica Sinica,* 16 (1974): 372–80; Guo Fu, Xia Bangying, and Liu Jinglong, "The Action and Source of the Gonadotropic Factor in the Male Migratory Locust," *Kexue tongbao (Science Bulletin),* 17 (1966): 220–23; T. L. Zou, "The Distribution of the Migratory Locust and Ecological Study of Its Breeding Ground in China," *Agricultura Sinica,* 1 (1935): 239–72; Yu Jijing, Chen Yonglin, and Ma Shijun, *Acta Entomologica Sinica,* 4 (1954): 1; and Yu Duanshu and Ma Shijun, "The Choice of Oviposition Site and the Hatching of Eggs of the Oriental Migratory Locust in Relation to the Salt Content of the Soil," *Acta Phytophylactica Sinica,* 3 (1964): 333–44.

[37] C. L. Metcalf, W. P. Flint, and R. L. Metcalf, *Destructive and Useful Insects,* 4th ed. (New York: McGraw-Hill, 1962), pp. 61–62, 412–15.

the practical utilization of coleopterous predators such as *Coccinella septempunctata,* and in microbial control with the bacterium *Bacillus thuringiensis* and its exotoxin, and with the fungus *Beauveria bassiana.* Much of the groundwork for the success of these programs was probably developed during the period of close association with the U.S.S.R. during the 1950's when many Chinese made extended visits to the Soviet Union.[38]

*Trichogramma.* Chinese entomologists have identified 12 species of these egg parasites, which are tiny wasps about 0.4 mm in length weighing about 4 μg.[39] *Trichogramma* have been characterized as living insecticides, since the adult female wasps have great mobility and highly developed searching capacity for the eggs of lepidopterous pests. Four species of *Trichogramma, T. dendrolimi, T. australicum, T. ostrineae,* and *T. japonicum,* are mass-produced in special institute laboratories and communes. The parasitic wasps are cultured on the large eggs of the giant silkworm moths, *Antheraea perniyi* or *Samia* (=*Philosamia*) *cynthia,* or in some cases on the eggs of the rice moth, *Corcyra cephalonica.* As many as 175 *Trichogramma* have been reared from a single giant silkworm egg, and the average is about 60. In commercial production large numbers of giant silkworm cocoons are placed on an automated belt that is rotated to place newly emerged virgin female moths in a convenient position for collection. The moths are passed through an electric meat grinder, which squeezes out the intact eggs. These are stored and can be used over a period of time for parasitoid production. The eggs are cleaned, dried, and glued onto paper cards, either for immediate use or for short-term storage. The cards are placed in large glass cages or alternatively in outdoor cages containing literally millions of adult wasps, whose eggs are laid in the silkworm eggs on the cards.[40]

After parasitism of the eggs on the cards, the latter are ready for release at carefully calculated dosages. It is important that the emergence of the *Trichogramma* wasps from the egg cards coincides with the oviposition patterns of the pests to be controlled. Release rates of *Trichogramma* in China are high compared with those used in previous attempts in other parts of the world, which undoubtedly contributes to the high degree of control obtained.

Several major insect pests are controlled with *Trichogramma,* which are an integral part of IPM strategies. European corn borer, *Ostrinia nubilalis,* in Jilin province was controlled by *Trichogramma* releases in programs that covered over 34,000 ha in 1974. The release rate was about 230,000 to 310,000 per ha at about 90 points per mou (1 mou = 0.066 ha) beginning just before the corn borer eggs were laid and during the peak egg-laying period. The cost was stated to be slightly more than 2 yuan per ha.[41]

[38] NAS, *Insect Control.*

[39] Pang Xianfei and Chen Tailu, "*Trichogramma* of China (Hymenoptera: Trichogrammatidae)," *Acta Entomologica Sinica,* 17 (1974): 441–54.

[40] NAS, *Insect Control,* pp. 90–91.

[41] "The Control of European Corn Borer Using Trichogrammatid Egg Parasites," *Acta Entomologica Sinica,* 18 (1975): 7–9.

Rice leaf roller, *Cnaphlocrocis medinalis,* is controlled in Hunan and Guangdong provinces as well as in other areas by release of *Trichogramma,* with about 70–85 percent parasitism. The rate of release is correlated with the egg density of the pest and ranges from 150,000 to 600,000 per ha. As much as 20 percent of the total rice acreage in Guangdong province is treated in this way.[42]

The spiny pine caterpillar, *Dendrolimus sibericus,* is controlled in the Jilin Forest Preserves by *Trichogramma* release at levels of 10,000 wasps at 10 points per ha. Control ranges from 80 to 90 percent parasitism at a cost of about 1.4 yuan per ha. The method is ideally suited for use in the densely wooded mountainsides where reforestation is practiced.

*Trichogramma* also play an important part in other IPM programs developed in China for managing the pink bollworm, *Pectinophora gossypiella;* the sugarcane borer, *Diatraea saccharalis;* and other lepidopterous pests. A high degree of technology has been developed for the rearing and mass release of the four important species of *Trichogramma.* The generally successful results come from suitable production and release methods, adequate short-term prognosis of weather conditions and target insect status, and use of the most effective *Trichogramma* species and/or strain for the area and crop involved. In general, stable control as good as or better than that from the use of chemical insecticides was reported. Obviously, the successful use of *Trichogramma* is incompatible with the regular usage of broad-spectrum insecticides. The successful employment of *Trichogramma* in Guangdong province, for example, has cut insecticide usage by half or more. This was reported to be welcomed enthusiastically by farm workers because personal hazard from pesticide poisoning was substantially reduced.[43]

*Coccinella septempunctata.* This large and aggressive ladybird beetle has been widely transferred from natural vegetation into cotton and citrus for the control of aphids and leaf rollers.

*Dibrachys cavus.* This indigenous wasp parasitizes overwintering pink bollworms and has been mass-reared and liberated in barns in Shanghai and Shaanxi province to control pink bollworms pupating in stored cotton seeds.[44] Use of this parasite has been an important part of the cotton IPM program.

*Ducklings.* A particularly innovative program of biological control using young ducklings as predators has been developed by the Zhongshan University Institute of Entomology. Ducklings are herded through rice fields in flocks of 1,000 or more and voraciously consume a variety of rice pests, including the rice grasshopper, *Oxya chinensis;* the brown plant hopper, *Nephotettix cincticeps;* and the rice stem borer, *Chilo suppressalis.* Al-

[42] NAS, *Insect Control,* pp. 92–98.

[43] *Ibid.,* p. 97.

[44] Zhu Hongfu, "Strategies and Tactics of Pest Management with Special Reference to Chinese Cotton Insects," *Acta Entomologica Sinica,* 21 (1978): 297–308.

though this is an ancient Chinese practice, it is supported by meticulous research on predator/prey ratios and duck biology. For example, a single hungry duckling was found to consume about 200 insects per hour. In the Daxia Commune of Guangdong province, about 3,900 ha of rice are protected each year by about 220,000 ducks that are herded through the early rice crop by 853 duckherders. The pest population is reduced by 65–70 percent. This basic program is supplemented by the liberal use of *Trichogramma* and *Bacillus thuringiensis.* The amount of insecticides applied on early rice decreased from 75,427 kg in 1972 to 34,112 kg in 1974 and to only about 6,700 kg in 1975. This program is very popular with peasants working in the field who strongly prefer a pesticide-free environment and presumably with the ducks as well.[45]

*Microbial controls.* The utilization of insect diseases in control programs is practiced on a large scale in China. Millions of metric tons of formulations of the spore-forming bacterium *Bacillus thuringiensis* (BT) are applied to rice, corn, cotton, vegetables, and forests for the control of lepidopterous larvae. BT is produced by both liquid fermentation and semisolid lower-yield fermentation methods using damp peanut and soybean meal in shallow baskets. The latter is cheaper and better suited to local commune production, and the output is standardized by bioassay before use. The production of BT insecticide has been supported by substantial scientific research on various serotypes of BT, and 12 distinct strains have been characterized. A new and effective form now used is *B. thuringiensis ostrineae.*[46] The beta-exotoxin yields from various types or strains have been evaluated in reference to strain of bacterium and methods of culture. BT applications were stated to produce 90–100 percent kill of pine caterpillar and 80–100 percent kill of cabbage armyworm, *Barathra brassicae.* BT is highly compatible with other biological control methods, such as mass release of *Trichogramma,* and because of its safety is popular with peasants.

The fungus *Beauveria bassiana* has been described for many years as a possible microbial control, but its use has not been developed in the West. However, a strain attacking the European corn borer has been successfully used in Jilin province in conjunction with *Trichogramma* on more than 60,000 ha.[47] The fungus is cultured in flasks, crocks, or shallow pits in the ground, using wheat and rice bran as the culture medium. *Beauveria bassiana* is applied as a granular or powdered product to piles of corn stalks and stubble collected in the fall to kill the overwintering larvae and to the tassels and silks of the summer crop. The control obtained was described as 83–96 percent for the overwintering larvae and 80–90 percent for the spring generation.[48] The use of *Beauveria bassiana* has the advantages of efficiency,

[45] NAS, *Insect Control,* pp. 106–8.

[46] *Ibid.,* pp. 91, 99.

[47] Xiu Zhengfeng et al., "Field Application of *Beauveria bassiana* (Bals) Vuill. for European Corn Borer Control," *Acta Entomologica Sinica,* 16 (1973): 203–6.

[48] NAS, *Insect Control,* pp. 100–2.

low hazard, simplicity of production, and low cost. It is fully compatible with all other IPM techniques.

Among other approaches, nuclear polyhedrosis viruses (NPV) are being developed for the control of such pests as tussock moth, the cotton leafworm, and cabbage caterpillars.

*Light Traps*

Insect forecasting and control in China involves the use of light traps on a scale much greater than that in any other country. In Hunan province there are said to be about one million "blacklight" traps in an area of about 210,500 km$^2$. Networks of electric power lines intersected most of the rice fields, and at night the eerie blue-violet glow of numerous blacklight traps was visible from the train traveling between Changsha and Guangzhou. The spacing of these blacklight traps varied from one trap per 1 to 2 ha in Jiangsu province, one trap per 2 to 3 ha in Guangdong province, and one trap per 3 to 4 ha in Hunan province.

A variety of traps were seen; all consisted of a 20-watt fluorescent tube emitting a continuous spectrum of light from 300 to 500 nm, but with peak emission at 365 nm.[49] The fluorescent tube is centered between four vanes of glass, and the whole apparatus is commonly mounted over a permanent brick water basin or bucket. The water basin usually contains sticky rice water to trap the stunned insects, and the nutritious gruel thus obtained is fed to fish or hogs. For forecasting, electric grids or cyanide jars were used as the killing device beneath the fluorescent tube.

The development of these blacklight traps was apparently the result of careful study of the response of night-flying lepidoptera to monochromatic light. Ding et al. studied the phototactic response of *Heliothis armigera* and *H. assulta* to 13 monochromatic lights ranging from 333 to 656 nm.[50] The 333-nm light was most attractive to these cotton bollworms, but armyworm adults responded best to 375 nm. At the Nantong Institute of Agriculture, the attractive power of 20-watt blacklight traps and 200-watt incandescent traps was evaluated. During 1973 a combination of the two light sources caught 63,097 destructive insects, compared with 11,035 by the blacklight alone. In 1972 the blacklight–incandescent light combination caught 38,940 pests, the blacklight alone caught 14,956, and a combination 20-watt blacklight plus a 20-watt white fluorescent light caught 40,573 pests.[51]

*Insecticides*

The origin of the use of indigenous Chinese plants containing insecticidal constituents is lost in the mists of 5,000 years. However, their use was ex-

[49] *Ibid.*, pp. 133–36.

[50] Ding Yanjin, Gao Wuqing, and Li Dianmo, "Study on the Phototaxic Behavior of Nocturnal Moths: The Response of *Heliothis armigera* (Hubner) and *Heliothis assulta* Guenee to Different Monochromatic Lights," *Acta Entomologica Sinica*, 17 (1974): 307–17.

[51] "Effects of Light Traps Equipped with Two Lamps on Capture of Insects," *Acta Entomologica Sinica*, 18 (1975): 289–94.

tensive in pre-Revolutionary China and continues to be substantial today. In 1958 over 500 native plant and mineral products were processed into more than 10 million tons of insecticides and fungicides.[52] Among the principal products derived from native plants were alkaloids such as nicotine from *Nicotiana* spp., anabasine from *Anabasis aphylla,* rotenoids from *Derris* spp., and extractives from *Kalanchoe* (effective against aphids) and garlic. Research in this area is still continuing, and in conjunction with the renewed interest in Chinese medicinal plants, useful discoveries may be made.[53]

The worldwide publicity given to organochlorine insecticides, such as DDT, BHC, and other new synthetic organic insecticides, after World War II had an immense impact on pest control in China. Domestic production of DDT and BHC began after 1949, and about 60 million kg of BHC and 4.3 million kg of DDT were produced in 1958. The introduction of BHC was especially rapid and increased by 11,000 percent between 1952 and 1958. It was used throughout China for the control of major pests. Airplane applications were widely used for control of the migratory locust; between 1954 and 1958 more than 5.0 million ha of crops were saved from locust outbreaks in this way.[54] At first this widespread use of the organochlorines was widely acclaimed. In 1963, Tsien-hsi Cheng said that "the mass production and wide application of BHC is a milestone in the history of Chinese agriculture." However, further usage in China produced the same series of problems as found elsewhere in the world—widespread insect pest resistance, development of secondary pests into major pest problems, and environmental pollution. By 1975 the use of BHC and DDT was rapidly being phased out and was no longer permitted on fruits and vegetables.[55]

BHC preparations of high gamma-isomer content were still used as granular applications to control the European corn borer, *Ostrinia nubilalis,* and the cotton thrips, *Thrips tabaci,* and BHC dusts were used on rice and cotton. DDT was used on cotton in a limited way, as a single application to control the pink bollworm, *Pectinophora gossypiella,* and the bollworms, *Heliothis armigera* and *H. assulta.* DDT is also used to a minor extent on tobacco and grapes in conjunction with Bordeaux mixture to control leaf rollers.

Early in the 1950's a national decision was made not to produce the cyclodiene insecticides (aldrin, dieldrin, heptachlor, endrin, and chlordane) for agricultural use because of their long-term persistence and environmental toxicity. Small amounts of chlordane and mirex are still used for termite control in soil and timbers. Toxaphene is used in China in a limited way to control bollworms, but its use is forbidden near rice paddies and fish ponds

[52] Cheng, "Insect Control."

[53] Shin-feng Chiu, Lin Shih-ping, and Ching-yung Hu, *Toxicity Studies of Insecticidal Plants of Southwestern China* (Guangzhou: National Sun Yatsen University, College of Agriculture, 1944); and *Zhongguo tunongyao zhi* (Manual of Chinese Native Pesticides) (Beijing: Kexue Chubanshe, 1959).

[54] Cheng, "Insect Control."

[55] NAS, *Insect Control,* pp. 66–67.

because of its high toxicity to fish. The independent Chinese experience with the persistent organochlorine insecticides is particularly interesting in view of the controversy produced by their regulation and ban in many Western countries.

Organophosphorus insecticides are the basic synthetic insecticides in use in China today, and more than 30 different compounds are produced commercially. Production of organophosphorus insecticides at the principal factory in Shanghai was more than 18 million kg in 1974. The order of production and use was described as trichlorfon $>$ dichlorvos $>$ dimethoate $>$ phosmet $>$ fenitrothion $>$ phosphamidon $>$ malathion.[56] Parathion and methyl parathion were used widely in the late 1950's as replacements for BHC and DDT, but problems with insect resistance and human and animal toxicity have greatly curtailed their use, which is largely limited to dusts for control of rice pests. The use of methyl parathion and demeton is now forbidden on fruit-bearing trees, and the latter is used only on cotton to control the cotton aphid, *Aphis gossypii.*

The emphasis today is entirely on organophosphorus insecticides of low mammalian toxicity. Trichlorfon, introduced in 1957, is apparently the insecticide produced on the largest scale in China today.[57] It is publicized as the "enemy of a hundred worms." Its popularity is due to its safety, lack of persistence, specificity against chewing pests, and environmental degradability. Trichlorfon is the main deterrent against chewing insects in IPM programs in cotton and rice, and on cabbage and other vegetables and deciduous fruits.

Dichlorvos (DDVP) is widely used as a residual fumigant for crop pests, e.g. for the soybean pod borer, *Grapholitha glycinivorella.* Sorghum stems are soaked in the technical material and placed among the beans at a rate of 300 stems per ha, using 1 kg of dichlorvos. Dichlorvos is also used in cotton fields, absorbed on a porous carrier, to control bollworms and as an aphicide on cotton, tobacco, and cabbage, and is the principal insecticide used to control flies and mosquitoes.

Dimethoate, another safe insecticide, is widely used on field and fruit crops for general pest control and is the basic insecticide for IPM programs on deciduous fruits, citrus, rice, and cotton. Phosmet is the basic insecticide used on tea to control scale, *Leucaspsis japonica.* Extensive research has demonstrated that phosmet does not result in detectable residues or off-flavors in tea picked seven days after application.

Use of parathions has been greatly curtailed, and fenitrothion, a much safer derivative of methyl parathion, has been substituted as a general insecticide for control of rice pests. Multiple applications are widely used for the control of green rice leafhopper, *Nephotettix cincticeps,* and brown plant hopper, *Niloparvata lugens.* Malathion, another safe organophosphorus

[56] *Ibid.*, pp. 77–79.
[57] *Ibid.*, pp. 77–78; and Wang Nenwa, "Characteristics and Application of Dipterex," *Agricultural Science Bulletin* (Beijing), 1958, 4.

ester, has limited use for citrus pest control and is used as a household insecticide.

Phoxim, a very safe insecticide, has been evaluated in massive field programs on more than 50 pests of rice, cotton, tea, tobacco, fruits, vegetables, and stored grains. Its great safety and short residual action make it popular with peasants.[58] Carbamates have not been widely used for pest control in China. Carbaryl is the only carbamate extensively applied; it is used for the European corn borer and for bollworms and aphids on cotton. There is much concern about its toxicity to bees, and its use involves official warnings to agriculturists not to let bees out. The highly toxic carbofuran, aldicarb, and methomyl have apparently not been considered for development. Chlordimeform is widely used in IPM programs, especially for red spider mites on deciduous fruits and for bollworms on cotton.

Insecticide resistance has proved as vexatious in China as elsewhere. The housefly and numerous species of mosquitoes have become resistant to DDT and BHC, and to various organophosphorus insecticides. Fruit-tree red spider mites and the cotton aphid, *Aphis gossypii*, have developed resistance to parathion and methyl parathion and to a wide range of other organophosphorus insecticides. As a result chlordimeform is now used for control in many areas. Serious resistance problems occur in rice; the paddy borer, *Tryporyza incertulas*, is highly resistant to both DDT and parathion, and the rice leafhopper is highly resistant to trichlorfon and BHC.[59]

Impressive efforts are being made to delay the onset of insect pest resistance by alternating insecticide treatments on apples and citrus; three to five different insecticides are used in rotation. Determinations of resistance and cross-resistance are given high priority. However, the prevailing scientific attitude about resistance is pessimistic, and the increase in the resistance problem is a major factor in encouraging pest management programs.

Experience in China with the widespread use of broad-spectrum insecticides has been generally unfavorable, partly because of the emergence of secondary pests. The experience with deciduous fruits has been typical. The early use of DDT gave good control of a wide variety of leaf rollers, but nearly eliminated the natural enemies of many pests, including red spider mites. The mites were controlled with parathion at first, but rapidly became resistant. The use of insecticides increased until they accounted for 25 percent of the production cost of the fruit. The leaf roller, *Leucoptera scitella*, became a rampant secondary pest. Intensive ecological study led to complete abandonment of broad-spectrum insecticides and to the development of a pest management program emphasizing orchard sanitation; handpicking of overwintering cocoons; interplanting orchards with legumes to encourage the ladybird beetle, *Coccinella septempunctata*, and other predators as well as the parasite *Aphytis proctia;* together with selective spraying using lime sulfur, trichlorfon, and chlordimeform.[60]

[58] NAS, *Insect Control*, pp. 86–87.
[59] *Ibid.*, pp. 83–85.
[60] *Ibid.*, p. 64.

Much the same applies to citrus pests. The use of DDT and BHC eliminated the principal chewing insect pests and those sucking pests with weak adaptive capacity. The substitution of organophosphorus insecticides produced resistant races of pests, especially of the citrus red mite, *Panonychus citri,* which became the dominant pest. More recently, new lepidopterous and coleopterous pests have invaded citrus orchards as a result of the effects of pesticides and changes in cultural practices.

Rice is China's most important crop and suffers from the attacks of at least 89 species of insect pests. The regular use of DDT and BHC in the 1950's gave excellent control of some of these pests, including the rice grasshopper, *Oxya chinensis,* the rice green stink bug, *Nezara viridula,* and the rice leaf weevil, *Lema oryzae.* However, the rice stem borer, *Chilo suppressalis,* remained the most important pest, destroying an average of 20–30 percent of the crop in Hunan province. The use of broad-spectrum pesticides and changing crop practices caused the rice leafhopper, *Nephotettix cincticeps,* and the brown plant hopper, *Niloparvata lugens,* to become serious pests (the former transmits the virus diseases yellow stunt, yellow dwarf, and common dwarf). Secondary pests, the grass leaf roller, *Cnaphalocrocis medinalis,* and the rice thrips, *Thrips oryzae,* developed into major pests causing great damage to yields. The change to double cropping of rice decreased the importance of the rice stem borer, but the paddy borer, *Tryporyza incertulas,* became the major pest of late rice and is now resistant to DDT and parathion. Other secondary pests, the purplish stem borer, *Sesamia inferens,* and the rice case butterfly, *Parnara* spp., are increasing in severity.

There is much concern in China about the overall effects of the intensive use of pesticides on human health and environmental quality. The use of such highly toxic insecticides as parathion and methyl parathion is very closely regulated, and handling is limited to highly trained and experienced personnel. The Ministry of Agriculture has established strict rules about application equipment, protective masks and clothing, and the availability of antidotes. Rules at the production brigade level prevent the storage of concentrates of dangerous insecticides, e.g. rat oral $LD_{50} < 10$ mg per kg. A special person controls these materials in each production brigade. Every effort is being made to replace these dangerous materials with safer pesticides.

At the national level there are a number of restrictions on pesticide use. These control organomercurials and arsenicals and forbid the use of parathion and demeton on fruits and vegetables. DDT and BHC are being phased out. New insecticides that have adverse health and environmental properties must have governmental approval before use. An Institute of Environmental Protection was established in Beijing in 1978, and a national system of pesticide residue tolerances is being developed along with a system of restricted periods between application and harvest.

Insecticides play important roles in China's pest management programs, in its remarkable progress in producing food for more than one-fifth of the world's population, and in its exceptional standards of hygiene and public health. More than 100 individual insecticide compounds are produced commercially in China, and the government aims to become totally self-sufficient in this area. Insecticides, we were told, are to be used with the objectives of "guaranteeing good harvests and diminishing plagues, not for the benefit of certain persons." There are no patents or secret techniques, and production methods are circulated immediately throughout China.

The manifold disadvantages of the unrestricted use of broad-spectrum pesticides have been recognized, and efforts are being made to develop procedures for the judicious use of insecticides in a sound ecological framework. There is almost no aerial application of insecticides; selective treatment of hot spots of infestation in field or orchard is the general rule. Biological and cultural control measures are used wherever feasible, and supplemental insecticide controls are planned so as to be minimally disruptive to the total ecology of field, orchard, and forest.

### *Pheromones*

As elsewhere, the potential use of insect sex pheromones for monitoring insects and trapping them has captured the enthusiasm of Chinese entomologists. Active research projects enjoying considerable support are concerned with the identification and production of the pheromones of such diverse pests as the pine caterpillar moth, *Dendrolimus punctatus;* the European corn borer, *Ostrinia nubilalis;* the pink bollworm, *Pectinophora gossypiella;* the oriental fruit moth, *Grapholitha molesta;* and the paddy borer, *Tryporyza incertulas*. This research has included excellent analytical and synthetic chemistry and detailed biological studies.[61]

Field studies to investigate the utility of these pheromones for insect control are being conducted on a large scale. Mass trapping with the pink bollworm pheromone, *gossyplure,* in water basin traps positioned through a 27-ha cotton field at 30 traps per ha captured 290,000 male pink bollworms. However, this was estimated to be only 25 percent of the total number present, and it was concluded that the insect population was too large to control by mass trapping. Attempts were made in 1974 to disrupt male courtship by permeating the air with *gossyplure* in 2-ha test plots using the pheromone on string wicks or in polyethylene bags. The results were encouraging, and both methods resulted in about a 90 percent decrease in male response. As an indication of the effort involved, field research on the pink bollworm involved 139 production teams in 44 production brigades in Hanhu county, Hebei province.[62]

### *Host Plant Resistance*

Breeding crop plants for resistance to insect pests appears to be in its infancy in China. However, progress is being made in selecting and distribut-

[61] *Ibid.*, pp. 116–24.
[62] *Ibid.*, pp. 122–23.

ing two varieties of soybean resistant to the soybean pod borer, *Grapholitha glycinivorella.* The cotton aphid, *Aphis gossypii,* is a most harmful pest, and a large number of resistant varieties have been selected. In 1975, it was estimated that 47,000 ha of aphid-resistant cotton were being grown in Shaanxi province.[63] Host resistance studies for insect pests of corn, wheat, and rice were just beginning in 1975.

## METHODS OF PLANT DISEASE CONTROL

### Control by Exclusion of Diseased Plant Material and Pathogens

One of the foundation stones of disease control in the PRC is the intensive commitment to restrict the movement of any seed or plant propagation material that may carry plant pathogens.[64] Efforts are made at all agricultural administrative levels to ensure that seeds of major food crops are pathogen-free. This is well illustrated by the quarantine restrictions enforced in 1974 in Shanghai that prevented the unloading of several American ships with cargoes of wheat found to contain spores of the dwarf bunt fungus, *Tilletia controversa.* However, not all quarantine restrictions are justified. A quarantine against dwarf bunt fungus in Jilin province has been in effect since 1966. But only spring wheat is grown in this province, and it should not be affected.[65] A similar quarantine exists against wheat seed from fields affected by the "take-all" pathogen, *Gaumannomyces graminis* var. *tritici;* it is based on the possible presence of root pieces from infected plants that may be carried in seed shipments. This results from the practice in some areas of harvesting wheat by pulling plants from the soil.

Since many communes produce their own seed and propagate other planting stock, the possibility of spreading a pathogen over a large area by the use of contaminated or infected seed from a single major source is minimized. In the case of certain virus diseases, communes probably do not have the capability of using relatively sophisticated techniques for detecting viruses in seed, such as immunofluorescent staining and related procedures. Thus, they may not be able to ensure that material used for propagation is virus-free. However, if the seed supply on an individual commune or groups of communes within a county becomes contaminated, the flexibility exists to obtain a supply from disease-free areas. One of the important functions of each county government is to maintain a "seed station." In each station, one individual is responsible for preparing certification slips for shipments of seed out of the county. Several cases can be cited in which rigid quality seed control resulted in control or elimination of specific diseases. For example, the use of wheat seed only from the areas free of flag smut, *Urocystis tritici,* is claimed to have been the major factor in control of that disease in Shaanxi

[63] *Ibid.,* pp. 113–16.

[64] *Plant Quarantine Subjects,* Handbook for Plant Protection Workers, no. 7 (Shanghai: Shanghai Renmin Chubanshe, 1971).

[65] Kelman and Cook, "Plant Pathology."

province. In Jiangsu province careful studies on transmission of the bacterial streak pathogen, *Xanthomonas translucens* f. sp. *oryzicola*, demonstrated that it was spread mainly by rice seed. Strict quarantine measures were imposed; communes on which the disease appeared were not permitted to sell their seed to other communes and were required to obtain pathogen-free seed. All seed lots shipped from one county to another had to be certified free of bacterial streak. As a result of a rigidly enforced quarantine and use of disease-free seed for new plantings, incidence of this disease has been markedly reduced in Jiangsu and Guangdong provinces.

Vegetable seed and vegetatively propagated planting material produced in a commune are usually used only in that commune to reduce the risk of spreading pathogens. Similarly, the transport of vegetables is restricted to short distances from points of production; this also minimizes long-distance spread of pathogens.[66]

A commitment to clean seed has meant a willingness to use hand labor to an extent that probably has few parallels in modern agriculture. This is illustrated by efforts to remove seed infected by the scab fungus, *Gibberella zeae*, from healthy wheat and barley grain. Scabby seed are separated from healthy seed by repeated hand-winnowing or by flotation in 20 percent salt solutions.[67] In Shaanxi province large grain storage centers refuse to accept wheat unless it is completely free of scab-infected seed. As a result, the communes have to accept the cost of cleaning seed.

Years ago the wheat gall nematode was a severe problem in northern growing areas. Clay-silt emulsion and salt-flotation methods were used to ensure seed free from nematode-infested "cockels" or galls. A machine for removal of the cockels was developed by the late Professor V. M. Zhu of Jiangsu Academy of Agricultural Sciences and widely used for screening out cockels. This disease is no longer considered to be a large-scale problem in China.

Until recently, large-scale seed potato production was restricted to Heilongjiang, Jilin, and Inner Mongolia.[68] When attempts were made to grow seed potatoes in southern provinces, high yields could not be maintained. At one time over 200,000 tons of seed potatoes were shipped annually from Heilongjiang and neighboring provinces to southern regions. Since 1960 widespread and marked changes in cultural practices have made it possible to reduce the impact of viruses on crop yield in the southern provinces and to increase the use of locally grown seed potatoes.[69]

Early varieties are used to produce two crops a year, with the first crop of

[66] P. H. Williams, "Vegetable Crop Protection in the People's Republic of China," *Annual Review of Phytopathology*, 17 (1979): 311–24.

[67] V. A. Johnson and H. L. Beemer, Jr., eds., *Wheat in the People's Republic of China*, Report no. 6 (Washington, D.C.: Committee on Scholarly Communication with the People's Republic of China, 1977).

[68] Chinese Academy of Sciences, Institute of Genetics, Tuber Group, *Struggle Against Potato Degeneration* (Beijing: Kexue Chubanshe, 1973).

[69] *Ibid.*

presprouted tubers planted as early as possible in the spring. One month before the normal harvest time, disease-free plants are carefully selected and marked in the plots from which seed tubers are to be harvested. Diseased plants are rogued whenever symptoms of virus diseases (leaf roll, PVX, and PVY) appear. The tubers to be used for seed are in some instances removed carefully by hand from growing plants without digging the entire plant. Plants from which these tubers are removed are often left to continue growing and produce additional tubers at the normal harvest period. The seed tubers from the spring crop are then chemically treated to break dormancy, presprouted, and planted for the second crop. Before frost, any plant showing symptoms of virus infection or of other diseases is rogued. Small- and medium-sized tubers are selected as seed for the spring crop the following year, and larger tubers are reserved for consumption. In areas where the ring rot pathogen, *Corynebacterium sepedonicum,* may be a hazard, only small tubers from healthy plants are selected for seed. The elimination of cutting of seed tubers reduces the prospect of dissemination of the ring rot bacterium. Apparently seed potatoes are the only vegetables subject to quarantine in movement across county lines. This restriction is concerned mainly with the spread of seed potatoes infected by *C. sepedonicum.*

A number of well-adapted varieties of potatoes with short growth periods that can be used in this double-cropping procedure are currently available. Although there are local variations in the application of this system, it appears that the procedure is effective and has resulted in a large increase in potato production in the southern provinces.

In Guangdong province, these procedures have been adapted for production of three potato crops per year. Seed tubers come from early varieties planted in the fall and spring. The winter crop is planted mainly for local consumption. However, in areas of southern China where brown rot or bacterial wilt, *Pseudomonas solanacearum,* is present, plantings are made only in the late fall and early spring to avoid the high summer temperatures that enhance wilt development. So far, no wilt-resistant varieties have been developed for this region.

In addition to developing procedures for obtaining virus-free seed in the southern provinces, projects are under way at the Cytology and Tissue Culture Laboratory of the Institute of Botany in Beijing, as well as in certain other institutes, to produce virus-free seed potatoes using meristem culture techniques and serological procedures. Virus-free clones developed from tissue cultures are planted in cool regions in Inner Mongolia, where aphid transmission of viruses is unlikely to occur. Plants are indexed using serological procedures to determine whether they are virus-free. Tests for the presence of potato viruses X and Y are made. In addition, tuber-indexing procedures are followed at a number of experimental stations. Virus-free seed potatoes are then sent to other provinces as a source of foundation seed and for further multiplication. At present only a limited number of provinces may be directly involved in this program.[70] Potato seed growers in some areas prefer to use true seed to obtain virus-free potatoes. The use of botanical seed for planting of potatoes has been perfected further in the PRC than in any other country.

*Control by Avoidance*

Certain diseases that are intensified under overhead irrigation, in particular bacterial diseases, are not favored by the extensive use of ditch and furrow irrigation practiced in China. The relatively infrequent occurrence of many of the seed-borne bacterial diseases of legumes, crucifers, and solanaceous crops in North China may in part be attributable to this method of irrigation.[71] In Central and South China, the severity of bacterial soft rot of Chinese cabbage has been reduced by the use of deep furrows and high ridges to drain excess soil moisture after heavy rains. This was first recommended by W. F. Chiu in 1952 to control this disease in North China.[72] This procedure is not recommended for use on saline soils.

Early or late planting to avoid disease has become a common practice for certain vegetable crops. Eggplants and peppers are planted early under plastic so that they start producing fruit in July, one month ahead of normal; this helps avoid losses from *Phytophthora, Pythium,* and *Phomopsis.* Changes in planting dates to avoid the peak dissemination of viruses have been recommended for the control of mosaic diseases of Chinese cabbage and wheat yellows caused by the barley yellow dwarf virus and the wheat rosette dwarf.[73]

The practice of planting two vegetables in the same plot, which can increase efficiency of land use in many instances, also serves to minimize intensification of damage from certain foliar pathogens. This is applied most effectively if unrelated species are used, such as cabbage, eggplant, or chives. However, downy mildew, *Peronospora parasitica,* and bacterial soft rot, *Erwinia carotovora,* are difficult to control when Chinese cabbage, leaf mustard, cauliflower, and related species are used in intercropping sequences.

Double or triple cropping wheat with rice or corn has favored the intensification of head blight and scab, *Gibberella zeae,* to the point that it is considered a major problem on wheat.

Sexual spores (ascospores) that can be disseminated by the wind are released from perithecia formed on rice, corn, or wheat straw. The greatest damage from this disease occurs in the lower Chiang Jiang area, where soils are moist at the surface during most of the period when wheat is growing;

[70] Williams, "Vegetable Crop Protection."

[71] *Ibid.*

[72] W. F. Chiu, "The Theoretical Basis of Integrated Control of Soft Rot of Chinese Cabbage," in *China's Plant Protection Science* (Beijing: Kexue Chubanshe, 1961).

[73] W. F. Chiu, "Studies on Kuting, a Virus Disease of Chinese Cabbage," *Acta Phytopathologica,* 3 (1957): 31–43; idem, *Virus Diseases of Cereal Crops* (Taiyuan, Shanxi: Agricultural Bureau, 1974); and idem, "On the Development of Virology of Agricultural Plants," *Scientia Agricultura Sinica,* (1978): 68–69.

prior to anthesis, temperatures are warm and humidity high because of air currents from the South China Sea. The severity of this disease can be minimized by improving soil drainage so that the water table is maintained below 1 m during the wheat-growing period. It is also necessary to use resistant varieties and, in some cases, fungicides to obtain effective control.

*Eradication*

The direct removal of weed hosts and diseased plants is one of the key cultural practices that are employed effectively in China to reduce losses from virus diseases. For example, in an effort to reduce wheat rosette dwarf, it has become a standard practice on communes in northern China to encourage removal of all grasses along the ditches, furrows, and adjacent areas to destroy the source of virus inoculum and hibernating insect vectors before planting. This labor-intensive practice, in conjunction with insecticidal control of the plant-hopper vector, *Delphacades striatellus,* seems to have controlled this virus. Similarly, wheat streak mosaic has been reduced in Shaanxi province by not locating wheat fields next to fields where host grasses (*Panicum*) are prevalent. A number of communes have attempted to eradicate those grasses, such as *Panicum miliacerim,* that serve as hosts for the vector, the mite *Aceria tulipae.* These grasses grow during the wheat-free period of June through September. In the absence of the grass hosts, the cycle for transmission can be effectively broken. An effort of this type would be beyond practical application in the United States and Western European countries, but eradication of a given weed host may be possible when a large and well-organized labor force is willing to make the necessary effort.

Sanitation, which is practiced to an unusual degree in vegetable production, may be an important factor in reducing the severity of certain pathogens that can overwinter in plant debris.[74] For example, as soon as such crops as cabbage are harvested, all fallen leaves and other residue are removed by hand from the field and used as animal feed. Leaves and vines of various other vegetable crops are also removed to be added to compost heaps or fed to the ever-present pigs.

Because of the limited land area, it is often impossible to practice crop rotation for major crops such as rice or wheat to the degree that might be desirable. However, crop rotation is recognized as an essential facet of disease control on communes involved in vegetable production near large population centers. The variety of vegetables allows the rotation of individual fields with nonrelated crops. Relatively complex rotation sequences have been developed so that a given species is not replanted on the same plot for three to five years. The increase in damage from soil-borne pathogens has been attributed to a failure to maintain an adequate time lag between planting of closely related vegetable crops.

The practice of flooding fields for paddy rice may be one of the key fac-

[74] Williams, "Vegetable Crop Protection."

tors in the control of certain soil-borne pathogens in China. Paddy rice is grown on a larger acreage than any other crop, and each rice field is flooded for two to four months each year. The anaerobic conditions associated with flooding are known to reduce the survival of many soil-borne pathogens.[75]

Composted organic fertilizer is added to the soil in China on a scale unequalled by any other country. The organic matter includes various proportions of animal manure, crop refuse, and human waste; approximately 25 percent (by weight) is silt dredged from the bottom of canals, ponds, or streams. Plant material (wheat and rice straw) may account for as much as 60 percent of the total. After a period of anaerobic fermentation, the aerated and partially dried product is applied at rates as high as 75–80 metric tons per hectare. Green manure crops capable of fixing nitrogen are grown in rotation with such crops as cotton or sweet potatoes. The heavy use of organic wastes may be one of the key factors in the relatively low incidence of such soil-borne pathogens as plant parasitic nematodes. Continuous application of great amounts of organic material may foster populations of bacteria and fungi that are antagonistic to many common soil-inhabiting fungi.[76] It should be emphasized that severe damage from certain root pathogens on vegetable crops can and does occur, but in general the major field crops appear to be relatively free of damage from such pathogens as *Pythium*, damping-off, root rot, and *Phytophthora* root rots.

### *Breeding for Disease Resistance*

Breeding programs with emphasis on yield and disease resistance receive major attention in the research efforts of all agricultural colleges, provincial research institutes, and the Institutes of Botany, Genetics, and Atomic Energy in Beijing. In general, conventional breeding programs are used. Most breeding work at the colleges and institutes involves cooperative effort between plant breeders and plant pathologists. However, the truly unique feature of the breeding programs is the intense effort given to breeding, selection, and varietal evaluation at the commune level, which involves thousands of workers with relatively little academic training. Apparently research workers from the institutes and agricultural colleges have been able to teach peasants the basic techniques. As a result, there are few, if any, countries in the world today in which so many individuals at all levels of professional training are actively engaged in varietal improvement. Much of this effort does not involve the use of well-designed plots employing appropriate statistical analyses. Nevertheless, the total effort expended on major crops is truly enormous. Since a breeding or crop improvement group has been trained in almost every commune, it is possible for breeders in each province to provide these teams with experimental lines for testing under a

[75] R. J. Cook, "Management of the Associated Microbiota," in J. G. Horsfall and E. G. Cowling, eds., *Plant Disease* (New York: Academic Press, 1977), pp. 145–66.

[76] K. F. Baker and R. J. Cook, *Biological Control of Plant Pathogens* (San Francisco: W. H. Freeman, 1974).

wide range of environmental conditions. Once a particular line or variety has been obtained and tested, it is possible to expand the seed of the new line and distribute it over many millions of hectares in a relatively short period of time.

It should be recognized that as a result of these organizational capabilities, a "Green Revolution" in rice had occurred in China, unheralded and unpublicized, well before dwarf rice varieties were developed at the International Rice Research Institute (IRRI) in the Philippines. Early in the 1960's, improved dwarf varieties were released in China and gained wide acceptance. By 1965 these varieties had been planted on over 3,000,000 ha, and by 1973 the planted area had doubled. In addition to their other attributes, these new varieties were quick to mature (in contrast to the IRRI varieties first released). The relatively short growing period was important in multiple cropping.[77]

The most impressive aspect of the program to improve crop productivity was the intensive effort to exploit available scientific knowledge and to integrate this with peasant farming practices known to work from past experience. It is unlikely that the production increases necessary for food supplies to keep pace with population growth would have occurred without this effort.

The evaluation of lines for resistance to specific diseases is based mostly on the exposure to pathogens under natural field conditions. Field and greenhouse inoculations to evaluate cotton for resistance to *Fusarium* wilt have been used at the Northwest College of Agriculture at Wugong, Shaanxi province, as part of an intensive effort to improve the levels of resistance to *Fusarium* wilt in cotton varieties. In 1974 a cooperative project involving 30 scientists in 17 provinces was in progress. A large collection of isolates from the regions involved had been obtained, and the isolates were being used in the screening process. A large project for evaluating resistance to wheat scab was under way at Jiangsu Agricultural Science Research Institute in Nanjing. The scab fungus was cultured on autoclaved barley seeds, and these were spread on the soil in varietal test plots in March. By early April the perithecial stage had formed, and the ascospore inoculum necessary for uniform infection of susceptible lines was produced. Several varieties with good field resistance to scab have been obtained from this research. It is interesting to note that these varieties may not be inherently resistant to the pathogen, but may escape infection by flowering early while still in the boot or within a tightly closed lemma and palea.

Potato breeders are actively involved in the development of virus and late blight-resistant locally adapted potato varieties. Breeders cooperate with the Potato Research Institute in Heilongjiang, where lower temperatures favor flowering and seed set. Seed produced at the Institute is returned for further selection under local conditions.

[77] International Rice Research Institute, *Rice Research and Production in China* (Los Banos, Philippines: IRRI, 1978).

In Jilin province the severity of rice blast has been minimized by the practice of planting varieties with four different sources of resistance. Under supervision from the Institute for Agriculture at Gongzhuling, this procedure was implemented, with the aim of reducing the appearance of new races.

One of the major contributions of breeding disease resistance in crop plants has been the development of lines of wheat with high resistance to stem, stripe, and leaf rusts. The physiological races have been studied intensively, and their distribution throughout China has been mapped. This has involved the determination of the regions where these rusts overwinter and oversummer. Both vertical and horizontal resistance to literally thousands of cultivars has been studied, and inoculations have been made with known races under controlled conditions in greenhouses.[78]

In parts of Shaanxi and adjacent provinces, over 600 different wheat cultivars are being grown as part of an effort to interfere with the ideal situation for autumn movement of stripe rust from the cool mountains of north-central China eastward to the main winter wheat areas.

Because of the organizational network at the county and commune level, such practices can be implemented over a large area once the applications have received general approval. A similar procedure was being followed in the planting of clones of poplar with different sources of resistance to poplar rust. Apparently at least four different clones were being used in the propagation of cuttings.

In the case of vegetable crops, it is likely that selection by peasant farmers over many years has resulted in lines or cultivars with high levels of resistance in many vegetables. This is particularly true of Chinese cucumbers. (Incidentally, the "China Long" cucumber cultivar was introduced to the United States from China by Dr. H. Porter, who worked as a plant pathologist at Nanking University in the mid-1920's. The primary source for cucumber mosaic resistance that has been used extensively by plant breeders in America is this "China Long" cultivar.)

One of the practices dependent on hand labor for success is the grafting of watermelons on root stocks of a pumpkin, *Cucurbita pepo,* that is resistant to root rot and *Fusarium* wilt. This is done in Shanghai and in Jiangsu and Guangdong provinces. Seedlings are grown in beds or protected frame shelters to the first leaf stage, and the watermelon plants are grafted (cleft or side-grafts) to the pumpkin root stocks. After two weeks, successful grafts are selected for transplanting to the field.

### *Chemical Control*

Fungicides are in widespread use for disease control, in particular for seed treatment and certain foliage pathogens. Treating seed with chemicals to en-

[78] China Wheat Rust Control Research Coordinating Committee, *Progress in the Study of the Prevalence of Wheat Stripe Rust in China and Future Work* (n.p.: CWRCRCC, 1978), 12 pp.; and Y. S. Wu, *Identification and Prevalence of the Physiologic Races of Wheat Stem Rust in China (1956–1965 and 1973–1978)* (Shenyang: Shenyang Agricultural College, Plant Immunity Laboratory, 1978), 38 pp.

hance seedling germination and growth is an ancient practice in China. In a book on farming written in the first century B.C., a variety of seed-treatment materials were discussed.[79] This may be the earliest documented report in the world of the value of seed treatment for disease control.

Seed-treatment materials such as organic mercurials are still used to a limited extent. However, current plans call for the phasing out of these compounds as supplies are depleted. Effective control of the Bakanae disease of rice caused by *Gibberella fujikuori* was obtained with organic mercury seed treatments, but formaldehyde is currently recommended as a substitute; when it is properly applied, the disease can be effectively controlled. Common bunt, *Tilletia foetida,* and flag smut of wheat, *Urocystis tritici,* as well as other cereal smuts, are apparently effectively controlled by using seed-treatment chemicals, or hot-water treatment in the case of loose smut of wheat, *Ustilago tritici.* However, in most countries the chemicals currently in use in China have been replaced by other, more effective systemic compounds.

Application of pesticides on communes is generally dependent on hand labor (usually teams of three to eight young women) and the use of small hand-operated sprayers. Other types of spraying equipment are being manufactured and used on a small scale.

Fungicides are commonly used in the lower Chang Jiang area for control of wheat scab. Two sprays are applied, one at the time of flowering and one a week later. The fungicide of choice is Bavistin (methyl 2-benzimidazole carbamate). In an effort to improve the effectiveness of fungicide applications, ascospore release patterns were studied by the Plant Protection Laboratory of Shanghai Academy of Agricultural Sciences. Spore traps were placed on many communes, and readings were taken by plant protection workers at the brigade and team level. When the numbers of ascospores trapped reached a certain number per field, applications of the fungicidal sprays were made. This is one of the few instances known in which epidemiological data were used directly for control of a field crop disease by timed application of a fungicide.

Intensive work has been under way for over a decade to develop antibiotics for control of foliage diseases. The main search has involved screening of antibiotics for compounds effective in the control of the rice blast disease. In the laboratory of the Microbiology Institute of Plant Physiology in Shanghai, thousands of cultures of *Streptomyces* have been tested for production of antibiotics that might be used as fungicides. One antibiotic, qingfengmycin, that is reported to be highly effective in controlling rice blast has been obtained. The production process for qingfengmycin has been simplified so that the crude antibiotic can be made at the commune level. The strain of *Streptomyces* (*S. qingfengmycin* sp.) is cultured on autoclaved wheat, rice bran, or cornmeal; the antibiotic is then extracted, concentrated, and ap-

[79] S. Shi, tr., *On "Fan Sheng-chi Shu," an Agriculturalist Book of China Written in the First Century B.C.* (Beijing: Science Press, 1974), 68 pp.

plied as a foliage spray.[80] Several other antibiotic sprays have been developed that can be produced at the commune level.

Bordeaux mixture is still widely used; it is usually prepared in the field in large glazed earthenware containers. The cost of the copper sulfate ($0.30/kg) is low, and the lime is readily available since it is produced in kilns on most communes. Bordeaux mixture is commonly used for control of foliage blights—late blight, *Phytophthora infestans,* and early blight, *Alternaria solani,* of potato; and early blight and Septoria leaf spot, *Septoria lycopersici,* on tomato.

The dithiocarbamates have been used more extensively in recent years, particularly for controlling downy mildews on vegetable crops and cucurbits; leaf mold, *Cladosporum fulvum;* and early tomato blight. Daconil is recommended for cucumber downy mildew control in Jiangsu and Guangdong provinces.[81] The shift to other organic fungicides will probably occur at a rapid pace in the immediate future, if costs can be kept low.

Nematicides are used on a limited scale at present; this reflects the general view that plant parasitic nematodes cause relatively little economic loss to Chinese agriculture. Few researchers are working solely on the biology and control of plant parasitic nematodes. In 1977 an infestation of rice by the root knot nematode was reported on the island of Hainan. It is considered to be the first report of this nematode on rice in China. This discovery, as well as other evidence of damage, may encourage a greater level of interest in research on nematode diseases and their control in China. The strain of *Meloidogyne* apparently has adapted to the standard cultural practices that have kept other strains under control in the past.[82] It is possible that flooding of fields for paddy rice, the intensive use of organic matter for fertilizer, the use of intercropping, and shifting of different crops in the growing of vegetable crops may combine to minimize the impact of nematodes and thus reduce the need for their control with nematicides, as is almost essential for the production of certain high-value crops in other countries.

## *METHODS OF WEED CONTROL*

Research on weed control has not been emphasized at research institutes or agricultural colleges in China as much as it has in the United States. This reflects the availability and commitment of the commune labor force to keeping fields weed-free by manual labor and the fact that weeds collected along roadsides or field borders constitute a major source of animal feed. Indeed, some species of weeds are harvested for their medicinal value and for human consumption. Thus, weed populations are generally kept at rela-

[80] "Studies on Qingfengmycin. III: Solid Media Fermentation," *Acta Microbiologica Sinica,* 15 (1975): 315–19.

[81] Williams, "Vegetable Crop Protection."

[82] "A New Paddy Rice Meloidogyne Disease," *Guangdong's Agricultural Science,* 74 (1977): 51.

tively low levels, and the number of weed plants reaching maturity and producing seed is greatly reduced.

In general, the weed species in upland vegetable crops are similar to those that occur in the United States. In the vicinity of Shanghai, the major weeds in wheat were listed as *Poa* spp., chickweed, field bindweed, foxtail, water foxtail, goosetail, and wild oats. Manuals that illustrate in color the characteristics of the major weed species and list the species affecting production of important crops are available. Despite the constant and unremitting effort involved in hand removal of weeds, there is ample evidence of the continuing need for weed control in many major crops. Furthermore, the common use of weeds for animal feed and composting may be a factor in maintaining populations of certain weed plants.

A wide range of weed control chemicals is listed in the manuals on pest control and fungicides, but the scale on which these materials are used is limited compared with the West. The members of the Wheat Studies Delegation reported that they had observed only one instance in which 2,4-D was actually used for weed control in wheat production. There is evidence that the emphasis on chemical control will expand in the future. However, the use of chemicals will be more complicated in China, particularly in areas where strip cropping or the scheme of mixed plantings followed on many vegetable communes is practiced. Cropping sequences and multiple cropping also present problems with the type of residues that may be involved and the adverse effects that may result to crops planted in succession. The variation in sensitivity to injury of the various crops makes it difficult to apply weed control chemicals.

In the Xian area, where wild oats are sometimes a problem in wheat, the wild oat panicles are hand-stripped in the seed fields to ensure that the oat seeds do not contaminate the wheat seeds at harvest.

Many agricultural research institutes in China have laboratories that evaluate herbicides. In most of these laboratories, the primary objectives are to determine the relative effectiveness and proper application rates, rather than the mechanisms by which these compounds affect plant metabolism adversely.

## *EDUCATION AND COMMUNICATION*

As of the late 1970's education in plant protection was provided at ten major agricultural universities and at numerous small colleges and provincial agricultural research institutes. At the agricultural universities the basic programs in plant protection did not differ significantly from those that were required for other students in agriculture. All students were expected to master courses in general science, as well as entomology, plant pathology, and weed science. As in other disciplines, students were required to spend as much as one-third of their time in studies under field conditions at research substations or communes. Projects away from the university were usually under the direct supervision of a professor who traveled and worked with the students. Mathematics, statistics, and related basic science courses were not emphasized to the same degree at the agricultural colleges as they were in other disciplines. Recent changes in curricula should result in more training in the basic sciences, especially since plant protection programs now require four years of undergraduate training. Training in foreign languages is provided; the number of students registered in English courses has increased markedly in recent years, whereas enrollment in Russian courses has dropped.

Revised teaching manuals and textbooks in plant pathology are becoming available, but many courses and student reading assignments include old texts that provide basic source material in plant pathology, entomology, and weed science. The primary emphasis in the 1970's has been to print extension manuals on plant protection. These describe the specific stages in life cycles of insects and pathogens in remarkable detail, and provide information on various methods for control, including cropping practices, resistant cultivars, and chemical and biological control practices.[83]

In general, teachers do not use projected material and colored slides as much as their Western counterparts. However, beautifully illustrated posters with color paintings of disease symptoms and key facets of the life cycle of important fungal pathogens and insect pests are available in teaching departments and at most research institutes. Much emphasis has been placed on preserved specimens and collections of mounted and labeled diseased plant specimens and insects. Certain provincial agricultural research institutes have initiated the practice of mounting the preserved specimens of the adults, immature stages, and eggs of insect pests, as well as their predators. These are then distributed on a large scale to communes.

University professors, plant protection specialists from provincial and county institutes, and commune representatives may be involved in drafting and revising material for insect and disease control manuals, which are usually issued by municipal or provincial administrative units. Most are concerned with pest problems on single crops, such as rice, wheat, or cotton; others are more general and concerned with a broad range of crops, such as vegetables. Copies of these manuals, which include numerous color plates, can be purchased at most bookshops at very low cost.

Continuing education is also strongly emphasized. Individuals who work in research stations or plant protection units in counties or communes have many opportunities to learn more about their specialties. Short courses are taught by experts either from the local provincial and county plant protec-

[83] *Rice Pests and Their Control* (Shanghai: Shanghai Renmin Chubanshe, 1974); Agricultural Press, Editorial Committee, *Illustrations of Chinese Crop Diseases and Insect Pests. Vol. 4, Insects and Diseases of Cotton and Jute* (Beijing: Nongye Chubanshe, 1972); *Mass Forecasting of Crop Pests* (Shanghai: Shanghai Renmin Chubanshe, 1973); *Control of Insects and Diseases of Fruit Trees* (Shanghai: Shanghai Renmin Chubanshe, 1972); and *The Control of Cotton Pests* (Shanghai: Shanghai Renmin Chubanshe, 1973).

tion institutes or from nearby agricultural colleges. Many communes now have practical one- or two-year course programs that provide instruction in crop and plant protection and animal science. In addition, the teachers at these commune schools may be experienced workers who provide primary instruction to children at a practical level.

In the past, advanced training in entomology, plant pathology, and weed science at the colleges and institutes was limited to individuals chosen by teachers and fellow students and the revolutionary committee of the college. They were appointed as assistants and received on-the-job training under the direct supervision of senior professors. This did not involve any formal course work, but these assistants could initiate special research projects in consultation with one of the professors, and this served as the equivalent of thesis research experience. In the late 1970's, however, this pattern of advanced training was being revised, and more formal requirements for graduate study were being developed.

Because of rapid rotation sequences, heavy use of organic fertilizers, extensive rice culture that involves flooding of fields, care in the selection of disease-free planting material, strict quarantine regulations, careful roguing of diseased plants and weeds, and the intensive effort to select improved varieties for disease resistance, many common diseases in China have been either brought under control or reduced in severity. However, certain foliage diseases of corn and sorghum, wilt of cotton, and blast and bacterial blight of rice, for example, continue to be a cause for concern.

In terms of basic research, relatively little work was conducted on the physiology of disease or the biochemical basis for disease resistance in plants during the Cultural Revolution. Research did not involve emphasis on the structure, multiplication, or molecular biology of plant viruses, on plant parasitic nematodes in general, or on the ecology of soil-borne plant pathogens. Pressing, immediate needs justified the primary emphasis on solving problems in the field.

The intensity of land use in China has few parallels in the world; thus, one would have expected severe nematode problems on many crops. It is difficult to determine whether the limited number of individuals concerned with nematode diseases reflects an absence of these problems or a lack of awareness of the subtle impact of plant parasitic nematodes on crop yields.

It is evident that the practice of assigning research scientists, technicians, and students for extensive work periods at the communes and field stations has had a twofold effect. The scientists are better informed about the specific disease problems that need to be controlled in their area. The peasant farmers are educated to understand the nature of disease agents and are themselves directly involved in developing and improving control practices. However, the practical gains of this pattern of assignment to the communes must be weighed against the disruptive effects.

The administrative networks that involve plant protection teams make it possible to institute regional controls on movement of seed and plants, as well as to determine the varieties to be planted. However, it should be recognized that on occasion this has been accompanied by setbacks; certain plant varieties recommended for large-scale plantings, for example, have had to be withdrawn because of susceptibility to disease.

The most impressive aspect of current programs to reduce disease losses in China is the intensive effort to utilize fully the available knowledge in plant pathology, as well as in related fields of agriculture, and to integrate this knowledge with those practices that peasant farmers have found to work through years of experience. It should be emphasized, however, that the success of many of these control programs is totally dependent on labor-intensive practices. This pattern of assigning scientists to communes and field stations for extensive periods will probably be modified in line with other changes made in China's educational system. The availability of a large labor force has made it possible to develop patterns of crop and soil management in China that are too costly in those countries where mechanization has been adopted.

Multiple cropping and the use of organic matter as fertilizer are two practices that have an indirect impact on many soil-borne plant pathogens. As a means of increasing food production, double and triple cropping, intercropping, and catch-cropping (use of a short-growth-period crop planted between two other crops with longer growth periods) have been utilized more intensively since 1949 than before.

Not all efforts for disease control have been successful, and new diseases that have appeared in recent years have jeopardized production of certain crops in local areas. Peanut rust, *Puccinia arachidis,* for example, has spread rapidly in Guangdong province since 1970. By 1972 over 60 percent of the area planted to peanuts had been affected. The loss in yield ranged from 20 to 30 percent in the Zhanjiang region. In 1977 intensive work was in progress to develop means for controlling this disease.[84]

Although Chinese farmers are able to feed a vast population that continues to grow rapidly, opportunities still exist for production to be increased. Chinese agricultural scientists may be able to learn some useful things from foreign scientific journals and texts, especially those reporting on agriculture in other developing countries.

We, in turn, have an opportunity to learn from China certain management practices in disease and insect control and in the use of organic waste for fertilizers. Furthermore, because of the widespread, intensive efforts being made to select improved high-yielding varieties in China, we should also be interested in the prospect of exchanging cultivars of plant materials of common interest. There would also be value in making a careful assessment of those Chinese cultural practices that seem to have had a moderating effect on the severity of soil-borne pathogens, in particular plant pathogenic nematodes.

[84] "Origin of Infestation of Peanut Rust and a Study on the Biological Characteristics of the Pathogens," *Guangdong's Agricultural Science,* 74 (1977): 40–45.

*Thomas B. Wiens*

# *ANIMAL SCIENCES*

Animal sciences in the PRC have long been a weak sister to the other branches of agricultural science. The production of animals has been less emphasized than crop production in the priorities of Chinese leaders because of the currently low living standard of the Chinese people. Also, animal agriculture has remained largely within the private economy and therefore outside the area of greatest concern to socialist planners. As far as draft animals were concerned, their eventual replacement with machinery was a foregone conclusion, and the maintenance of the draft animal stock was intended only to buy time.

As a result, animal sciences have not attracted the best talents or been granted much publication space. Animal specialists and their publications have emphasized the practical problems of husbandry. The impact of the Cultural Revolution and its aftermath (1966–76) reinforced these existing tendencies, rather than nipping a growing scientific capability in the bud as was the case in other scientific disciplines. Given the renewed emphasis on science and technology in the PRC evident in the late 1970's, animal specialists in China will have to scramble to catch up.

This seems to be their intention. On August 2, 1978, a national society for animal research was reestablished (following its dissolution during the Cultural Revolution), and a conference of specialists drafted plans through 1985 for scientific research on this subject.[1] Although the content of these plans was still unknown by mid-1979, greater scientific rigor and less extension orientation were already evident in the literature.

It should not be inferred that the Chinese can claim no scientific accom-

[1] "National Animal Husbandry Forum Closes," *Foreign Broadcast Information Service (FBIS)*, Aug. 8, 1978, p. E14.

TABLE 1. *Midyear Stocks of Live Animals in China, 1949–78*
(in millions)

| | *1949* | *1952* | *1957* | *1962* | *1965* | *1972* | *1973* | *1976* | *1977*[a] | *1978*[a] |
|---|---|---|---|---|---|---|---|---|---|---|
| Pigs | 57.8 | 89.8 | 115.3 | 122.0 | 167.0 | 260.0 | 260.0 | 280.0 | 292.0 | 301.0 |
| Large animals | 59.8 | 76.2 | 83.5 | 71.0 | — | 95.0 | — | — | 94.0 | 94.0 |
| Oxen | 33.8 | 45.0 | 50.5 | — | — | — | — | — | — | — |
| Buffaloes | 10.2 | 11.6 | 13.1 | — | — | — | — | — | — | — |
| Horses | 4.9 | 6.1 | 7.3 | — | — | — | — | — | — | — |
| Asses | 9.5 | 11.8 | 10.9 | — | — | — | — | — | — | — |
| Mules | 1.5 | 1.6 | 1.7 | — | — | — | — | — | — | — |
| Camels | 0.2 | 0.3 | 0.4 | — | — | — | — | — | — | — |
| Sheep and goats | 42.3 | 61.8 | 98.6 | 113.0 | — | 148.0 | — | — | 161.0 | 170.0 |
| Goats | 16.1 | 25.0 | 45.1 | — | — | — | — | — | — | — |
| Sheep | 26.2 | 36.9 | 53.4 | — | — | — | — | — | — | — |

SOURCES: For 1949–57: Nai-ruenn Chen, *Chinese Economic Statistics* (Chicago: Aldine, 1967), p. 340; 1957 midyear pig stocks are from *Zhongyang hetso tongxun* (Central Cooperative Bulletin), 1959, 3 (Mar.): 31. The 1957 end-of-year pig stock was 146 million. For 1962: growth rates between 1962 and 1971 given in *Foreign Broadcast Information Service (FBIS),* Oct. 3, 1972; 1971 figures for sheep and goats are not known, and the 1972 figure is used under the assumption that no growth in stocks occurred in 1971–72; the 1962–71 trend in the growth rate was assumed to have continued through 1972 for large animals and pigs; 1962 figures are thus derived from 1972 estimates. For 1965: growth in pig stocks over 1949 given in *FBIS,* Sept. 19, 1973; and *How to Raise Pigs* Editorial Group, *Zenyang yangzhu* (How to Raise Pigs) (Shanghai: Shanghai Renmin Chubanshe, 1975), p. 2. For 1972: growth rates over 1949 given in *FBIS,* Sept. 19, 1973. For 1973: growth rates of pig stocks over 1965 and 1949 given in *How to Raise Pigs* Editorial Group, *Zenyang yangzhu,* p. 2. For 1976: from diplomatic sources. For 1977–78: State Statistical Bureau communiqué, June 27, 1979, in *FBIS,* June 27, 1979, p. L11.

[a] End-of-year stocks.

plishments in the field of animal sciences. In some aspects of immediate importance to production—particularly in breeding and vaccine development—Chinese scientists have been active in the past and can cite substantial achievements. Nevertheless, the main thrust of Chinese efforts in the 1970's was to improve the techniques of animal production, rather than to conduct research that added significantly to world knowledge.

This situation is likely to change in the future, not only because of China's renewed commitment to scientific research, but also because of the higher priority given to animal production evident in recent political decisions. It is now recognized that the process of economic modernization implies greatly increased consumption of meat products and that this can be accomplished only by establishing animal agriculture as a specialized industry, rather than as a peasant side occupation. It has also become clear to Chinese leaders that mechanization contributes more to the productivity of animal agriculture than it does to plant agriculture. For these two reasons, animal sciences are likely to find themselves at the cutting edge of China's modernization program in agriculture.

*Chinese Animal Agriculture Since 1949*

In quantitative terms, animal agriculture has been rather successful since the founding of the PRC in 1949. It is not possible to be absolutely precise about the details, but Table 1 shows the estimated stocks of animals in China in various years between 1949 and 1978. The number of pigs grew from about 146 million in 1957 to 301 million in 1978, an average annual growth rate of 3.5 percent. In the same period, the number of cattle, horses, asses, mules, and camels grew 0.6 and the number of sheep and goats 2.6 percent annually. The share of animal products in the gross value of all agricultural produce rose from 12.9 to 13.9 percent over the two decades ending in 1978.[2]

Growth has, however, been sporadic. In the 1950's, for example, the number of sheep and goats expanded steadily and rapidly, pig numbers grew moderately, but large-animal numbers increased by only a small percentage. But when the Great Leap Forward of 1957–58 ended in major crop failures, numbers of both pigs and large animals fell precipitously, although the number of sheep and goats continued to grow. Pig stocks recovered smartly and displayed impressive growth rates through 1978, but the number of large animals recovered far more slowly and then stopped growing. There is evidence that production of sheep and goats peaked in the mid-1960's, but then stagnated or declined until the late 1970's.[3]

Quantitative information about poultry is scarce. Averages based on

[2] Wang Gengjin, "On a Few Issues in Agricultural Planning Work," *Jihua yu tongji (Planning and Statistics)*, 1959, 13 (June): 16; and Li Xiannian, "Speech at National Agricultural Conference," *FBIS*, Aug. 18, 1978, p. E3.

[3] "Sinkiang's Wang Feng Speaks About Animal Husbandry," *FBIS*, July 21, 1978, p. M1.

poultry numbers for five provinces in 1957–58 suggest that nationally there were about 1.3 poultry per capita, or 4.1 poultry per ton of grain produced, or 6.5 poultry per pig. Extrapolation of these ratios yields possible totals of 850 million poultry in China in 1957 and 1.4 billion in 1978. Rough estimates from Chinese sources are about 18 percent higher than both figures.[4] Between 1953 and 1957, in the relatively well-off municipality of Shanghai, average per capita annual consumption of poultry and eggs was 1.74 and 2.78 kg, respectively.

*Consumption Patterns*

Per capita consumption of meat in China is extremely small. In 1955, peasants consumed an average of 4.6 kg of meat per year per capita, of which 3.3 kg were pork and the remainder primarily poultry. Urban residents consumed somewhat more—5.9 kg of pork.[5] Judging from production trends, these figures have since nearly doubled, but it is probably still true that peasants eat meat mostly around harvest time, the New Year, and during the busy farming seasons, except in rich agricultural areas.[6] Demand in urban areas was held in check during the 1960's and 1970's by strict rationing.

Despite the demand for meat in China, beef and mutton are not consumed in large quantities, except in pastoral areas and by Muslims. Since large animals are needed for their draft power, they are rarely slaughtered until their work capacity is virtually exhausted. Eggs are an important source of animal protein, but annual per capita consumption was still probably well under one hundred in 1978.

Dairy products are not a part of the traditional Chinese diet, but a dairy industry has been established since 1949. By the early 1960's, the town of Hailaer (in Heilongjiang province) had become the largest dairy center in China, and there were dairy farms and processing plants on the outskirts of major municipalities.[7] Powdered and condensed milk, butter, and cheese were produced for export or consumption in pastoral areas and fresh milk for the use of urban children.

China produced roughly 103,000 metric tons of wool in 1974, double the amount produced in 1954. If the proportion of sheep to goats remained unchanged during these 20 years, then wool yields per sheep have increased from about 1 kg to 1.3 kg.[8]

Most of China's animal produce is consumed in China, but in 1975

[4]H. V. Henle, *Report on China's Agriculture* (Rome: Food and Agriculture Organization (FAO) 1974).

[5]Nai-ruenn Chen, *Chinese Economic Statistics* (Chicago: Aldine, 1967).

[6]"Problems Reflected by Peasants' Buying Meat to Eat," *Dagongbao* (Beijing), Feb. 18, 1965.

[7]"China's Biggest Dairy Center Expands," New China News Agency (NCNA) dispatch in English, Oct. 30, 1961, in *Selections from China Mainland Press (SCMP)*, 2,612 (Nov. 3, 1961); and "Record Milk Output in Shanghai," NCNA dispatch in English, Jan. 15, 1964.

[8]*Jingji daobao*, 528 (July 22, 1957): 4; and "PRC Woolen Industry Makes Headway Since Cultural Revolution," *FBIS*, Feb. 4, 1975, p. E8.

exports (mainly to Hong Kong) of live animals, meat, and crude animal materials brought in $720 million, or 10 percent of the country's total export earnings.[9]

## *Draft Animals*

Most of China's animal stock of large animals consists of draft animals, used primarily as pack animals or for hauling, plowing, irrigation, and traditional industrial processes. Using traditional implements, an unusually good breed of ox, such as the Qinchuan, can plow 2 ha of dry field per day, and a pair of oxen can plow 0.3–0.4 ha of clay field. However, the plowing capacity of the average ox is probably not more than 0.2 ha per day. Good breeds of water buffalo can plow an average 0.25–0.3 ha of paddy field per day. The average hauling force of oxen is 80 kg, compared with 110 kg for water buffalo. The Mongolian horse can commonly carry 100–120 kg and cover 100–120 km per day in grazing areas.[10]

Mechanized transport and plowing grew rapidly in the 1970's and began to displace draft animals. But at the same time new devices, such as improved plows and rubber-tired carts, greatly increased the work capacity of draft animals and hence their economic viability. In some cases, such as plowing muddy paddy fields and mountain plots or transportation in areas without roads or under severe weather conditions, draft animals remained indispensable.

## *Manure Use*

Animal production is also valuable in China for the manure it produces. This organic fertilizer, rich in potassium and phosphate relative to plant requirements, is required in intensive farming to preserve and restore soil conditioning. As Table 2 shows, animal manures were the major source of fertilizer in China between 1957 and 1971. Their significance in crop cultivation can be seen from the fact that the traditional price paid for a slaughtered pig in China barely covered its production costs; a farmer's profit derived from the use of the pig's manure.

## *Geographical Distribution of Animal Raising*

Pig raising is geographically distributed across China roughly in proportion to per capita availability of grain and other feed sources. It is relatively unimportant only in pastoral areas, which are concentrated in the far north and northwest. China's grazing region in Heilongjiang, Inner Mongolia, Xinjiang, Qinghai, Tibet, Gansu, and northwestern Sichuan provinces ac-

[9] United States, Central Intelligence Agency, *Handbook of Economic Statistics, 1977* (Washington, D.C.: CIA, 1977).

[10] People's Republic of China, Ministry of Agriculture, ed., *Basic Knowledge of Agricultural Production Techniques*, tr. Joint Publications Research Service (JPRS) of *Nongye shengchan jishu jiben zhishi* (Shanghai, 1965), JPRS no. 61,746 (Washington, D.C.: JPRS, Apr. 15, 1974).

TABLE 2. *Estimated Supplies of Fertilizer by Source, 1957–71*
(in million tons)

| *Source* | *1957* | *1962* | *1965* | *1971* |
|---|---|---|---|---|
| Large animals | 4.92 | 4.21 | 4.56 | 5.45 |
| Pigs | 3.15 | 3.01 | 4.63 | 6.77 |
| Night soil | 1.86 | 2.23 | 2.34 | 2.61 |
| Green manure | 0.27 | 0.44[a] | 0.58 | 0.58[a] |
| Bean cakes | 0.53 | 0.44[a] | 0.39 | 0.39[a] |
| Other nonchemical | 1.65 | 1.65[a] | 1.65[a] | 1.65[a] |
| Chemical fertilizers | 0.38 | 0.84 | 2.10 | 4.35 |
| Total nutrient | 12.76 | 12.81 | 16.25 | 21.79 |
| kg/ha cultivated | 114.0 | 114.6 | 152.3 | 194.9 |
| Percent chemical | 3.0 | 6.5 | 12.9 | 19.9 |
| Total N (kg/ha) | 51.1 | 53.5 | 69.8 | 92.8 |

SOURCES: Estimates are based on S. Ishikawa, "Factors Affecting China's Agriculture in the Coming Decade," working paper (Tokyo: Institute of Economic Affairs, 1968), Table 12, pp. 60–61, and Appendix 3. Ishikawa's estimates of the human and animal population have been replaced with estimates based on recent official statements (see Table 1). Estimates are constructed by applying growth rates of population series to 1957 nutrient yield estimates (based on Chinese data; utilization rates for pig manure are assumed to rise from 0.7 in 1957 to 0.8 in 1962 and 0.9 in 1965–71). Chemical fertilizer nutrient based on A. L. Erisman, "China: Agriculture in the 1970's," in U.S., Congress, Joint Economic Committee, ed., *China: A Reassessment of the Economy* (Washington, D.C.: Government Printing Office, 1975), Table 2, p. 333, adjusted to reflect differing estimates of total domestic production in 1971.

Cultivated acreage based on CIA, *Agricultural Acreage in Communist China, 1949–1968: A Statistical Compilation* (Washington, D.C.: CIA, 1969); acreage is assumed constant after 1962.

Manure from sheep, goats, and poultry has been ignored. It should also be noted that night soil use is largely confined to private plots and suburban vegetable production.

[a] Interpolated or assumed constant.

counts for 38 percent (3.2 million km$^2$) of the country's land area and about one-third of its total livestock (excluding pigs and poultry).[11]

Although this pastoral region dominates in the production of horses, cattle, sheep, goats, and other large animals, most provinces maintain sizable stocks of draft animals, sheep, and goats. Also Heilongjiang specializes in horse breeding, and Guangxi and Guizhou in breeding water buffalo.

### *Breeds of Animals*

China has developed numerous native breeds of pig, some of which possess excellent characteristics (see Table 3). The better breeds may be crudely classified as North and South China types. The former have big frames, large litters, ample milk, strong maternal characteristics, early sexual maturity, plenty of subcutaneous and belly fat, strong disease resistance, low fattening and dressing rates, thick and loose skin, and inferior meat quality. The latter have a relatively high fattening rate, thin skin and tender meat, and early maturity, but smaller frames and litters.[12] The

[11] Zhi Ming, "Economic Construction in People's Communes of Pastoral Areas Must Center on the Development of Livestock Breeding," tr. from *Minzu tuanjie* (Nationalities Unite), 1962, 1, *Selections from China Mainland Magazines (SCMM)*, 308 (Apr. 9, 1962): 4.

[12] Shanghai Municipality, Jinshan County, Xingta Commune, Revolutionary Committee, ed., *Mosuo guilü yanghao zhu* (Discovering the Methods for Good Pig Raising) (Shanghai: Shanghai Renmin Chubanshe, 1975).

TABLE 3. *Characteristics of Native and Crossbred Pig Breeds*

| *Variety* | *Daily weight gain (kg) (under good feeding conditions)* | *Feed/unit weight gain (kg/kg)* | *Dressing (percent)* | *Litter size at birth* | *Lard as percentage of carcass weight* | *Skin thickness (cm)* |
|---|---|---|---|---|---|---|
| *North China:* | | | | | | |
| Northeast Popular (Large variety) | .30 | — | 81 | 12–15 | 4.5 | .56 |
| Xiangcheng (Henan) | .34 | — | 69 | 15 | — | — |
| *South China:* | | | | | | |
| Meihua (Guangdong) | — | — | 65–70 | 8–12 | — | — |
| Ningxiang (Hunan) | — | — | 65–70 | 8–9 | 6.6 | .45–.51 |
| Jinhua (Zhejiang) | .39 | 4.2 | 72 | 15 | 7.1 | .33 |
| Neijiang (Sichuan) | .44 | 5.1 | 60–74 | 10–11 | 5.3 | .56 |
| Huai (Fujian) | — | — | — | — | 7.1 | .34 |
| Luchuan (Guangxi) | — | — | — | — | 4.7 | .41 |
| *Hybrids:* | | | | | | |
| Xinjin (Liaoning) | .35–.48 | — | 75+ | 9 | — | — |
| Dingxian (Hebei) | .38 | — | 72 | 10 | — | — |
| Haerbin White (Heilongjiang) | .66 | 4.3 | 78 | 11 | — | — |

SOURCES: Zhang Zonghui, *Zhongguo yangzhufa* (Chinese Methods of Pig Raising) (Nanjing: Animal Husbandry and Veterinary Publications Press, 1957), pp. 24–53, 281; *How to Raise Pigs* Editorial Group, *Zenyang yangzhu* (How to Raise Pigs) (Shanghai: Shanghai Renmin Chubanshe, 1975), pp. 49–50; Chen Runsheng, "Discussion of Some Problems of Breeding of Certain Characteristics in China's Pig Breeds," *Zhongguo nongye kexue (Scientia Agricultura Sinica),* 1977, 4 (Nov.): 83–86; PRC, Ministry of Agriculture, ed., *Basic Knowledge of Agricultural Production Techniques,* tr. Joint Publications Research Service of *Nongye shengchan jishu jiben zhishi,* Shanghai, 1965, JPRS no. 61,746 (Washington, D.C.: JPRS, Apr. 15, 1974), pp. 244–47; and Tianjin Municipality, Baoti County Agricultural Research Institute, *Qunzhu siyang* (Feeding of Pig Herds) (Tianjin: Tianjin Renmin Chubanshe, 1976), pp. 4–14.

Chinese consider their native varieties superior to foreign breeds in their ability to process coarse feeds, but this alleged capability has not been systematically measured.[13]

The major breeds of other animals in China are summarized in Table 4. About 20 percent of China's cattle stock are water buffalo, which are limited almost entirely to southern paddy rice areas.[14] The remainder consists mostly of common oxen, although the Qinchuan and Sanhe oxen have been

[13] Chen Runsheng, "Discussion of Some Problems of Breeding Certain Characteristics in China's Pig Breeds," *Zhongguo nongye kexue (Scientia Agricultura Sinica),* 1977, 4 (Nov.): 81.

[14] W. R. Cockrill, "Report of a Visit to China, April/May 1974," mimeographed (Rome: FAO, 1974).

TABLE 4. *Major Breeds of Large and Small Animals in China*
(superior breeds in italics)

| *Animal* | *Origin* | *Adult male average* | |
|---|---|---|---|
| | | *Height (cm)* | *Weight (kg)* |
| Common ox: | | | |
| *Qinchuan* | Shaanxi | 140 | 574 |
| Nanyang | Henan | 142 | 424 |
| Luxi | Shandong | 142 | 525 |
| Yanbian | Northeast | 131 | 480 |
| Danjiao | Shanghai | 132 | 500 |
| Yak | Tibet, Qinghai | 114 | 370 |
| Water buffalo: | | | |
| Binhu | Hunan | 135 | 495 |
| Wenzhou | Zhejiang | — | — |
| Horses: | | | |
| Mongolian | Inner Mongolian A.R. | 128 | 290 |
| *Sanhe* | Heilongjiang | 146 | 351 |
| *Yili* | Xinjiang | 148 | 393 |
| Hequ | Qinghai | 139 | — |
| Asses: | | | |
| *Guanzhong* | Shaanxi | 140 | — |
| *Shandong* | Shandong | — | — |
| Sheep: | | | |
| *Xinjiang fine-wool* | Xinjiang | 73 | 90 |
| *Northeast fine-wool* | Northeast | 78 | 89 |
| Han | Henan, Hebei | — | 50 |
| Tong | Shaanxi | — | 50 |
| Mongolian | North, Northeast, and Northwest | 69 | 52 |
| Tibetan | Tibet, Qinghai | 66 | 65 |
| Tan | Ningxia | — | — |
| Hu | Zhejiang | 67 | 40 |

SOURCES: PRC, Ministry of Agriculture, ed., *Basic Knowledge of Agricultural Production Techniques*, tr. Joint Publications Research Service of *Nongye shengchan jishu jiben zhishi*, Shanghai, 1965, JPRS no. 61, 746 (Washington, D.C.: JPRS, Apr. 15, 1974), pp. 218–21, 231–35, 254–61; Wu Yunfeng, ed., *Jianming xumu shouce* (Simplified Animal Husbandry Handbook) (Huhehot: Nei Menggu Renmin Chubanshe, 1974), pp. 157–61; and *How to Raise Plow Oxen* Editorial Group, *Zenyang yang gengniu* (How to Raise Plow Oxen) (Shanghai: Shanghai Renmin Chubanshe, 1975), pp. 26–28.

identified as superior breeds.[15] The yak and its hybrid with the common ox, the "pianniu," are found in the plateaus and mountain areas of Tibet, Qinghai, and Sichuan, where they have adapted to cold weather and high altitudes.

The Mongolian horse is the most numerous, oldest, and most widely distributed variety in China and is known for its endurance. But because of its small frame, it is not very useful for heavy transportation work, for which the Sanhe (western Heilongjiang) and Yili (Xinjiang) varieties are preferred. Asses outnumber horses in China and are widely used in the north for plow-

[15]Tang Yijin, "A Brief Discussion on Problems of Developing Livestock Production," tr. from *Xin jianshe* (New Construction), 1964, 12 (Dec.), *SCMM*, 456 (Feb. 15, 1965): 22.

ing and as pack animals. Superior breeds include the Shandong variety, which has been exported abroad for breeding purposes.[16]

China's sheep population may be categorized, according to their major economic functions, as fine-wool, semi-fine-wool, coarse-wool, "fur-skin" wool, and lamb-skin varieties. China originally had no native fine-wool breeds of sheep; the overwhelming majority were coarse-wool types, such as the Mongolian sheep, which in the early 1960's accounted for over half of China's stock. The best breed of Mongolian sheep, the Wuzhumuqin variety, was characterized by rapid weight gain and high meat and fat yield.

The Northeast fine-wool sheep was China's first such breed. It resulted from crosses of imported Rambouillet fine-wool sheep with local Mongolian varieties, made under Japanese auspices between 1913 and 1924. However, the Xinjiang fine-wool sheep, bred in 1954, is the most highly prized and widely distributed fine-wool variety in China and the mainstay of the country's stock improvement program. There were about 1.5 million in 1977.[17] Since fine-wool sheep constitute roughly 20 percent of China's stock, there must be several million more crossbreeds.[18]

Information on goats is hard to find, mainly because they are worth less than sheep and have not attracted much government attention. Information on native poultry breeds other than the world-famous Peking duck is also scanty. A 1964 survey of domestic breeds of chickens, ducks, and geese by the Chinese Academy of Agricultural Sciences' Domestic Research Institute in Jiangsu recommended 16 good breeds for widespread adoption.[19] Breeds observed by foreign visitors include:

1. The Peking duck, which weighs 2.6 kg after 65 days of force-feeding. Layers produce 180 eggs a year, with 80 percent hatching rate by incubator.
2. The Ma, or sparrow, duck, a small but efficient forager observed in Hunan province, said to lay 200–250 eggs a year.
3. Muscovy ducks, seen in Guangxi and Guangdong provinces.
4. Lionhead (Shifo) geese, a large grey-brown breed, which is slaughtered at 100 days, weighing slightly over 3 kg. Breeders produce 35–40 fertile eggs a year.[20]

### *Pig Raising*

Household pig raising, which accounts for the majority of pigs produced in China, usually involves rather crude shelters, built near or against residences, frequently with mud-brick walls and thatched roofs. Weather protection and drainage are poor, and pigs often wander loose, foraging or gleaning, despite government efforts to promote penning and manure col-

[16] Ministry of Agriculture, ed., *Basic Knowledge.*
[17] "Improved Animal Breeds," *FBIS,* Mar. 28, 1978, p. E23.
[18] Cockrill, "Visit to China."
[19] "General Survey of Domestic Fowls Concluded in China," NCNA dispatch in English, July 13, 1964; in *SCMP,* 3,259 (July 16, 1964).
[20] Cockrill, "Visit to China."

lection. In more prosperous areas, brick-walled and floored pens with troughs for collecting manure and urine are common. In the south, however, pig-raising manuals recommend bamboo and mud pens with undrained mud floors, the straw, ash, and pig droppings accumulating on the floor until needed as fertilizer.[21]

Pig-raising manuals also provide good, simple designs for collective pens run by communes. (Collective pig raising accounts for less than a third of the pigs raised in China.)[22] Designs are often tailored to the specific requirements of farrowing sows, boars, and meat pigs. The pens, which house 100–200 pigs each, are provided with partially open shelters, exercise yards, baths, feeding troughs, and drainage systems that combine sanitation with conservation of manure. Protection against the cold may be provided by a simple heating system, but is more commonly accomplished through wind protection, provision of ample straw, and/or a dugout pit under the shelter. Collective pens are typically located at the edges of fields or ponds to save labor in the transport of manure and fodder.[23]

The traditional Chinese pig production process consists of two generations: a virtually universal first generation, with births in the spring and slaughter about ten months later, and a second generation, which is confined mainly to the south, with births in the late summer and slaughter in the succeeding year. The process is determined by feed availability and the desired slaughter date. The weaning of the first generation in May–June coincides with the summer harvest so that the main pig-feeding period occurs at the time of peak crop supply. This first generation of pigs is then slaughtered about eight months later, just when feed and food become scarce and meat is needed for the New Year celebration. Similarly, the birth of the second generation coincides with the autumn harvest, and its slaughter takes place in spring, the agriculturally busy season when extra food is most necessary, and at summer harvest time. The traditional pig production process is economic in other ways as well since it taps resources that would otherwise go to waste: family food scraps; the labor of wives, old people, and children, as well as the leisure time of regular workers; and family capital.

The Chinese desire to exploit limited resources to the maximum has led to some ingenious interactions between pig raising and other productive activities. In South China, for example, collective pens are often sited next to

[21] E. B. Shaw, "Swine Industry of China," *Economic Geography,* 14 (Oct. 1938): 381; and *How to Raise Pigs* Editorial Group, *Zengyang yangzhu* (How to Raise Pigs) (Shanghai: Shanghai Renmin Chubanshe, 1975).

[22] U.S., Department of Agriculture, Economics, Statistics, and Cooperatives Service, *People's Republic of China Agricultural Situation: Review of 1977 and Outlook for 1978* (Washington, D.C.: Department of Agriculture, 1978).

[23] *How to Raise Pigs* Editorial Group, *Zenyang yangzhu;* Tianjin Municipality, Baoti County Agricultural Research Institute, *Qunzhu siyang* (Feeding of Pig Herds) (Tianjin: Tianjin Renmin Chubanshe, 1976); and Yan Chen, *Zenyang gaohao dongji yangzhu* (How to Handle Properly Winter Season Pig Raising) (Tianjin: Tianjin Renmin Chubanshe, 1976).

ponds, into which liquid wastes are washed. These provide nutrients for aquatic plants grown as pig feed and support cultured fish. The pond mud (fertilized by fish waste) can later be applied as base fertilizer.[24] Similarly, in some piggeries, trellised grapevines are used to provide shade, with grapes as a by-product.[25] In addition, the high-protein feed from by-products of grain milling, wine making, and other food-processing industries is used to feed pigs.[26]

Side by side with the traditional production process, there is a commercial pig-raising industry, located near cities and in relatively prosperous grain-producing areas. This produces a more continuous flow of pigs, fattened more quickly (eight months from birth to slaughter) on better-quality feeds. The scale of operation is usually small (less than 100 pigs per pen) in rural areas, but can run into several thousand pigs on commercial farms near major municipalities.

Government policy in the late 1970's was concerned primarily with improving the supply of meat for urban areas and export. Chairman Hua Guofeng called for the establishment of large numbers of pig (and chicken) farms in the suburbs of major cities.[27] These were to be modern, large-scale commercial operations raising up to 10,000 pigs each, run by municipalities, state farms, and communes. Their mission was to improve urban food supplies without imposing additional burdens on the peasantry.

### *Pig Feed*

A cursory reading of popular literature and visits to model pig-raising units suggest that Chinese pigs are essentially scavengers, depending on a combination of food scraps and field wastes with no important alternative uses and imposing no burden on human food supplies. However, the government's tight control over the use of feedgrains, the voluminous propaganda promoting substitutes, and the content of China's pig-raising literature indicate that pigs do pose vigorous competition for the use of foodgrains and foodgrain lands.

In fact, feeding habits vary widely from foraging (common in the poorer areas—usually mountainous and low population-per-unit areas) to pen raising on grain and various by-products (in the fertile lowland areas and near cities). Most pigs have a mixed diet consisting of (1) concentrates, which include coarse varieties of grain (barley, kaoliang [sorghum], peas, and beans), the bran or millings of grain, soybean, cotton, and peanut cakes, and wine dregs (these "fine feeds," as they are called in China, are low in fiber content and high in digestible nutrients); (2) forage, which includes squashes and melons, the leaves of vegetables, such as cabbages or beans,

[24] Henle, *Report.*
[25] Cockrill, "Visit to China."
[26] Liu Ruilong, "Summary Report of the National Keypoint Pig-Raising Counties Discussion Meeting," *Zhongguo nongbao* (Chinese Journal of Agriculture), 1957, 11 (Nov.):1.
[27] "People's Daily Urges Mechanized Animal Husbandry," *FBIS,* Feb. 15, 1978, p. E11.

and various trees, green manure crops, such as purple vetch, and various grasses and weeds; and (3) roughage, which includes more fibrous materials, such as the stalks or straw of grains and other plants.

Animal feeds in China are generally high in forage and roughage content and low in concentrates. Official policy favors minimum use of feeds that humans can eat as well. In 1977, for example, the government's recommended pig-feeding strategy was to skimp on concentrate in the middle period of growth and to increase it sharply in the final fattening stage, leading to a prolonged fattening period and a rather low slaughter weight (see Table 5). By 1979 just under 10 percent of grain produced in China was used as animal feed. However, there was some flexibility. More concentrate was used in North China, for example, where short growing seasons result in a scarcity of forage. By 1978 model collective pig-raising units were still using about 82 kg of grain per pig fattened to 60 kg.[28]

To counteract declining sources of forage and fodder in crop-growing areas (caused by such factors as absorption of land by irrigation works and replacement of traditional crop varieties by semidwarfs with low yields of straw), China has sought new sources of fodder and forage and improved on traditional means of processing them.

Beginning in the early 1960's, the growing of water hyacinth (*Eichhornia* spp.) and similar crops on water surfaces was heavily promoted. By the late 1970's, they had become a major source of pig forage in South China.[29] Extension literature publicized the existence and possible uses of these and hundreds of other wild and cultivated plants, which were carefully catalogued in the 1950's.[30] To improve the nutritional value of such plants as pig feed, this literature campaigned against the common custom of cooking feed and advocated the use of raw, or preferably fermented, mixed feeds.[31] These typically consist of finely chopped straw, stalks, and weeds mixed with water and supplemented with small amounts of grain (carbohydrate), ammonia (or other sources of usable nitrogen), edible salt, and mineral nutrients. Rumen fluid taken from cattle provides the enzymes for the fermentation, which breaks the coarse fibers down into usable protein, thereby increasing the nutritional value of fodder. Apparently pigs can normally digest only 15–20 percent of cooked coarse fibers, but they can digest 45–55 percent of fermented fibers.[32] As of 1978, the fermentation technique was not

[28] Yan Chen, *Zenyang gaohao dongji yangzhu;* Xingta Commune, Revolutionary Committee, ed., *Mosuo guilü yanghao zhu; How to Raise Pigs* Editorial Group, *Zenyang yangzhu;* Baoti County Agricultural Research Institute, *Qunzhu siyang;* Ministry of Agriculture and Forestry, Animal Husbandry Bureau, *Dali fazhan yangzhu shiye;* and *Yangzhuye yao dafazhan* (We Must Greatly Develop the Pig-Raising Industry) (Beijing: Agricultural Publications Press, 1976).

[29] Cockrill, "Visit to China."

[30] PRC, Ministry of Agriculture, ed., *Zhongguo siliao zhiwu tupu* (Chinese Fodder Plants Handbook) (Beijing: Kexue Puji Chubanshe, 1959).

[31] Li Fuxing, "Raw, Dry Pig Feed," *Kexue shiyan (Scientific Experimentation),* 1974, 12 (Dec.):24.

[32] *Artificial Stomach Fermented Feed* Scientific Cooperative Group, ed., *Rengong liuwei faxiao culiao* (Artificial Stomach Fermented Feed) (Beijing: Science Publications Press, 1976).

TABLE 5. *Pig-Feeding Method Designed to Minimize Concentrate Consumption, 1974*

| | *Weight of pig (kg)*[a] | | | |
|---|---|---|---|---|
| *Measure* | *9–30* | *30–40* | *40–50* | *50–75* |
| Feeding period (days) | 90 | 70 | 30 | 50 |
| Daily weight gain (kg) | .18–.32 | .10–.30 | .20–.40 | .40–.70 |
| Daily feed requirements: | | | | |
| Concentrate (kg)[b] | 0.5–0.6 | 0.35 | 0.35 | 1.8 |
| Forage (kg)[c] | 0.7 | 10.0 | 15.0 | 3.6 |
| Fodder (kg)[d] | 0.35 | 0 | 0 | 1.8 |
| Feed units[e] | 0.8–0.9 | 1.0–1.4 | 1.3–1.5 | 3.5–4.2 |
| Crude protein (percent) | 18–19 | 17–18 | 13–15 | 11–12 |

SOURCE: Wu Yunfeng, ed., *Jianming xumu shouce* (Simplified Animal Husbandry Handbook) (Huhehot: Nei Menggu Renmin Chubanshe, 1974), pp. 73–74.

[a] Animal weight stage: from weaning (at approximately 9 kg) to slaughter (at over 50 kg live weight).

[b] Concentrate: including grain, grain processing wastes (e.g. bran), and other high-nutrient feed sources.

[c] Forage: including leaves, weeds, vegetables, and other plant products, at fresh weight (except for some dried grasses).

[d] Fodder: including ground-up stalks, straw, etc.

[e] Feed units: nutrient-equivalent as compared with a kilogram of oats. Concentrate is generally in the range 0.7–1.4 feed units; forage 0.05–0.25

widespread; it required investment in equipment and considerable labor, and its economic viability had not been demonstrated.

Work has been under way since 1971 to replace the fermentation process by using cellulase molds to break coarse fibers down into less complex carbohydrates.[33]

### *Feed and Fodder of Draft Animals and Sheep*

Feeding practices for draft animals in China remain largely traditional. In the crop-raising areas, the general practice is to allow grazing on grasses and weeds during the seasons when they are abundant. During the agriculturally busy season and winter, animals depend largely on hay feed, supplemented by appropriate amounts of concentrate.

In addition to wild grasses and tree leaves, green manure crops, such as *Astragalus sinicus* L. and *Tecoma grandifiora* Loisel, are favored green feeds in South China. In winter, rice and millet straw are the predominant feeds. Green storage of silage is a recommended but not widely followed practice that minimizes nutritional losses; shredded raw materials are tightly packed in specially constructed cellars and covered with wet clay to seal out air. Because of the low nutritional content of dried stalk feeds, they are generally supplemented with such concentrates as bran, bean cakes, and soybeans and such green feeds as are available.[34] Standards are of course higher on breeding and dairy farms, where concentrate supplies the majority of nutrients.

[33] Guangdong College of Agriculture and Forestry, Department of Animal Husbandry and Veterinary Medicine, "Isolation Selection of Cellulase Molds and Breeding of Induced Mutants," *Guangdong nongye kexue (Guangdong's Agricultural Science)*, 1974, 2: 46.

[34] *How to Raise Plow Oxen* Editorial Group, *Zenyang yang gengniu* (How to Raise Plow Oxen) (Shanghai: Shanghai Renmin Chubanshe, 1975); and Ministry of Agriculture, *Basic Knowledge*.

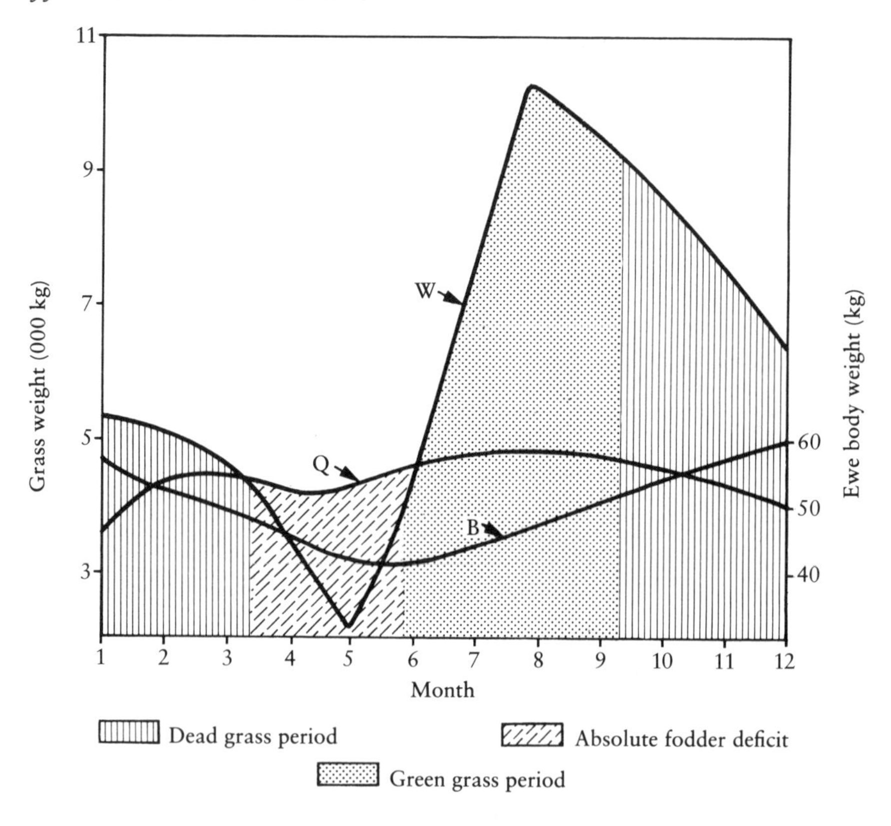


W Dry weight of grass yield per 12 hectares per month (000 kg)

B Average live body weight of adult ewe (kg)

Q Dry weight of grass required to feed a flock of 55 sheep for a month (000 kg)

SOURCE: Ren Jizhou et al., "Pastoral Production Process and Seasonal Animal Husbandry," *Zhongguo nongye kexue* (*Scientia Agricultura Sinica*), 1978, 2 (May): 90.

Draft animals generally have heavy nutritional requirements and, due to the shortage of fodder, may collectively consume up to four times as much concentrate as pigs. The stagnation in draft animal production consequently has been a necessary condition for the diversion of increasing proportions of feed supplies to pigs. This in turn has facilitated the rapid growth rate of the pig stock in the 1960's and 1970's.

Sheep raising in the pastoral areas suffers from a fodder-animal imbalance. The accompanying figure summarizes the results of a one-year experiment in grazing a flock of Xinjiang fine-wool sheep in the Gansu highlands on 0.22 ha each. The supply of grass over the year is the area shown under line *W;* the flock's consumption requirement is the area under line *Q*. The average weight of an adult ewe over the year is indicated by line *B*. The active grass-growing period lasts less than four months, from mid-May to early September; in the remaining months, the grass is withered. The flock's nutritional requirements are covered for all but 2.5 months of early spring, but the lower nutritional value of dead grass plus the absolute deficit in spring result in a downtrend in animal weight in winter and spring that is reversed only in mid-May, leading to a low net weight gain over the year. (The winter weight loss of ewes is partly due to lambing, however.)

In the pastoral areas, the fodder-animal imbalance is generally handled by cutting and storing grass fodder during the surplus season to be fed to the animals during the deficit season. As in the pig cycle, however, the animal population is seasonally adjusted in such a way as to relieve the fodder-animal imbalance; meat consumption is usually greatest between September and November, and in the colder areas, sheep are slaughtered in November to provide meat to last through spring.[35]

In 1978 it was proposed that sheep raising should approach the extent of seasonal adjustment characteristic of pig raising. The authors of an article in *Zhongguo nongye kexue (Scientia Agricultura Sinica)* argued that the efficiency of fodder utilization is the key to increasing animal products. They noted that the method of dry-storing fodder resulted in the loss of at least 75 percent of its nutritional value and that the cost of fodder consumed (in relation to the value of the eventual animal products) was lowest if the sheep were slaughtered at ten months of age. Consequently, they argued that instead of overwintering the majority of the animal stock, almost all animals should be slaughtered at 10–12 months of age, well before the spring food deficit season. The best animals would be retained for breeding to permit further expansion of flocks to take advantage of surplus, fresh forage in the summer months.[36]

This proposal has some obvious problems: in particular, fodder is the only unit cost considered. As the authors note, an expansion of numbers of breeding ewes might be required; and in view of the reported low breeding rate of sheep, this might well be prohibitive. (The authors report that to replace the slaughtered animals, breeding ewes would have to be maintained at a ratio of 0.8 or more per animal slaughtered.) In addition, cold-storage facilities would have to be greatly expanded to accommodate the larger number of animals slaughtered at one time. (These proposals, which did not reflect official policy, are an example of the resumed late-1970's policy of letting a "hundred schools of thought contend.")

A considerable effort was made in the 1950's and early 1960's to expand fodder supplies in the pastoral areas. Surveys of grassland resources were

[35] Peng Zhenqiu, "A Consideration of the Problems of China's Animal Husbandry Statistics and the Planning Year," *Tongji gongzuo tongxun (Statistical Work Bulletin)*, 1956, 21 (Nov.): 20.

[36] Ren Qichou et al., "Pastoral Production Process and Seasonal Animal Husbandry," *Scientia Agricultura Sinica*, 1978, 2 (May): 87.

carried out, and Chinese hydrologists and geologists mapped the distribution of underground water resources. A program of well-sinking was launched to overcome problems of aridity. This program accelerated in the 1970's because of improvements in well-drilling and pumping technology and increased pump production.[37]

In the early 1960's, seed farms were set up in Inner Mongolia and other pastoral areas to provide high-quality forage seed to the herdsmen. Herdsmen were encouraged to settle down and supplement pastoral activity with crop cultivation, and acreage of cultivated fodder crops expanded. Cultivated fodder was used primarily during the winter-spring deficit season, but the extent of its current importance relative to natural grasses cannot be assessed.

Common fodder crops, mostly introduced since 1949, include *Medicago denticulata* Willd., *Melilotus suaveolens* Ledeb., Sudan grass, Dzungarian alfalfa, clover, and oats.[38] Recently, attempts have been made to popularize *Lithospermum officinale* L., a grass introduced from Europe.[39] As a harbinger of future efforts in this area, a Chinese translation of a leading Western text on grasslands management for meat production was published in 1977.[40]

According to recent Chinese press discussions, the stagnation of animal production in pastoral regions since 1966 has been partly due to competition over land use between grain production and pasturage. Under the Maoist policies practiced during the Cultural Revolution and its aftermath, each locality sought grain self-sufficiency, often through preemption of the best pasture land. This was accompanied by an assault on the right of herders to retain a small private flock or herd. Government policy in the late 1970's was to restore both the primacy of animal agriculture in the pastoral regions and the right to retain some private animals. Whether these measures are sufficient to restore dynamism to the sheep- and goat-raising industries remains to be seen.

### *Breeding*

The history of exotic pig breeds in China dates to around 1800 when the original Yorkshire breed was imported and crossbred with South China pig varieties.[41] Three North China "improved breeds," which resulted from

[37] "Successful Research in Pasture and Livestock in Inner Mongolia," NCNA dispatch in English, Jan. 10, 1964; and "Efforts to Improve Grasslands in Northwest China," NCNA dispatch in English, July 27, 1963, in *SCMP,* 3,030 (July 31, 1963).

[38] "Efforts to Improve Grasslands"; and "Sudan Grass for China's Leading Stock-Breeding Region," NCNA dispatch in English, May 14, 1963, in *SCMP,* 2,982 (May 20, 1963): 18.

[39] Beijing Academy of Agricultural Science, Institute of Animal Husbandry and Veterinary Science, "*Lithospermum officinale,* L.: A Superior Fodder Grass," *Scientific Experimentation,* 1976, 10 (Oct.): 20.

[40] J. M. Wilkinson and J. C. Tayler, *Beef Production from Grassland* (London, 1973), Chinese tr. North China Agricultural University, Animal Husbandry Department (Beijing: Science Publications Press, 1977).

[41] *How to Raise Pigs* Editorial Group, *Zenyang yangzhu;* and Zhang Zonghui, *Zhongguo yangzhufa* (Chinese Methods of Pig Raising) (Nanjing: Xumu Shouyi Tushu Chubanshe, 1957).

crosses with Yorkshire, Poland-China, and Berkshire exotics before 1940, benefited from further breeding and promotion work after 1949. In addition, the Russian Large White was introduced in 1950 and became the mainstay of breeding work. The Danish Landrace was imported in the 1960's and has since been widely tested. According to Chinese breeding manuals, the Berkshire is easily and quickly fattened and has a sturdy frame, but small litters; the Yorkshire accepts coarse feed and has large litters; the Landrace has a high feeding efficiency in terms of weight gain per unit of feed, but requires high-quality feed in large amounts each day; the Large White (a derivative of the Yorkshire) has most of the advantages of both the Berkshire and the Yorkshire including acceptance of coarse feeding or foraging, but its initial weight gain is relatively slow.[42]

The Chinese began testing Duroc boars imported from the United States in the 1970's and may soon test Canadian Lacombe boars.[43]

Although there is no information on attempts to upgrade China's common ox with foreign genetic stock, there has been a major effort to improve dairy cattle through infusion of exotic blood from the Soviet Union and the Netherlands.[44] The primary exotic breed has been the Holstein-Friesen. By the late 1970's there were several hundred thousand inbreds and crossbreds in China.[45]

Some efforts have been made to cross exotic water buffalo breeds with the native varieties; Murrahs were imported from India beginning in 1957, and a small herd of Nili buffaloes was brought in from Pakistan in the 1970's. Crossbreeding of native horse varieties with Soviet stallions of fine breeds was conducted on a large scale from 1956 on.[46] Because of the inferiority of China's native breeds of sheep, exotic fine- and semi-fine-wool sheep breeds were also imported from the Soviet Union early on. Finally, the exotic variety of poultry most commonly observed by foreign visitors is the White Leghorn.[47] China has imported poultry breeding stock from Canada, the Netherlands, and the United States.

Artificial insemination in pig breeding was introduced to China in the mid-1950's, but probably became significant only in the 1970's. Most pig-raising manuals published since 1972 provide detailed instructions about the procedure and the simple equipment required for it.[48] By 1974, 60 per-

[42] Wu Yunfeng, ed., *Jianming xumu shouce* (Simplified Animal Husbandry Handbook) (Huhehot: Nei Menggu Renmin Chubanshe, 1974); *How to Raise Pigs* Editorial Group, *Zenyang yangzhu;* and Ministry of Agriculture and Forestry, Animal Husbandry Bureau, ed., *Dali fazhan yangzhu shiye.*

[43] Paul Findley, "Report of Congressman Paul Findley on Illinois Agricultural Mission to the P.R.C., March–April 1978," *Congressional Record,* 124, 79 (May 25, 1978): H4,637.

[44] "Eighteen State Farms on Peking Outskirts," NCNA dispatch in English, Aug. 24, 1962; in *SCMP,* 2,810 (Aug. 31, 1962).

[45] "Improved Animal Breeds."

[46] "Heilungkiang Breeds More Horses," NCNA dispatch in English, Apr. 5, 1956.

[47] M. E. Ensminger, "The People's Republic of China That We Saw in 1977," mimeographed (Clovis, Calif.: Agriservices Foundation, 1977); and H. C. Champeau, "Illinois Agricultural Mission to the People's Republic of China," mimeographed (1978).

[48] *How to Raise Pigs* Editorial Group, *Zenyang yangzhu.*

cent of the communes in Guangxi province had used the technique; about 30 percent of sows and gilts in the province were artificially inseminated by 1975.[49]

The average conception rate claimed in Guangxi was 80 to 90 percent. It was not clear how many impregnations were achieved per boar, although the targeted number is nearly ten times the natural impregnation potential.[50] Difficult transport conditions and lack of refrigeration equipment limited the use of artificial insemination; unrefrigerated sperm lasts only 24 hours.[51] However, work in Guangxi has been proceeding on means of freezing pig sperm, using a mixture of sucrose, egg yolk, and glycerol as a diluent. In 2,595 sows impregnated through 1976, the conception rate was 42 percent. There were 185 litters, with an average of nine shoats per litter. Refrigeration during transport was provided by liquid nitrogen containers.[52]

Use of artificial insemination in the breeding of horses, cattle, and sheep began very early; it was extensively practiced in sheep breeding in Inner Mongolia and Xinjiang as early as 1955.[53] But by the late 1970's it was used widely only on state horse-breeding farms. Its use in sheep and cattle breeding was quite high in some areas, but low in others.[54]

Since at least 1962, research has been under way on the use of frozen semen in cattle breeding (pellet freezing with liquid nitrogen). The conception rate with dairy cattle after the first artificial insemination is about 61 percent, using frozen semen. Successful freezing of stallion semen has not been matched by successful freezing of sheep semen, however.[55]

Research on embryo transplants in sheep and cattle under way at the Chinese Academy of Science's (CAS) Institute of Genetics in Beijing has begun to open the way for more rapid expansion of improved breeds. Gonadotrophic hormones, extracted from sheep and pig pituitaries or (more economically) contained in injections of pregnant mare serum, are used to stimulate increased ovulation in ewes or cows of quality breeds; the fertilized ova are extracted and implanted in the reproductive tract of ordinary animals. In one particularly successful experiment carried out in the mid-1970's, 11 transplanted ova and one nontransplanted ovum from a single ewe (out of a total of 19 fertilized ova passed) were successfully carried to maturity. In general, the fertilization rate of the induced ova is about 90 per-

[49] Guangxi-Zhuang Nationality Autonomous Region, Institute of Animal Husbandry, "A Representative Survey of Pig Artificial Insemination in Kwangsi," *Yichuan xuebao (Acta Genetica Sinica)*, 3, 1 (Jan., 1976): 45.

[50] Tianjin Municipality, Middle and Primary School Teaching Materials Research Section, *Yangzhu* (Raising Pigs) (Beijing: People's Education Press, 1976).

[51] *How to Raise Pigs* Editorial Group, *Zenyang yangzhu.*

[52] Guangxi-Zhuang Nationality Autonomous Region, Institute of Animal Husbandry, "The Study of Pig Boar Semen Freezing Technique," *Scientia Agricultura Sinica*, 1977, 2 (May): 85.

[53] "Better Sheep to Be Bred in Inner Mongolia," NCNA dispatch in English, Oct. 13, 1955; in *SCMP*, 1,150.

[54] "Report of Veterinary Delegation from Health of Animals Branch, Agriculture Canada, That Visited China from October 8–29, 1975," mimeographed (1975).

[55] *Ibid.*

cent, and the rate of embryonation of transplanted ova ranges from 40 to 70 percent.[56]

The endocrinology of reproduction is also studied at the CAS Institute of Zoology in Beijing, using prostaglandins, luteinizing-hormone–releasing factors (LRF), and pituitary gonadotrophins. Prostaglandins ($PGF_{a2}$ and $PGE_2$) and LRF have been synthesized by the Institutes of Biochemistry (Shanghai) and Materia Medica (Beijing).[57]

Chinese breeding objectives, for pigs as well as other animals, are dominated by considerations of economic efficiency and adaptability to Chinese conditions. In general, a maximum growth rate of meat and lard is sought, subject to severe constraints on the rate and quality of feeding. Pigs are expected to be amenable to cold or hot weather, disease resistant, and strong in frame. Sows must bear and be able to nurture large litters; shoats should have high weights at birth and weaning. In most areas, pigs are raised for both lean meat and lard, which must therefore be a high proportion of carcass weight. In some localities, other objectives—high yield of bristles or ability to forage—may also be important.[58]

In the past, Chinese breeding strategy has emphasized commercial crossbreeding or "commercial hybridization"; that is, $F_1$ outcrosses between native and exotic breeds for meat production, as opposed to development of new and stable inbreeds. This strategy stems from the 1950's Michirinian theory of animal breeding, reinforced by the suppression of Mendelian genetics under the Gang of Four.

One of China's leading texts on pig raising for college students and technical workers, published in 1957, was unequivocal in describing the theoretical base of China's breeding work: Mendelian genetics were termed "metaphysical," "antiscientific," and "lacking a rational basis in experiment"; Michirinian theories of inheritance were described as having a "complete experimental basis" and representing the "scientific road of dialectical materialism."[59]

This choice of Michirinian theory as the basis for pig breeding has practical consequences. For example, in accordance with its emphasis on the heritability of environmental influences, breeders were told that the beneficial results of crossbreeding depended on excellent feeding and management.[60]

[56] CAS, Institute of Genetics, Laboratory No. 203, "A Method for Increasing Ovulation in Domestic Animals," *Yichuan yu yuzhong (Genetics and Breeding)*, 1976, 4 (July): 19; "Genetics in Sheep Breeding," *FBIS*, Jan. 27, 1978, p. E15; Lin Tongyong, "Production of Pregnant Mare Serum Under Rural Conditions," *Genetics and Breeding*, 1976, 4 (Nov.): 24; and CAS, Institute of Genetics, Laboratory No. 203, "Borrowed Pregnancy: Transplantation of Livestock Fertilized Ova," *Scientific Experimentation*, 1976, 6 (June): 16.

[57] "Report of Veterinary Delegation from Health of Animals Branch."

[58] Shanghai Municipality, Leap Forward State Farm, Revolutionary Committee, *Yangzhu shengchan jishu wenda* (Questions and Answers on Pig Production Technique) (Shanghai: Shanghai Renmin Chubanshe, 1976); Chen Runsheng, "China's Pig Breeds"; and *How to Raise Pigs* Editorial Group, *Zenyang yangzhu*.

[59] Zhang Zonghui, *Zhongguo yangzhufa* (Chinese Methods of Pig Raising) (Nanjing: Animal Husbandry and Veterinary Publications Press, 1957).

[60] *Ibid.*

Chinese breeders now consider that crosses evaluated under good feeding conditions do not necessarily preserve their superiority under poor feeding conditions and that to ensure adaptability pigs should be bred and tested under ordinary production conditions.[61]

More significantly, the Chinese accepted that animal breeding should take advantage of "hybrid vigor" (heterosis). Heterosis results from $F_1$ crosses of breeds with extremely different characteristics and is diluted by subsequent crossings. Thus, Chinese breeding aimed to take advantage of the disparate characteristics of native and exotic breeds for production of $F_1$ meat pigs.[62] (The alternatives would be to continuously increase the proportions of exotic blood through repeated crossings or to develop new and superior inbred lines.) However, evidence produced to demonstrate heterosis has been able to show only that $F_1$ outcrosses exhibit characteristics superior (in quantitative terms) to those of one parent (usually the native). Only rarely is it demonstrated that the $F_1$ outcross is more vigorous than either parent.[63]

Surprisingly, Chinese breeders were just beginning to sort out which economic characteristics are dominant in crosses by the late 1970's. Recent research on this subject has made good use of statistical techniques and shown concern with standardization of measurement and conceptual clarification. It points the way to further research that could be of considerable value to breeders.[64]

Since the native-exotic hybrids have characteristics generally intermediate between the two parents, the weaker economic characteristics of native breeds can be substantially improved through outcrossing. As a result, commercial production favors the outcrosses in areas where feed supplies are adequate. Native breeds do persist, however, partly because of the breeding strategy discussed above, partly because they frequently demonstrate overall superiority under conditions of poor feeding, and partly because of their fecundity.[65]

Despite the emphasis on native-exotic commercial outcrossing, at least two new inbreds have been developed since the 1950's, specifically the Shanghai and the Haerbin White varieties. The Shanghai is well adapted to eat coarse and green feed and food-processing industry by-products, and it has relatively rapid growth, a high dressing percentage, strong disease resistance, and a large litter size (around 12 shoats). The Haerbin White variety offers cold tolerance, quick growth, and high birth and weaning weights. Six months after weaning it gains about 0.60 kg a day with high-quality

[61] Chen Runsheng, "China's Pig Breeds."

[62] *How to Raise Pigs* Editorial Group, *Zenyang yangzhu.*

[63] Chen Runsheng, "China's Pig Breeds"; and Heilongjiang Sanjiang White Pig Breeding Group, "The Study of the Form of Inheritance and Hybrid Vigor in the Body Form of Pigs," *Scientia Agricultura Sinica,* 1978, 1 (Feb.): 85.

[64] Chen Runsheng, "China's Pig Breeds"; Heilongjiang Sanjiang White Pig Breeding Group, "Body Form of Pigs"; and Jiangsu Institute of Agricultural Science, Department of Animal Husbandry and Veterinary Science et al., "Heredity of Some Quantitative Traits in New Waiying Swine," *Acta Genetica Sinica,* 2, 4 (Nov. 1975): 287.

[65] Chen Runsheng, "China's Pig Breeds."

feeding and 0.56 kg a day with poor-quality feeding. When fed fermented mixed feed, it requires about 3.7 kg to gain 1-kg weight.[66] Neither breed has sufficiently stabilized inherited characteristics, however.[67]

Most generalizations about Chinese pig-breeding programs also apply to the breeding of other animals. For animals with low reproductive efficiency, the stud's ability to service many females each year makes the crossing of imported males with native females the most economical strategy; only for animals with a high reproductive efficiency, such as poultry, can inbred flocks be multiplied rapidly enough to affect total stocks. Thus, the general breeding objective in China is to eliminate the weaknesses of native varieties while retaining their strengths through outcrossing. In the case of dairy cattle, Holstein exotics are crossed with native black-and-white cattle, which are small and have a low milk yield but good heat tolerance. The crosses are apparently heat tolerant and provide a good milk yield without the high concentrate requirement and nutritional demands of Holsteins.[68] The yield of crossbred milking cows per lactation ranges from 4,000 to 6,000 kg and averaged 5,800 kg in the north.[69]

The average size of native breeds of water buffalo is small, although individual animals vary greatly. Crosses between Murrahs and native swamp buffaloes yield an animal of uniform size and appearance with outstanding propensities for both work and milk production.[70]

In horse breeding, the objective is to retain the endurance of the Mongolian breed while increasing size and hauling strength through crossbreeding.

Sheep-breeding objectives are more complex, depending on the type of wool required, the priority attached to meat production, climatic conditions, and the type of feeding method to be used. However, an increase in quality and quantity of wool yield are normally the primary breeding objectives, with increased meat and fat yields secondary goals.

There have been attempts to interbreed the Peking duck with local varieties to improve growth rates, egg production, meat quality, and "living habits." A cross between Jinding and Peking ducks reportedly combines the former's egg productivity with the latter's meat quality and was being popularized in the 1970's.[71]

Finally, an unlikely example of an attempt to take advantage of heterosis was the crossbreeding on a Qinghai farm of the local wapiti deer with the spotted deer of northeastern China to produce a better yield of antlers, from which a medicinal tonic is extracted and exported.[72]

[66]"The Creation of a New Breed: The Haerbin White Pig," *Scientia Agricultura Sinica,* 1977, 2 (May): 95.

[67]*How to Raise Pigs* Editorial Group, *Zenyang yangzhu.*

[68]Cockrill, "Visit to China."

[69]"Improved Animal Breeds."

[70]Cockrill, "Visit to China."

[71]Xiamen University, Department of Biology, Zoology Research and Teaching Group, "Utilizing the Hybrid Vigor of Domestic Ducks: Comparison of the 'T'u-pei' Duck and the 'Pan-fan' Duck," *Yichuanxue tongxun (Bulletin of Genetics),* 1973, 4 (Nov.): 1.

[72]"More Deer Antlers," *Peking Review,* 21, 27 (July 7, 1978): 47.

*Veterinary Medicine*

China's program of prevention and treatment of disease in pigs could well serve as a model for other less developed countries. The death rate among weaned pigs was roughly 7.6 percent in 1953, fell to about 4 percent in 1956, and to near zero in the 1970's. Programs to reduce the death rate of pigs from hog cholera have also been successful following Chinese development and large-scale use of a vaccine against this disease in the early 1960's.

Chinese veterinary medicine emphasizes prevention, rather than cure. There are veterinary stations in all communes, staffed by full-time cadres with middle-level technical training. Semitrained "barefoot veterinarians" practice at the team level; their duties include inspection and reporting of disease outbreaks and quarantine of infected animals; promotion of sanitary measures, such as maintaining clean and dry pens or disinfecting equipment and housing; ensuring that breeding boars are disease-free; practice of artificial insemination; inoculation of all pigs twice yearly against hog cholera, erysipelas, and plague; and promotion of the inclusion of antibiotics in pig feed. Inspection and quarantine are also practiced at markets where shoats are purchased. Veterinary care is provided to private pig raisers and collective owners alike—either free or at low blanket rates per pig. A "life insurance" system offers partial compensation in the event of a pig death. (This helps enforce the prohibition against the sale of dead pigs.)[73]

The veterinary program for other animals is similar. By the early 1960's, there were large numbers of trained or semitrained veterinary personnel in the pastoral areas. Prevention rather than cure and traditional rather than Western medical techniques were emphasized. By 1957 rinderpest and foot and mouth disease in cattle were under control; by the 1970's they were no longer a problem.[74] Sheep and cattle pneumonia, anthrax, and intestinal parasites were eliminated by the 1960's, and sheep scabies was largely under control by the mid-1970's.[75] Even allowing for some journalistic exaggera-

[73] Gaozhou County CCP Committee, Agricultural Office, "Pig Raising Develops Rapidly in Kaozhou County," tr. from *Nonglin gongzuo tongxin (Agriculture and Forestry Bulletin)*, 1964, 6, *SCMM*, 429 (Aug. 10, 1964): 30; Chen Xinghui and Ding Siyin, "An Investigation into the State of Simultaneous Development of Grain and Animals, Etc.," *Jingji yanjiu (Economic Studies)*, 1965, 8 (Aug.): 26; Cockrill, "Visit to China": R. G. Nash and T. H. Cheng, "Research and Development of Food Resources in Communist China," *Bioscience*, Nov. 1965, p. 707; "Report of Veterinary Delegation from Health of Animals Branch"; *How to Raise Pigs* Editorial Group, *Zenyang yangzhu;* Ministry of Agriculture and Forestry, Animal Husbandry Bureau, *Dali fazhan yangzhu shiye; Yangzhuye yao dafazhan;* and Hunan Province, Ningxiang County CCP Committee, "Preserve the Party's Basic Line: Diligently Develop Collective Pig Raising," *Scientia Agricultura Sinica*, 1977, 1 (Feb.): 88.

[74] Nash and Cheng, "Food Resources"; and "Report of Veterinary Delegation from Health of Animals Branch."

[75] "Northwest China National Minority Area Increases Famous Sheep," NCNA dispatch in English, July 28, 1964, in *SCMP*, 3,270 (July 31, 1964); "Achievements and Prospects in Animal Breeding," NCNA dispatch in English, Dec. 19, 1957, in *SCMP*, 1,678 (Dec. 24, 1957); "Continuous Increase of Livestock in Tibet," NCNA dispatch in English, in *SCMP*, 3,087 (Oct. 25, 1963); "Rapid Livestock Increase in Southwest China National Minority Areas," NCNA dispatch in English, July 20, 1964, in *SCMP*, 3,264 (July 23, 1964); "Report of Veterinary Delegation from Health of Animals Branch"; *Summary of World Broadcasts (BBC), Far East*, W847 (Oct. 8, 1975): A20.

tion, the achievement was substantial. In Inner Mongolia, for example, the percentage of animals dying of traditional diseases dropped from 5 percent in 1958 to 1 percent in 1963.[76]

Vaccination programs in the 1970's emphasized use of brucellosis strain-19 vaccine for calves and vaccination against Newcastle disease for chickens. Vaccination against eradicated diseases had stopped except in areas bordering other countries. Animals were also treated for external and internal parasites. Equines were tested for glanders, chicken for pullorum disease, and cattle for tuberculosis. Certain diseases that are common elsewhere, including scrapie, atrophic rhinitis in swine, rabies, trichinosis, bovine cysticercosis, anaplasmosis, and piroplasmosis, are allegedly not found in China.[77]

Recent research on preventative medicine has concerned theileriasis in cattle, encephalitis in equines, and liver flukes in cattle.[78]

Chinese veterinary medicine combines basic Western cures with herbal remedies and acupuncture. Various common antibiotics are recommended for treatment of erysipelas, swine plague, asthma, and influenza. Herbal medicines are sometimes injected in combination with antibiotics as the former are usually cheaper and prescriptions can be made up locally.[79] A combination of common chemical compounds or herbal decoctions are prescribed for diarrhea, intestinal parasites, eczema, and food poisoning. The only effective drug prescribed for hog cholera is still very limited in supply. Acupuncture is not mentioned in pig-raising manuals, and a recent article on the subject admitted that there is no evidence that it cures any pig disease.[80] The large volume of publications offering new prescriptions for pig diseases suggests considerable practically oriented research in this area, with a distinct emphasis on herbal remedies.[81]

[76] "Five Years of Livestock Increases in Inner Mongolia," NCNA dispatch in English, in *SCMP,* 3,259 (July 16, 1964).

[77] "Report of Veterinary Delegation from Health of Animals Branch."

[78] Gansu Provincial Institute of Veterinary Medicine, "Cell Culture Studies of *Theileria annulata* Schizonts and Their Immunogenicity," *Dongwu xuebao (Acta Zoologica Sinica),* 21, 3 (Aug. 1975): 243; Wang Yongyi et al., "Maternal Antibody Against Japanese B Encephalitis Acquired by Foals and Its Effects on Immunization by Live Vaccine," *Weishengwu xuebao (Acta Microbiologica Sinica),* 16, 1 (Feb. 1976): 17; and Xiamen University, Department of Biology, Parasitology Laboratory, "Investigation on the Epidemiology and Prevention of Liver Fluke in Cattle," *Dongwuxue zazhi (Journal of Zoology),* 1974, 3 (Aug.): 32.

[79] *How to Raise Pigs* Editorial Group, *Zenyang yangzhu.*

[80] Yang Yongdao, "Some Problems amid the Expansion of Acupuncture Technique in Swine Diseases," *Scientia Agricultura Sinica,* 1977, 2 (May): 90.

[81] Jiangxi Province, Linchuan County, Bureau of Agriculture, "Treatment of Arthritis in Domestic Animals with a Combination of Western and Chinese Traditional Medicine," *Journal of Zoology,* 1974, 4 (Nov.): 23; Fujian Normal University, Parasitological Research Laboratory, "On the Development of *Gnathostoma Hispidum* in the Intermediate Host with Special Reference to Its Transmission Route in Pigs," *Acta Zoologica Sinica,* 22, 1 (Feb. 1976): 45; Jiangxi Province, Agricultural Research Institute, Department of Animal Husbandry, "Treatment of Pig Asthma with Lei-kung-t'ung," *Journal of Zoology,* 1974, 3 (Aug.): 35; Shanghai Municipality, Jinshan County, Fengwei Commune, Animal Husbandry and Veterinary Station, "Herbal Prescriptions for Hog Diseases," *Kexue puji (Popular Science),* 1974, 12 (Nov.): 31; Shaanxi Province, Yanan District, Animal Husbandry and Veterinary Control Station, "The Use of Selenium as a Preventative Measure in Controlling Certain Pig Diseases," *Acta Zoolog-*

At the grass-roots level, curative medicine in large animals is concerned mostly with common and noncontagious ailments, such as digestive problems, heat stroke, arthritis, strained muscles, etc. Manuals recommend both Chinese herbal prescriptions and Western treatments, with no judgments about their relative efficacy.[82]

Use of acupuncture anesthesia in animal surgery has been growing since 1970 and has been demonstrated to two delegations of foreign veterinarians.[83] Effectiveness is high except in operations on the urinary tract or hernias where, as in humans, acupuncture is not used. The main problems are occasional failure to achieve full sedation and inability to find suitable pressure points for the head. Research into the mechanism of acupuncture anesthesia has barely begun.

### *Mechanization*

In 1978 Chairman Hua Guofeng called for mechanization of pig raising and said that research in this area had high priority.[84] The main emphasis was on "mechanizing" water supplies (i.e. introducing piped water supplies to pig pens) and using machines to harvest and mix feed and fodder. For example, the fodder shredder, which has been in use since 1965, has made it possible to use crop by-products in fermented feeds.

Mechanization began relatively early, in the 1960's, in pastoral areas because of labor scarcity and, in the case of state farms, of fixed, high wage rates. By 1963 Inner Mongolia, the forerunner in this as well as other aspects of stockbreeding, claimed that one-third of its forage collection was accomplished by tractors, horse-drawn harvesters, and rakes, that electric shears had replaced hand shears in some communes, and that powered pump wells were providing water for a significant number of livestock. The local Research Institute for Stockbreeding Mechanization had designed, among other things, a "multiduty automobile equipped for milking, shearing, and water pumping."[85] Since then, most dairy farms in Inner Mongolia and around major cities have been equipped with milking machinery. Incubation with artificial heat is also common on major poultry farms and the force-fed Peking duck is provided regulated doses of feed by an electric-powered machine.[86] Despite these examples of the use of machinery, animal husbandry in China remains labor intensive.

---

*ica Sinica,* 22, 1 (Feb. 1976): 53; Fujian Province, Fuqing County, Dongzhang Commune Animal Husbandry and Veterinary Station, "Treatment of Diarrhea in Pigs with Chinese Traditional Drug, Chih-li-ling," *Journal of Zoology,* 1974, 4 (Nov.): 22; and "Single-Injection Treatment of Pig Paralysis," *Scientific Experimentation,* 1976, 8 (Aug.): 33.

[82] *How to Raise Plow Oxen* Editorial Group, *Zenyang yang gengniu.*

[83] "Report of Veterinary Delegation from Health of Animals Branch."

[84] He Gang, "Speech by Vice Agriculture Minister Ho Kang," *FBIS,* Mar. 29, 1978, p. E24.

[85] "Inner Mongolia Begins to Mechanize Stockbreeding," NCNA dispatch in English, Dec. 10, 1963.

[86] Cockrill, "Visit to China."

In the late 1970's, Chinese researchers were developing an innovation to save labor by replacing machinery: a subcutaneous injection of an unspecified substance that causes wool to fall off naturally, obviating manual sheep shearing.[87] This technique is probably similar to the experimental use of cyclophosphamide in the United States.

*Research Centers and Publications*

At the National Science Conference in March 1978, President Jin Shanbao of the Chinese Academy of Agricultural and Forestry Sciences reviewed the achievements of Chinese research in agriculture, which he said had reached "advanced world levels." He listed two achievements in animal sciences—the preparation of vaccines for cattle plague and hog cholera.[88] These vaccines were already widely used in the early 1960's, however, and there lies the point: most Chinese research achievements in animal science were made before the Cultural Revolution and its aftermath (1966–76). With a few notable exceptions, such as work on transplantation of sheep and cattle ova and efforts to breed efficient cellulase molds, research activity was moribund during these 11 years.[89]

It is not clear what institutions were conducting research on animal sciences in the 1970's. Because of extensive reorganization during the Cultural Revolution, many institutions formerly in the field had been moved, abolished, or their functions significantly changed. The few foreign delegations that have visited animal sciences research institutes in the 1970's found that about half the nominal staff was away at "grass-roots levels" and that little research was under way. University programs were suffering from deteriorated quality of students, abbreviated programs, and courses concentrating exclusively on practical matters.[90]

The only research of any quality that was published in the 1970's came from the CAS Institutes of Genetics and Zoology in Beijing, which were responsible for research on ovum transplantation and animal endocrinology, the Institutes of Animal Husbandry and Veterinary Science in Beijing, Inner Mongolia (Huhehot), Heilongjiang (Fuyu), Shanghai, Guangxi-Zhuang Autonomous Region (Nanning?), Gansu, Guangxi, and Xinjiang, the Poultry Science Research Institute (Jiangsu), and the Gansu Provincial Institute of Veterinary Medicine (Lanzhou). The only institutions of higher education showing signs of life were Guangdong College of Agriculture and Forestry (cellulase mold research), Jiangsu Agricultural College (development of some vaccines at their microbiological laboratory), and two universities in Fujian.

[87] "Heilungkiang Wool-Shearing Method," *FBIS*, Aug. 16, 1977, p. L3.
[88] Jin Shanbao, "Chin Shan-pao Speech," *FBIS*, Mar. 27, 1978, p. E18.
[89] CAS, Institute of Genetics, Laboratory No. 203, "Borrowed Pregnancy"; and Guangdong College of Agriculture and Forestry, Department of Animal Husbandry and Veterinary Medicine, "Isolation Selection of Cellulase Molds."
[90] "Report of Veterinary Delegation from Health of Animals Branch."

National journals devoted most of their space to papers collectively authored by laboratories, institutes, experimental units, or farms. The majority of papers came from grass-roots organizations and typically concerned breeding experiments, herbal or acupuncture remedies, or the experience of a production unit in increasing its livestock numbers. In content and style, they can be characterized as extension literature, rather than research. They lacked the kind of quantitative data, analysis of statistical significance, and comparison of treatment with control groups that would persuade a Western-trained scientist to accept the results claimed.

In 1978–79 there was a spectacular change in the quality of national agricultural journals, such as *Dongwu xuebao (Acta Zoologica Sinica), Dongwuxue zazhi (Journal of Zoology), Zhongguo nongye kexue (Scientia Agricultura Sinica), Weishengwu xuebao (Acta Microbiologica Sinica), Yichuan yu yuzhong (Genetics and Breeding)*, and the popularizing *Kexue shiyan (Scientific Experimentation)*. These journals published a large number of papers with real scientific content, signed by individual authors, giving details of the conditions under which experimental data were gathered, adequate quantitative data on treatment groups and controls, standard deviations, correlation and regression coefficients, and, in many cases, measures of statistical significance. It was encouraging to find that the Chinese understood scientific methodology and tools of analysis, even if it had been politically unfashionable to use them during the Cultural Revolution. Most of the provincial agricultural journals had also resumed publication by 1978, but unlike the national journals, were not available for subscription outside the PRC.

Animal scientists in China emerged from the Cultural Revolution and the overthrow of the Gang of Four to pick up where they left off more than a decade previously. This point of departure found them backward relative to their colleagues in the other agricultural sciences, as well as to those in the developed world. It is hard to predict how rapidly these gaps can be narrowed, but the prospects are good. Many Chinese specialists in animal sciences have the basic training necessary to assimilate the results of foreign research. In addition, the practical orientation of Chinese research in the natural sciences will continue to ensure crucial support from such disciplines as genetics, zoology, and chemistry. This advantage is not necessarily enjoyed by the animal sciences in other developing countries.

Despite China's success in increasing animal production within a socialist economic framework, animal agriculture in the 1970's remained predominately a labor-intensive, household-based activity or side activity. Labor productivity and feed conversion rates were not high, and it was difficult to maintain a high rate of production growth at stable or decreasing costs. The drive to increase production of meat and other animal products rapidly as a part of the four modernizations program probably implies future research emphases on the practical problems of grassland development and assimilation of modern, concentrate-intensive production techniques for pigs and poultry. Such problems of adapting known technologies are likely to absorb the energies of most of China's animal scientists in the near future. But considerable scope remains for advanced research in the areas of breeding and veterinary medicine, which could also contribute to productivity.

*Leo A. Orleans*

# *A NOTE ON AGRICULTURAL MECHANIZATION*

Agricultural mechanization has been the subject of controversy in China for almost 30 years. Despite Mao's early proclamation that "the fundamental way out for agriculture lies in agricultural mechanization," the implementation of this goal was impeded by disagreements on the most basic questions: not only "how" or "how fast" but, given China's terrain and ample manpower, "whether" agriculture should in fact be mechanized. The recurring debates resulted in shifting priorities and changing definitions. In general, however, the definition of agricultural mechanization progressed from soil turning, raking, and rolling—now referred to as the "three olds"—to the extremely broad definition that came into use in the late 1970's. Now, in addition to conventional activities like plowing, agricultural mechanization includes irrigation, rural electrification, farmland construction, the production and application of chemical fertilizers and pesticides, and the processing of farm products. There is also no doubt that discretional agricultural mechanization is an integral part of the current drive to modernize agriculture.

This note is intended to provide some perspective on China's overall policies of agricultural mechanization, the new policies that Beijing began implementing in the late 1970's, and the recurring problems that China has encountered and struggled with since the 1950's.

## *Agricultural Mechanization Policies*

Given the backwardness of Chinese agriculture in 1949, the initial goals were fairly obvious. It was necessary to improve the primitive tools and implements on which Chinese peasants had relied for centuries and then to introduce new, more modern, and efficient equipment. Peasants stood to benefit enormously, for example, from having rubber tires rather than wooden wheels on their horse- and man-drawn carts, by using metal rather than wooden plows, and by substituting mechanical irrigation pumps for human-powered waterwheels. The serious debates started in the mid-1950's when, influenced by the Soviet experience, the goal for mechanization of agriculture was to have tractors plowing 70 or 80 percent of the country's total cultivated farmland. Although this goal was remote, it raised several key questions. What kind of economic priority should be assigned the production of heavy agricultural equipment? Was mechanization applicable to all of China's diverse agricultural regions? Who was to own and manage agricultural machinery: communes or state-run tractor stations? Where was the skilled manpower needed to operate and maintain the tractors to be found, and would the widespread introduction of labor-saving machinery create serious labor utilization problems in the countryside?

During the Great Leap Forward (1958–59), the diversion of millions of peasants into nonagricultural activities quickly turned China's rural labor surplus into an artificial labor shortage, and by 1959 it seemed appropriate for Mao to proclaim that "basic mechanization" of agriculture should be achieved in ten years. Just a year later, however, the Great Leap was in shambles, and agricultural mechanization lost its priority. When discussion about how best to improve efficiency in agriculture resumed, the emphasis seems to have shifted to a more modest goal, referred to as "semimechanization." By the mid-1960's the Chinese had clearly reached a compromise position: the major emphasis was on producing walking tractors, water pumps, electric threshers, and a variety of small powered tools and implements. With minor vacillations this policy lasted through the Cultural Revolution period until 1977 when China announced its "four modernizations" program.

By the late 1970's the modernization of agriculture was at the foundation of China's development plans. In a report to the Fifth National People's Congress in June 1979, Chairman Hua Guofeng stressed the need to give greater material and technical support to agriculture and promote mechanization of farming. He revealed that government investment in agriculture had increased from 10.7 percent of the state budget in 1978 to 14 percent in 1979. Total agricultural expenditure in 1979 was expected to be 17.4 billion yuan.[1]

The detailed plans announced in a three-week National Conference on Agricultural Mechanization held in Beijing in January 1978 showed the far-reaching scope of Chinese plans to modernize agriculture. Mechanization of farm work was defined as including machines for cultivation, farmland capital construction, drainage and irrigation, plant protection, transport, harvesting, and processing farm and sideline products, in addition to chemical

[1] New China News Agency (NCNA) dispatch, June 25, 1979; in *Foreign Broadcast Information Service*, July 2, 1979.

fertilizers, insecticides, plastic sheets for nursing seedlings, machinery for forestry, animal husbandry, and fisheries, the building of small power stations, and semimechanized farm implements.[2] The nation was instructed to proceed "on all legs"—to combine modern machinery produced in major industrial plants with innovations and improvements developed at the local level. The established targets prescribed that machines should be used in 70 percent of the "main work" in agriculture by 1980 and 85 percent by 1985. By late 1979, adjustments were already being made in these overly ambitious targets. Furthermore, there were many complaints about the extreme unevenness between the mechanization of plowing and the other field activities:

> In order to achieve the set target of 70 percent mechanized farming, some places have simply sought to increase the number of tractors and the percentage of tractor-plowed land. This, however, has resulted in the disproportionate situation in which 46 percent of the country's farmland is now tractor-plowed, while mechanically seeded, harvested, and transplanted farmland is only 10 percent, 2–3 percent, and 1 percent, respectively.[3]

Chinese leaders recognize that national policies must be adapted to varied local conditions. For example, the comparatively sparsely populated, flat, grain regions in the north and northeast are especially suited to mechanized farm work; observers during the 1970's reported considerable progress in mechanized plowing, sowing, harvesting, and threshing. But in the rice-growing areas of the south, where most of the paddy fields are located on terraced slopes in tiny patches, it is almost impossible to use modern machines, and the wet paddy fields are mostly worked by manual labor. In the pastoral areas the conditions differ again. The recommended emphasis is on mechanizing fodder cultivation, raking, and transport facilities.

China's leaders maintain that they are not abandoning Mao's dictum that "we live by intensive farming" by introducing new machinery.[4] They argue that increased farm yields can be achieved only in conjunction with mechanization in sparsely populated areas and that close integration between mechanization and intensive farming is essential. It is expected that the labor power saved by mechanization will go into "broader and deeper economic construction in various fields." Mao's lifelong dream—the ultimate integration of industry with agriculture and a reduction of differences between manual and mental labor—is to be achieved with the gradual industrialization of the people's communes. In fact, many localities should be able to use additional manpower, given China's increasing emphasis on livestock, fishing, and orchard operations and the development of commune and brigade industries.

Individuals, however, are not necessarily interchangeable. Mechanization relies heavily on young people educated at middle-level agricultural schools. Semiliterate peasants who have spent half their lives working in the fields

[2] *Peking Review*, 1978, 8 (Feb. 24).
[3] NCNA dispatch, Oct. 14, 1979.
[4] NCNA dispatch, Oct. 13, 1977.

TABLE 1. *Inventory and Production of Tractors*

| Year | Garden Tractors | | | | Conventional Tractors | | | |
|---|---|---|---|---|---|---|---|---|
| | Production | | Inventory | | Production | | Inventory | |
| | *Physical Units* | *15-HP Units* | *Physical Units* | *15-HP Units* | *Physical Units* | *15-HP Units* | *Physical Units* | *15-HP Units* |
| 1958 | | | | | 957 | 1,150 | 22,700 | 45,330 |
| 1959 | | | | | 4,871 | 9,400 | 29,500 | 59,000 |
| 1960 | | | | | 11,900 | 23,800 | 40,700 | 81,300 |
| 1961 | | | | | 8,100 | 16,200 | | |
| 1962 | | | | | 6,550 | 13,100 | 51,700 | 103,000 |
| 1963 | | | | | 7,850 | 15,700 | 56,700 | 113,300 |
| 1964 | 600 | 150 | | | 9,420 | 18,840 | 61,600 | 123,100 |
| 1965 | 3,500 | 875 | 18,000 | 4,500 | 9,700 | 19,470 | | |
| 1966 | 5,520 | 1,380 | 25,000 | 6,250 | | | 75,000 | 150,000 |
| 1967 | 8,400 | 2,100 | 35,000 | 8,750 | | | | |
| 1968 | 10,700 | 2,675 | 47,000 | 11,750 | | | | |
| 1969 | 12,800 | 3,200 | 59,000 | 14,750 | | | | |
| 1970 | 25,000 | 6,250 | 85,000 | 21,250 | | | | |
| 1971 | 38,500 | 9,625 | 125,000 | 31,250 | 36,875 | 88,500 | | |
| 1972 | 8,400 | 21,000 | 225,000 | 56,250 | 40,560 | 97,350 | | |
| 1973 | 11,200 | 28,000 | 350,000 | 87,500 | 56,790 | 136,290 | | |
| 1974 | 126,000 | 31,500 | 600,000 | 150,000 | 48,240 | 120,600 | | |
| 1975 | 175,000 | 43,750 | 900,000 | 225,000 | 56,790 | 136,290 | | |
| 1976 | 248,500 | 62,125 | 1,000,000 | 250,000 | 54,260 | 135,650 | | |
| 1977 | 320,000 | 80,125 | 1,090,000 | 272,500 | 99,300 | 238,320 | 467,000 | 1,120,800 |
| 1978 | 324,200 | 81,050 | 1,370,000 | 342,500 | 113,500 | 272,400 | 557,000 | 1,336,800 |
| 1979 | 308,000 | 77,000 | | | 124,000 | 297,600 | | |

SOURCE: National Foreign Assessment Center, *China: Economic Indicators* (Washington, D.C., 1980).

usually require considerable training before they can perform more demanding jobs. Furthermore, since China's rural population is not mobile, it will be necessary to synchronize the mechanization of field activities with local industrial development. Because such planning is extremely difficult, it is all too easy for one county or commune to have both a surplus of unskilled labor and a shortage of trained labor. As the New China News Agency explained in 1979: "The peasants will buy tractors to replace manpower, but a diversified economy will not yet be developed, and the use of the labor force will not be planned. As a result, the collective income will not be increased, but will be reduced."[5] This problem of absorbing released farm labor may help to explain why the Chinese primarily stress the need for each laborer to farm more land (which is scarce) and only secondarily the need to increase per acre production (which is the real priority). "We should follow the principle of keeping the agricultural labor force employed right there at their posts," they say.

*Some Production Aspects of Agricultural Mechanization*

As of early 1979, there were 1,900 agricultural machinery manufacturing enterprises in China, of which fewer than 30 were able to produce large- and medium-sized tractors. The rest were county-level units producing a wide range of small- and medium-sized farm machinery.*

According to the State Statistical Bureau, there were in 1978 557,000 large- and medium-sized tractors in China (90,000 more than in 1977), and 1,370,000 garden tractors (280,000 more than in 1977). Total horsepower of powered irrigation equipment was 65,580,000, an increase of 5,530,000 HP over 1977.[6] Table 1 puts these figures into historical perspective. It should be added that in 1979 there were about 2,400 county-, commune-, and brigade-level agricultural machine repair shops to maintain the machinery available in China.

The accomplishments reflected in these statistics of machinery production must be qualified, however. First of all—as Beijing is the first to admit—most agricultural machinery in China is low in quality, poorly managed, and difficult to maintain and repair. Second, not all agricultural machinery is used for directly agricultural purposes. Foreigners visiting China nearly always comment that most tractors they see are used for transport rather than field activities. This was confirmed, for example, by some "responsible members" of the All-China Agricultural Society who, traveling outside Beijing, counted 173 tractors, of which 170 were being used for transport.

Third, there is a shortage of rolled steel and oil, which will worsen as mechanization progresses. In the northern and northeastern provinces, for example, significant advances in mechanization have not reduced the use of

* Some of the specifics that relate to production are discussed in other chapters of this volume. See Vaclav Smil, for example, on rural electrification and Jack R. Harlan on pest control.

[5] NCNA dispatch, Apr. 20, 1979.

[6] *Beijing Review*, 1979, 27 (July 6).

horses for farming and transport. The Chinese refer to this as "double expenditure": horses eat fodder and farm machines consume oil, so that while grain production may increase, peasant income does not. They estimate that if motor vehicles replaced horse carts in Heilongjiang province, the amount of grain saved would equal the total quantity of commercial grain delivered by the province to the state each year.[7]

Many of the problems China faces in agricultural mechanization are the result of the past emphasis on local self-reliance. Since the mid-1960's, thousands of county-level production units have been turning out a wide range of small- and medium-sized farm machines, using their own resources and all too often their own designs as well. The consequences of this mass innovation were discussed at the May 1979 Beijing Conference held by the Ministry of Agricultural Machinery. The participants pointed out that the past overemphasis on self-sufficiency had resulted in too many "we-make-everything-you-need" enterprises. These were extremely inefficient, producing small quantities of poor-quality, expensive goods. They also created an imbalance in the national economy: an excess of factories producing relatively simple products and a shortage of factories capable of producing the more complex, precision products that the country needed.[8]

China's policy of self-reliance also resulted in an absence of national standards and coordinated policy in producing spare parts. Ever since the 1960's, China has been racked with complaints about the lack of spare parts and the inadequate (or nonexistent) interchangeability between parts and components of tractors, diesel engines, and other agricultural machinery. One unidentified province, for example, was accused in 1978 of having 46 varieties of tractors and over 110 varieties of farm diesel engines—a nightmare for any repair unit.[9] Similarly, the "Hongqi No. 100" tractor was reported in 1978 to have 1,700 parts, of which over 300 were manufactured by nearly one hundred plants in other localities.[10] The manufacturer of each part, rather than the tractor assembly plant, was held responsible for the part's quality and performance, creating almost insurmountable problems for any production unit faced with a malfunctioning tractor. Because of the poor quality of machinery and the unavailability of spare parts, the Chinese estimated in 1979 that only 70 percent of tractors, hand-drawn tractors, and power machinery for farm use in China were serviceable. This could mean that 150,000 tractors were out of action at any given time. Since peasants must purchase machinery ("it is ineffective to present peasants with farm machines gratis," say the authorities), it is understandable why poor serviceability "dampens mass enthusiasm for farm mechanization."

Beijing was attempting to correct some of these problems in 1979 by initi-

[7] NCNA dispatch, Apr. 20, 1979.

[8] NCNA dispatch, May 15, 1979.

[9] *Hongqi (Red Flag)*, 1978, 2 (Feb.); tr. Joint Publications Research Service (JPRS), *JPRS*, 70,869 (Mar. 31, 1978).

[10] NCNA dispatch, Aug. 28, 1978.

ating an intensive campaign to achieve "standardization, serialization, and versatility." This effort was aimed at upgrading the quality of machinery, reducing the number of models designed to perform identical functions, increasing the production of adequate replacement parts, and generally improving the efficiency of the agricultural engineering industry.[11] Chinese leaders have also adopted some short-term measures to help the peasants. One report from Beijing said: "In 1978, 358 factories belonging to the central authorities and Beijing municipality sent out 1,200 workers and technicians . . . to help rural technicians repair 21,000 pieces of agricultural machinery and improve or produce 5,300 machines for summer farm work, build 156 mechanized and semimechanized pig sties, 44 chicken farms, 297 small water conservancy projects . . . and 10 small hydroelectric power stations."[12]

In attempting to introduce mechanization across a broad range of agricultural activities, China has had both successes and setbacks. The Chinese have done quite well in irrigation and rural electrification—sectors that received early priority (see Table 2). Progress in the mechanization of crop cultivation, however, has been patchy and shows great regional variation. As of 1979, many aspects of agricultural engineering were still very new to China or nonexistent.

In general, the Chinese believe that they have a reasonable mechanical base from which to start the modernization of agriculture. They reported in 1978 that the total power of all farm machinery was 170 million HP, or approximately 0.7 HP for every 10 mou (1.66 acres) of arable land.

The leaders see planning and coordination within the agricultural sector as their main practical problem. To improve on this, a State Agricultural Commission was established in 1978, under the direct control of the State Council. The latter oversees the ministries of Agriculture, Forestry, Water Conservancy, Land Reclamation, and Agricultural Machinery and almost certainly cooperates with other ministries concerned with agriculture, such as the newly reestablished Ministry of Chemical Industry. The ministry most directly involved in agricultural engineering is the Ministry of Agricultural Machinery. As of 1979, it probably controlled China's 1,700 agricultural machine research institutes, and its job was to ensure increased production of better-quality agricultural machines and parts and implementation of "standardization, serialization, and versatility" within the factories. It was also charged with training technical forces for rural areas and persuading factories to satisfy their customers and take full responsibility for all their products by a system of guaranteed repair, exchange, and refund.

In the final analysis, however, the degree of success China will achieve in agricultural mechanization will only partly depend on the Ministry of Agricultural Machinery's effectiveness. In the long run, China's real challenge is

[11] *Red Flag*, 1978, 2 (Feb.); tr. JPRS, *JPRS*, 70,869 (Mar. 31, 1978).
[12] Beijing City Service, Apr. 23, 1979.

TABLE 2. *Powered Irrigation Equipment Production and Inventory*
(1,000-HP Units)

| *Year* | *Production* | *Inventory* | *Year* | *Production* | *Inventory* |
|---|---|---|---|---|---|
| 1959 | 1,790 | 3,300 | 1969 | | |
| 1960 | 2,300 | 5,960 | 1970 | | |
| 1961 | 1,000 | | 1971 | 4,080 | 18,600 |
| 1962 | 1,360 | 5,800 | 1972 | 5,100 | |
| 1963 | 640 | | 1973 | 7,624 | 29,260 |
| 1964 | 715 | 6,700 | 1974 | 11,436 | 38,500 |
| 1965 | 953 | 7,700 | 1975 | 17,544 | 50,800 |
| 1966 | 1,078 | 10,260 | 1976 | | 54,000 |
| 1967 | | | 1977 | | 60,050 |
| 1968 | | | 1978 | | 65,580 |

SOURCE: National Foreign Assessment Center, *China: Economic Indicators* (Washington, D.C., 1980).

the integration of agricultural mechanization with the country's overall economic development and coping with the social changes attending agricultural modernization. Although the ambitious targets set for agricultural mechanization as part of the four modernizations have been dropped, the competition for natural, human, and capital resources will place serious strains on the economy. China will have to allocate substantial resources for diversifying the rural economy in order to absorb the labor force displaced by mechanization and simultaneously find jobs for as many as 15 million young people entering the work force each year.

In addition, Beijing has a serious political challenge in the rural areas. The requirements for larger numbers of educated workers in agriculture and subsidiary industries are apparently being met by the educational system. The question now being asked is how to meet the rising expectations of this vital segment of society. In short, one of the crucial issues China faces in the 1980's and 1990's is how to "mechanize" the countryside while maintaining social and political stability among the rural masses.

*Chang-Lin Tien*

# *ENGINEERING*

The role of engineering in the modernization of a nation has always been crucial, but in the case of the People's Republic of China, this role will be more important in the 1980's and 1990's. China's immediate national goal is to modernize its agriculture, industry, national defense, and science and technology by the end of this century. The modernization of science and technology is basic, and engineering is, of course, a key element in science and technology. The importance of engineering is underscored by the rapid, ambitious pace envisioned in the official plans, since basic science alone cannot sustain such a pace.

A close analysis of the eight top-priority areas announced at the National Science Conference in March 1978 (agriculture, energy resources, materials, computers, lasers, space science and technology, high energy physics, and genetic engineering) immediately reveals the prominent role engineering is to play. In agriculture, for example, the three essential problems are mechanization, irrigation, and fertilizer production, all of which require a strong engineering base and involvement. Even in high energy physics and genetic engineering, advanced-level precision engineering instruments and devices are essential.

Engineering is a broad and general discipline. A proper view of its scope and functions is fundamental to the understanding of its contributions to the general development of science and technology. This is particularly germane in the case of China's development, since priorities between 1949 and 1979 have vacillated between fundamental science and production technology.

Like many other scientific disciplines, engineering has been expanding rapidly, but the growth rate in engineering has generally been much higher due to the demands of an increasingly technological society. Subdisciplines like civil, electrical, and mechanical engineering are now broad disciplines in themselves and contain further well-structured subdivisions. In fact, many of these subdivisions are independent academic departments in Chinese universities, a lingering effect of the Russian influence of the 1950's. Specialization or overspecialization is perhaps more a characteristic of engineering than of other basic scientific disciplines and in a certain sense is ironic, since the increasingly fast pace of technological development often renders a specialized engineer obsolete within a short time. In addition, advanced technology requires engineers with more general and fundamental backgrounds, well versed in basic scientific disciplines.

These necessary but conflicting demands for specialization and generalization as well as the increasingly broadened scope of engineering result in two domains distinguished by their philosophical emphasis and scientific content. In the United States, these are called engineering science and engineering technology, while in China they are often referred to as technical sciences and application technology. The boundaries of these two domains are not well defined and overlap in many instances. Technical sciences may serve as a bridge between basic sciences and application technology, but more often than not, the technical sciences are independent, fundamental subjects. The latter point, while important, is frequently overlooked or forgotten, resulting in a gross misunderstanding of the fundamental elements in technical or engineering sciences.

In terms of fundamental content and application potential, science and technology can be divided into three general domains: basic sciences, technical sciences (engineering or applied science in the United States), and application technology (engineering technology in the United States). As indicated above, this division is general and imprecise and may cause considerable confusion in the case of certain borderline disciplines. There are, for example, spirited discussions among Chinese scientists working in the field of mechanics whether mechanics should be classified as a basic or a technical science. This classification, which may appear to be simply a semantic exercise, assumes a certain degree of importance in China's centrally planned and regulated development. Six disciplines are clearly marked as basic sciences: mathematics, physics, chemistry, astronomy, geology, and biology. The category of technical sciences includes such disciplines as chemical, civil, electrical, and mechanical engineering as well as many other engineering specializations.

The Chinese Academy of Sciences (CAS) has research institutes in both basic and technical sciences, while research and development work in application technology is performed in institutes run by the appropriate industrial ministries or local governments. Each large factory has its own research and development unit, directly concerned with improving production technology.

The following discussion of engineering research and education in China emphasizes their unique features and developments, particularly when viewed from the prevailing American standpoint. Specific examples are employed for illustration, but no comprehensive statistics for either support or evidence are given. Some observations and interpretations are personal, based on my interactions with China's scientists and engineers during several visits in the last few years.

## *HISTORICAL PERSPECTIVE AND DEVELOPMENT*

Before the People's Republic was founded in 1949, China experienced two decades of foreign and civil war and a century of foreign interference and domination. Despite its rich cultural background, China in 1949 was industrially and scientifically underdeveloped. In a country with an estimated population of 400–600 million, the production of crude steel was a trifling 150,000 metric tons (compared with the production of well over 100 million tons each in the United States, the Soviet Union, and Japan in the 1970's with their much smaller population bases). A small group of trained engineers—about 25,000 with college degrees in engineering—helped operate factories whose machinery was entirely foreign made. Fewer than 5,000 engineers were graduated in 1949, approximately ten per one million population.[1] When the CAS was founded in 1949, it had between 16 and 22 research institutes and units. None was devoted to engineering.

Moreover, a subtle negative social attitude toward engineering and technology existed in old China. The overwhelming majority of the population was illiterate, with strong beliefs in empiricism and tradition, and suspicious of and resistant to any changes brought by the adoption of new technologies. These attitudes caused considerable resistance to modern engineering and technology. Among intellectuals, craftsmanship or engineering skills had always been regarded as lowly arts due to the centuries-long domination of neo-Confucianism. This influenced the training of engineers more toward theory than practice, and engineers shied away from design and manufacturing.

Over the past 30 years, the development of Chinese engineering has been highly impressive. There have been some serious setbacks at different periods, due largely to ideological and political disruptions. Indeed, the development of engineering and the society and economy as a whole is intimately tied to ideological and political development. In general, the first 17 years (1949–66) were marked by rapid, impressive achievements in engineering, while the period of the Cultural Revolution and its aftermath (1966–76) was characterized by stagnancy in engineering development. The period since 1976, often called the New Period, is to be guided by the ambitious plan of the "four modernizations."

[1]Data on China's scientific and engineering manpower obtained from Chu-yuan Cheng, *Scientific and Engineering Manpower in Communist China, 1949–1963*, NSF 65–14 (Washington, D.C.: National Science Foundation, 1965), chaps. 2 and 4; Leo A. Orleans, "Scientific and Technical Manpower," in *Science and Technology in the People's Republic of China* (Paris: Organization for Economic Cooperation and Development, 1977), pp. 93–112.

During the first 17 years, there were also distinctive periods of rapid and slow development. The first three years (1949–52) were largely a period of political consolidation and economic stabilization, rather than of any engineering and scientific development. The following five years, the period of the First Five-Year Plan (1953–58), however, saw great and rapid development. The plan's aim was to maximize China's rate of industrial growth, and achievements in engineering were particularly impressive. For example, steel production in 1958 reached eight million metric tons, 50 times higher than that in 1949. Enrollment in engineering colleges jumped from 30,300 in the 1949–50 academic year to 177,600 in 1957–58, a sixfold increase in nine years. In research, the number of research institutes in the CAS rose from 16 in 1949 to 91 in 1958; most of the additions were newly established institutes in the technical sciences. Substantial assistance was also provided by the Soviet Union during this period, with, at one time, approximately 10,000 Soviet personnel in China and 4,000 Chinese students in the Soviet Union. This period of rapid growth, however, was followed by a period of severe cutbacks.

During the following three-year period (1958–61), three major independent factors crippled industrial and agricultural development. First, worsening Sino-Soviet relations resulted in a total withdrawal of Soviet specialists and the end of Soviet supplying of machinery and replacement parts in 1960. The heavy reliance on Soviet technical aid in the 1950's left China unprepared to operate its infant but fast-growing industries. Industrial production dropped sharply, and numerous factories were closed. Perhaps this traumatic experience influenced China's overemphasis on self-reliance and resulted in an overcautious attitude toward utilizing foreign technology in the next 15 years. Second, the lack of adequate planning and the excessive stress on production statistics during the Great Leap Forward, a massive experimental program to accelerate growth in agriculture and industry through ideological, political, and institutional means, caused a sharp downturn in agricultural and industrial development. Third, three consecutive years of bad weather did not help China's agriculture either.

The Great Leap Forward did, however, accelerate the training of engineers, a factor that became quite significant in subsequent engineering development. During the 1962–63 academic year, for example, college graduates in engineering reached 77,000, an amazing 38 percent of the total number of college graduates that year and an increase of 16 times over the annual number of engineering graduates during the period of 1948–51. This corps of young engineers contributed greatly to the smooth and speedy recovery of industrial production and engineering research during the post-Leap period (1961–66). In fact, owing to their age and the disruption of the educational system during the Cultural Revolution, this group of engineers can be expected to assume an increasingly important role in the coming years in the modernization of Chinese engineering and technology.

The magnitude of the adverse effects of the Cultural Revolution on industry and education in science and engineering has gradually become known. Industrial production dropped significantly, and numerous factories either completely shut down or operated at minimal production levels. Qinghua (Tsinghua) University in Beijing, one of China's leading technical colleges, enrolled 12,000 students in the fall semester of 1963, five times more than the 1947 enrollment, but it admitted no students for several years at the height of the Cultural Revolution and had only a small student body afterward. The academic quality of even this select group of students is subject to serious question because of interruptions in high-school programs and the laxness of academic selection criteria. In engineering, however, a number of positive elements, such as the balance of theory versus experiments, the linkage of fundamental knowledge with practical applications, and the interaction of universities and factories, were stressed during this period. Unfortunately, in actual practice, the prevailing chaotic atmosphere and the administrative and political abuse probably negated most beneficial results.

The sharp reverse in direction after 1976 highlighted the significance of engineering and resulted in a deep commitment to the rapid development of engineering. Its future success, nevertheless, will depend largely on the continuation of this course over an extended period of time.

## *RESEARCH DEVELOPMENTS IN ENGINEERING*

Engineering is concerned with the conception, planning, design, manufacture, operation, synthesis, and management of devices, structures, and systems for applications. Much of engineering work is dependent on existing knowledge in the basic sciences, the purposes of which are pure and simple, i.e. to acquire new knowledge irrespective of practical implication. Engineering research, however, bears directly on applications. This application aspect dictates intimate interaction with social and economic needs, which in turn impose certain constraints on engineering with regard to efficiency, the economy, and ecological impact. As technology evolves along with the society, the imposition of new and additional constraints requires improved or innovative technological concepts and systems in which engineering research plays a key role.

Engineering research is especially crucial to the development of new and advanced technologies. The acquisition or transfer of technology from advanced nations is often difficult because of restrictions that originate from commercial and proprietary, as well as political and military, considerations. In China's ambitious "four modernizations" program and its expressed priority on new and advanced technologies, engineering research must be greatly emphasized. Chinese leaders and scientists have repeatedly called for an emphasis on research in the basic sciences.[2] Indications that the importance of engineering research was understood and perhaps debated by

[2]See, for example, Zhou Peiyuan, *Guangming ribao (GR)*, Oct. 16, 1972.

Chinese leaders were apparent in a number of articles published in 1978 and 1979. In particular, a 1978 article in the policy-oriented official magazine, *Hongqi (Red Flag),* was entitled "Great Emphasis on Research Work in Technical Sciences."[3]

*Emphases and Priorities*

A clue to the emphases and priorities in engineering research during the 1980's can be found in the December 1977 meeting called by the Ministry of Education in conjunction with other related governmental agencies[4] to plan education in the applied sciences and new technologies in colleges and universities. Specifically, the meeting adopted plans of development in 14 areas: mechanical engineering, electrical engineering, civil and architectural engineering, hydraulics and hydraulic engineering, chemical engineering, electronics, computer science, semiconductors, automation, mechanics, optics, environmental science, materials science, and engineering thermophysics. It is interesting that the first five represent relatively classical areas of existing technology, while the last nine belong to the broad class of new and advanced technology. Moreover, such specialized topics as semiconductors and engineering thermophysics were elevated to the same level of prominence as the first five general, broad disciplines. Indeed, possibly as a consequence of this directed emphasis, two national scientific meetings were held on the subject of thermophysics in 1978: one on engineering thermophysics at Mogangshan, Zhejiang province, and the other on space thermophysics at Guangzhou.[5] The size of these two meetings on a relatively specialized subject, as well as the short period of time between them, was quite amazing, even by the standards of advanced industrial nations, and clearly signaled China's determination to develop the ability to handle new, advanced technologies.

Since China lags far behind advanced nations in the existing, well-developed technical disciplines, the development of research is perhaps not as urgent as the rapid and broad development of both personnel training and production. Indeed, most research activities in these areas, predominantly handled through ministerial research institutes, are hardware- and production-oriented and best classified as developmental work. As China becomes increasingly devoted to the acquisition of Western technology and withdraws from its isolationist role, the developmental work in these ministerial research institutes will face a major adjustment. Difficult questions remain, however, regarding the amount of research effort to be allocated to selected domains in conjunction with imports of Western technology. The industrial models of postwar Japan and West Germany can perhaps be used as a guide, but the different social systems and institutional characteristics must be considered.

[3] Wu Zhonghua, *Hongqi,* Apr. 1978, pp. 58–61.
[4] *China Daily News* (New York), Dec. 27, 1977.
[5] For further details, see *GR,* Dec. 12, 1978, and Dec. 23, 1978.

*CAS Research Institutes*

The most important scientific research organization in China is the CAS, which consists of four departments: Physics, Chemistry, and Mathematics; Biology; Earth Sciences; and Technical Sciences.[6] During the Cultural Revolution, basic scientific research was significantly downgraded, and production-related research and development work was emphasized. To have a close liaison with production activities, many research institutes were transferred from CAS jurisdiction.[7] Owing to its applied orientation, the Department of Technical Sciences suffered the most. Only one unit remained under central direction, and five were under joint direction with provincial and municipal governments, which was a drastic reduction from the previous total of 24 institutes.

After the downfall of the Gang of Four in October 1976, the CAS gradually returned to its earlier institutional structure and emphasis on basic scientific research. The state of the CAS in 1979 must still be regarded as transitional, and more changes can be anticipated. It is impossible at this stage to present accurate organizational information for the engineering research institutes. Sizes of institutes may vary greatly, but most probably have a staff ranging from 200 to 1,000, with about half of the staff being research personnel.

The group of research institutes under the jurisdiction of the Department of Technical Sciences is oriented more toward existing, well-established technologies with direct application potential. For instance, the six institutes under its direction in 1975 were the National Institute of Metrology, Beijing Institute of Automation, Beijing Institute of Chemical Engineering and Metallurgy, Beijing Institute of Electrical Engineering, Shanghai Institute of Silicate Chemistry and Technology, and Shanghai Institute of Metallurgy. Other institutes, which had presumably been transferred from CAS jurisdiction before 1975, but may have been moved back since, include the Haerbin Institute of Civil Engineering and Architecture, Beijing Institute of Mechanical Engineering, Beijing Institute of Radio Engineering, Shanghai Institute of Metallurgy and Ceramics, and Changchun Institute of Optics and Precision Instruments. In general, research in these institutes is more applied than in other CAS institutes, but still more basic compared with that in the factory- and ministry-run institutes.

Engineering research in new, advanced technologies is mostly carried out in the institutes under the Department of Physics, Chemistry, and Mathematics. For example, the Beijing Institute of Mechanics is heavily involved in aeronautical engineering research, and the Haerbin Institute of Engineering Mechanics in earthquake engineering. Judging from the names, ad-

[6] More comprehensive information can be found in Susan Swannack-Nunn, *Directory of Scientific Research Institutes in the People's Republic of China* (Washington, D.C.: National Council for U.S.–China Trade, 1977).

[7] See Richard P. Suttmeier, "Scientific Institutions," in *Science and Technology in the People's Republic of China*, pp. 53–67.

vanced engineering research is also the major function of the following institutes: Institute of Semiconductors, Institute of Computing Technology, Institute of Acoustics, Institute of Chemical Engineering and Metallurgy, and Institute of Applied Chemistry.

*Beijing Institute of Mechanics*

The Institute of Mechanics in Beijing can serve as an example of a CAS institute. I visited this Institute and had scientific exchanges with its research staff during my three visits in 1973, 1974, and 1978.

The Institute was established in 1956, when the Chinese government mapped out a comprehensive 12-year plan for scientific and technological development for the period 1956–67. Its director in 1979 was Qian Xueshen (known in the United States as H. S. Tsien), a world-famous aeronautical scientist who received his Ph.D. from the California Institute of Technology in 1939 under Theodore von Karman. Before returning to China in 1956, Qian had been von Karman's protégé and a professor at the Massachusetts Institute of Technology and the California Institute of Technology. As of 1979, Qian appeared to have relinquished the Institute directorship and was serving as vice-chairman of the Science and Technology Commission for National Defense, to which he has devoted most of his time for many years. A former deputy director was the late Guo Yonghuai, also a Caltech Ph.D. (1945) and a professor at Cornell in the early 1950's. Guo made notable research contributions while in the United States and was well known for the Poincaré-Lighthill-Kuo (PLK) singular-perturbation method in applied mathematics.

The Institute expanded in 1966 by merging with the Institute of Power Engineering in Beijing, which was established in 1956 with major research tasks in thermodynamics and fluid mechanics of turbomachinery. The first director of the Institute of Power Engineering was Wu Zhonghua, who received his Ph.D. from MIT (1947) and held the positions of research scientist at the National Advisory Committee for Aeronautics' Lewis Research Center in Cleveland (now under the National Aeronautics and Space Administration) and professor at the Polytechnic Institute of Brooklyn (now the Polytechnic Institute of New York). Wu became a deputy director of the Institute of Mechanics after the merger of the two Institutes and appeared to be in full charge of the Institute's activities in the late 1970's. Wu became increasingly influential as a leading spokesman for the technical sciences community in the 1970's. He was the author of the *Red Flag* article mentioned earlier, was on the cover of the July 1978 issue of *China Reconstructs,* and was depicted as a model scientist in a long article in *Renmin ribao (People's Daily)* of March 24, 1978.

The Institute has about 1,000 personnel, including a technical staff of 500, divided into over ten technical divisions and two technical service units. It covers the major disciplines of engineering thermophysics and fluid and solid mechanics. In the general area of engineering thermophysics, there are at least three departments: Heat and Mass Transfer, Combustion, and Aerothermodynamics of Turbomachines. Each has approximately 30 to 50 technical staff. The head of the Heat and Mass Transfer Department is Ge Shaoyan, who received his Ph.D. in 1955 from the Illinois Institute of Technology and worked for two years at Minneapolis-Honeywell before returning to China. Other departments in the Institute, covering the broad area of solid and fluid mechanics, are Explosion and Soil Mechanics; Plasmadynamics and Magnetohydrodynamics; Lasers; Fatigue and Solid Mechanics; and Supersonic Shock Tubes. The two technical service units are for instrumentation and computers. In addition, there exist central administrative units.

Many of the senior research staff (roughly those over 50 years old) were educated and did research in the West, particularly in the United States. They occupy most leadership positions and are highly capable and knowledgeable in their respective fields. They are, however, somewhat hampered by relatively obsolete supporting facilities (instrumentation, computing, library, photocopying). Some of the mid-level staff (those between 40 and 50 years old) received their training in the Soviet Union during the 1950's. The younger research staff (mostly 30 to 40 years old) were educated in China and appear to be quite competent technically. Few research workers are under 30 because of the interruptions in higher education during the Cultural Revolution period. In general, despite their high technical capability and potential, the expertise of the research staff could be greatly enhanced through more widespread and frequent communication with the international technical community.

A close look at the active research topics in the Institute vividly reveals the mixed objectives of supporting research for existing advanced technologies while leapfrogging into new research endeavors. For example, in the area of existing technology, original research of excellent quality in the field of engineering thermophysics is being conducted under Wu Zhonghua's direction on three-dimensional turbomachine fluid mechanics, a topic to which Wu made pioneering contributions while in the United States in the early 1950's. For turboblade cooling, impressive analytical and experimental work has been carried out under the guidance of Ge Shaoyan on turbulent heat transfer over a flat plate with coolant injection. Finite-element numerical techniques have been developed for the calculation of turboblade temperature fields. This concentrated research on turbomachines is obviously due to the need to improve jet engine design. In the area of new research, the Chinese determination to remain in the forefront of research is demonstrated by their investigations of the heat pipe, a novel thermal device invented in the United States during the mid-1960's; the Wankel (rotary) engine, which generated worldwide excitement in the automobile industry a few years ago; and solar energy utilization.

*Ministerial Research Institutes*

Engineering research is also performed in a variety of other organizations. A significant portion of applied or industrial research is carried out in the research institutes operated by various industrial ministries. Many of these institutes are under joint jurisdiction with provincial or municipal governments. Most military-oriented engineering research is classified, and little is known about it other than its existence. Research and development units exist in almost all medium- and large-size factories, but their main functions are to handle problems and improve existing production lines or machinery. Some substantial industrial complexes may have their own research institutes. For example, Wuhan Heavy Machinery Works has a research institute operated jointly with the First Machine Building Ministry.

The major industrial ministries are the seven Ministries of Machine Building, whose tasks are to develop and produce basic machine tools and equipment (the First Machine Building Ministry); nuclear weapons (Second); aircraft (Third); electronic equipment (Fourth); conventional armaments (Fifth); ships (Sixth); and missiles and aerospace systems (Seventh). Some of these ministries are totally military oriented, but most provide machines and equipment for both civilian and military applications. Many other industrial ministries are strongly involved in technology in addition to their primary service functions. These include the Ministries of Chemical Industry, Coal Industry, Light Industry, Metallurgical Industry, Petroleum Industry, Post and Telecommunication, Railways, Textile Industry, Transportation, and Water Conservancy and Electric Power. Each industrial ministry has about ten to twenty research institutes under its own jurisdiction and others under joint jurisdiction with local governments and large factories. For example, the First Machine Building Ministry has under its central jurisdiction over 15 general research institutes with specific research objectives on abrasives and grinding tools; agriculture machinery, bearings, castings, and forgings; construction machinery, cranes, and transport; general machinery; heavy machinery; hydraulic pressure, instruments, and meters technology; machine tools (several institutes); material, motor vehicle, and scientific and technical information; and a number of research academies (design, electrical machinery, and machinery science).[8]

Some ministerial research institutes are heavily involved in new, advanced technology. An excellent example is the Research Institute of Space Technology in Beijing, which is under the jurisdiction of the Seventh Machine Building Ministry. Most of this Institute's work is probably classified, and little information is available about it. My interactions with the research staff in space thermophysics indicate that their expertise is considerable. About 50 research personnel in the space thermophysics group work on high-temperature and cryogenic heat pipes, cryogenic thermal insulation, spacecraft modeling and similitude theory, and surface adsorptivity and emissivity ratios, which are all very up-to-date topics, even in advanced nations. Judging from the size of this group, an intelligent estimate of the size

[8] Jack Craig et al., "A Survey of China's Machine Building Industry," in U.S., Congress, Joint Economic Committee, ed., *Chinese Economy Post-Mao* (Washington, D.C.: Government Printing Office, 1978), 1: 284–322.

of the Institute would be from 500 to 1,000. It has also been speculated that a new research institute for nuclear reactor technology has been established near Wuxi in Jiangsu province, probably under the jurisdiction of the Ministry of Water Conservancy and Electric Power. This move appears to be in concert with the 1979 negotiations on the acquisition of two 900-megawatt French nuclear power plants with pressurized water reactors (PWR) licensed by U.S. Westinghouse.

*Chengdu Cutting Tool Research Institute.* Compared with the CAS research institutes, few ministerial research institutes have received foreign visitors. During the tour (September 8–October 2, 1978) of the U.S. Engineering Education Delegation to China, six research institutes were visited, of which five belong to the CAS. The sixth was a ministerial research institute, the Chengdu Cutting Tool Research Institute, under the jurisdiction of the First Machine Building Ministry.[9]

The production of machine tools, which are basic to all industrial production, increased from an estimated total of 1,600 units in 1949 to 90,000 units in 1975. From 1975 to 1978, the annual rate of increase was reported to be 10 percent, but definitive figures were lacking. About 2,000 plants manufacture approximately 1,500 types of machine tools. Major facilities are located in Shanghai, Beijing, Kunming, Haerbin, Chengdu, Nanjing, Shenyang, Tianjin, Jinan, Chenzhou, and Wuhan. In general, China's machine tool industry, although it has advanced rapidly since 1949, still lags behind the level of industrialized nations.

The First Machine Building Ministry operates several machine tool research institutes. The Beijing Machine Tool Research Institute emphasizes research on the lathe and is often called the Beijing Lathe Research Institute, while the Dalian (in Liaoning province) Machine Tool Research Institute concentrates on combination machine tools. There is also a research institute in Guangzhou.

The Cutting Tool Research Institute was founded in 1956 and moved from Haerbin to Chengdu in 1965. It is responsible for conducting basic and applied research on cutting tools, tool materials, and measuring instruments and for disseminating information relevant to metal-machining industries. The Institute has a total of 500 personnel, including 34 engineers and 186 technicians. During the Cultural Revolution, more than 30 of the Institute's engineers were transferred elsewhere for unexplained reasons. The Institute has not yet completely recovered from that setback, and productivity remains low.

The Institute has five research departments, one department for information dissemination, and three pilot workshops. The five departments conduct research on tool materials; new cutting tools, including research on

[9] Much of the following information is from a private communication from B. T. Chao, professor of mechanical and industrial engineering, University of Illinois, Urbana, who visited China in Sept. 1978 as a member of the U.S. Engineering Education Delegation.

basic theory; production technology; measuring instruments and gauges; and standardization and systemization of tools and gauges. The three pilot workshops conduct performance tests for superhard high-strength steel (HSS), cubic boron nitride, and ceramic cutting tools; high-speed carbide hobs, thread chasers, and thread-rolling tools; and special tool-manufacturing processes, including chemical vapor deposition of hard surface coatings and special purpose machines. Experiments on hard coating using chemical-vapor and ion-deposition techniques involved titanium nitride, titanium carbide, aluminum oxide, and titanium nitride carbide. Nitride coating on HSS was being explored. The Institute was also considering initiating research on plasma deposition, the purpose of which was not only to obtain a hard surface, but also to increase the toughness of the base material. The Institute was active in research and development work on cubic boron carbide. A special press was capable of exerting a 3,600-ton force in six directions on a blank having a surface area of about 1 $cm^2$. This was the only laboratory in China doing research on boron carbide tool material. The project began in 1973 and showed good promise for producing tools suitable for cutting superhard steels.

The inspection laboratory was equipped with a variety of high-quality instruments, including optical toolmakers, microscopes, optical comparators, and surface final analyzers. (Some were made in China, others imported from Europe.) This laboratory was air-conditioned and had controlled humidity. In one pilot workshop, carbide cutting blanks were sintered in small batches for use in threading high-strength steel pipes in the Daqing oil field. This shop produced 8,000 blanks per year, and tools were finish ground in the workshop. In another workshop, production performance tests were conducted for carbide hobs. Alloy steel gear blanks of 240 BHN were machined at 180 m/s. It was reported that 120 to 160 blanks were produced between resharpenings.

*University Research*

Largely because of Soviet influence in the 1950's, university research has never been emphasized as much as in American universities. Some faculty members in major technical universities and colleges do conduct engineering research, but on a smaller scale than their counterparts in the United States. During the Cultural Revolution, university research was drastically curtailed. Particularly devastating was the nearly total demolition of research equipment and facilities, which will take several years to rebuild. Public pronouncements in 1978 and 1979 indicated that research in Chinese academic institutions will be greatly expanded. This tilt toward research was also reflected by the admission of a large number (compared with the past) of graduate students, the establishment of new research units in these institutions, and the travel abroad by junior faculty members for advanced research and study.

Despite the generally modest efforts of university research, a few bright spots do exist in some research areas, largely through the strong, persistent efforts of individual faculty members. In the area of engineering thermophysics, for instance, notable research work is being conducted at Qinghua University (on plasmadynamics and high-temperature thermophysics), the Chinese University of Science and Technology (thermophysics research closely coordinated with the CAS Institute of Mechanics in Beijing), Chongqing University (on heat pipes), Nanjing Engineering College (on magnetohydrodynamics), Xian Jiaotong University (on thermophysical properties), and Tianjin University (on internal combustion engines and geothermal power).

*Research Communication*

Owing to institutional hierarchical structure, communication among various academic and research organizations has not been particularly strong and efficient. The common means of research communication, besides one-to-one direct communication in some special instances, are topical research conferences and publications, both of which were effectively eliminated during the Cultural Revolution. In 1978 and 1979, many topical research conferences were revived or newly initiated, and research journals gradually reappeared. Chinese participation in international exchanges accelerated, and attendance at international conferences greatly increased. Several papers were presented at these conferences, but no research papers were submitted for publication in research journals outside China.

The activity and conduct of topical research conferences can best be illustrated by reviewing two national conferences on thermophysics held in 1978. The Second National Engineering Thermophysics Conference was attended by 250 research specialists. (The First Conference had been held in 1965, before the Cultural Revolution, and there had not been any national conference activity for 13 years.) The conference lasted for ten days, two or three times the length of similar American conferences, and received 260 research papers, but owing to time limitations, only 129 papers were presented orally. Papers were grouped into five topical sections: thermodynamics, aerothermodynamics, heat transfer, combustion, and thermophysical measurements. About 30 percent of the papers were on heat transfer, including heat transfer in jet engines (film cooling, temperature distribution in turboblades); thermal control in spaceflight (heat pipes, active thermal control, thermal control coatings, modeling and similitude theory); heat transfer in nuclear reactors (two-phase flow and heat transfer, critical heat flux); industrial furnaces and heat exchangers (furnace heat transfer, radiation exchange); solar energy (solar collectors, natural recirculation); and thermal conductivity measurements (laser techniques, transient methods). Other surveys of heat transfer work dealt with recent advances in heat and mass transfer and on heat transfer problems in spaceflight. Proceedings of the conference were to be published.

During the conference, the Chinese Society of Engineering Thermophysics was founded, and Wu Zhonghua was elected president. Among the many activities planned was a research journal about engineering thermophysics.

A month after this conference, a related meeting on space thermophysics was organized at Guangzhou under the auspices of the Chinese Society of Aeronautics and Astronautics, with 140 participants from the Institute of Space Technology, related CAS research institutes, technical universities and colleges, and various industrial organizations. This conference was the second such meeting; the first one was held in 1974. The conference received nearly 100 papers on a variety of topics, including thermal control methods during spaceflight, heat transfer between exteriors and interiors of space vehicles, thermal modeling experiments and measurement techniques, thermal control materials, thermophysical properties, measurements, and other fundamental problems in heat transfer.

The large number of papers offered in these two conferences and the announced topics are indeed quite impressive, but as of 1979 these papers had not yet been published, and it was impossible to assess their quality. Some clues concerning research quality, however, can be extracted from the limited number of available journals. Basically, there are two kinds of Chinese journals, those published by a technical or scientific society, such as the *Chinese Journal of Mechanical Engineering* published by the Chinese Society of Mechanical Engineering and those published by a technical university, such as the journals of Xian Jiaotong University and Shanghai Jiaotong University. Papers in the former category originate from various research and industrial organizations; the latter is limited mostly to the research results of the particular university. Most of these journals are published quarterly.

Before the Cultural Revolution, journal quality was adequate when compared with international standards, although these journals were not well known outside China since they were published in Chinese and had limited circulations. For the most part, journals published in 1978 and 1979 contain results of research done during the Cultural Revolution under difficult working and communication conditions. Although the papers are generally of high quality, some are out of date to a certain degree in terms of topic selection and methods of computation and measurement. With strong governmental support and enhanced international communication, the timeliness of Chinese research work should improve notably in the 1980's. These technical journals are, however, published in Chinese, which poses a major obstacle to most non-Chinese researchers. Perhaps selected papers about significant research contributions should be made available in English, as is often done in Japan.

## *PRODUCTION TECHNOLOGY*

Although research is essential to China's long-term industrial development, the ambitious plan for immediate increases in industrial production entails a rapid expansion and improvement in China's production technology. The present level of China's technological capabilities is still far behind that of industrialized nations, despite a most impressive record of advancement in industrial development since 1949. China remains largely an agricultural country with about 80 percent of its population in rural areas, and agricultural development can be expected to have top priority in China's overall planning for years to come. Modernization of agriculture, however, depends on the development of a strong, broadly based production technology in order to meet extensive agricultural requirements in machinery, electricity, chemicals, transportation facilities, and construction materials and supplies (for further discussion of this subject, see Leo Orleans's note on agriculture in this volume).

### *The Nature of Chinese Industry*

Chinese industry is characterized by a high degree of variation and dispersiveness. Several factors have contributed to the development of these characteristics: the rapid development from minimal levels of industry after 1949, the strong infusion of Soviet technologies in the 1950's, self-development and the selective transfer of Western technologies between 1962 and 1976, the hierarchical structure in industry, the lack of standardization and automation, and the policy of developing small, dispersed, communal industries paralleling large central industries.[10] Technological levels, product quality, and production efficiency may vary widely from industry to industry, factory to factory, and locality to locality.

Communal industry is a uniquely Chinese creation. The widely publicized commune factories in rural areas and the neighborhood factories in urban areas are generally small in size and primitive in technology. They utilize local labor (including housewives for part-time labor) and locally available raw materials, and often work with self-designed tools and simple machinery. Their products are mostly nonstandardized and of low quality, and produced in small quantities to fill local agricultural or industrial needs. Communal industry also serves important functions in diffusing technology vertically and horizontally and in generating mass awareness and participation in the technological transformation of an underdeveloped society.

The state of Chinese technology is best exemplified by large central industries and their products as presented at its trade fairs and exhibits, of which the most representative are the semiannual Canton Trade Fair and the Shanghai Industrial Exhibition, which have exhibited such items as a 100-megawatt hydrogenerator set with water-cooled rotor and stator; a 12,000-ton (hydraulic) forging process; a 200,000 × electron microscope; precision thread-grinding machines; 32-ton dump trucks; electronic computers; and many other electronic and mechanical consumer products. Another indica-

[10] For an excellent discussion of the nature of Chinese industry, see Hans Heymann, Jr., "Industrial Structure and Technological Advancement," in *Science and Technology in the People's Republic of China*, pp. 143–55.

tion of China's technical capability is its rapid and successful development of nuclear weaponry and missile systems during the 1960's and 1970's.

China's large central industries are located mostly in urban areas in coastal provinces (e.g. Shenyang, Anshan, and Changchun in the northeast, and Nanking and Guangzhou in the south) and in the independent municipalities of Beijing, Shanghai, and Tianjin. Despite concerted efforts to relocate large industries inland, only one-third of industrial production occurs inland. Major inland industrial centers include Xian, Lanzhou, Wuhan, and Chongqing.

In China, large industrial complexes have been established for basic industries, including the petrochemical, metallurgical, machine-building, automotive, and electronic component industries. Many of these originated from the massive transfer of Soviet technology and equipment in the 1950's but have subsequently been modified and greatly expanded by the Chinese themselves. Their products often show features characteristic of Soviet and East European designs. Recent additions to large industrial plants have been imported primarily from Japan and Western Europe. For example, the Lanzhou Chemical Industrial Complex consists of a number of plants producing petrochemicals, chemical fertilizers, rubber, and chemical machinery, as well as research institutes, design academies, construction companies, and wastewater treatment plants. It has a total work force of about 31,000, including 3,000 scientists and engineers. The complex was originally built with Russian help, but subsequent expansion was made through Chinese efforts, combined with some infusion of West German technology and British equipment.

Notable independent Chinese efforts are best represented by the massive petroleum extraction complexes in the Daqing, Shengli, and Dagang oil fields. Indeed, the "Daqing spirit" or "Daqing experience" was widely publicized during the Cultural Revolution as a model of self-reliant development. Despite this most impressive record of development, Chinese leaders have recognized that advanced foreign technology and equipment are needed in order to achieve more efficient and greater petroleum production. In fact, signs in 1979 clearly indicated that Western technology will play an important role in Chinese petroleum extraction.

To illustrate the current level of China's production technology, three basic industries—iron and steel, machine tool, and automotive—are discussed in the following sections.

### *Iron and Steel Industry*

The iron and steel industry, which provides basic construction materials for all industries, is fundamental to technological development. Since the founding of the PRC, the government has emphasized the rapid development of the iron and steel industry.[11] Despite severe setbacks caused by the

[11] See also Rong Ye, "China's Steel Industry Recovers," *China Reconstructs*, Mar. 1979, pp. 56–59; and William W. Clarke, "China's Steel: The Key Link," *U.S.–China Business Review*, July/Aug. 1975, pp. 27–40.

Soviet withdrawal of technical aid and the Cultural Revolution, production of crude steel has increased dramatically from 150,000 metric tons in 1949 to over 31.7 million metric tons in 1978, which placed China at approximately the same production level as Great Britain, France, and Italy and behind the top four producers, the United States, the Soviet Union, Japan, and West Germany. This increase in steel output, however, still could not meet the growing demand for steel in various Chinese industrial sectors, and China remains a significant importer of iron and steel products and metallurgical production equipment from Japan and other industrialized nations.

China's potential for growth in the iron and steel industry is substantial due to its large supplies of iron and coal. China has the world's third largest coal reserves (behind the United States and the Soviet Union), and supplies of other necessary ingredients, such as manganese and limestone, are adequate. Most of China's iron ore and coaking coals are, however, low grade and must be beneficiated, or upgraded, to make them usable.

China's iron and steel industry consists of two parallel industries at very different technological levels: a modern industry with centrally managed, large, integrated facilities producing the bulk of steel output and a more primitive industry with many small locally operated and maintained plants using a simpler technology. This results in a number of imbalances, such as product nonstandardization, lack of quality control, and low efficiency, but it also effectively utilizes local raw material and labor resources to meet local needs.

China produces various types of both modern and native blast furnaces for making pig iron. The modern type is based primarily on earlier Soviet designs, with some Chinese improvements. The largest, which was put into operation at the Anshan steelworks in early 1978, has a working volume of 2,580 m$^3$, or an annual capacity of 1.5 million tons. China has not yet undergone the new technological transformation that has swept world steel making since 1960 as the open hearth furnace (OH) has been replaced by the basic oxygen furnace (BOF). Currently, OH still accounts for 60–70 percent of China's capacity, followed by BOF at 15–25 percent, electric furnaces at 5–10 percent, and side-blown converters at 10–15 percent. The last two types, no longer economic in the West, are widely used in China's small steel-making units. In the steel-finishing area, China has experienced considerable difficulty in producing high-productivity finishing mills, and the country is extremely short of finishing steels, especially seamless steel pipes.

The Anshan Iron and Steel Plant in Liaoning province continues to overshadow all other plants in China. It has 12 blast furnaces (including the recently installed 2,580-m$^3$ one), about 25 OH's (200- to 500-tonners), two large BOF's, various mines and sintering and beneficiation plants, plus blooming, plate, hot strip, rail, and pipe mills. Its annual steel production capacity is about 7 million tons. The Shanghai Steel Plant Complex has perhaps eight steel plants with a total annual capacity of approximately 5 million tons. The No. 1 Steelworks, which is the largest, has three 30- to 35-ton BOF's, a 120-ton BOF, and two 70-ton OH's. The Wuhan Iron and Steel Plant in Hubei province is the third largest, with an annual production rate of over 2.5 million tons. In 1979, the Plant was undergoing a major expansion, with $550 million worth of West German and Japanese production equipment being installed. Other major plants with production capacities of 1 to 2 million tons include the Beijing or Shijingshan Plant in Beijing, the Baotou Plant in Inner Mongolia, the Maanshan Plant in Anhui province, the Taiyuan Plant in Shanxi province and the Chongqing Plant in Sichuan province. Numerous medium- and small-sized plants are scattered throughout the country, including recent additions in Lanzhou and Ürümqi in the northwest.

*Machine Tool Industry*

The machine tool industry supplies the bulk of capital equipment for every component of machine building and construction. The level of production technology in the machine tool industry is often an excellent indicator of overall industrial development. Since 1949, China's machine tool industry has made tremendous progress, and annual production has risen from 1,600 machine tools in 1949 to about 100,000 in the late 1970's.[12] The country is now basically capable of meeting domestic demand for general purpose machine tools, although specialized and precision machine tools must still be imported from Japan and other industrialized countries. China has, in fact, developed an export market in the Western countries for several basic machine tools, largely because of their competitive prices. A 1977 exhibit of China's exportable machine tools in Hong Kong displayed a variety of gear-shaping machines, cylindrical grinders, optical profile grinders, lathes, and gear-grinding machines.

China produces most major types of metal-cutting and metal-forming tools, including lathes, boring machines, grinders, milling machines, shapers, gear cutters, thread-cutting machines, hydraulic presses, broaching machines, and planers. Major production centers are concentrated in urban areas of the east, north, and northeast, particularly in Beijing, Shanghai, Shenyang, and Haerbin. Typical of the largest machine tool factories are the Shenyang No. 1 Machine Tool Factory and the Shanghai Machine Tool Factory. The Shenyang plant manufactures nine types and 40 models of lathes, including single and multiple spindle, horizontal and vertical turret, and general and special purpose lathes. Customers are automobile manufacturers, agricultural machinery manufacturers, and other material-processing industries. Total annual production is around 3,500 units, of which about 15 percent are exported to foreign countries. Total personnel number 6,900, including 36 engineers, 640 technicians, and 2,400 production workers.

The Shanghai Machine Tool Factory, which was originally an agricultural

[12] For additional information, see Craig et al., "China's Machine Building Industry."

machinery factory, specializes in the production of precision grinding machines. It has about 6,000 personnel, distributed in ten production divisions, and manufactures 60 to 70 types of grinding machines (from 1 to 30 tons), including numerically controlled grinders. The factory operates a grinding machine research institute and the oldest "July 21 university," attended by about 400 technicians (the name is derived from the date of Mao's inspection tour in 1968 at the height of the Cultural Revolution).

The technological level of China's machine tool industry still lags far behind that of the most advanced industrial nations. Factors contributing to this lag include the imitative nature of Chinese development, the lack of innovative research, the gap between laboratory and factory capabilities, the low quality of raw materials, poor quality control, and low factory productivity. In newer technologies, China has begun limited production of numerically controlled machine tools; multiaxis tools, including numerically controlled two- and three-axis machines; and multispindle and multicutter machine tools, including two- and four-head models. It appears, however, that production of such equipment continues to be low and limited to single laboratory prototypes or small-batch factory production.

*Automotive Industry*

The automotive industry is another example of the unique Chinese combination of large, centrally directed, modern facilities and small, regional plants using simple technologies.[13] During the Cultural Revolution, provinces and municipalities throughout China began manufacturing motor vehicles for agricultural and transportation purposes. More than 30 provinces, autonomous regions, and municipalities now produce various types of motor vehicles. Most of these plants, however, are small, utilize low-level technology, and have small outputs and low productivity. The large motor vehicle plants were designated as the lead factories to coordinate the efforts of these smaller plants and to provide them with the necessary guidance, training, and technology. Since increasing production output demands better production planning and efficiency, more specialized component and subassembly plants, such as those for tires, engines, and chassis, have been developed to provide mass supply for large assembly plants.

Trucks are the major product of the Chinese automotive industry, constituting approximately 70 percent of the total output. The other 30 percent consists of jeeps, automobile sedans, and buses (not including small three-wheel motor vehicles, which are popular for cargo transport in rural areas). The Chinese automotive industry's rapid development is best reflected by truck production. China produced its first trucks (1,654) in 1956 at the Soviet-designed Changchun No. 1 Motor Vehicle Plant in Jilin province, but annual production exceeded 140,000 trucks in 1978. About 1.1 million trucks were then in service, a 25-fold increase from 42,000 trucks in 1952.

[13] For additional information, see Craig et al., "China's Machine Building Industry"; and Heymann, "Industrial Structure and Technological Advancement."

Most of the trucks China has produced are general-purpose cargo vehicles suitable for a wide range of transportation services, but limited numbers of special trucks used in mining, forestry, construction, and petroleum transport are also produced. The most common model of the general-purpose trucks is the 4-ton "Liberation" cargo truck, with a 110-HP engine, patterned after the Soviet ZIL-150 truck and ZIL-120 engine of the 1950's, which the Russians had previously copied from earlier American prototypes (GM and Hercules). Other truck models include the 2.5-ton "Leap Forward" copied from the Soviet GAZ-130 and 63, the 8-ton "Yellow River" 160-HP diesel modeled after the Czech SKODA 70 GRT, and a large 32-ton, 12-cylinder, 400-HP special diesel truck.

The first automobile sedans were the "Dongfanghong" (East is Red) produced in 1958 at the Beijing Motor Vehicle Plant. Other common models are the premium-class "Hongqi" (Red Flag) produced in Changchun and the "Shanghai" produced in Shanghai.

The largest automotive plant in China is the Changchun No. 1 Motor Vehicle Plant, which accounts for about half of China's output. With a work force of 30,000 (20,000 in production and 10,000 in auxiliary functions), the plant has a total work area of 1.5 million $m^2$. In addition to large-scale production, the Changchun plant plays a significant role as a center for technical training in the Chinese automotive industry. More than 10,000 skilled automotive workers have been trained at Changchun and then sent or returned to other motor vehicle plants throughout the country. The plant operates a July 21 university and a secondary-level technical school. With the late 1970's push for increased production and modernization of technology, the Changchun plant began negotiating with Japanese and Western automotive giants (including General Motors and Ford) for assistance in enlarging its capacity to produce 4- to 6-ton trucks from 70,000 to 150,000 units per year. Nanjing in Jiangsu province has the second largest plant in China with a staff of 9,000 and a production capacity of 10,000 trucks per year.

The Chinese automotive industry, while developing rapidly and impressively, is still relatively backward both in models produced and in production methods. China imports a large number of modern trucks (in 1976, for example, Japan and Europe supplied about 75 percent of the 11,700 imported trucks). On the other hand, China also exports about 10 percent of its truck production to various third-world countries.

## ENGINEERING EDUCATION

Realization of the "four modernizations" in China by the year 2000 depends on China's ability to produce a large pool of high-quality scientific and technical personnel in the 1980's.[14] Indeed, one of the main goals an-

[14] This section is based on the draft reports of the U.S. Engineering Education Delegation to China, Sept. 8–Oct. 2, 1978, and my three visits in 1973, 1974, and 1978.

nounced at the National Science Conference in March 1978 was to have at least 800,000 scientific research personnel by 1985. If the existing pattern of China's scientific and technical manpower utilization continues, engineers will constitute about 60 percent of the 800,000, a very large ratio. Therefore, engineering education and training can be expected to assume a significant role in the coming years.

Before 1949, Chinese higher education, despite its elitist nature, was greatly influenced by American practices. Many of the best Chinese students, a large number of whom were in engineering, were sent to the United States for study and assumed influential positions after returning to China. In the 1950's, however, Sino-Soviet cooperation made a major change in the education and training of engineers. Not only were a large number of engineers sent to Russia for training, but engineering colleges in China were patterned after the Soviet system, resulting in highly specialized departments with narrowly based, professionally oriented curricula. Universities and colleges concentrated on teaching; participation in research was minimal. The Cultural Revolution nearly destroyed Chinese higher education; and engineering education, both in quantity and quality, declined sharply during this period. The downfall of the Gang of Four in 1976 and the subsequent accelerated program of exchange with the West, however, have brought about another new era in Chinese engineering education.

In view of China's political, ideological, and institutional characteristics, animated discussion and debates regarding different aspects of engineering education will certainly continue in the 1980's. Particular issues of concern are such seemingly conflicting demands as quality versus quantity in education, red (i.e. politically sound) versus expert, general versus specialized education, technical science versus application technology, and teaching versus research. Although confusion and conflicts are bound to occur, further strain will be generated by the need to retrain existing engineering personnel, the competition among engineers for education, research funds, and industrial positions, and differences in background and preparation between engineering graduates of the 1970's and those of 1980's. It cannot be overemphasized that the 1980's will be a most stimulating and challenging period for engineering education in China, a period that may exert a long-lasting influence on China's industrial development.

*Technical Universities and Colleges*

As of 1979, there were about 460 full-fledged institutions of higher education in China, some of which had just recently been reopened after being shut down or merged with other institutions for several years during the Cultural Revolution. The total number is probably still increasing, as China puts more resources into higher education. China has returned to a pre–Cultural Revolution system and has designated 89 key institutions as recipients of special attention and priority funding in order to improve scholarship and education.

All Chinese institutions are fully supported and administered by the government and fall into three different types of governmental administrative jurisdiction. The Ministry of Education sets overall policy and coordinates the major common functions for all institutions, such as establishing nationwide or regional standards, entrance exams, and curricula. Only 35 institutions are under its direct jurisdiction, including such key comprehensive universities as Qinghua (in Beijing) and Xian Jiaotong, and Tongji (in Shanghai). Many polytechnic universities and engineering colleges are administered by the industrial ministries, and a few by the CAS. For instance, Shanghai Jiaotong University, which emphasizes marine engineering, is under the jurisdiction of the Sixth Machine Building Ministry (for shipbuilding), while the Beijing Post and Telecommunications College is operated by the Post and Telecommunications Ministry. In fact, each ministry probably administers between one and ten colleges. The CAS runs several universities, including the Chinese University of Science and Technology and its branch campuses, the Haerbin Polytechnic University, and Zhejiang University. The majority of Chinese universities and colleges are, however, under the direct jurisdiction of provincial and municipal governments.

Although Chinese universities do not grant academic degrees, college students must complete a rigorous and relatively rigid program before graduation. Except for the five-year program in the CAS universities, the curriculum is now four years in length and represents a return to the Western system predominant in China before 1949. The school year in China is 40 to 45 weeks long, considerably longer than the 32 to 33 weeks common in the United States. The 1979 curricula, which were being revised, for various specialties in engineering have the following approximate breakdown in course coverage: 20 percent in common core courses, including foreign languages, physical education, and political ideology; 20 percent in basic science courses, such as mathematics, physics, and chemistry; 35 percent in basic general engineering courses, for instance, mechanics, machine design, thermodynamics, fluid mechanics, materials, and electronics for mechanical engineering; and 25 percent in specialty courses, such as courses in mechanical manufacturing technology (metal cutting, tools, instruments, machining processes, automation). Elective courses are seldom offered.

Three representative examples of Chinese technical institutions are discussed below.

*Qinghua University.* Qinghua University in Beijing, a leading polytechnic university in China, was founded in 1911 as a preparatory school for students selected for college education in the United States. Qinghua became a comprehensive university in 1925 and was an important center for Sino-American scientific cooperation in China before 1949. In 1952, Qinghua was reorganized as a polytechnic university concentrating on various engineering disciplines. It has not only been an influential technical institution, but has also played a prominent role in spearheading various political and ideological movements, particularly during the Cultural Revolution.

As of 1979, Qinghua University had a teaching staff of 2,800, including 180 professors and associate professors, a larger number of lecturers (equivalent to assistant professors in the United States), and an even larger number of assistants. The small size of the professorial staff came about because no promotions were granted during the Cultural Revolution. Student enrollment was 8,000, still below the level of 12,000 students reached just before the Cultural Revolution. Projections call for 20,000 students in 1985, an increase of 150 percent. In 1978, 300 graduate students were admitted, and this number can be expected to increase sharply in the 1980's. The student-to-faculty ratio is considerably lower than the normal ratio in the United States.

During the Cultural Revolution, academic departments at Qinghua underwent a restructuring, resulting in 11 departments: Architectural Engineering, Automatic Control, Electrical Engineering, Electronics (including computer science and engineering), Engineering Chemistry, Engineering Mechanics, Engineering Physics, Hydraulic Engineering, Mechanical Manufacturing, Precision Instruments, and Radio Engineering. Apparently, this structure is being modified again, in terms of both consolidating existing departments and creating new ones. It appears that a new department of thermal power engineering was established in 1979, and other new departments covering industrial engineering were then being contemplated.

Under the direct jurisdiction of the Ministry of Education, Qinghua University in 1978 was headed by President Liu Da, who was previously in charge of the Chinese University of Science and Technology before moving to Qinghua. He was assisted by over six vice-presidents, who assumed different responsibilities, such as instruction, research, or administration. Zhang Wei, a professor of engineering mechanics who received his doctorate in engineering from the University of Berlin in 1944, was in charge of scientific research. Zhang had traveled abroad frequently, and had visited in the United States in 1977 and 1978. At the departmental level, the structure consists of one chairman, several vice-chairmen, and a number of teaching group leaders.

In 1979 Qinghua University became formally affiliated with the University of California at Berkeley in order to facilitate such activities as exchange of visiting scholars and students, cooperation on scientific research, and exchange of scientific publications, curricular information, and library facilities.

*Shanghai Jiaotong University.* Shanghai Jiaotong University, another leading polytechnic university in China, is administered through the Sixth Machine Building Ministry (for shipbuilding). Jiaotong University, which was founded in 1896 as Nanyang College in Shanghai and renamed Jiaotong University in 1921, is one of the oldest universities in China. It is particularly known for the high quality of its engineering graduates, many of whom have become leaders in various engineering disciplines both in China and abroad. In 1956, some of the faculty and students were transferred to

Xian in Shanxi province to establish the Xian Jiaotong University, which later became an independent university under the direct jurisdiction of the Ministry of Education. Shanghai Jiaotong University was originally designated the lead institution in naval architecture and marine engineering, but it gradually evolved into a full-fledged polytechnic university.

In 1979, the university had a faculty of more than 1,600, including about 70 full professors, 70 associate professors, and 830 lecturers, supported by a nonacademic staff of 2,500. Student enrollment in 1978 was 5,100 and was projected to increase to 12,000 in 1985. In 1978, 132 graduate students were admitted, and projections called for increases to 200 new graduate students in 1979 and 400 in 1980, reaching a level of 1,000 in 1985.

The governing body is the University Council, which was headed at the time of my last visit by Vice-Premier Wang Zhen of the State Council. He was known to be a strong advocate of science and technology, but his duties in Beijing probably prevented him from taking an active role in the University. Vice-chairman of the University Council was Deng Shuzhu, who led a delegation from the university during a 47-day tour of the United States in 1978. Zhu Wuhua, a Harvard-trained (Ph.D., 1926) electrical engineer was president of the university. He was assisted by several vice-presidents, including Associate Professor Zhang Shou, who was a member of the 1978 delegation visiting the United States.

The university had ten academic departments: Naval Architecture, Marine Engineering, Electrical Engineering and Computer Science, Electronic Engineering, Materials Science and Engineering, Mechanical Engineering, Precision Instruments, Applied Mathematics, Applied Physics, and Engineering Mechanics. Additional engineering programs or departments were contemplated in industrial engineering and operations research and in bioengineering. To strengthen the "two-center" (teaching and research) role of the university, five research centers were established in 1978: Naval Architecture and Ocean Engineering, Marine Engineering, Electronic Engineering and Computer Science, Materials Science and Engineering, and Mechanics.

The 1978 visit to the United States was a major undertaking by the university.[15] The delegation, consisting of 12 members in various engineering disciplines, visited 20 cities, 27 institutions of higher education, and 14 scientific research institutes. They met numerous leading scientists and engineers, including 200 alumni of Jiaotong University. Agreements have also been worked out to establish formal affiliation with four American universities: University of California at Berkeley and at San Diego, University of Michigan in Ann Arbor, and Washington University in St. Louis.

*Beijing Aeronautical College.* Beijing Aeronautical College is a specialized engineering college under the direction of the Third Machine Building

[15] *GR*, Dec. 19, 1978.

Ministry.[16] It was founded in 1952 by merging 8 of the 11 departments of aeronautical engineering then existing in China. The other three departments were joined to form Nanjing Aeronautical College, a smaller counterpart in the south.

The college consists of seven academic departments: Aircraft Engineering (aerodynamics and structures), Aero-engines, Automatic Control, Electronics, Computers, Technology, and Materials. In 1978, the vice-president of the college was Shen Yuan, an early graduate of Qinghua University who received his Ph.D. from the University of London in 1946. A specialist in aerodynamics, Shen had also been president of the Chinese Society of Aeronautics and Astronautics since 1964. The college had a faculty of about 1,200 and a student body of around 3,000.

Major facilities include a low-speed wind tunnel with a 1.5-m square cross-section and a maximum speed of 40 m/s (and a Reynolds number of 500,000) and a blowdown-type of high-speed wind tunnel with a Mach number range of 1.5 to 3.0. Other specialized laboratories are structured for instruction and research on engine combustion and structural design, aerodynamic heating, materials properties, and flight simulation.

It is difficult to paint a clear picture of the future development of engineering in China because engineering, like many areas of activity, depends strongly on what political and ideological course China takes in the coming years. Indeed, the history of the PRC reveals an unpredictable, vacillating pattern in China's course of development. In science and technology, this vacillation is best characterized by the "struggle of two lines"—"red or expert." However, China's deep commitment to the "four modernizations," as well as the expressed support of the Chinese people for such a commitment, appears to have charted an irreversible course of rapid economic and technological development.

China's potential for advances in engineering is good, despite the existence of many constraining forces. On the positive side lies the determination of Chinese leaders to modernize science and technology, buttressed with generous budgetary allocations and aggressive acquisition of Western technology. This has been manifested by purchases of complete factory facilities and equipment in many key industries and the initiation of an extensive scholarly exchange program with the West. Although some confusion and inefficiency will inevitably result from such a frantic pace of development, a noticeable gain in the general level of China's engineering expertise can be expected during the 1980's. This gain should appear particularly impressive when compared with the relatively weak and narrow technological base of the 1970's.

In the long run, however, to accelerate or even maintain a fast pace in

[16] For additional information, see Gerald Corning, "An Aeronautical Visit to China," *Astronautics and Aeronautics*, Apr. 1974, pp. 16–19.

China's engineering development would be an extremely formidable task. General constraints on development exist because of the lack of a strong financial base for extensive capital investment in technology and the limited supply of trained technical personnel. A particularly significant factor is the gap left by the near absence of engineering training during the Cultural Revolution. Moreover, the adaptation of Western technology will necessitate some adjustment in technical terminology and design methods for those engineers who were trained during the period of Russian influence in the 1950's.

Furthermore, actions taken to achieve a short-term gain in engineering development do not guarantee a rapid long-term advancement in engineering, for two reasons. First, China's acquisition of Western technology, although capable of exerting an immediate impact on the nation's technological development, represents only the beginning of the dynamic process of technology transfer. This process must continue with appropriate diffusion, digestion, and absorption of the new technology. Even successful technology transfers would result only in a technology of a derivative nature. Engineering progress at the advanced level must come from innovative research and technology, which requires a scientific and technological base of considerable breadth and depth. Second, selective development of key industries as currently practiced in China, while easily resulting in impressive short-term gains, may indeed become an impediment to a broad engineering development program in the long run. The intimate and complex interrelations among technologies in different sectors and at various levels should not be underestimated.

Engineering in China is presently undergoing a rapid transformation and the 1980's and 1990's will be a crucial period of great challenge. In the upward thrust of China's technology, difficult problems, seemingly insurmountable obstacles, and even severe setbacks will inevitably occur. The task is arduous, but the long-tested ingenuity and entrepreneurship of the Chinese people should prove an invaluable asset in achieving the goal of technological modernization. Such a feat would indeed mark a dramatic return to China's historical preeminence in technology.

*Vaclav Smil*

# *ENERGY*

In absolute quantitative terms, the progress of Chinese energetics has been quite impressive.[1] Between 1949 and 1978, raw coal production grew from 32.43 to 618 million metric tons (MT), elevating China from eighth to third place in the world; crude oil extraction expanded from a negligible 0.12 to 104 million MT, putting the country into a class with Libya, Kuwait, and Nigeria and changing it from an importer to a rising exporter of petroleum; natural gas flow reached about 52 billion cubic meters ($m^3$), making China the world's fifth largest producer.

But in comparative terms, the situation is rather different. Per capita consumption of primary energy is currently about 600 kilograms of coal equivalent (kgce) annually, a value characteristic of a poor developing country (India's consumption is about 250, Brazil's 750, and Mexico's 1,300).[2] And quantitative comparisons of Chinese energy technology with current industrialized world standards reveal serious gaps and deficiencies in virtually all exploration, extraction, transportation, and conversion activities.

The Chinese are, of course, aware of this technological lag, which they themselves estimate to be at least 15 to 20 years, and they recognize the relative weakness of the whole energy sector in the national economy. Not surprisingly then, accelerated development of energy resources was placed second (after agriculture) among the eight spheres singled out for special attention in the eight-year plan (1978–85) for scientific modernization.[3]

[1] V. Smil, *China's Energy Achievements, Problems, Prospects* (New York: Praeger, 1976); and idem, "China's Energetics: A System Analysis," in *Chinese Economy Post Mao* (Washington, D.C.: Government Printing Office, 1978), pp. 323–69.

[2] U.N., Department of Economics and Social Affairs, *World Energy Supplies* (New York: U.N., annual publication).

[3] "Abridgement of Fang Yi Report to the National Science Conference," New China News Agency (NCNA), Mar. 28, 1978; in *Foreign Broadcast Information Service,* Mar. 29, 1978, p. 8.

This chapter attempts to provide as comprehensive an evaluation of the current state of China's energy technology as the available information permits. By presenting the maximum number of specific facts and placing them in an international perspective, it aims at enabling readers to appreciate the extent to which Chinese energy technology lags behind that of the most advanced nations and the magnitude of the future tasks facing the country.

## *COAL*

Coal is China's most important fossil fuel, providing some 70 percent of the nation's primary energy. The PRC is one of the world's three greatest coal producers; with more that one-half billion tons of raw coal extracted annually, it is surpassed only by the Soviet Union and the United States. And yet the coal industry has been repeatedly labeled as a "weak link" in China's economy, and an inadequate coal supply has been one of the country's most serious and intractable economic problems.

This has been caused by chronic underinvestment and lack of modern technology, resulting in low rates of mine mechanization and coal processing. In 1977–78, new policies designed to modernize the coal industry came into effect, but it will take China at least 15 years to elevate its coal mining to the current technological level of the other two coal superpowers.

### *Production*

The single most important characteristic of China's coal production is its low level of mechanization. In the words of the minister of coal industry, coal extraction "remains a weak link of the national economy with comparatively backward techniques," whose "labor productivity and some major economic and technical standards . . . today compare unfavorably with the world's advanced level."[4]

Unfortunately, there are no recent official figures on the degree of mechanization, but a good estimate can be derived by considering a variety of relevant information. Mechanized extraction in 1958, the last year for which a reliable nationwide breakdown is available, accounted for 31.76 percent of total production in the mines operated by the Ministry of Coal Industry.[5] In Kailuan, China's largest and most modernized coal-mining region (annual output of 25.2 million MT before the devastating July 1976 earthquake), nearly 50 percent of production came from mechanized workfaces in 1975.[6] And *Renmin ribao (People's Daily)* editorialized in February 1978 that "our coal industry is not mechanized to any extent."[7]

[4] Xiao Han, "Developing Coal Industry at High Speed," *Peking Review (PR)*, 1978, 8 (Feb. 24): 7–8.

[5] I. I. Bazhenov, I. A. Leonenko, and A. K. Kharchenko, *Ugolnaia promyshlennost Kitaiskoi Narodnoi Respubliki* (Moscow: Gosgortekhizdat, 1959), p. 254.

[6] NCNA in English, in *Summary of World Broadcasts (SWB)*, 978 (May 3, 1978): 10.

[7] Beijing home service, in *SWB*, 970 (Mar. 8, 1978): 7.

Considering that two-fifths of China's coal production comes from small, local mines (run by counties and communes) where mechanization is either very low or nonexistent (on the average certainly no more than 20 percent), the highest conceivable share of currently mechanized coal extraction in China is about 40 percent, but is more likely about one-third of the total. In comparison, extraction is virtually 100 percent mechanized in the U.S.S.R. and the United States.[8]

While manual extraction with picks or pneumatic drills, blasting with dynamite, and loading with shovels into carts remain the predominant coal-mining methods in Chinese mines, many individual collieries use advanced, efficient technologies. The room-and-pillar method, which has been traditionally used in both conventional and continuous mining and leaves about half the coal in place as pillars to support the roof, has been mostly displaced in large, modern mines by the longwall retreating method introduced by the Soviets in the 1950's.[9]

In longwall mining, huge cutting heads (shearers) travel the length of the coal seam's face and dump coal onto a conveyor that hauls it away; the miners are protected by a steel canopy supported by jacks or props. After a complete pass, the whole assembly (conveyor, canopy, and shearers) is moved forward, and the roof behind is left to cave in. Longwall retreating is not only faster, cheaper, and safer than other underground extraction techniques, but also commonly allows recovery rates of 70–80 percent.

The longwall technique in China is certainly most advanced in Kailuan. For example, at Fangezhuang, a large mine opened in 1964, the pre-earthquake (1975) output reached 80,000 MT per month from a single longwall operation in a 275-m deep, 4.2-m thick, 13° inclined seam with wall length up to 180 m, and daily progress of about 3 m.[10] Another modern method used in Kailuan is hydraulic mining. The Lujiaduo mine was equipped with Polish machinery in the late 1960's, but now uses hydraulic guns designed by Tangshan Coal Machinery Research Institute (*Kailuan* 4 type, weighing only 130 kg and working with a pressure of 142 atmospheres). Its pre-earthquake, solely hydraulic output reached 3 million MT annually.[11] Outside Kailuan, there are only two other hydraulic mines—in Henan and Shandong—and several experimental hydraulic workfaces elsewhere in the country.

Most equipment in Chinese coal mines is domestically produced, although only a few plants specialize in producing coal machinery. Serially produced equipment, which includes drills, cutter-loaders, scraper convey-

[8] Tsentralnoie Statisticheskoie Upravleniie, *Narodnoe khoziaistvo SSSR* (Moscow: Statistika, annual publication); and National Coal Association, *Bituminous Coal Facts* (Washington, D.C.: National Coal Association, biennial publication).

[9] Yuan-li Wu and H. C. Ling, *Economic Development and the Use of Energy Resources in Communist China* (New York: Praeger, 1963), pp. 44–48.

[10] H. Harnisch and H. G. Gloria, "Eindrücke von chinesischen Steinkohlenbergbau in Kailaner Revier," *Glückauf,* 111, 21 (Nov. 6, 1975): 1,012.

[11] *Ibid.*, p. 1,013.

ors, belts, winches, hoisting machinery, carrying cars, air compressors, water pumps, and a variety of automatic control devices, is largely based on Soviet and East European designs of the 1950's and on even older Western prototypes.[12] Electric mine locomotives have been made in China since 1958, and small power shovels and conveyors for surface mining are also produced. However, the coal-mining machinery industry had a rather low development priority until at least 1971, and even now, with the current drive to mechanize, it continues to suffer from shortages of high-quality steels.[13]

Consequently, there were major purchases of foreign equipment in the 1970's. In 1973 and 1974 the Chinese bought at least $116 million worth of advanced equipment from Poland, Great Britain, West Germany, and the United States, and during 1978 they initiated negotiations that could ultimately lead to purchases totaling nearly $6 billion.[14] The greatest part (about $4.2 billion) should eventually go to Krupp, Demag, Thyssen, and Ruhrkohle of West Germany to set up five new deep mines, modernize another one, and open two large surface mines; Britain's Dowty, Anderson Strathclyde, Gullick Dobson, Powell Duffryn, and National Coal Board and Japan's Mitsui Mining will develop other mines and deliver various machinery.

Current Chinese plans are "to basically mechanize work in the mines in 10 years, with the major mines equipped with coal-cutters and tunneling machines, continuous transport facilities, automatic coal lifting, washing and dressing machines and computerized communications and dispatching systems."[15] This will require further imports and substantial expansion of domestic machinery production in addition to mass training of technicians and miners to handle the sophisticated equipment.

The planned expansion of Chinese coal production (1977 output is to be doubled in a decade) will necessitate the opening of a large number of new shafts. This is an area of coal technology in which the Chinese have become fairly experienced; in 1977 they put 42 pairs of new shafts into production, in 1976 more than 50, and in 1975, 22.[16] A few examples give a good glimpse of their capabilities. The deepest mine in China, with 925-m main and 894-m auxiliary shafts, 20 km of passages, and 600,000 tons per year capacity was built by the Beipiao Mining Bureau (Hebei) between August 1966 and autumn 1973.[17] Liangwa (Hunan) colliery's two 250-m shafts and 600,000 tons per year mine were completed in two years.[18] For a new

[12] J. Craig, J. Lewek, and G. Cole, "A Survey of China's Machine-Building Industry," in *Chinese Economy Post Mao*, p. 314.

[13] U.S., Central Intelligence Agency, *China: The Coal Industry* (Washington, D.C.: CIA, 1976), pp. 10–11.

[14] *Ibid.*, p. 16; CIA, *China: Post-Mao Search for Civilian Industrial Technology* (Washington, D.C.: CIA, 1979), p. 5.

[15] Xiao Han, "Coal Industry," p. 7.

[16] NCNA in English, in *SWB*, 950 (Oct. 12, 1977): 15; and *SWB*, 965 (Feb. 1, 1978): 9.

[17] NCNA in English, in *SWB*, 786 (July 31, 1974): 5.

[18] NCNA in English, in *Survey of People's Republic of China Press (SPRCP)*, 6,154 (Aug. 10, 1976): 77.

Dadun (Shansi) colliery, 17,076 m of tunneling and 58,300 $m^2$ of floor space were prepared in less than four years.[19] Again, the technology used in opening new mines is mostly outdated; Fushun Coal Mine Machinery Plant produced China's largest tunnel-boring machine (3.8-m diameter) in 1971, a quarter century after such devices became commercially available in the West.[20]

Chinese experience and capabilities in opening and operating large surface mines are much more limited. Although over 50 percent of American coal and some 30 percent of Soviet production originate in open mines, the Chinese share was only 7 percent in 1958 and does not surpass one-tenth of the total today.[21] (Only modern mechanized operations are included in these figures; small-scale rural outcrop mines are excluded.)

The low share of efficient surface mining, the low level of deep-mine mechanization, and the high reliance on small rural mines result in low labor productivity. In 1957 and 1958, the last two years for which official figures are available, the average output per man per day in the Ministry of Coal Mines was between 1 ton (including above-ground labor) and 1.8 tons (underground only), values roughly comparable with West European performance at that time, but much below U.S. production (8.91 tons including above-ground workers at underground mines and 21.64 tons in surface mines).[22]

Recent Chinese disclosures of raw coal output and labor force enable the productivity for Kailuan and Shanxi, China's foremost coal-producing province, to be calculated. Kailuan output per man per day before the earthquake was below 1 ton when all 100,000 workers on the mining bureau's payroll are included and about 2 tons counting underground miners only.[23] The Shanxi figure, depressed by a large output from inefficient small mines (which account for nearly 40 percent of the total), is only 0.69 tons per man per day.[24] In contrast, the current U.S. rates are 10.75 tons for underground mines and 30.50 tons for surface operations; West European deep-mine figures float around 3 to 4 tons per manshift underground.[25]

These comparisons clearly show the potential for major improvements in the years ahead. Another part of coal technology where the Chinese need to make substantial advances is coal preparation.

### *Preparation*

Coal preparation usually involves removal of the associated rocks by crushing and washing the raw coal, gravity separation of coal and incom-

[19] NCNA in English, in *SWB*, 938 (July 20, 1977): 10.

[20] Craig, Lewek, and Cole, "China's Machine-Building Industry," p. 314.

[21] Smil, "China's Energetics," p. 341.

[22] Bazhenov, Leonenko, and Kharchenko, *Ugolnaia promyshlennost*, p. 446; and National Coal Association, *Bituminous Coal Facts*, p. 83.

[23] Beijing home service, in *SWB*, 860 (Jan. 14, 1976): 7; "A Heroic People, a Heroic Army," *PR*, 1976, 34 (Aug. 20): 6; and NCNA in English, in *SWB*, 978 (May 3): 10.

[24] CIA, *China: The Coal Industry*, p. 8; and Shanxi provincial service, in *SWB*, 970 (Mar. 8, 1978): 10.

[25] National Coal Association, *Bituminous Coal Facts*, p. 39; and Association of the Coal Producers of the European Community, *Energy in Western Europe: Vital Role of Coal* (London: National Coal Board, 1977): p. 39.

bustible foreign material (sometimes followed by dewatering of smaller sizes), and sorting into fractions of consistent size and quality for different markets. Although virtually all solid fuels extracted in developed nations undergo some processing before marketing, coal preparation has been a consistently lagging segment of China's coal technology.

In 1949 only four plants, with an annual capacity of 5.2 million MT, were in operation; a decade later 46 installations could handle 41.4 million MT, and the 1970 output was estimated at 83.6 million MT.[26] Even when assuming (overoptimistically) a sustained 10 percent growth since 1970, the 1977 coal preparation capacity would be no more than 200 million MT. This means that the share of China's coal output prepared before marketing rose from 10 percent in 1949 to 20 percent in 1959, 30 percent in 1970, and about 40 percent in 1977. It is surprising that even such a large and modern mine as Fangezhuang, with an annual production of 3.6 million MT, does not have its own washery, but just dry sorts.[27]

Recent Chinese reports on the state of coal preparation are rare. In Shaanxi there is a new, large modern coal washery (600,000 tons annual capacity) designed and manufactured in China. The equipment for crushing, transporting, feeding, washing, flotation, filtering, and dewatering took three years to build. China's first jet turbulent flow flotation machine was produced in the Nanshan Coal-Dressing Plant (Hegang Mining Bureau, Heilongjiang).[28] The new flotation assembly, used for dressing finely pulverized coal and for retrieving coal powder from the sludge, is claimed to be much more efficient and easier to maintain and repair than the previously imported machines.

One important recent development concerns the unavoidable by-products of coal extraction and of the preparation process—the large amounts of coal waste and mine tailings. A nationwide effort is now under way to make these wastes useful in a variety of ways.[29] Tailings with high-energy content (up to 2,500 kcal/kg) are burned in mine boilers and by local industries or made into briquettes for household use. Combustion of wastes and tailings leaves large quantities of slag and ash, and these may in turn become useful commodities; reports mention their use in producing cement, building materials, fertilizers, and molding powder and in the extraction of many valuable elements and compounds (germanium, gallium, indium, vanadium, molybdenum, nickel, copper, lead, sodium silicate, aluminum chloride, and ferrous sulfide).[30] In the Nanbiao Mining Bureau (Liaoning), the economic value of wastes now equals one-third of the raw coal price.[31]

[26] A. B. Ikonnikov, "The Capacity of China's Coal Industry," *Current Scene,* 11, 4 (Apr. 1973): 1–9.

[27] Harnisch and Gloria, "Kailaner Revier," p. 1,012.

[28] Shaanxi provincial service, in *SWB,* 876 (May 5, 1976): 7; and NCNA in English, in *SWB,* 891 (Aug. 18, 1976): 9.

[29] Zhejiang provincial service, in *SWB,* 927 (May 4, 1977): 2; and NCNA in English, in *SWB,* 970 (Mar. 8, 1978): 8.

[30] NCNA in English, in *SWB,* 956 (Nov. 23, 1977): 7; and *SWB,* 959 (Dec. 14, 1977): 10–11.

[31] NCNA in English, in *SWB,* 970 (Mar. 8, 1978): 10.

*Transportation*

Due to large regional disparities in coal production, China has always had to move large amounts of coal from the north to the south. The task has been somewhat eased with the development of new mines in the nine provinces south of the Chang Jiang (Yangtze), but the region still produces less than a quarter of China's raw coal. It accounts for some 40 percent of the country's population and industrial production and has to import one-third of its coal demand.[32]

Coal is thus easily the most important commodity hauled on China's railroads, amounting to nearly 40 percent of the total tonnage.[33] Many Chinese railroads and much of the rolling stock are relatively new; almost 60 percent of China's estimated inventory of 253,000 freight cars was built after 1965.[34] But there is no special coal transportation technology akin to the American or European system of unit trains. (Here, assemblies of large cars, in trainloads exceeding 10,000 tons, are pulled by powerful diesel engines and move continuously between mines and consumption points, both of which have automated loading and unloading facilities.)[35]

The Chinese move coal in ordinary gondola or hopper cars, with an average load of around 40 tons per car (newer cars of modern design carry 50 to 65 tons, the older ones 33 tons or less). The total train load does not exceed 3,000 tons.[36] Moreover, they still pull most trains with steam locomotives, which are energy inefficient (typical conversion efficiencies are less than 10 percent compared to more than 30 percent for diesels) and perform less well than modern diesel or electric engines.

As part of the nationwide technological modernization, railway transportation is to achieve considerable advances by 1985. Major north-south lines are to be double-tracked and electrified, electrical and diesel locomotives should haul over 60 percent of the total load, bigger freight cars should be introduced, and automation of traffic control and mechanization of loading should be gradually extended.[37] If accomplished, this modernization will go a long way toward improving the status of China's coal transportation.

Coastal freighters and river barges are the second most important way of shipping coal. Several 25,000-ton ships move regularly between Qinhuangdao and Shanghai, and barge assemblies of several thousand tons ply the Chang Jiang.[38] Highway transport of coal is insignificant, and so far there have been no attempts to experiment with coal slurry pipelines.

[32] NCNA in English, in *SWB*, 910 (Jan. 5, 1977): 6.

[33] Kang Chung-mou, "A Study of the Energy Sources on the China Mainland," *Issues and Studies*, 11, 4 (Apr. 1975): 60.

[34] Craig, Lewek, and Cole, "China's Machine-Building Industry," p. 309.

[35] J. C. Kneiling, *Integral Train Systems* (Milwaukee: Kalmbach, 1969); and R. C. Rittenhouse, "Fuel Transportation," *Power Engineering*, 81, 7 (July 1977): 48–56.

[36] Craig, Lewek, and Cole, "China's Machine-Building Industry," p. 309; and P. W. Vetterling and J. J. Wagy, "China: The Transportation Sector, 1950–71," in U.S., Congress, Joint Economic Committee, ed., *People's Republic of China: An Economic Assessment* (Washington, D.C.: Government Printing Office, 1972), pp. 159–60.

[37] NCNA in English, in *SWB*, 979 (May 10, 1978): 24.

[38] Kang Chung-mou, "Energy Sources," p. 73.

## OIL AND GAS

Since the mid-1960's, hydrocarbon production has been the most successful branch of Chinese energetics. Soviet-aided oil and gas exploration during the 1950's, disappointing for many years, finally resulted in the discovery of the giant Daqing field in Heilongjiang in September 1959.[39] Discoveries of other large fields followed shortly: Shengli in Shandong between 1960 and 1964, Dagang near Tianjin between 1964 and 1967, and Panshan in Liaoning in 1964. Major natural gas and oil strikes were also made in the 1960's in the Sichuan Basin.

As a result, Chinese hydrocarbon extraction expanded tremendously; between 1967 and 1977 crude oil production grew at an annual exponential rate of nearly 19 percent, while natural gas output climbed at 16.6 percent per year.[40] Most of this fast expansion was achieved with technology based on antiquated Soviet designs. Now that the exploration and production is moving into deeper strata, less accessible locations, and offshore waters, the industry faces the need for extensive technological modernization and acquisition of advanced foreign equipment.

### *Exploration and Production*

Modern hydrocarbon exploration and production technology, practically nonexistent in China before 1949, was introduced by the Soviets during the 1950's. Substantial imports of equipment were later made from Romania, and recently China has benefited from purchases of advanced Western and Japanese equipment. Largely Soviet-oriented technology has been the source of important limitations. Inadequate seismic equipment and lack of computerized field units hinders the location of deep structures. Reliance on old turbodrills and shortages of high-quality drill bits, drilling and casing pipes, mud pumps, and gas-treating facilities have resulted in inefficient field operations.[41] But the Chinese have at least mastered serial production of most essential components for shallow drilling in relatively soft formations and are now trying to improve on their performance and quality.

Virtually all basic types of onshore hydrocarbon exploration equipment are now produced in China. The Xian Geophysical Instruments Plant, set up in 1955, is the principal supplier of gravimeters, seismographs, radiometers, magnetometers, geophones, well-logging devices, isotope well-testing instruments, and gas analyzers.[42] The sophistication and performance of these devices are invariably behind Western levels. For example, the Xian plant turned out the first experimental digital seismograph in spring 1976, about a decade after its introduction in the West.[43]

[39] A. A. Meyerhoff and J. O. Willums, "Petroleum Geology and Industry of the People's Republic of China," *CCOP Bulletin*, 10 (1976): 178.
[40] Smil, "China's Energetics," p. 367.
[41] Meyerhoff and Willums, "Petroleum Geology," pp. 204–5.
[42] NCNA in English, in *SWB*, 828 (May 28, 1975): 10–11.
[43] NCNA in English, in *SWB*, 876 (May 5, 1976): 13.

The complex of Lanzhou (Gansu) plants, established with Soviet aid, is China's main producer of oil- and gas-drilling equipment; the Petrochemical Machinery Plant raised its annual production of medium-depth (up to 3,200-m) drilling rigs from 40 to 100 in 1977.[44] The latest Lanzhou-built experimental oil-fracturing machine works with the pressure of 1,200 atmospheres, while the largest mobile (truck-mounted) units pump liquid into the hydrocarbon formations under the pressure of 850 atmospheres.[45]

Baoji Petroleum Machinery Plant (Shaanxi) has been producing small drilling rigs, drill pipes, collars, and tool joints since the 1950's and is now making lightweight links (150 tons capacity), elevators (250 tons capacity), and pipe tongs (10 tons capacity).[46] Various Shanghai and Beijing-Tianjin area plants supply light drilling rigs, drilling bits, oil valves, control equipment, engines, and pumps. Chongqing Machinery Plant produced China's first experimental hydraulically controlled blowout preventer in May 1977.[47] In recent years the Chinese have devoted more attention to the introduction of improved drilling bits. The first fine-grain synthetic diamond drill bit was developed in Beijing's Institute of Powder Metallurgy between 1969 and 1974 and has been widely used since; official claims ascribe to it "doubled work efficiency" and "reduction of rolled steel consumption by 80% in hard formations."[48]

Improved drilling technology led, obviously, to deeper wells. Exploratory drilling was limited to 1,000–2,000 m in the 1950's; depths of 4,000–5,000 m were reached in the early 1970's, and in 1978 exploratory drilling in Sichuan reached 7,175 m, two decades after wells over 7,000 m deep were completed in the United States.[49] The Chinese have been using their deep experimental wells to gain experience in mud injections, logging, coring, and well-reinforcing techniques under high temperatures and pressures.

As for field production, the Chinese have repeatedly claimed that they "devised the method of injecting water into different oil-bearing formations in the early stage of production to maintain the oil pressure."[50] In fact, waterflooding was introduced in China by the Soviets, who started injecting their Tuimazy field in Bashkiria in 1948 and now produce more than four-fifths of their oil from water-driven wells.[51] Not surprisingly then, the Chinese used in their fields the line-drive waterflooding predominant in the Soviet Union. They have only recently turned toward the five-spot or nine-spot

[44] NCNA in English, in *SWB*, 949 (Oct. 5, 1977): 14.

[45] NCNA in English, in *SWB*, 871 (Mar. 31, 1976): 10.

[46] NCNA in English, *SPRCP*, 5,749 (Dec. 9, 1974): 26.

[47] Craig, Lewek, and Cole, "China's Machine-Building Industry," p. 316.

[48] NCNA in English, in *SWB*, 979 (May 10, 1978): 7.

[49] NCNA in English, in *SWB*, 999 (Sept. 27, 1978): 19; NCNA in English, in *SWB*, 938 (July 20, 1977): 11; and J. E. Brantly, *History of Oil Well Drilling* (Houston: Gulf Publishing, 1971), pp. 305–23.

[50] NCNA in English, in *SWB*, 774 (May 8, 1974): 10.

[51] V. D. Shashin, quoted in *Oil and Gas Journal*, 73, 21 (May 26, 1975): 80–82.

arrangements commonly used in the West, which increase recovery efficiency but slow down the production rate.[52] A recent field production innovation has been the development of computerized oil-flow measuring devices at the Daqing oil field.[53]

Offshore exploration started from earthen causeways in the Bohai Gulf in the late 1960's. China's first offshore drilling rig, *Bohai I,* drilled its first well in May 1971; the rig, a small jack-up with four legs (73 m long and 2.5 m in diameter) capable of drilling up to 6,000 m in 30 m of water, was built in Liaoning's Hongqi shipyard.[54] In 1972 the Chinese bought a secondhand jack-up, *Fuji* (renamed *Bohai II*), from the Japan Drilling Company, a rig capable of working in up to 54 m of water.[55] During 1974, Shanghai's Hudong shipyard pieced together the hulls of two old Liberty cargo vessels to build China's first drill ship, a catamaran (*Gandan I*), capable of working in waters 80 m deep.[56]

In June 1974 the Chinese began a large-scale aerial magnetic survey of their coastal waters, which lasted nearly four years, covered more than 250,000 km, and resulted in a series of maps at scales of 1:500,000 and 1:200,000.[57] This survey and the purchases of seismic instruments and of an advanced geophysical survey ship (*Lady Isabelle*) from the French Compagnie générale géophysique, as well as French and American geophysical data-processing equipment (704 Raytheon computer), have enabled the Chinese to start assessing their offshore hydrocarbon resources accurately.[58]

The potential in the Bohai Gulf and in the East and South China seas has been promising enough to effect a steady expansion of China's offshore drilling fleet, which included a dozen drilling rigs by the summer of 1978. Besides *Bohai I, Bohai II,* and *Gandan I,* there were two other smaller Chinese-built jack-ups capable of drilling 2,000 to 2,500 m in 20 to 25 m of water, another Shanghai-built catamaran (*Gandan II*), two semisubmersibles (*Bailong* and *Borgny Dolphin*), a drilling barge, and three large modern jack-ups (*Nanhai I, Nanhai II,* and *Bohai IV*) built by the Robin Loh yard in Singapore, with drilling depths of up to 8,000 m in water depths of up to 100 m.[59]

Recent purchases have included a large semisubmersible Aker H-3 rig from Norway, two National Supply rigs, and a Bethlehem jack-up.[60] The

[52] Meyerhoff and Willums, "Petroleum Geology," p. 205.
[53] NCNA in English, in *SWB,* 959 (Dec. 14, 1977): 11.
[54] NCNA in English, in *SWB,* 830 (June 11, 1975): 10–11.
[55] Meyerhoff and Willums, "Petroleum Geology," pp. 205–6.
[56] NCNA in English, in *SWB,* 810 (Jan. 22, 1975): 15.
[57] NCNA in English, in *SWB,* 987 (July 5, 1978): 2.
[58] W. R. Scheidecker, "Petroleum Developments in Far East in 1976," *American Association of Petroleum Geology Bulletin,* 61, 10 (Oct. 1977): 1,835.
[59] Meyerhoff and Willums, "Petroleum Geology," p. 206; and "Active Rigs in the Asian Region," *Petroleum News South East Asia,* monthly supplement.
[60] H. J. Mugar, "China's Offshore Development," speech given to the Rotary Club of Hong Kong, Apr. 27, 1978.

Japanese have won orders for two semisubmersible Hitachi rigs to be delivered in January 1980.[61] Clearly, the Chinese are very serious about swiftly expanding their offshore drilling and production capacities.

*Transportation*

Development of modern crude oil transportation in China—by large, long-distance pipelines and tankers—started in the early 1970's. Until the late 1960's, virtually all China's crude oil was moved in an expensive, wasteful, and clumsy manner, by small railway tank cars and trucks. There was only one major crude oil line, built in 1958–59, which linked Kelamai oil field with Dushanzi refinery in Xinjiang.[62] China's largest domestically built oil tankers used for coastal shipping were two 15,000 deadweight tons (dwt) Daqing series vessels (numbers 27 and 28) launched in 1969, and most of the ships built after 1949 were only 1,000 to 4,500 dwt—true midgets by world standards.[63]

Although an extensive oil pipeline construction program started in 1970, by mid-1976 only 3,200 km of new trunk lines were complete, and a further 2,000 km were under construction. The total length of China's long-distance oil pipelines in 1977 was still less than 6,000 km, or about 42 km per million MT of extracted crude, while the corresponding relative values were about 120 km in the U.S.S.R. and 675 km in the United States.[64] According to the New China News Agency, just half of China's crude oil was moved by pipelines in 1977; the rest had to be transported from the oil fields by more than 30,000 railroad tank cars.[65]

Chinese long-distance pipelines are 20, 25, 30, and 61 cm in diameter. (The largest oil pipes are now 80–150 cm in diameter in the United States and 90–170 cm in the U.S.S.R.) The pipelines are electrically welded from 8–12 m sections and wrapped in six alternating layers of plastic cloth and bitumen before emplacement. Chinese-made pipes are mostly under 30 cm in diameter, and the larger, spiral-formed sizes do not withstand high pressures.[66] Another difficulty is imposed by the high-pour point and high-paraffin content of many Chinese crude oils, which must flow in heated pipelines in winter.[67] Chronic shortages of high-quality steel and the inadequate capacity of domestic pipe factories have necessitated continuous imports

[61] Kyōdō in English, in *SWB*, 991 (Aug. 2, 1978): 20.

[62] B. A. Williams, "The Chinese Petroleum Industry: Growth and Prospects," in United States Congress, Joint Economic Committee, ed., *China: A Reassessment of the Economy* (Washington, D.C.: Government Printing Office, 1975), p. 246.

[63] Smil, *China's Energy*, p. 39.

[64] U.S., Central Intelligence Agency, *China: Oil Production Prospects* (Washington, D.C.: CIA, 1977), p. 25; and V. Smil, "China Reveals Long-Term Energy Development Plans," *Energy International*, 15, 8 (Aug. 1978): 25.

[65] NCNA in English, in *SWB*, 963 (Jan. 18, 1978): 13.

[66] CIA, *China: Oil Production Prospects*, p. 25.

[67] S. Marsden, "Properties of Natural Gas and Crude Oil and Their Impact on PRC Energy Development," in Thomas Fingar, ed., *China's Energy Policies and Resource Development* (Stanford, Calif.: U.S.–China Relations Program, 1976), p. 36.

of large-diameter seamless pipe from Japan, West Germany, France, and Italy.[68]

Pipe laying has recently benefited from imports of modern machinery, and the Chinese are able, through arduous labor, to build the lines in difficult terrain and harsh weather in a relatively short time. The extent of human exertion and the low level of mechanization is well illustrated by China's most important long-distance pipeline, which connects Daqing with two ports on the Bohai Gulf; digging the trench for the 1,152-km section to Qinhuangdao required 14 million $m^3$ of earthwork and stonework and was done without machinery in less than half a month by amassing tens of thousands of laborers along the line.[69] The most difficult line of all, even by international standards, is the construction of a 1,100-km strategic pipeline from Qinghai to Tibet.[70]

Natural gas pipelines are much less extensive than crude oil lines and have been primarily concentrated in Sichuan, where more than 1,000 km of gas pipelines were laid by 1975 to supply local iron and steel, fertilizer, and power generation industries.[71] China's first domestically designed gas-pumping station was completed in 1976, and major imports of pipes and pumps are likely with the increasing importance of natural gas as petrochemical feedstock.[72]

Tanker transportation is less advanced than pipeline technology. After the first two 15,000-dwt vessels were launched in 1969, it took another five years to design and build China's first 24,000-ton tanker—*Daqing No. 61*.[73] A number of identically sized ships (178 m long and 25 m wide) have since been built by Hongqi shipyard in Lüda (Liaoning). In August 1976, this shipyard also produced and launched China's first 50,000-ton tanker, *Xihu,* which is powered by the country's largest marine diesel engine (six cylinder, 18,000 HP).[74] However, vessels of 15,000–50,000 dwt are still too small to take the best advantage of economies of scale, which rise greatly between 20,000 and 200,000 dwt.

The Chinese lack the necessary shipyard facilities and construction know-how to build larger ships (i.e. of more than 100,000 dwt). Until recently, they did not even have any deep-water ports to accommodate them. The only harbor capable of handling tankers up to 100,000 dwt is the new Lüda pier (Nianyu Bay in Liaoning) completed in May 1976 by welding together nine steel bridges, each with a span of 100 m.[75] It is a terminal of the Daqing

[68] CIA, *China: Oil Production Prospects,* p. 25.

[69] "Building Oil Industry Through Self-Reliance," *PR,* 1975, 2 (Jan. 10, 1975): 13–16; and NCNA in English, in *SWB,* 809 (Jan. 15, 1975): 19–20.

[70] CIA, *China: Oil Production Prospects,* p. 25.

[71] NCNA in English, in *SWB,* 814 (Feb. 19, 1975): 14.

[72] NCNA in English, in *SWB,* 876 (May 5, 1976): 7.

[73] Liaoning provincial service, in *SWB,* 776 (May 24, 1974): 13.

[74] NCNA in English, in *SWB,* 893 (Sept. 1, 1976): 10–11; and Radio Beijing for Southeast Asia, in *SWB,* 888 (July 28, 1976): 7.

[75] Liaoning provincial service, in *SWB,* 877 (May 12, 1976): 11.

pipeline and loads large foreign vessels with export oil, while the old Lüda port handles smaller domestic tankers destined for eastern refineries, especially in Shanghai and Nanjing.[76]

Another new oil terminal, on Huangdao island opposite the Qingdao port, handles Shengli crude and has an onshore storage facility of about 110,000 tons. The 1,260-m-long pier can serve tankers up to 70,000 dwt.[77] Qinhuangdao, the other port terminal of the Daqing pipeline, was reconstructed after 1973 to receive 35,000-dwt ships at three wharves and 20,000-dwt vessels at two smaller berths.[78] The only deep-water dock in the south is Zhanjiang in Guangdong, whose 138-m berth and four mooring dolphins can receive ships up to 70,000 dwt.[79] China's remaining major ports, including Shanghai and Tanggu, can accommodate only 10,000- to 30,0000-dwt tankers. Since 1974 the Chinese have been expanding their tanker fleet by purchases of secondhand ships (in the 50,000- to 100,000-dwt category) from Norway and Japan. Negotiations to buy more Japanese ships were under way in 1979.

Little is known about China's oil storage. Strategic oil supplies stored in anticipation of armed conflict with the Soviet Union must be relatively large, but are most likely dispersed in thousands of small oil storage tanks of about 60 $m^3$ throughout the country.[80] China's first large—10,000 $m^3$—underground "watertight stone-cave oil tank" was completed at an undisclosed location in April 1977, and the first large above-ground steel tank to store oil for large power plants was finished in Lüda in December 1977.[81]

*Refining*

Modern Chinese crude-oil refining had its origins, as did the other principal sectors of the hydrocarbon industries, in Soviet technology, which in turn derived from earlier American experience. Before the Soviets withdrew their technical assistance in 1960, they designed and built China's first large modern refinery, in Lanzhou, and helped to reconstruct and expand other small installations. They also introduced basic techniques of atmospheric and vacuum distillation (the latter is necessary for the production of heavy gas oil, lubricating oil, and bitumen fractions), thermal cracking (processing that breaks down and rearranges the molecular structure of hydrocarbon chains by applying high temperature and pressure), and coking (relatively severe cracking to completely convert residual pitch and tar into gas, liquid fractions, and coke).[82]

[76] CIA, *China: Oil Production Prospects*, p. 26.
[77] NCNA in English, in *SWB*, 899 (Oct. 13, 1976): 10.
[78] Hebei provincial service, in *SWB*, 875 (Apr. 28, 1976): 12.
[79] NCNA in English, in *SPRCP*, 5,702 (Sept. 11, 1974): 76.
[80] NCNA in English, in *SWB*, 770 (Apr. 10, 1974): 1.
[81] NCNA in Chinese, in *SWB*, 927 (May 4, 1977): 1–2; and Lüda city service, in *SWB*, 960 (Dec. 21, 1977): 2.
[82] G. D. Hobson, ed., *Modern Petroleum Technology* (New York: Wiley & Sons, 1973), *passim*.

After the Soviet withdrawal, the Chinese bought complete refineries from Italy (ENI and Snam Progetti in 1963 and 1965) and a heavy-oil cracking installation and olefin extraction unit from West Germany (Lurgi Gesellschaft) in 1964.[83] They gained access to the technology of American catalytic cracking (where high-boiling hydrocarbons are converted into lower-boiling substances under high temperature and pressure in the presence of a catalyst) and platforming (catalytic reforming of paraffins and naphthenes for the production of high-octane-number hydrocarbons, first introduced by United Oil Products in March 1949) via the Esso-Shell refinery in Havana.[84] These acquisitions enabled them to start work on the first two fully Chinese refineries: Daqing, which opened in 1966, and the Beijing General Petrochemical Plant (Dongfanghong), which began operating in 1969.

By 1973 oil-refining capacity was 3.7 times that of 1965, and most refineries had catalytic cracking, platforming, and delayed coking (long-residence cracking) installations, which only a few had had in the mid-1960's.[85] In the early 1970's the first Chinese direct computer-controlled distillation columns were put into operation at the Shanghai refinery. They are well designed and laid out.[86]

Technologies introduced in 1974 included molecular sieve separation for normal paraffin (a procedure made commercial in the West in 1954 to enhance the value of oil distillates as raw materials for fuels and lubricants) and hydrofining of lubricating oils ("hydrofining," a term introduced by Exxon, means hydrodesulfurization, the removal of sulfur from distillates in the form of hydrogen sulfide).[87] The first domestic equipment for transfer line catalytic cracking was installed at Yumen refinery in 1975, and the first large, all-Chinese catalytic cracking unit went on stream in January 1978.[88] In general, Chinese refining technology of the late 1970's resembled that of the West in the late 1950's.

Advances in the Chinese refining industry can be illustrated by the increasing number of final products: only ten basic varieties were made in the early 1950's, 80 were added by 1960, and more than 100 new kinds were introduced between 1965 and 1974. By 1979 China's refineries turned out a nearly complete range of important fuels, lubricants, and petrochemical feedstocks.[89] However, the depth and the quality of products from a typical Chinese refinery still lagged behind the output of a similar Western or Japanese installation.

[83] Chu-yuan Cheng, *China's Petroleum Industry* (New York: Praeger, 1976), p. 134.

[84] Hobson, *Modern Petroleum Technology*, p. 332; and Williams, "The Chinese Petroleum Industry," p. 245.

[85] NCNA in English, in *SPRCP*, 5,710 (Oct. 7, 1974): 33.

[86] S. Yuan, "An Inside Look at China's HPI," *Hydrocarbon Processing*, 53, 4 (Apr. 1974): 106.

[87] Hobson, *Modern Petroleum Technology*, p. 416.

[88] NCNA in English, in *SWB*, 888 (July 28, 1976): 5; and Hubei provincial service, in *SWB*, 970 (Mar. 8, 1978): 11–12.

[89] Williams, "The Chinese Petroleum Industry," pp. 244–45; and Cheng, *China's Petroleum Industry*, p. 95.

With the growth of crude oil production, the size of refineries has been steadily increasing. The Soviet-built Lanzhou refinery had an annual capacity of only 0.5 million MT. The largest installations in the late 1960's reached 2–3.5 million MT, and in 1978 Daqing and Lüda refineries had annual capacities in excess of 5 million MT, and at least seven more refineries processed over 2 million MT a year.[90] But even the largest Chinese refineries are very small in comparison with the world's giants, such as Baytown in Texas, Baton Rouge in Louisiana, Venezuela's Amuay Bay, or Italy's Milazzo, which can process over 500,000 barrels of crude oil a day, or more than 25 million MT a year.[91]

Large refineries are potentially among the worst sources of air and water pollution, but the Chinese do not seem to have put much effort into controlling effluents. Although a Chinese report describes "clear skies" and "fresh air everywhere" in Beijing's oil refinery, where off-gas is not burned but used to fuel furnaces, sulfur-containing gases in Shanghai refinery are simply vented.[92] It was only in 1977 that Zhangding refinery, a medium-size enterprise on the southern bank of the Chang Jiang in Hunan, built China's first purifying installation, using activated carbon absorption to treat 600 tons of wastewater an hour as well as an ozone purifier with a capacity of 200 tons of water per hour.[93]

## *ELECTRICITY*

Considering China's large production and consumption of primary energy (in absolute terms), the country's power generation is undoubtedly very underdeveloped. In the mid-1970's the United States produced over 800 kilowatt hours (kwh) of electricity for each ton of coal equivalent energy consumption, the Soviet ratio was about 740, and the Indian around 640, but Chinese generation was just below 400.[94] The Chinese are producing an inadequate amount of electricity in relation to the magnitude of their total energy use. According to the Chinese government, power generation is "a weak link" in the national economy; the number of installed generating units has fallen short of plan, much of the equipment is obsolete, overloaded, and in bad repair, and generation does not meet the growing needs of industry and agriculture.[95]

Major technological advances are needed in the manufacture of generating equipment, and the construction of large power plants, as well as the upgrading and extension of transmission lines.

[90] Williams, "The Chinese Petroleum Industry," p. 263.

[91] A. Cantrell, "Worldwide Refining," *Oil and Gas Journal,* 75, 53 (Dec. 26, 1977): 148–90.

[92] "Visit to an Oil City," *China Reconstructs (CR),* 25, 1 (Jan. 1976): 15; and Yuan, "China's HPI," p. 107.

[93] NCNA in English, in *SWB,* 983 (June 7, 1978): 12–13.

[94] Smil, "Energy Development Plans," p. 25.

[95] Qian Zhengying, "Electric Power Should Play a Pioneering Role," *SWB,* 956 (Nov. 23, 1977): 4.

*Generating Equipment*

Chinese capacity to produce modern power-generating equipment remained very limited until the late 1960's. The largest thermal turbogenerator produced in Shanghai in 1952 rated a mere 240 kw, and only the acquisition of power technology from the U.S.S.R. and Czechoslovakia during the 1950's enabled the relatively fast growth of generator and boiler ratings.

A 3-megawatts (Mw) set was made in 1953, a 6-Mw unit was put into operation in 1955, and a 12-Mw unit a year later; a 50-Mw turbogenerator was completed in Haerbin during 1959 and became operational in 1962.[96] It was not until the late 1960's that the ratings surpassed 100 Mw; the first 100-Mw unit, designed in 1965, was put into operation in 1968, and a prototype 125-Mw set with a 400 tons per hour (t/hr) reheat boiler was manufactured between November 1968 and September 1969.[97] In comparison, the first 100-Mw unit was introduced in the United States before 1930 and in the U.S.S.R. in 1939.[98]

In the two decades between 1949 and 1969, the ratings of the relatively less sophisticated hydrogeneration equipment increased much faster. The Haerbin plants produced a 6-Mw set in 1953, 10 Mw in 1955, 15 Mw in 1956, and 72.5 Mw in 1959.[99] The first 125-Mw unit came in 1966, a 150-Mw set in 1968, and China's first turbogenerator over 200 Mw—a 225-Mw unit at Liujiaxia (Gansu)—started to operate in 1969.[100]

Construction of the first 200-Mw thermal unit commenced in 1971, but the 670 t/hr boiler for it was completed only in 1974, and the equipment went on stream at Beijing's Jingxi station in 1975.[101] In 1971 work also started in Shanghai on China's first 300-Mw thermal turbogenerator; it was installed at the Wangding coal-fired power plant (Jiangsu) between 1973 and 1974 and achieved full capacity in January 1976.[102] The only rating advance in hydrogeneration in the 1970's was the completion of the first 300-Mw unit installed at Liujiaxia.[103] Thus, in both types of generation, 300-Mw sets are now China's largest domestically designed power equipment, while the world's top capabilities are 1,380 Mw for thermal and 700 Mw for hydroturbogenerators.

Comparisons of the highest power ratings do not, of course, reveal the tech-

[96] R. Carin, *Power Industry in Communist China* (Hong Kong: Union Research Institute, 1969), pp. 84–85.

[97] I Fan, "Industries in Mainland China," in R. Carin, ed., *Communist China 1969* (Hong Kong: Union Research Institute, 1970), p. 312.

[98] U.S., Federal Power Commission, *The 1970 National Power Survey* (Washington, D.C.: Government Printing Office, 1971), pp. I-5-2–I-5-3; and P. S. Neporozhnyi, ed., *Elektrifikatsiia SSSR 1917–1967* (Moscow: Energiia, 1967), p. 33.

[99] PRC, State Statistical Bureau, *Ten Great Years* (Beijing: Foreign Languages Press, 1960), pp. 76, 104–5; and Carin, *Power Industry*, pp. 84–85.

[100] W. Clarke, "China's Electric Power Industry," in *Chinese Economy Post Mao*, p. 411.

[101] NCNA in English, in *SWB*, 808 (Jan. 8, 1975): 7.

[102] NCNA in English, in *SWB*, 883 (Jan. 23, 1976): 11–12.

[103] "China's Biggest Hydro-Power Station," *PR*, 1975, 7 (Feb. 14): 11–12, 21.

nological gap in typical unit sizes. Median ratings of newly installed units in large power systems are usually two to five times smaller than a handful of larger sets, and the average size of all units in service is smaller still. Typical sizes of new thermal units in China were 10–25 Mw in the late 1950's and the early 1960's, 25–50 Mw in the late 1960's, and 50–125 Mw in the 1970's. The median ratings of new U.S. thermal units are 500–750 Mw; large Soviet power plants have serialized 400-or 800-Mw sets, and ratings of about 400 Mw are common in Europe.[104] Chinese units are still too small to take full advantage of economies of scale, but the 600-Mw turbogenerators now under development in Shanghai cannot be used widely until high-voltage interconnections link the major power systems and raise the limit put on the largest unit size by inevitable reserve and safety considerations.[105]

In aggregate terms, China produced only 30 Mw of power-generating equipment in 1952, less than 1,500 Mw in 1958, 2,300 Mw in 1970, and about 6,600 Mw in 1976.[106] These are impressive increases, but the largest figure just about equals the same year's total for Italy. Production appears to be nowhere near the available plant capacity, and much of the equipment is derived from outdated Soviet and Czech designs of the 1950's. Moreover, a comparison of annual production totals and actual capacity increases leads to the conclusion that about one-third of all new units are used to replace overloaded, inadequately maintained, damaged, and obsolete equipment.[107]

Not surprisingly then, potential import needs are enormous. Between 1972 and 1976, about 4,500 Mw of power-generating machinery was bought abroad, consisting mostly of 100- and 200-Mw Soviet and Czech units and including two 320-Mw oil-fired sets from Italy—so far the largest turbogenerators in China—now operating in the Beidagang power plant.[108] The goal of the eight-year modernization plan in electricity generation—to construct 30 large hydro and thermal power stations—will require enlargement of existing power equipment factories in Shanghai, Haerbin, Deyang, and Beijing, and most likely, further imports of complete power plants or large generating units.

Chinese capabilities are also considerably behind world levels in gas turbine generation. As of 1979, the largest Shanghai-made prototype unit had 10 Mw, ten times less than the serially available Western and Soviet gas turbines. The Chinese have had to import 8.5- to 25-Mw sets to cover their peak-load demand in large cities.[109] Yet another area of deficiency is in pumped storage generation, which has become an essential and fast-grow-

[104] E. V. Abraham et al., "Trends in Power Boilers," *Power Engineering*, 82, 2 (Feb. 1978): 44.

[105] Smil, *China's Energy*, p. 48.

[106] Craig, Lewek, and Cole, "China's Machine-Building Industry," p. 292.

[107] *Ibid.*, p. 307.

[108] Tianjin city service, in *SWB*, 1,019 (Feb. 21, 1979): 6.

[109] Clarke, "China's Electric Power Industry," p. 408.

ing component of modern power systems. Although the largest European storage capacities surpass hundreds of megawatts and the two largest U.S. projects top 1,500 Mw, the first Chinese pumped storage installation at Miyun reservoir near Beijing has, since 1973, only 11 Mw.[110]

However, there is at least one area of power-generating technology where the Chinese claim world primacy: direct water-cooling of both thermal and hydraulic generators. They constructed the world's first 12-Mw thermal generator with inner rotor and stator water-cooling in 1958 and over the years applied the technology in their larger sets, including the 300-Mw ones.[111]

The advantages of water-cooling hydraulic units, as opposed to air-cooling, include an increased output of about 60 percent, a more uniform temperature distribution in the machine parts, and longer insulation life.[112] Consequently, some of the world's largest hydraulic units outside China have been built with water-cooling, including the 508-Mw Soviet turbogenerators at Krasnoyarsk (water-cooling of stator only) and the 400-Mw vertical Francis-type machines at the U.S. Raccoon Mountain station (water-cooling for both stator and rotor winding).[113] On the other hand, the world's largest generators (600 Mw) built by Westinghouse for Grand Coulee are all air-cooled.[114] Water-cooling will also be essential for the new generation of very large thermal machines (1,300–1,500 Mw),[115] but as with the hydro units, there is no compelling engineering need or persuasive reason to use it for relatively small machines, as the Chinese do.

*Power Plants*

In 1977 China had at least 178 thermal and hydro power plants of 30 Mw and higher, several thousand small fossil-fueled installations, and over 70,000 small rural hydro stations. The aggregate installed capacity of these stations amounted to 40,500 Mw, equal to about 8 percent of the American total, 18 percent of the Soviet capacity, and about 1.8 times India's total capability.[116]

About 62 percent of China's generation capacity and 71 percent of production capacity is in thermal power plants. In 1977 at least 69 percent of

[110] NCNA in English, in *SWB*, 761 (Feb. 6, 1974): 13.

[111] NCNA in English, in *SWB*, 795 (Oct. 20, 1974): 13; and NCNA in Chinese, *SWB*, 883 (Jan. 23, 1976): 11.

[112] E. N. Foster, "The Design of Generators for Hydropower," *Water Power and Dam Construction*, 29, 1 (Jan. 1977): 49.

[113] J. M. Gamous, M. F. Krasilnikov, and V. J. Brygalov, "Krasnoyarsk Is Proving Ground for World's Largest Hydro Sets," *Energy International*, 11, 7 (June 1974): 14–16; and H. Wirschal, "Raccoon Mountain Pump/Turbines Are World's Largest," *Energy International*, 10, 1 (Jan. 1973): 18–20.

[114] P. D. Johrde, "Power Generation at Grand Coulee," *Water Power and Dam Construction*, 29, 4 (Apr. 1977): 44–47.

[115] J. Papamarcos, "Large Turbine-Generators: Straining at the Leash," *Power Engineering*, 76, 9 (Sept. 1972): 28–33.

[116] U.N., *1977 Statistical Yearbook* (New York: U.N., 1978), pp. 388–407.

the total thermal rating was met by 119 fossil-fueled stations with ratings of 30 Mw and over.[117] Most of these stations did not exceed 100 Mw. China's largest thermal power plant in 1979 was the 1,100-Mw coal-fired Jinghe station in Kaiyuan county in Liaoning. Other large stations included the 750-Mw Daohe near Tangshan (Hebei), the 713-Mw Wangding (Jiangsu), which has China's first two 300-Mw units, and 600-Mw plants in Liaoning, Shijingshan (Beijing), and Shandong.[118] In contrast, the largest American and Soviet thermal power plants have topped 3,000 Mw. There are scores of stations over 2,000 Mw in the world, and work is under way on several 4,000- to 6,000-Mw installations.

Coal was the only Chinese power plant fuel during the 1950's and much of the 1960's. In the late 1970's, it generated some 75 percent of thermal power. Subbituminous coals with heat values of 3,000–5,000 kcal/kg are burned most frequently, but lignites and anthracites are also used. Expanded construction of mine-mouth power plants is one of the main goals for the 1980's. The only large natural-gas-fired stations are in Sichuan; oil-fired plants (burning unrefined crudes from Daqing, Shengli, and Dagang) are mostly located at or near oil fields.

Relatively small-sized thermal stations equipped with outdated machinery lead to considerable fuel waste. The last reliable average national heat rate—604 g of standard coal (7,000 kcal) per kwh in 1957—was some 20 percent above the comparable Soviet value and nearly 50 percent larger than the average American consumption.[119] Since then, this wide gap has been somewhat narrowed by introducing more efficient units and also by increasing attempts to utilize waste heat to preheat the feedwater.[120] However, the new policy of using low-grade fuels (rock coal, sapropelic coal, oil shales) will cause combustion inefficiencies that may negate the better thermal performance of new units.

No recent figures—nationwide or for individual stations—are available for power plant load factors. The best capacity and production estimates result in an average annual usage of 3,849 hours per installed thermal kilowatt (kw) in 1977, i.e. a load factor of 43.9 percent. This is about equal to Indian load factors in the late 1970's, and 10–25 percent lower than, respectively, the comparable American and Soviet usages.[121] Power plant controls and safety are also behind Western standards. China's first computer-controlled 100-Mw steam turbogenerator was installed only in 1975, after three years of development at Gaojing power plant at Shijingshan near Beijing.[122]

The first large hydro stations in China—Suping in Liaoning and Fengman

[117] Clarke, "China's Electric Power Industry," pp. 427–33.
[118] *Ibid.*; and "Huge Power Plant," *PR*, 1978, 49 (Dec. 8): 31.
[119] Smil, "China's Energetics," p. 313.
[120] NCNA in English, in *SWB*, 4,660 (July 25, 1974): C1.
[121] U.N., *1977 Statistical Yearbook*, pp. 388–407.
[122] NCNA in English, in *SWB*, 813 (Feb. 12, 1975): 16.

in Jilin—were built by the Japanese during their occupation of Manchuria. Fengman remained by far the largest Chinese hydro station until 1965 when Xinanjiang in Zhejiang reached 652.5 (9 × 72.5) Mw.[123] Of the more than 30 large- and medium-sized hydro projects started with Soviet and East European aid during the Great Leap Forward, Xinanjiang was one of the handful that were actually completed more or less according to plan. Most of the large projects were simply abandoned when the Great Leap Forward collapsed and were finished only in 1973–75.

These delayed projects include several of China's largest power stations. Liujiaxia on the Yellow River in Gansu with 1,225 Mw is the largest of all and was the first Chinese power plant to top 1,000 Mw.[124] Its concrete gravity dam is 147 m high and 100 m long, spanning the narrowest part of a steep gorge and creating a reservoir of about 5.7 billion $m^3$; the lake fluctuates between 75 $km^2$ in spring and summer and 125 $km^2$ in winter, and the spillway discharges up to 7,400 $m^3$/sec in flood. The powerhouse contains five units, including China's only 300-Mw hydroturbogenerator, and annual electricity production averages 5.7 billion kwh, implying a high load factor of 53 percent. In comparison, however, the station does not even rank among the world's top 20 structures.[125]

Other large stations completed between 1973 and 1975 are the 900-Mw Danqianggou (Han Shui in Hubei); and the 300-Mw Yenguojia, the 272-Mw Jingdongxia, and the 180-Mw Babanxia, all on the Yellow River, which has become the largest water-power source in China (although its basin contains only 6 percent of the nation's hydro potential).[126] Six completed or partially operating Yellow River stations had a capacity of 2,130 Mw in 1977, and about 2,000 Mw will be added in several existing plants and in two new large stations under construction, including 1,600-Mw Longyangxia. All together, 66 hydroelectric stations of 30 Mw and over were identified as of 1977, including seven under construction. Only nine had capacities over 300 Mw.[127]

About one-third of China's hydro power capacity is installed in small, rural power plants, whose number grew from a mere 50 in 1949 to 50,000 in 1973 and 87,000 in 1979.[128] Although the available provincial figures show these stations to average about 50 kw, most of these village projects are smaller than 25 kw. The dams for small stations are built by mass labor with traditional tools from local materials.[129]

Most of these structures are earth-fills. The national dam statistics for

[123] Smil, *China's Energy*, p. 82.

[124] "China's Biggest Hydro-Power Station," *PR*, 1975, 7 (Feb. 14): 11–12, 21.

[125] International Commission on Large Dams, *World Register of Dams* (Paris: ICLD, 1973), *passim*.

[126] Wu and Ling, *Economic Development*, p. 167.

[127] Clarke, "China's Electric Power Industry," pp. 427–33.

[128] Smil, *China's Energy*, pp. 85–86; and NCNA in English, in *SWB*, 1,025 (Apr. 14, 1979): 4.

[129] Smil, "China's Energetics," p. 335.

1949–72 show that 12,517 dams higher than 15 m were completed. Of these, 11,877 were earth-fills, 438 stone masonry, 129 rock-fills, and 73 concrete.[130] Small dams (15–30 m) were in the overwhelming majority (12,321); only 46 structures were higher than 60 m. Consequently, only 254 storages surpassed a capacity of 100 million $m^3$, and 10,596 lakes stored less than 10 million $m^3$.

As with the thermal stations, no exact figures are available on plant load factors, but the estimated capacity and production values for 1977 implied an average annual use of 2,592 hours per kw. This is considerably lower than in India, the United States, and the U.S.S.R.[131]

*Transmission*

In 1949 the PRC inherited only 11,410 km of high-voltage (110-kv) transmission lines.[132] The basic domestic technology developed fairly quickly with Soviet, Czech, and Hungarian aid. Nearly 7,500 km of new high-voltage links were added between 1949 and 1957, and the first 220-kv, 370-km-long tie started operating in 1954.[133] Transformer production advanced from the first 44 kv/20 Mva device in 1953 to 220 kv/40 Mva single-phase, 110 kv/60 Mva three-phase, and 110 kv/60 Mva three-phase units in 1958.[134]

More than 7,000 km of high-voltage links were added during 1958 and 1959, the first two years of the Great Leap Forward, but afterward little progress occurred until the mid-1960's, except for the completion of the first fully Chinese-made 220-kv tie from Xinanjiang hydro station (Zhejiang) to Shanghai.[135] Technical advances in the latter half of the 1960's and the early 1970's included 1,000-kv condensors, silicon-controlled rectifiers, 110-kv underwater oil-filled cables, serial manufacture of 250-Mva transformers, and the limited introduction of 300-, 330-, and 360-Mva units. Air-blast breakers with interrupting capacities of 15,000 Mva at 330 kv and the first domestic aluminum-wrapped 330-kv steel cable were also produced during this period.[136]

The first transmission of power at 330 kv over short distances started in June 1972, but China's sole 330-kv line—a 534-km link between Liujiaxia hydro station (Gansu) and the Guanzhong plain (Shaanxi) capable of carrying 420 Mw—was completed only in early 1975. This line, for which all cables, wires, switches, transformers, lightning arresters, and porcelain insulators were made in China, was China's top achievement in transmission technology as of 1979.[137] China's alternate-current transmission technology

[130] Chinese Society of Hydraulic Engineering, "Construction of Dams for Water Conservancy in the People's Republic of China," in *Onzième congrès international des grands barrages, comptes rendus 5* (Madrid: ICOLD, 1973), pp. 363–64.

[131] U.N., *1977 Statistical Yearbook*, pp. 388–407.

[132] Carin, *Power Industry*, p. 111.

[133] *Ibid.*, pp. 111–12.

[134] Clarke, "China's Electric Power Industry," pp. 410–11.

[135] Carin, *Power Industry*, p. 115.

[136] Clarke, "China's Electric Power Industry," p. 411.

[137] NCNA in English, in *SWB*, 814 (Feb. 19, 1975): 13.

lags between 15 and 25 years behind the West. Small European countries have been linked with 440-kv interconnections since the 1960's; the United States and the U.S.S.R. have operated long-distance lines of 765 and 750 kv, respectively, and plan 1,200- to 1,500-kv lines.[138] Furthermore, the Chinese do not have any of the direct-current transmission links that have become essential for transferring large blocks of power over long distances.

Yet another weakness of China's transmission system is the lack of interconnection among the regional and provincial power grids. The two largest regional power grids in the northeast and the north are only connected by a single 220-kv line. Provincial networks in the east and south are completely separated, except for the Anhui-Zhejiang-Jiangsu system. The main provincial transmission lines are 110 kv and are being upgraded to 220 kv. Long-distance lines are carried on steel towers or concrete pylons. By 1979, subprovincial lines were built at 110 kv, and typical country distribution lines were 35 kv, mostly carried on locally made concrete poles. Chinese engineers were developing a 500-kv alternate-current transmission system at Shenyang Transformer Plant and Wire and Cable Plant, China's two leading transmission technology centers. A 1978 Beijing home broadcast stated that 500- to 750-kv power systems would soon appear in China.[139]

## ALTERNATIVE TECHNOLOGIES

Alternatives to fossil fuels and hydroelectricity have become a major preoccupation of energy technology research in Western nations trying to ease their dependence on increasingly scarce and costly traditional resources. Nuclear energy played a key role in this search for about 25 years, but direct and indirect uses of solar radiation (ranging from passive heat storage to biomass fuels and ocean thermal currents), as well as geothermal energy, became more prominent in the late 1970's.

In comparison, the Chinese, with their abundant and largely undeveloped fossil fuel resources, have not had any pressing need to search for alternatives. Nevertheless, the eight-year modernization program called for "great efforts" in "exploring new sources of energy in order to change China's energy pattern gradually."[140] By 1979 the Chinese had started a very modest geothermal energy program, were using solar stoves and water heaters in some localities, and were experimenting with small wind-power generators. But they had no nuclear power installations. Their greatest success in alternative energetics was with rural biogas generation.

### *Biogas*

The PRC can claim world leadership in small-scale biogas energy technology. There were more than 5.76 million methane digesters in over 1,000

[138] Smil, *China's Energy*, p. 181.
[139] Beijing home service, in *SWB*, 1,006 (Nov. 15, 1978): 3.
[140] "Abridgement of Fang Yi Report," p. 10.

counties in May 1978.[141] The national total has grown by hundreds of thousands each year since this relatively simple yet demanding technology started to spread from two Sichuan Basin counties in the early 1970's.[142]

Most of the digesters are family units with a capacity of 10 m$^3$ (2 m$^3$ per person is considered sufficient to provide enough biogas for cooking and lighting).[143] They are fed by a mixture of pig manure, human excreta, crop residues or grasses, and water. (The most frequently used ratios are 30, 10, 10, and 50 percent, respectively.) Various digester designs have been tried in China, but by far the most popular type is the water-pressurized tank developed in central Sichuan; it is a rather simple structure consisting of intake (loading), fermentation, and outlet (sludge) chambers.[144] Fermentable materials pass from the loading compartment into the main chamber, where the anaerobically generated biogas displaces the fermenting liquid as it gathers below the rigid cover plate. Spent material is removed from the sludge compartment at least once a year and is applied on fields as an excellent organic fertilizer.

As long as the digester is tightly sealed, a variety of construction materials can be used. Cheap, locally available materials are always preferred, and the digesters are built from mortared rocks, bricks, and cement. A typical 10-m$^3$ capacity unit requires about 100 kg of lime, 100 to 200 kg of cement, 1 m of steel or PVC pipe (1 cm in diameter) to serve as the gas outlet, and a few meters of plastic or bamboo tubes to deliver the biogas to the nearby house (the gas is under low pressure and cannot be moved far).[145] A simple glass safety valve (pressure gauge) is connected to the digester gas outlet to monitor tank leaks, estimate the quantity of stored gas, and protect the digester from cracking.

The waterproof and gasproof structure is gradually filled with aerobically decomposed material. When gas production starts, proper care must be taken to maintain suitable liquidity, the right carbon-to-nitrogen ratio of newly added digestible matter (about 1:25), and the optimum temperature and pH for methanogenic fermentation. Any organic acids released by the decomposing phytomass can easily be neutralized by adding the necessary amounts of lime solution or grass ash to maintain pH at 7–8.[146] Retaining the optimum temperature of around 30°C during cold seasons without heating the digester content is impractical; biogas production either stops alto-

[141] NCNA in English, in *SWB*, 983 (June 7, 1978): 13.

[142] Sichuan provincial service, in *SWB*, 760 (Jan. 30, 1974): 11; and *SWB*, 792 (Sept. 11, 1974): 6.

[145] "Construction of the Fixed Top, Fully Enclosed Biogas Plant," in M. G. McGarry and J. Stainforth, eds., *Compost, Fertilizer, and Biogas Production from Human and Farm Wastes in the People's Republic of China* (Ottawa: International Development Research Center, 1978), pp. 56–57.

[143] "Questions About Marsh Gas Tank," *Kexue shiyan (Scientific Experimentation)*, 1974, 7 (July): 30.

[144] "Crude Method of Manufacturing and Utilizing Marsh Gas," *Scientific Experimentation*, 1973, 5 (May): 32.

[146] "Questions About Marsh Gas Tank," p. 30.

gether or is greatly reduced.[147] Consequently, a backup energy source is needed, except in the warmest regions. Biogas is burned in simple stoves and lamps made of a clay and carbon mixture; gas from larger digesters (up to 100 $m^3$) is also used to power crop-processing machinery and irrigation pumps or to generate electricity.[148]

The diffusion of biogas energy technology from Sichuan has been organized by the State Planning Commission, the Chinese Academy of Sciences (CAS), and the Ministry of Agriculture and Forestry. There has also been applied research at local universities and publication of practical construction and operation booklets.[149]

Although not all digesters are used for methane generation (some were built simply to decompose manure), China's edge in rural biogas technology is considerable. In 1977 India had some 15,000 digesters, or about 0.1 per 1,000 families, South Korea had about 25,000 digesters in the mid-1970's (3.6 per 1,000 families), and Taiwan had 8,000 small family units (2.4 per 1,000 families). The ratio in the PRC in 1978 was nearly 30 digesters per 1,000 families.[150]

### *Solar Radiation*

According to *Kexue shiyan (Scientific Experimentation),* solar radiation reaching China could provide energy equivalent to 100,000 electricity-generating plants with a total capacity of 12 million kw.[151] Naturally, as everywhere else, only a tiny fraction of this immense flow can be practically harnessed, and as of 1979, China's use of solar radiation was much less advanced than its indirect solar energy uses (hydropower, biogas).

Infrequent Chinese reports have mentioned small solar stoves and relatively large solar water heaters. The first series of 1,000 small solar stoves was manufactured experimentally in Shanghai in 1974, to be used by peasants in the surrounding communes. The simple device had a parabolic collector and boiled 3 liters of water in 20 minutes and cooked a kilogram of rice in 15 minutes. Its performance was comparable to that of a small coal stove.[152] A reflector-type solar stove for boiling water used in Tibet had 20-$m^2$ mirror reflectors and an average capacity of 200 liters of boiled water a day when both summer and winter performance was taken into account.[153] Similar devices were used in several other provinces, but no nation-

[147] *Ibid.*

[148] NCNA in English, in *SPRCP,* 5,864 (June 2, 1975): 24.

[149] Sichuan provincial service, in *SWB,* 812 (Feb. 5, 1976): 6; and E. El-Hinnawi, *China Study Tour on Energy and Environment* (Nairobi: United Nations Environment Program, 1977), p. 37.

[150] A. K. N. Reddy and K. K. Prasad, "Technological Alternatives and the Indian Energy Crisis," *Economic and Political Weekly,* special issue (Aug. 1977): 1,484; and National Academy of Sciences, *Methane Generation from Human, Animal, and Agricultural Wastes* (Washington, D.C.: NAS, 1977), p. 5.

[151] "Umbrella-Shaped Solar Disks," *Scientific Experimentation,* 1974, 10 (Oct.): 31.

[152] "Solar Energy Stoves," *PR,* 1974, 40 (Oct. 4): 38.

[153] NCNA in English, in *SWB,* 866 (Feb. 25, 1976): 10.

wide figures or technical details were available. Folding-type solar stoves for boiling and cooking manufactured in Gansu province cost 15 yuan (about U.S. $7), much less than comparable devices in India.[154]

There have been reports of large-scale solar water heaters from Beijing and Tibet. Beijing Electric Bureau, for example, had a collector with a 108-$m^2$ surface area and 12-ton capacity, sufficient to provide daily warm bath water for 400 to 500 people.[155] Beijing Technical Exchange Center organized a citywide seminar on solar water heaters in 1975, and many experimental devices have been set up in hotels, barbershops, and bathhouses.

In Lhasa, where annual sunshine averages 3,000 hours in clean mountain air, a 280-$m^2$ glass absorber heats daily bath water for 300 people in summer and 100 in the cold winter; similar baths have been built in more than ten locations in the Lhasa area.[156] The Tibet Industrial Architectural Institute has also introduced solar energy equipment in factories and has distributed a "solar energy oven with slanting surface" for use in farming and pastoral areas.[157]

### *Wind Energy*

The only information available on wind-harnessing technology comes from the Inner Mongolian Autonomous Region, where the Prairie Research Institute built a small wind-powered 100-w electric generator in cooperation with a Shanghai factory.[158] This simple and cheap device has been used increasingly by nomadic Mongolian herdsmen because it can be disassembled and packed in a wagon for moving. The Inner Mongolian Agricultural and Animal Husbandry Mechanization Research Institute has also developed a larger 2,000-w generator for stationary use. Electricity generated by wind in Inner Mongolia is mainly used for well-water pumping, lighting, radio communication, and movie projection. There have also been some unique local applications in wool shearing and electrifying fences against wolves. Research institutes in the region are trying to develop a relatively large-scale standardized generator for multipurpose settlement uses.

### *Geothermal Energy*

The ongoing collision between the Eurasian and Indian plates has given rise to numerous geothermal manifestations through much of China, above all on the Tibetan plateau. China's undoubtedly large geothermal energy potential has yet to be systematically assessed, but the first small experimental power stations indicate a serious Chinese interest in using the Earth's

[154] Gansu provincial service, in *Joint Publications Research Service,* 66,191 (Nov. 21, 1975): 15.

[155] "City of Peking Actively Begins Work on Use of Solar Energy," *Chung-kuo hsin-wen* (Hong Kong), Aug. 30, 1975, p. 2.

[156] NCNA in English, in *SWB,* 866 (Feb. 25, 1975): 9–10.

[157] NCNA in English, in *SPRCP,* 6,028 (Feb. 5, 1975): 76.

[158] "Use of Wind Energy in Inner Mongolia Explored," *Guangming ribao (Guangming Daily),* Aug. 10, 1977, p. 2.

heat. The first one started operating in Fengshun county (Guangdong province) in 1970.[159] Hot water from an artesian well (103.5°C at the bottom of a 800.81-m-deep shaft) flows to the surface at a rate of 80 t/hr, with a temperature of 91°C at the opening. After degassing (to remove $H_2S$, $CO_2$, $NH_3$, and N so they will not affect evaporation, increase pump load, and cause corrosion), hot water is vaporized in two 5.5-$m^3$ low-pressure expansion containers and the steam is introduced to an 86-kw turbogenerator rotating at 1,500 rpm. The condenser is made of 800 copper tubes with a total cooling surface of approximately 78 $m^2$. Naturally, the overall thermal efficiency of such a plant is rather low, internal electricity consumption is quite high, and as the Chinese admit, problems with gas elimination and corrosion have yet to be solved.

China's second small experimental power station was designed and built by Beijing's Department of Power Industry and Beijing University in Huailai county, about 90 km northwest of the capital, in 1974.[160] Hot water pumped from a well vaporizes an intermediate fluid of low boiling point (chloroethane or isobutane), which then runs a low-pressure turbine. No technical details have been released regarding this project. The third station was built between 1972 and 1975 at Huizhang spa, Ningxiang county, Hunan province. It uses low-pressure vaporization of the 92°C water to propel a 300-kw generating set.[161]

Between 1974 and 1976, the Qinghai-Tibet Plateau Multipurpose Scientific Survey Team (sponsored by the CAS) surveyed a large steam field in the Yangbajan area, 4,300 m above sea level, 90 km northwest of Lhasa.[162] By 1978 the field had become the site of the country's first geothermal steam power plant, with an underground steam temperature of more than 300°C and generating capacity of 1,000 kw. A photograph in *China Reconstructs* shows a medium-size power station.[163]

This installation is very small in comparison with the world's largest geothermal stations. The Geysers in California, now rating in excess of 500 Mw, was projected to have nearly 1,000 Mw by the 1980's, and power plants of several hundred megawatts were either operating or planned in about a half-dozen nations in the late 1970's.

## *Nuclear Energy*

In December 1978 China agreed to purchase two 900-Mw nuclear power plants from France, but shortly afterwards, the deal was postponed indefinitely. By mid-1979, China had no nuclear power station, although it has maintained a relatively large research and development program for nuclear

[159] Guangdong, Geothermal Power Generation Experimental Group, "An Experimental Geothermal Power Station," *Scientific Experimentation,* 1971, 6 (June): 36–37.

[160] "Geothermal Power Station," *PR,* 1974, 20 (May 17): 31.

[161] Hunan provincial service, in *SWB,* 886 (July 14, 1976): 7.

[162] Ko Tze-yuan, "Tibet's Abundant Geothermal Resources," *Ta kung pao,* Apr. 8, 1976, p. 13; and Zhang Mingdao, "Tibet's Geysers," *CR,* 26, 11 (Nov. 1977): 44–46.

[163] "Geothermal Power Station," *CR,* 27, 5 (May 1978): 11–12.

weapons since the early 1960's and has also engaged in advanced thermonuclear fusion research. Domestic development of nuclear power is theoretically quite feasible; China has the research experience, adequate supplies of both natural and enriched uranium, and an opportunity to rely on substantial foreign experience in designing and operating various nuclear power systems.

Moreover, the Chinese have no qualms about safety problems. They see nuclear power as having "outstanding merits," with "accident rates . . . lower than [those] in other sectors of industry . . . basically a clean energy source."[164] It was hardly surprising that the eight-year modernization program envisioned an acceleration of scientific and technical research in nuclear energetics and contained plans to "speed up the building of atomic plants."[165] Although operation of China's first small domestic experimental power reactor may not be far off, the contribution of fission power to China's twentieth-century energy balance will be very small.

The Chinese are also expanding their thermonuclear research. This started in 1973–74, with the construction of the first small experimental Tokamak-type device by the CAS Institute of Physics in Beijing.[166] According to a Japanese report in 1976, work was under way on the construction of a four-story structure to house a medium-sized Tokamak facility at the Xianan Institute of Physics, one of China's newest nuclear fusion research centers, in Luoshan (Sichuan).[167] Furthermore, the Beijing Institute of Physics was developing a new plasma-generating device, a belt pinch with a rectangular cross-section, and a condenser capacity of 2.5 Mjoule, and the Institute of Optics and Precision Instruments in Shanghai was experimenting with plasma generation by laser beams in 100-micron diameter glass beads.[168]

As in other areas of science and technology, China intends to catch up with the most advanced nations in nuclear fusion research by the end of the century.

Accelerated development is to propel China's economy to the front ranks of the world by 1999. "Catching-up and surpassing" timetables of Communist leaders—great favorites of both Khrushchev and Mao in the late 1950's—have shown a uniform tendency to go awry. If taken literally, Zhou Enlai's call, which has been adopted by Hua Guofeng and Deng Xiaoping as one of their main slogans, is destined to meet a similar fate. One key reason for this almost inevitable outcome is the state of China's energy technology.

Energy is, naturally, the principal mover of economic development, and as I have endeavored to show, China's capabilities for extracting, transport-

[164] Feng Zezhun, "Nuclear Power," *Scientific Experimentation*, 1976, 12 (Dec.): 29–31.
[165] "Abridgement of Fang Yi Report," p. 10.
[166] Radio Beijing for Southeast Asia, in *SWB*, 869 (Mar. 17, 1976): 2.
[167] Kyōdō in English, in *SWB*, 963 (Jan. 18, 1976): 4.
[168] *Ibid.*

ing, and converting its energy resources in 1979 were some 20 years behind advanced world performance. It will not be enough to close this gap in the 1980's and 1990's; research and development of new energy technologies have come to be among of the most intensive intellectual and engineering undertakings in industrialized societies, and major breakthroughs are inevitable.

Moreover, there are several important commercial energy technologies with which the Chinese have no practical experience: power generation with fission water reactors, direct current high-voltage transmission, and liquefied natural gas transportation. In addition, there are other promising energy technologies that have attracted considerable research and development attention in the West for which the Chinese have no comparable serious research programs. Coal gasification to produce synthetic "pipeline" gas, coal-fired magnetohydrodynamic power generation, and fast-breeder reactors are examples of these.

The eight-year plan of scientific modernization aims to narrow some of these gaps. Gasification and liquefaction of coal and high-voltage transmission are explicitly singled out as important research and development areas. Research efforts are to focus on hydrocarbon drilling and recovery, large-dam construction, efficient energy use in power generation, iron and steel, and chemical and metallurgical industries. Improvements in these areas are certainly imperative; Chinese steel production is more than three times as energy intensive as its Japanese counterpart, and the small nitrogen fertilizer industry uses nearly twice as much energy per unit of output as modern large factories.[169]

However, the gaps are too large and too numerous to be eliminated or substantially narrowed in the time available. Although China will remain one of the world's top three coal producers, will move forward in the ranks of oil- and gas-producing nations, and will greatly enlarge its power generation, the variety, sophistication, and performance of its energy technology will almost certainly not attain the world's top level.

[169] Smil, "China's Energetics," pp. 356–57.

*Bohdan O. Szuprowicz*

# *ELECTRONICS*

The importance of electronics and computing in China today stems from the need to provide vital equipment and technology to implement the "four modernizations" program. The central role of computing and electronics was first underscored in an editorial in the *Beijing ribao (Beijing Daily)* in December 1977, which characterized the level of development of the electronics industry as the hallmark of modernization. The editorial stated that electronics, as a result, should have first priority in modernization.[1] Then, in 1978, Vice-Premier Fang Yi, in a speech to the National Science Conference, articulated the goals of the "four modernizations" with respect to science and technology and placed heavy emphasis on electronics and computing (see Appendix A).

## *THE TECHNOLOGY: PAST, PRESENT, AND FUTURE*

### *Electronics*

Although China has a large electronics industry, accounting for over 1 percent of its gross national product, with an annual output comparable to that of Britain, Canada, and East Germany, its technological base is at least five to ten years behind. The lag in the computer and software industry is even greater. The Chinese recognize that failure to maintain currency in electronics causes an ever-widening technology gap in this and other industries.

Priorities established to modernize China's electronics industry include improvement of the underdeveloped telecommunications system, more efficient transport, modernization of the armed forces, increased utilization of computers, and even politically prestigious space activities. Finally, in elec-

[1] *BBC Summary of World Broadcasts,* FE/5688/B11/2 (Dec. 5, 1977).

tronics China may have a potential export industry that could, in the long run, outproduce and outsell other countries and provide China with much-needed foreign currency.

The electronics industry was insignificant in China before 1949. When the Communists came to power, electronics did not immediately receive top priority. Small plants were consolidated into larger enterprises, however, and some new equipment was imported from the Soviet Union. The equipment was used in manufacturing basic electronics components and served as prototypes for domestic copies.

Initially, electronics was developed not as an independent discipline in China but as a planned program designed to duplicate existing technology previously developed in the Soviet Union. Not until the first Twelve-Year Plan for the development of science and technology in 1956 was electronics recognized as an important discipline and specific goals established for its development.

The Twelve-Year Plan called for the development of an independent semiconductor industry as a basis for the electronics industry in general and computer manufacture in particular. However, primary concentration was given to importation of five major plants for the manufacture of conventional electronics components, radios, and telephone equipment.[2] Initial production of components included basic elements such as resistors, connectors, tubes, capacitors, rectifiers, cables, wires, insulating materials, and various instruments and testing equipment—all indispensable tools in electronic equipment assembly operations.[3]

The first electronic-tube manufacturing plant began operation in Beijing in 1956. At that time over 200 major materials required in electronics had to be imported. By 1959 a domestic network of enterprises was established and was able to supply up to 97.6 percent of all required inputs.[4] By then the Beijing Electronic Tube Plant was producing 70 percent of all the tubes in China. The gross output value of radio and electronics production in 1958 was reported to be 200 times larger than in 1949.[5] By 1960 about 60 major electronics plants were in operation, and China was already manufacturing 160 different models of radio receivers, using both tubes and transistors. During this period, China produced its first television receivers and transmitters, computers, telephone exchanges, magnetic core memories, and transistors.

By 1960 China had developed a well-rounded base for its electronics industry, but the withdrawal of Soviet assistance stopped the supply of new basic research results and manufacturing know-how. As a result, China was forced to obtain new electronics equipment from noncommunist countries;

[2] Chu-yuan Cheng, *The Machine Building Industry in Communist China* (Edinburgh: Edinburgh University Press, 1972), p. 190.

[3] *Ibid.*, p. 307.

[4] *Ibid.*, p. 197.

[5] *Ibid.*, p. 307.

between 1960 and 1970 over U.S. $200 million worth of electronic products and manufacturing equipment was imported from Japan, West Germany, Britain, France, Switzerland, and Scandinavia.[6]

This revised acquisition policy increased the pace of development and reduced the technological gap between China and the West. Experts now agree that, despite some short-term setbacks, China saved itself considerable time in developing more modernized industrial and military electronic equipment than would have been possible otherwise.

The impressive advances in the late 1960's would not have occurred had not the industry been among those given special priority and protection from the disturbing effects of the Cultural Revolution (1966–69). These were given, said a prominent scientist writing in the *Bulletin of the Chinese Academy of Sciences,* primarily to "satisfy the demands of national defense."[7] Since in the 1960's the electronics industry was vital to China's effort to build up its defense establishment, it received not only sufficient domestic investment funds, but also the foreign exchange needed to purchase advanced Western equipment as prototypes for domestic production.

Thus, during the 1960's electronics production expanded greatly, and new products were introduced in growing numbers each year. Indeed, the number of plants expanded rapidly, from 60 major plants in 1960 to 200 in 1971.[8] Major products introduced during the 1960's included five types of computers, an electron accelerator, and digital controllers of vital importance to progress in nuclear sciences and plant automation.[9]

Until recently, relatively little basic research was done in electronics; adaptation was the main theme. But now research is needed to extend the practical limitations of production processes and to suggest alternative development paths. The National Plan for the Development of Science and Technology between 1978 and 1985 set forth important goals for increasing basic research in electronics. Semiconductor science is the basic sector in electronics, and the plan particularly stressed the need to "lose no time in solving the scientific and technical problems in the industrial production of large-scale integrated [LSI] circuits, and make a breakthrough in the technology of ultra-large-scale integrated [ULS] circuits" (see Appendix A).

### *Computers*

The development of computers and computer technology in China lags behind that in the Western world by 10 to 20 years, particularly in universities and research institutes. Because computer usage in China is still in its infancy, military and industrial priorities tend to limit computer use in other

[6] P. D. Reichers, "Electronics Industry in China," in U.S., Congress, Joint Economic Committee, *People's Republic of China: An Economic Assessment* (Washington, D.C.: Government Printing Office, 1972), pp. 86–111.

[7] Cheng, *Machine Building Industry,* p. 211.

[8] *Ibid.,* p. 120.

[9] *Ibid.*

areas such as economic planning and statistical data processing and retrieval. Reichers estimated in 1972 that at least half the computers operating in China were in military-oriented industrial research and applications.[10] The strong emphasis placed on computers in the National Plan for the Development of Science and Technology, however, can be expected to bring substantial increases in computer use in other industries.

The precise number of computers in use is difficult to estimate but can be deduced from information about production and model numbers to be roughly 1,500. Between 1964 and the end of 1978, a substantial number of computers were imported by China from at least a dozen different manufacturers in France, Britain, West Germany, the United States, Japan, East Germany, and Hungary. Until recently, almost every imported computer was different and in most cases included software and peripherals selected to solve specific problems. Many of these machines were intended for use in seismic oil exploration, industry, or banking. The pattern of these purchases suggests that China had a policy of "prototype purchasing" and tried to acquire the most advanced hardware and software possible, choosing the systems for specific applications.

Research and development in the Chinese computer industry began in 1956 with the establishment of the Institute of Computer Technology in Beijing under the Chinese Academy of Sciences (CAS). By 1959 additional institutes were established in Shanghai, Shenyang, Jinan, Nanjing, and Chengdu.

The first computer, the "August 1," was built at the Institute in Beijing in 1958. Based on the Soviet URAL-2, it was a first-generation vacuum tube machine. Many of the subsequent machines also used vacuum tubes, primarily because the Soviet assistance program was terminated in 1960, two years before the Soviet second-generation transistorized computers appeared. At that time China was only a year or two behind the Soviet Union in computer hardware development.

The first Chinese second-generation (transistorized) computer was developed at the Institute in Beijing in 1965, about five years after second-generation computers became operational in the United States. That machine led to a series of computers designed and produced during the Cultural Revolution. The fastest was the Model 109C, which became operational in 1967 and was rated at 115,000 operations per second (OPS) as compared with contemporary American and Soviet computers capable of a million instructions per second (MIPS).

In 1968 the Institute began construction of the first integrated circuit computer in China, the Model 111, which became operational in 1971. This machine, which used TTL logic (transistor-transistor logic), provided a 32K memory of 48 bits. But not until 1974 did Chinese computers first reach processing speeds of a million operations per second. In late 1977 visitors to

[10] Reichers, "Electronics Industry in China."

the Beijing Institute reported observing a computer, the Model 013, capable of speeds of 2 million OPS, using Chinese MSI circuits.[11]

However, most computers presently used are of modest speeds by Western standards. The DJS series ranges from the Model 210, with processing speeds of 100,000 OPS, to the Model 260, whose speed is less than 1 million OPS. Another of the DJS series, the Model 130, is properly classed as a minicomputer, the most powerful identified in China. This machine was announced in 1973 and uses a 16-bit word with a processing speed of roughly 500,000 OPS. A radio broadcast from Hefei, Anhui province, in April 1977 announced trial production of the first Chinese microcomputer, the DJS 050, which uses MOS LSI circuits. It is reported to be destined for use in network terminals, industrial automation, communications, banks, warehouses, and commerce.[12]

Input/output equipment and storage peripherals are regarded as the weakest area in Chinese computer hardware development. Virtually all input equipment to Chinese computers is based on paper tape. Visitors consistently report that punched card equipment is not used in China at all. By inference one can conclude that computers in China are not widely used for data-processing applications.

CRT terminals, graphic displays, and disc drives are only at prototype stages (the Model 111 is one of few to provide displays for Chinese and roman characters), but line printers, magnetic tape drives, magnetic drums, and typewriter terminals, all of antiquated design, are being produced and are in operation. Line printers and teletypewriters use primarily the roman alphabet and numerals for computer output.

Disc drives, a fast electrostatic matrix printer, and a selectric-type teletypewriter were seen under development at the Beijing Institute in 1977. More recently a color character display device with editing functions based on LSI circuits was trial produced by the Wuhan Industrial Machine Building Bureau.

Numeric printers with speeds up to 1,200 lines per minute have been reported in use.[13] But alphanumeric printers, such as Model JY-80, operating at 600 lines per minute with 80 to 120 characters per line, are more common. The Model DL-2 magnetic tape drive uses one-inch tape with 16 tracks and a density of 500 bits per inch, typically recording in block mode with 2,048 characters per block. The Model G-3 magnetic drum is reported to have 2 million bytes of capacity. Other peripheral devices seen in China include an X-Y plotter and a graphics terminal with paper tape input based on a 16-bit minicomputer.

The most frequently mentioned programming language used in China is ALGOL-60, but FORTRAN-IV and BASIC are also used, for example, in the DJS-130 minicomputer and the homemade computer at Fudan Univer-

[11] New China News Agency, Dec. 14, 1976.

[12] Anhui Provincial Service, "Electronic Microcomputer," Apr. 14, 1977; in *BBC Summary of World Broadcasts*, FE/W926 (Apr. 27, 1977): A10.

[13] Reichers, "Electronics Industry in China."

*Summary of Characteristics of Chinese Computers*

| *Year* | *Computer model* | *Word length (bits)* | *Rated speed (operations per second)* | *Maximum memory* |
|---|---|---|---|---|
| 1958 | August 1 | na | 2,000 | na |
| 1962 | DJS-1 | na | 1,800 | 2K |
| 1963 | DJS-2 | na | 10,000 | 4K |
| 1964 | Model 103 | na | 50,000 | 16K |
| 1965 | DJS-7 | 21 | 3,000 | 32K |
| | X-2 | 32 | 52,000 | 8K |
| | C-2 | 32 | 25,000 | 8K |
| 1966 | DJS-6 | 48 | 100,000 | 32K |
| | DJS-21 | na | 60,000 | 4K |
| | TQ-1 | na | 25,000 | na |
| 1967 | Model 109C | 48 | 115,000 | 32K |
| 1968 | Model 441 B3 | 24 | 70,000 | 32K |
| 1970 | Model 111 | 48 | 180,000 | 32K |
| 1971 | TQ-3 | 24 | 100,000 | 8K |
| | TQ-11 | 36 | 50,000 | 8K |
| 1972 | Model 709 | 48 | 250,000 | 32K |
| | DJS-17 | 24 | 100,000 | 16K |
| 1973 | TQ-6 | 48 | 1,000,000 | 128K |
| | Model 719 | 48 | 130,000 | 32K |
| 1974 | DJS-11 | 48 | 1,000,000 | 128K |
| | DJS-18 | 48 | 150,000 | 64K |
| | Model 702 | 24 | 100,000 | 8K |
| | Model 7756 | na | 920,000 | na |
| 1975 | TQ-16 | 48 | 110,000 | 32K |
| 1976 | Model 013 | 48 | 2,000,000 | 128K |
| | DJS-130 | 16 | 500,000 | 64K |
| | DJS-120 | 16 | 200,000 | 32K |
| 1977 | TQ-5A | 48 | 160,000 | 32K |
| | TQ-31 | 24 | 100,000 | 16K |
| | DJS-050 | 8 | na | na |
| | DJS-154 | 16 | 200,000 | 32K |
| 1978 | DJS-155 | na | na | na |
| | DJS-140 | na | na | na |
| | DJS-210 | 16 | 100,000 | na |
| | DJS-220 | 32 | 200,000 | 64K |
| | DJS-240 | 64 | 400,000 | na |
| | DJS-260 | 64 | 1,000,000 | na |
| | DJS-300 | na | na | na |

sity, which is equivalent to a Data General Nova 1200. Several universities and institutes expressed interest in APL, but thus far there seems to be no operating processor. The only time-sharing system reported thus far is the minicomputer at Fudan University. With that exception, which apparently uses a version of the Data General software, Chinese scientists seem to prefer to design their own software for the computers they manufacture, rather than designing hardware to take advantage of existing software.

The operating systems used on the various Chinese computers are for the most part relatively rudimentary. Most of the computers mentioned have relatively simple interrupt structures, very limited data channel facilities, and simple peripherals. The machines typically run one program at a time, with modest operator control. Thus the opportunity to design complex systems is not generally available. Further the burden of developing operating systems and applications programs tends to be placed on the users, since in China the hardware manufacturers provide only diagnostic programs and basic compilers.

Although development of computers at certain institutes was not, apparently, significantly disrupted by the Cultural Revolution and the Gang of Four, that is not at all the case with the work at universities. There everything stopped. As a result, universities now have outdated equipment and only the beginnings of computer science programs. The faculties tend to be fairly well read in some areas of current literature (and badly in need of current information in others), but their practical experience is limited. In addition, some faculty members received their degrees during the hard times, and their training is said to be quite weak.[14]

However, the problem of training is being attacked by computer science curricula and by development of short-term courses and computer symposia. During 1974, for example, the Science and Technology Bureau of the Shaanxi Provincial Revolutionary Committee held a symposium to exchange experiences in computer technology. Institutions of higher learning such as Xian Jiaotong University, North Shaanxi Industrial University, and North Shaanxi University held 15 short-term courses to train about 2,000 computer operators in one year.[15]

A curious and unique campaign to impart principles of linear programming and use of computers to the working masses in all walks of life was undertaken during the Great Leap Forward in 1958–60. Known as "Yunchouxue" (operations research), this campaign involved many mathematicians and students who instructed about 300,000 workers, peasants, and technicians in linear programming techniques. A further 8 million people were apparently exposed to various training courses on the merits of computer-oriented mathematics.[16]

[14] R. J. Robinson, "Findings Concerning Computer Development in the PRC," International Report, State University of New York at Albany, July 1979, p. 6.

[15] Shaanxi Provincial Service, "Application of Computers," Jan. 1, 1975; in *BBC Summary of World Broadcasts*, FE/W809 (Jan. 15, 1975): A18.

[16] Donald G. Audette, "Computer Technology in Communist China, 1956–1965," *Communications of the Association for Computing Machinery*, 9, 9 (Sept. 1966): 655–61.

Most visiting Western professionals express respect and admiration for what Chinese scientists have been able to accomplish despite poor planning, inadequate management, and insufficient coordination, while at the same time stressing that there are innumerable fields that would greatly benefit from more advanced computer technology.

*Telecommunications*

Until 1953 China's telecommunications effort was primarily devoted to reconstructing a limited and badly damaged telephone and telegraph system. Basic equipment plants were built with Soviet and East German assistance, and in 1958 a modernization plan proposed microwave radio relays and coaxial trunk lines. The failure of the Great Leap Forward delayed its implementation, and the Cultural Revolution produced yet further delay.

Significant expansion of telecommunications took place after 1970 with the construction of a network of long-distance facilities providing a transmission base for telephone, telegraph, and television. This expansion was accompanied by significant increases in the deployment of wired loudspeakers for the mass distribution of radio programs in rural China, the development of radio and television broadcasting facilities, and the design of a national television network. Production of radios, loudspeakers, and television sets increased. New facsimile transmission devices were also produced.

Since 1972 China has been using the International Telecommunications Satellite (INTELSAT) and operates 23 full-time half-circuits mostly for international communications. In 1977 it formally signed an agreement to become the 98th member of the INTELSAT organization.

Telephone transmission uses standard technology. Open wire systems, almost entirely replaced in the United States in the late 1960's, are still widely used. Most open wire carriers are 12-channel systems.

The entire network of transmission media consists of open wire lines, cables, radio, and microwave relays. A domestic satellite communications network is a future possibility. Millimeter waveguides or optical fiber communications are believed to be at exploratory stages.

Multiconductor, coaxial, and submarine cables form part of the networks but are not used uniformly throughout the country. Multiconductor cable is most widely used for municipal feeder and long-distance lines. Coaxial cable, capable of carrying one television and 3,600 telephone channels, was first produced in China in 1962. Since 1976, a 1,800-channel, coaxial cable has been placed to provide the main trunk line between Beijing, Shanghai, and Hangzhou. A low-capacity submarine cable connects China to many offshore islands. The high-capacity, 120-channel coaxial submarine cable installed between Tianjin and Lüda was imported from Japan. More recently, a 60-channel coaxial cable between Guangzhou and Hong Kong was installed by Cable and Wireless Ltd., and Fujitsu of Japan supplied a 850-km submarine coaxial cable capable of carrying 480 two-way telephone channels between Shanghai and Japan, thus linking China to transoceanic cable networks.

Many remote areas are not adequately served by the networks, and point-to-point HF radio (2–30 MHz) is widely used for transmitting telephone and telegraph traffic and as a standby system when wire lines are disrupted. Radio telephone also plays a role in international communications with several continents.

During the 1950's, Soviet and East European microwave relay systems with 12- and 24-channel capacities were introduced. Further development resulted in the production of 60-channel microwave equipment and, in 1964, a 600-channel system. In 1972 a transistorized prototype of a 960-channel system was developed.[17] Microwave trunk lines now connect Beijing with 23 of the other 28 provinces, municipalities, and autonomous regions. These are used to transmit television broadcasts, telephone, telegraph, and facsimile services.

Transistorized frequency division multiplexing equipment with up to 60 channels is manufactured, although most of the equipment consists of 24-channel systems comparable to those used in America in the late 1950's and early 1960's. Time division equipment with 32 channels is only in limited use. Pulse code modulation and digital networks are at exploratory stages. Chinese telecommunications scientists show considerable interest in advanced computer networks and have attended Consultative Committee on International Telegraph and Telephone (CCITT) sessions on data networks, text transmission, and modems.

Telephone switching equipment based on Soviet and East European designs with 200 to 900 lines is manufactured locally, but more modern equipment was imported in the late 1970's from Japan. The Chinese have shown interest in Western switching equipment and manufacturing facilities. This is clearly an important area for development if the telephone network is to expand significantly.

The huge unsatisfied demand for electronic communications in China is dramatically illustrated by the telephone system. It is estimated that there are about 5 million telephones in China. Automatic long-distance dialing is available only in Beijing, Tianjin, Jinan, Shanghai, Hangzhou, and a few other cities. Nevertheless, more than 98 percent of the communes are believed to have some form of primitive telephone service. Priority is assigned to Party, military, government, industrial, and commune officials, and except for a few high-ranking persons, telephones are not installed in private residences. Telephone conference calls are common and provide China's extensive bureaucracies with an economic method of exchanging information, receiving instructions, and coordinating activity without the need for travel and costly meetings.

The telegraph service uses a four-digit numeric code for faster transmission of Chinese characters, requiring coding and decoding of Chinese characters into telegraphic code. To speed up this process, automatic character-decoding equipment that can store up to 10,000 characters has been devel-

[17] *China Reconstructs,* June 1977.

oped.[18] The telegraph service provides a relatively small portion of China's communications, but the development of fast electrostatic matrix printers to work with decoding equipment may change this. Telex service is primarily used for international communications and is available only to government agencies, press services, and foreign officials.

Facsimile transmission service is another way of overcoming the problem of transmitting Chinese characters and was already available in 1957. Major Chinese cities were connected to Moscow, Berlin, Warsaw, New Delhi, and Stockholm. The Beijing Telegraph Bureau and Beijing Long Distance Telecommunications Bureau have been developing facsimile transmission devices since 1956, and a full page of the *Renmin ribao (People's Daily)* was first transmitted in 1958. A facsimile device announced in 1972 transmits a full page of newsprint in 3.7 minutes. Research is under way on laser beam color facsimile systems. China's meteorological service has been transmitting weather maps since 1970.

The Chinese broadcasting system consists of wired loudspeakers and radio and television transmission facilities controlled by the Broadcasting Affairs Administrative Bureau of the Central Committee of the Chinese Communist Party. Operation and possession of private transmitters is not permitted.

The national wired loudspeaker system is a vast network estimated to consist of 140 million loudspeakers throughout rural China. It is said to provide direct service to 90 percent of the production brigades and 65 percent of the rural households. The service originates at the county level from local broadcasting stations and is relayed by communes to production brigades, teams, and individual households. Even with the recent proliferation of individual radio receivers, the wired system is expected to continue in use as a key public information medium.

Radio broadcasting facilities in China developed from a single transmitting station in 1945 to over 251 transmitters in 1976. There are national-, provincial-, regional-, and county-level broadcasting stations, and at least 62 shortwave transmitters are employed in a vast international broadcasting service, with additional relay transmitters in Tirana, Albania.

Plans for a national television broadcasting system were first announced in 1958, but because of the economic setback caused by the Great Leap Forward, only 12 cities had television broadcasting stations by 1962. Japan and Western Europe provided television broadcasting equipment during the 1960's, and by 1969 all provincial capitals, except Lhasa, had stations. Trial color broadcasts began in 1973.

Observers say that if China is to become a modern industrial state, a more modern telecommunications system is required. But there is no consensus among them, nor apparently have the Chinese decided, whether this system should be predominantly based on domestic satellite, terrestrial microwave, cable, or other means. Because telecommunications technology is fairly well

[18] Radio Beijing, "Post, Telecommunications Services Modernized," Sept. 28, 1977; in *FBIS, People's Republic of China National Affairs,* Oct. 4, 1977, p. E12.

known, China's main problem lies in developing an adequate manufacturing capacity for modern equipment.

Specific plans for telecommunications research and development were discussed during the National Posts and Telecommunications Conference in August 1977. Unfortunately, almost no details of the outcome are available. However, during the National Conference on the Electronics Industry of November 1977, China announced plans to build experimental lines of laser communications and to use satellites, possibly for transmitting radio and television signals.[19] The most immediate priority, though, was to expand China's long-distance telephone network to all cities and eventually to over 2,000 rural counties.

*Satellites*

The Chinese spacecraft launched in January 1978 weighed between three and five tons and was successfully returned to earth. At that time, informed sources speculated that preparations for China's first manned spaceflight might be in advanced stages.[20]

The National Plan for the Development of Science and Technology between 1975 and 1985 singled out space science and technology as one of the eight sectors to receive priority allocations of resources. Quite specifically, the Plan called for the use of satellites, the development of remote sensing techniques, and the launching of space laboratories and space probes. The emphasis appeared to be on civilian space applications, such as meteorological observation, resource survey systems, and communications.

But in fact most Chinese satellites are military intelligence vehicles. By virtue of their orbital altitudes and inclinations to the equator, they are believed to be photo-intelligence systems.[21] It is logical to expect an intense surveillance satellite development program because this has been a strategic weakness.

China's satellite program—whether military or civilian in orientation—will require a significant effort in developing long-distance tracking and telemetering. Possible Chinese tracking stations in Zanzibar and Sri Lanka have already been reported. In addition, several oceanological research ships of the *Xiang Yang Hong* class appear to be suitably equipped for missile and satellite tracking. But indications of China's priorities and deficiencies in this area emerged when a Chinese oceanographic research team toured American and Japanese equipment suppliers in mid-1978. The team expressed interest in specific electronic equipment, including Doppler sonar, satellite-positioning systems, integrated navigation systems, inertial navigation equipment, sidescan radars, and other instruments that use microprocessors and minicomputers. Clearly, the Chinese planned to use a large, cen-

[19] Beijing Domestic Service, "Developments in Satellite Communications Discussed," Nov. 8, 1977.

[20] *China Business Review,* 5, 2 (Mar./Apr. 1978): 39.

[21] *Strategic Survey, 1975* (London: International Institute of Strategic Studies, 1976), p. 102.

tral shipboard computer to process data for all experimental stations. (A more flexible approach used in the West is to install a series of independent microprocessors and minicomputers.)[22]

*Lasers*

The special attention given to lasers in the National Plan (another of the Plan's eight priority sectors) may be closely connected with the space program. High-powered lasers are now regarded as effective antisatellite weapons, and in the future satellites may be equipped with them. The use of spaceborne lasers against ground targets has also been suggested. Chinese space strategists could not have missed the 1975 reports of a reconnaissance satellite operating over the Indian Ocean that was "blinded" by a Soviet laser system on the ground.[23] With several hundred Soviet satellites flying over Chinese territory daily, an antisatellite laser weapon might be attractive to China.

Laser telemetry would appear to be an area of mandatory development for the Chinese if they contemplate reaching the moon before the Russians. Their capabilities to do so already include adequate rocket power and await only the development of suitable electronic guidance systems and a political decision to proceed with such a program.[24]

*Semiconductors*

The Chinese are well aware of the problems they face in the key semiconductor industry. These were admitted at the National Conference on the Electronics Industry, held in Beijing in November and December 1977, and have been independently reported by many visitors who have seen Chinese semiconductor research and production facilities. One of the most comprehensive assessments was made by the members of the IEEE Delegation, who spent a month in China in 1977 and were shown numerous electronics factories producing telecommunications, radio, and computer equipment.

The IEEE scientists reported limited use of integrated circuits (ICs), mainly in computers and calculators. Although they were told that analog ICs were still at the design and discussion stage, they concluded that China's present demand for ICs and LSIs could be met by one or two factories. The know-how, however, was not widely available, and the actual production yield on standard ICs was on the order of 25 percent. For simple LSIs, yields were less than 10 percent.

Even though recent ion implantation equipment and minicomputers were seen in use, it seems that large-scale production is not yet feasible. Present forms of the so-called "clean" areas of semiconductor manufacture, where extreme care is required to control particulates, were reported only margi-

[22] "Making a Decision on Purchase of Foreign Technology," *China Business Review*, 5, 3 (May/June 1978): 9.

[23] *Strategic Survey, 1976* (London: International Institute of Strategic Studies, 1977), p. 27.

[24] James Oberg, "China in Space," *Current Scene*, Aug./Sept. 1977, p. 12.

nally effective. Minicomputers used to test semiconductor wafers, for example, were not utilized for the accumulation of quality control statistics. Photolithographic layouts requiring a minimum 10-micron line-width resolution were made by hand although laser-controlled step-and-repeat equipment did exist.

The state of the art of the Chinese semiconductor industry is perhaps best illustrated by the IEEE members' report on the Shanghai Radio Factory No. 14, possibly the most sophisticated of China's semiconductor manufacturing facilities, which they visited in October 1977. The plant, which started manufacturing transistors in 1972, employs about 800 workers, of whom 100 are technical personnel, including about 10 engineers. Ion implantation equipment manufactured elsewhere in China has reportedly been used at this plant since 1975. The state of the art of lithography was assessed at 10 microns line-width. Silicon wafers in use were 40 mm in size, but there were plans to introduce three-inch wafers in the near future. Each wafer was divided into 200 chips, but the yield was reported to be only 25 percent for ICs containing 20 to 100 active elements. Overall medium-scale integrated circuits (MSI) and LSI yield was reported at 10 percent.

The IEEE visitors saw furnaces for LSI fabrication and reported experimental work under way on plasma etching and silicon gates. LSI random access memory (RAM) chip testing was performed with a special DJS-1 minicomputer that identified failed parts. No design automation on circuit layout was observed.

Total production in 1976 was reported as 2 million circuits and devices, with a value of 10 million yuan, achieved by 600 workers. The output is primarily consumed by computer and process control equipment manufacturers, but total demand for LSI circuits was reported much larger than the production capacity of the plant.

In view of the importance of semiconductors to the country's modernization and China's problems with equipment, materials, and the high rejection rate, it is not surprising that China has been shopping for Western and Japanese technology to upgrade its capabilities. The desire to catch up at all costs in the manufacture of ICs is illustrated, for example, by the recent purchase of a complete IC plant from Toshiba Ltd. of Japan at about U.S. $150 million.[25] Although this bipolar manufacturing facility is limited to the production of television receivers, the Chinese are actively investigating the purchase of other, more sophisticated technology. Here, however, because of advanced production technology, sales from the West must be approved by the Coordinating Committee for Export to Communist Areas.

### *Instruments*

In discussing how to mobilize resources to implement scientific development policies at the National Science Conference in March 1978, Fang Yi

[25] John Hataye, "Hitachi, Toshiba to Build China TV Tube, IC Plants," *Electronic News*, July 17, 1978.

stressed the need to modernize laboratory and research facilities. He called not only for the modernization of existing facilities, but also for "emergency measures" to be taken in design and production of instruments and equipment. "It is now considered essential to strengthen the management of the design, production, distribution, and use of scientific instruments and bring them under an overall national plan," he said (see Appendix A).

The origins of the Chinese instruments industry go back to the late 1950's, when China began manufacturing simple laboratory and test instruments with Soviet and East German assistance. The Haerbin Electric Instrument Plant began production of ammeters, voltmeters, wattmeters, frequency meters, and potentiometers, while the Xian Thermo-Technical Instruments and Meters Factory produced precision apparatus required by the expanding oil, chemical, metallurgical, power, and electronics industries. By 1977 more than 70 plants, employing at least 90,000 workers, were reported to be manufacturing instruments. Two-thirds of the total output came from plants in Beijing, Shanghai, Nanjing, and Tianjin. Other instrument-producing areas included Jilin, Shatou, Qingdao, and Wuhan. Many Chinese instruments closely resemble Western models, so it is widely believed that this industry follows a "prototype purchasing" policy to speed up its own research and development programs and to keep up with new developments in the most advanced countries.

China imported Western instruments worth U.S. $80 million in 1975 and $60 million in 1976. Imported supplies provide between 30 percent and 50 percent of China's annual demand for instruments—the highest import dependency in any manufactured products category in Chinese foreign trade. Major suppliers of instruments to China are Japan, Switzerland, West Germany, Britain, France, the United States, and the Soviet Union, all of which are leading instrument manufacturers.

During the 1960's Chinese manufacturers transistorized many of their instruments and attempted to upgrade the industry. Innovation is extremely rapid in the instrument industry, however, and microcomputers have created a whole new generation of sophisticated instruments that China has yet to copy. Its emergency instrument development policies may in fact result from concern over the growing gap between Chinese achievements and the state of the art in leading industrialized countries.

Nonetheless, China appears to be making significant progress in developing its own instruments. Because of rapidly growing domestic demand, China has an opportunity to engage in large-scale economic production of various instruments and should be able to export such products at competitive prices.

### *Consumer Electronics*

Consumer electronics in China dates back to 1960. Although this sector does not have high priority, it does benefit from achievements in industrial electronics. China manufactures a vast number of transistor radios, transistorized phonographs, magnetic tape recorders, and black-and-white and color television receivers. Popular electronics magazines such as *Wuxian-dian* (Radio) constantly describe simple electronic devices, including transistor radios and television sets, and provide circuit diagrams for them. Small local groups of factories appear to use this information to establish small-scale production, based on availability of components and local demand.

The drive to expand radio and television receiver production began in 1971, and all major administrative regions in China appear to have sponsored the establishment of local manufacturing facilities. By 1975 at least 260 different types of transistor radios were being manufactured in China; between 1969 and 1975 the number of radio receivers produced increased dramatically from 1 million to 18 million.

Television set manufacture was estimated at 205,000 units in 1975. Production is growing by about 25 percent each year. A dramatic increase in output was expected to occur when a color television plant became operational in Xian in 1980. In 1979, China was importing an integrated color television factory with an annual capacity of a million sets from Hitachi Ltd. in Japan at a cost of $95 million.[26] This could make China one of the leading color television manufacturing countries in the world. In the meantime, some television sets were imported from Japan and Hungary.

### *Military Electronics*

Military procurement probably accounts for one-half to three-quarters of the total value of electronics production in China. This means that most electronics research and development reflects the strategic priorities required for modernizing the armed forces and maintaining space programs. Any estimate of the electronics component in China's annual military expenditure is necessarily speculative, but a method demonstrated by Carlson[27] suggests that it probably approaches 5 percent of the total, or about $1.5 billion. (This compares with about 11–13 percent for the Soviet Union and 14–18 percent for the United States.)

During a 1977 interview, Wang Zhen, then head of the Fourth Ministry of Machine Building, a defense production ministry specializing in electronics, said that the Chinese electronics industry "still cannot meet the needs of national defense." He specifically mentioned the importance of radars and computers and indicated the value of electronic devices in detecting very distant signals, their use in arms control, and their potential as an important combat tool.[28] These remarks suggest the general areas in which military electronics research and development receive priority.

The direction of China's research and development in military electronics can also be discerned from reports about China's military achievements, such

[26] Hataye, "Hitachi, Toshiba," *Electronic News*, July 17, 1978.
[27] John McNickel, "The Great Electronics Plot: The Carlson Curve," *Electronic Engineer*, Aug. 1972, p. 16.
[28] *FBIS, People's Republic of China*, Nov. 21, 1977, p. E1.

as nuclear tests and satellite and missile launchings. Trends are also discernible from descriptions of Chinese aircraft, ships, submarines, and communications equipment. On the basis of this fragmentary evidence, it is apparent that China has developed and manufactured radars for early warning, ground control interception, missile control, and naval use. China probably also produces airborne radars, avionics equipment, missile guidance equipment, sonar, lasar range finders, and nuclear instrumentation, as well as electronic countermeasures and infrared homing devices for missiles.[29] China has made progress in nuclear weapons, missiles, and satellite programs by duplicating early American, Soviet, French, and British achievements.

China has a substantial but antiquated radar development program and manufactures early-warning, ground control intercept, missile control, sonar, and airborne radars. It is not believed to have developed satellite-borne radar, antiballistic missile systems,[30] or terrain-following radars.

Although general military electronics priorities appear to be centered on aerospace, complex missiles, and nuclear weapons, China is believed to have devoted major efforts toward ensuring the adequacy and efficiency of its national and regional long-distance communications networks for military command and control. Certainly the technology displayed in military communications equipment is considerably more advanced than that of civilian equipment. For example, a portable military Chinese transceiver was compared with similar American equipment in 1970 and found comparable in terms of weight and battery life, although inferior in quality. The four modernizations policy will require considerable sophistication in both civilian and military communication systems and will pose a serious challenge to the ingenuity of the Chinese.

There are indications that Chinese military planners are concerned about the various techniques that might be used by an enemy in what is referred to as electronic warfare, which, in conjunction with reconnaissance and fire power, discovers and destroys electronic control systems. Although little information on this is available, there is evidence that China is in fact taking steps to protect the electronic equipment of its armed forces by developing effective countermeasures.[31]

Electronics also makes up an increasingly large part of the world armaments market. China participates in that market, and between 1966 and 1976 exported weapons valued at about $2.2 billion, equivalent to 3 percent of total world arms exports (estimated at $70.3 billion), placing it in fifth place among world exporters. In 1977, however, military exports declined to $57 million, moving China to seventh place behind Italy and West Germany.[32] This may indicate increasing domestic demand for electronic

[29] Reichers, "Electronics Industry in China."

[30] *Aviation Week and Space Technology,* July 18, 1977, p. 18.

[31] "Electronic Interference and Anti-interference," *Hangkong zhishi* (Aeronautical Knowledge), 1974, 8 (Aug.): 30–31.

[32] *U.S. News and World Report,* Aug. 1, 1977, p. 34; and David Fairhall, "Some Little Powers Are Making Big Arms Sales," *New York Times,* Apr. 3, 1977, p. E3.

material,[33] but China will have to modernize its electronic warfare equipment to remain credible as a military power and competitive as a supplier—even to the Third World customers that make up its primary market base.

## RESEARCH AND DEVELOPMENT: THE ORGANIZATIONAL FRAMEWORK

Electronic research and development is conducted at several levels in China:

1. The institutes of the Chinese Academy of Sciences (CAS) and the major universities. At least ten electronics institutes within the CAS and about 30 at the leading universities have been identified. The most important ones are discussed below.
2. Institutes and study groups in industrial colleges or operating under local authorities. Thirty such groups involving electronics have been identified.
3. The Fourth Ministry of Machine Building. This Ministry, one of the six production ministries that form the military-industrial complex, is believed to operate many electronics institutes and research centers throughout China.
4. Numerous small plants and workshops. These collaborate with universities and research institutes in trial production of new products and often undertake serial manufacture of successful designs.
5. Small, locally sponsored electronics "research" groups throughout China. Their employees investigate local industries and seek solutions to existing problems using simple electronics.

The following are some of the more important institutes and universities that have been identified.

### *The Institute of Semiconductors, CAS, Beijing*

Established in 1960, the Institute of Semiconductors is the leading organization engaged in semiconductor research. In 1975, 400 of its 900 staff were college graduates. Apparently their research program is dictated by a combination of industrial demand, their own interests, and those of the state.

The Institute has been involved in the production of gallium arsenide crystals, the development of masks for large-scale integrated circuits, and circuit design work. The Institute is one of the few in China using a computer (RJX-1) for circuit design.

Work has been reported on Gunn and IMPATT diodes made from locally available materials and on testing the power conversion loss and noise of Schottky mixer diodes, using conventional microwave equipment, some imported from Japan. Further work is reported on a room

[33] Richard Burt, "One Trouble with Arms-Control Talk: There's So Much of It," *New York Times*, Feb. 19, 1978, p. E3.

temperature and a single hetero-junction gallium-aluminum-arsenide laser in the infrared range, for which a 1,000-hour life is claimed at one-half power output.

Estimates have been made that 80 percent of the Institute's instrumentation was Chinese-made and that about 10 percent each came from Japan and the Soviet Union or East Europe.

It was noted that the Institute does not lead Chinese research in ion implantation. It had an ion implanter on order in 1976, but several other Chinese research institutes had built or operated such equipment as early as 1969.

### *The Institute of Metallurgy, CAS, Shanghai*

This Institute, established in 1928, has three branches: the Institute of Precious Metals in Kunming, the Institute of Materials Research in Changsha, and the Institute of Silicate Research (also known as the Institute of Ceramics) in Shanghai. Approximately 60 percent of the 1,000 staff is involved in research, with several older scientists having degrees from American and British universities.

The Institute uses the Czochralski technique to make gallium arsenide single crystals, producing materials for field effect transistor (FET) and Schottky diodes. Other projects include gas phase epitaxial growth of gallium arsenide and research on display technology using gallium phosphide. In late 1976, the Institute was planning research on lasers, light-emitting diodes, and liquid crystals displays.[34] Pilot production of emitter-coupled logic semiconductors was reported in 1973.

The Institute apparently conducts the largest ion implantation research program in China, using an implanter it built in 1969. The machine, some parts of which appear to have originated with nuclear research equipment, has a range of 22 to 200 keV with a possibility of expansion to 400 keV—comparable to American equipment available in the 1950's. Since the Institute is not a production facility, it is assumed that basic research is under way in this particular program.[35]

Work was observed at the Institute of Silicate Research on the manufacture of glass semiconductor material that can be used for information storage and retrieval.[36]

The Institute has double focusing atomic absorption and atomic fluorescence spectrometers, all built at the Institute; a crystal puller with automatic temperature control and a closed-circuit television monitor; and a small computer designed at the Institute for generating masks for integrated circuits.

[34] Genevieve C. Dean and Fred Chernow, *The Choice of Technology in the Electronics Industry of the People's Republic of China,* U.S.–China Relations Report no. 5 (Stanford, Calif.: U.S.–China Relations Program, 1978).

[35] *Ibid.*

[36] New China News Agency, Dec. 31, 1975; in *BBC Summary of World Broadcasts,* FE/W866 (Feb. 25, 1976): A4.

*Qinghua (Tsinghua) University, Beijing*

Qinghua University is the leading technological university in China and has Departments of Automatic Control, Electrical Engineering, Radio Electronics, and Computer Technology Sciences. It is noted for its design and production of early computers and operated one of the first laboratories in industrial electronics.

Its Integrated Circuits Laboratory apparently plays a key role in LSI research and development in China. Specific programs apparently are directed by electronics factories that manufacture computers and equipment for industrial process control. The University's Computer Technology Sciences Department prepares the required integrated circuit design using computer-aided design techniques. Products of the Laboratory include: 1,024-bit MOS LSI memories, CMOS circuits, shift registers, and JK flip-flops. As of 1978, reports indicated that all digital devices were built with aluminum gates, with fastest propagation times ranging from 20 to 50 nanoseconds. Silicon gate technology was just starting (apparently the only such program in China). P-channel–type integrated circuits were produced, using ion implementation techniques, and n-channel devices were in development stages.[37]

The University also manufactures semiconductor production equipment, including step and repeat cameras, reportedly with minimum dimensions of 10 microns. The ion implementation machine in use has an energy capacity of 20 to 100 keV, which is judged adequate for research and development work but unsuitable for large-volume, high-quality production on an industrial basis. Other equipment in the Laboratory was judged comparable to American equipment of 20 years ago.

The University's computer science program offers hardware and software courses, usually beginning in the second year and including ALGOL and BASIC. Machine language, programming, and computer design also are taught. Students specializing in computer hardware take a variety of electronics courses. In 1973, over 300 students specialized in computer science, with the majority taking hardware courses.[38]

*Fudan University, Shanghai*

The Integrated Circuits Laboratory at Fudan University is better classed as a training establishment for integrated circuit technicians than as a research center for semiconductor technology. The University does, however, operate an electronic instruments plant and 13 other factories, some of which are involved in the construction of computer equipment.

The Physics Department emphasizes microelectronics and trains students in the fabrication of integrated circuits. The Department has plans for the

[37] Dean and Chernow, *Electronics Industry.*

[38] New China News Agency, "Electronic Computer," Aug. 31, 1974; in *FBIS, People's Republic of China National Affairs,* Sept. 11, 1974, p. E4.

construction of an ion implementation machine in conjunction with the Shanghai Institute of Nuclear Physics.

The Computer Science Department, previously a part of the Mathematics Department, researches computer hardware design, computer software development, and English-language character recognition. The Department provides three specialty tracks: architecture of hardware and software, information processing, and artificial intelligence. In 1975 approximately 400 students were involved, plus 20 postgraduate students. Approximately one-third of the 110 teachers in the Department received their degrees during the Cultural Revolution and, it is said, are therefore not qualified. The Department operates an Optical Scanning Laboratory using a 24-bit, 8K minicomputer developed in Beijing. The Laboratory uses a hardwired device for initial matching, and the computer matches those characters it fails on! This project is being developed for the Postal Service, which wishes to be able to read handwritten zip code information. In 1979 the equipment could read typewritten characters from a specific typewriter with an error rate reported to be 0.01 percent.[39]

The central computer is a "719" manufactured in a nearby small factory (which previously made doorknobs). It is a 48-bit machine, not checked, capable of 125 KPIS, with a 32K memory. It has two drums, uses ALGOL 60 enhanced, and is quite unreliable.

The Department is reportedly in the process of designing and manufacturing a new computer. The central processing unit was to be operational by the end of 1979, with the software ready by 1982.

The Department also uses a homemade machine that is essentially identical to the Data General Nova 1200 and uses Data General software. Only one interactive terminal was connected, but plans are to add about 16 or more.[40]

The University's Physics Department was reported in 1974 to have developed a biomedical computer, Model JSY-1, which could also be used for seismic, paramagnetic, and nuclear work. The Shanghai Radio Plant No. 13 was to undertake full-scale production.[41]

*Nanjing University*

Nanjing University specializes in science and engineering. Its Radio Physics Department conducts research on ferromagnetism and electronic instruments, apparently including MOS semiconductor work, charge-coupled devices, TTL devices, and semiconductor materials analysis, as well as work on silicon veractor and IMPATT diodes. Several electronic components and telecommunications equipment factories in the area have relations with the University.

The Computer Science Department has been producing integrated circuits

[39] Robinson, "Computer Development," p. 10.
[40] *Ibid.*
[41] New China News Agency, Jan. 20, 1976.

and computers since 1973. It apparently produces machines in conjunction with local factories for use in banks, stores, government offices, schools, and research units.

### *Beijing University*

Beijing University apparently operates a special electronics factory that also conducts research and development programs on semiconductor products. These include 1,024-bit MOS memory devices for use in onboard computers or automatic control systems. If it produced advanced integrated circuits for use in satellites, missiles, and aircraft, it could become a sophisticated facility.[42]

The Computer Science Department is served by a home-built computer and, like Nanjing University, has many unqualified faculty and a modest student enrollment.

### *The Institute of Computer Technology, CAS, Beijing*

Much of the work of the Institute of Computer Technology is described in other sections of this essay. The Institute was formed in 1956 to develop domestic manufacturing of computers and to organize a large, sophisticated computing center for solving scientific and industrial problems. It operates its own factory and associates itself with advanced computer systems technology in specific industries. It probably has a staff of 500 to 600 engineers, technicians, and programmers, whose research activities include such industrial applications of computers as aerodynamic design, construction of dams, electric power systems, geodetic problems, machine tool automation, nuclear energy programs, oil exploration analysis, turbine blade design, and weather forecasting.

Active in computer systems development, the Institute was responsible for the 013, the "August 1," the 109C, and the 111 models. In 1977, disc drives of 10-megabytes capacity were in use on the Model 013 computer. Apparently the application, programming, and systems and computer development groups are the largest in the Institute. Reports in early 1970 mentioned work on real-time processing, multiprogramming software and hardware, computer graphics, computer-assisted instruction for teaching purposes, information retrieval, and the use of APL for interactive problem solving. However, this conflicts with reports that no time-sharing is operational on computers there, which implies that time-sharing is under development.

### *Shanghai Institute of Computer Technology*

Indications are that the Shanghai Institute of Computer Technology is more a large computer center than a full-scale research and development group (based on visitors' reports of hand testing of components). The Institute has been active in popularizing computers in industry, and has de-

[42] *Wuxiandian* (Radio), 4, 32 (Apr. 1976).

ployed three-fourths of its research workers to various production enterprises to study specific problems and to introduce computers. Special courses were established on the use of computers and optics design, shipbuilding, machine building, engines, aviation, power generation, and hydrology, with attendance of at least 1,700 workers and technicians from various cities.[43]

*The Hebei Institute of Semiconductor Research*

The exact location of this provincial research institute is not known, and no reports by foreign visitors have been made. However, it appears to be prominent in semiconductor research. Its activities are reported to include the design and fabrication of silicon and gallium arsenide devices, high-frequency low-noise transistors, high-power devices, field effect transistors, and research on vapor epitaxial growth of n-type layers. The latter devices are preferred for MOS products, suggesting specialization in LSI technology (further supported by reports of the existence of two ion implantation machines).

Other organizations and institutions that appear to be doing important research in electronics or computer development and research include:

Institute of Physics, CAS, Beijing
Beijing Normal University (Physics Department)
Jilin Institute of Applied Chemistry, CAS, Changchun
Zhongshan University, Guangzhou
Wuhan University
Xian Jiaotong University
Shandong University, Jinan
Institute of Computer Technology, CAS, Shenyang, Liaoning province
Institute of Computer Technology, CAS, Xian
Institutes of Automation, CAS, Xian, Jinan, Beijing, and Shanghai
Shanghai Institute of Heat Engineering Instruments and Meters
Tianjin Institute of Radio Technology
Chinese Universities of Science and Technology, Beijing and Hefei
Institute of Geophysics, Beijing
Institute of Petroleum Research, Beijing

*The Ministry of Posts and Telecommunications*

The Ministry controls all planning, research, and production of equipment and the operation of telecommunications systems. Several institutes reporting directly to the Ministry are responsible for planning the network and ensuring production of equipment. The Ministry approves plans, appoints factories to undertake manufacturing, and establishes production

[43] "Shanghai Computing Technique Institute Gets Good Results," *Ta Kung Pao* (Hong Kong), Nov. 14, 1974, p. 5.

targets. Services and products are conceived by committees representing end users, manufacturers, and designers. The Ministry's final approval is necessary before proceeding with important innovations. Apparently, three to four years are required for simple products to move from conception to final production.

Research and development in telecommunications is directed toward production technology and takes place within the manufacturing plants, supervised by the Ministry's Technical Department. At least 60 telecommunications equipment plants were identified in the early 1970's. The leading institutions in telecommunications are as follows.

*The Institute of Posts and Telecommunications Research, Beijing.* This leading telecommunications research organization, controlled by the Technical Department of the Ministry, was established in 1954 and was instrumental in formulating a 12-year plan, published in 1956. It apparently has several working sectors, including circuits, instruments, long-distance telephone systems, communication lines, telegraphy, and municipal telephone systems. It probably also has a satellite communications section. It has a subsidiary design institute and an experimental factory for developing facsimile transmission equipment, UHF radio transmitters and receivers, 100-line and automatic telephone switchboards, and up to 960-channel microwave equipment.

*Beijing College of Posts and Telecommunications.* Jointly controlled by the Ministry of Posts and Telecommunications and the Ministry of Education, the laboratories of this college work on electron and ion acceleration, radio fundamentals, reception, and transmission.

*The Institute of Broadcasting Research, Beijing.* This is a unit of the Broadcasting Affairs Administrative Bureau, a department of the Central Committee of the Chinese Communist Party. Cooperating closely with Beijing Broadcasting Materials Factory and Qinghua University, it assisted in developing the first television broadcasting station. The Institute is clearly important and influential in the telecommunications industry.

## DISSEMINATION OF TECHNOLOGY

The Chinese Society of Electronics (CSE) was established in 1956. Although it ceased functioning during the Cultural Revolution (as did all other professional societies), by 1977 it again claimed 100,000 members and in 1979 resumed publication of one of the two previously published quarterlies. The CSE is responsible for dissemination of electronics technology within China and cooperates with the major institutes. It organizes international travel and acts as host to foreign scholars and engineers visiting China.

Another major vehicle for disseminating advanced electronics technology from abroad is the industrial fair or technological exhibition. Of the 43 exhibitions held between 1971 and 1975, at least 23 showed electronic equipment. Of perhaps minor commercial value in the West, these events are extremely important in China. They not only provide a marketing forum for advanced products, but also perform a specific educational function for scientists and engineers. As such, careful control over the list of invitations is exercised, and attendance is usually limited to specialists from major manufacturing enterprises and research institutes.[44]

The universities are, of course, important in dissemination of technology. The leading universities in electronics education are Qinghua in Beijing and Fudan in Shanghai. Prior to the Cultural Revolution, many more universities featured electronics and related areas, and numerous polytechnic universities and technical colleges existed.

The Cultural Revolution and the Gang of Four virtually destroyed the university system, and it is only now being rebuilt. The Ministry of Education controls the universities, and the Fourth Ministry of Machine Building and local authorities often jointly operate the polytechnic and technical colleges.

Electronics is also taught on a part-time basis at universities and colleges and in special schools operated by manufacturing enterprises. Enrollments are large, reportedly 5,000 spare-time students in 1977.

Attempts are being made to disseminate basic electronics knowledge by special publications. For example, in Shanghai, the Municipal After-Work University of Engineering publishes a series entitled *Dialogues on Knowledge Concerning Transistorized Circuits*.[45] Material also appears in a Shanghai publication called *Materials for Popularizing Science,* and the city's Science and Technology and Exchange Station organizes public exhibitions and runs electronics courses. A six-month course on computer memories is also offered.

The Chinese media have also reported the establishment of small, local, county-run electronics research institutes involved in education and the design and manufacture of various devices meeting local and immediate needs.

## THE FUTURE OF CHINESE ELECTRONICS

Since the first electronic component plants were built in the 1950's, China has made remarkable progress in developing an electronics industry. Although China already has one of the ten largest electronics industries in the world, numerous factors suggest that in terms of potential it is only now at the "take off" point. One important factor is that in its drive toward modernization China can bypass many intermediate levels of electronics technology and can concentrate on the most advanced equipment available in the West. For example, the development of microprocessors has created a com-

[44] R. E. Klitgaard, *National Security and Export Controls,* Rand Corporation Report R-1432-1APRA/CIEP (Santa Barbara, Calif.: Rand Corporation, 1974), p. 47.

[45] *Kexue shiyan* (*Scientific Experimentation*), 1973, 11/12 (Nov.): 64–67.

pletely new situation in China's industry and bureaucracy, without precedent in other countries. Although success will depend on the ability to master the production of very large-scale integrated (VLSI) circuits, China may very well be able to bypass the minicomputer stage and move directly to microcomputers, just as it bypassed the black-and-white television stage in consumer electronics. By properly combining large-scale computers, appropriate minicomputers, and microcomputer technology, China can develop an intelligent blend, which in the West must wait until present equipment depreciates and can be replaced.

Another important reason for China's substantial potential in the electronics marketplace involves the size of its population. Regardless of the level of its electronics technology capability, China is reaching a stage where consumerism is bound to provide substantial impetus to growth. Because of the size of the population, no matter how fast the industry grows, the marketplace will be difficult to saturate. Thus, just as in the case of bicycles, sewing machines, and watches, China's electronic luxuries will provide the authorities with an excellent capital formation mechanism. For example, by keeping production of color television sets relatively limited and the prices high, Beijing will be able to siphon off vast amounts of savings for use in development of other sectors of the economy. When the Hitachi color television plant begins producing a million sets per year, these could easily be disposed of at $1,000 per set or more, thereby generating $1 billion in revenues in a single year at a high rate of profit to the state.

Apparently this is already practiced in other electronic product areas, such as Chinese calculators. Visitors to electronics factories in 1977 reported that a four-function calculator cost about $1,200 on the local market, while a similar calculator is now available in the West for $5 to $10 per unit.

Finally, further stimulus for China's electronics industry is bound to come from China's potential to compete successfully in world markets. Although still often ridiculed, China's potential to expand its industry into a global giant capable of influencing world markets to an extent greater than that of Japan should not be overlooked, for in a few short years, China will join Japan, the United States, and the Soviet block in possession of VLSI circuit technology.

China is already the second largest producer of radios in the world, having displaced Japan in 1976. It is true that Japan and other countries have reduced their production of radios in recent years, switching to television sets with larger value-added factors and potential as production costs escalated; but China is in a position to engage in this production at extremely competitive prices that cannot be matched in many parts of the world. In 1977 Chinese labor costs were estimated by the IEEE Delegation at only 12.5 cents per hour. Hong Kong, Taiwan, and South Korea have already shown the way for China to follow, which it can do on a much larger scale. What Japan has done to American electronics producers, China may be able to do to the Japanese and other electronics manufacturing countries in the future. This prospect is underscored by a statement from the director of the Max Planck Institute for Solid State Research in Stuttgart, Germany, after his visit to several Chinese electronic research and manufacturing facilities in 1977. He said that China will certainly challenge the Soviet and West German electronics industries in the future.[46]

The Chinese, however, are perhaps not quite so optimistic. Wang Zhen, then Minister of the Fourth Ministry of Machine Building, stated candidly in an interview in November 1977:

> Within the realm of the national economy, our electronics industry is a relatively weak link, the technical level of its products is not high, its production efficiency is low, and it still cannot meet the needs of national defense and the building of the national economy. There is still a considerable gap between the level of our electronic technology and the advanced world levels. We are not behind in the development of semiconductors, computers, and other specialized fields, but the gap between us and advanced world levels in other areas has widened.[47]

It is interesting to note that Wang Zhen considered China's progress in semiconductors and computers adequate, although assessments of Western scientists suggest that it is five to seven years behind in development of LSI circuits and ten or more years behind in developing high-speed computers. What his statement may mean, however, is that current Chinese semiconductor and computer achievements are sufficient to meet short-term objectives, such as placing a man in orbit.

The size of electronics production does not, of course, reflect the quality of the manufactured product, and here lies another difficulty for the Chinese. The Chinese electronics industry is noncompetitive and is almost entirely engaged in the production of specific products for a huge unsatisfied market, and the quality of the products is not always high. Electronics factories lack sufficient quality control, and the Chinese press has frequently referred to the need for better management in that industry. In the case of television manufacturing, for example, it has been claimed that as many as 8 percent of all television sets are in need of repair, and as a result bottlenecks exist in the factories, which cannot proceed with production. Admittedly, it is not known what percentage of defective units are produced in large modern plants as opposed to small local electronics workshops using possibly inferior components. But clearly the Chinese must improve the quality of their product if they are to satisfy their home market and win a large share of the export trade.

The computer industry, which may be able to "leap forward" across the minicomputer boundary, nonetheless has substantial challenges to overcome on the way to modernization. The first and most impressive challenge is the development of the expertise and capability to produce operating sys-

[46] Gosch, "Chinese Put Priority on Semiconductors."

[47] New China News Agency, "Minister Interviewed on Electronics Industry Prospects," Nov. 16, 1977.

tems for modern computers. Second, the Chinese must develop a communications system that can support computer innercommunication, although this might be minimized by using stand-alone microcomputers instead of time-sharing in many cases. Third, they must develop a broad base of application system design capability and programming. This in turn will require assimilation by a large percentage of Chinese professionals of some form of computer "metaphor" or basic understanding of the utility of the computer in various environments. Although this process of assimilation can undoubtedly be made faster than has been the case in the United States, the time required for widespread assimilation is difficult to estimate. That in turn will be closely tied to China's ability to resurrect its higher education system and to develop a broad base of education related to computing outside the university system.

*Baruch Boxer*

# *ENVIRONMENTAL SCIENCE*

It has been suggested that the longevity of Chinese civilization can be attributed partially to sound environmental practices.[1] Systematically defined responsibilities and limitations in relations between man and nature were described as early as the first century B.C.[2] These notions provided a natural philosophic rationale for environmental stewardship, which, despite some local and regional abuses, prevailed through the centuries. Since China represents perhaps the most illuminating example of the successful historical evolution of workable strategies for sustained productive use of human and natural resources, any attempt to characterize present-day environmental activities cannot overlook premodern contributions.[3]

Briefly, historical concern for the environment is reflected in the Chinese genius for water management—their ability to conserve minute amounts of moisture and distribute large volumes of water for irrigation; in their energy budgeting and waste recycling for maintaining the agricultural ecosystem; in their definition and application of environmental and landscape planning principles (codified in the *feng-shui,* or geomancy, locational belief system); and in their use of cartographic and topographic analysis in the context of hydraulic engineering to conserve soil and water. Another unique approach to environmental conservation can be seen in China's historical success in developing clusters of urban and rural communities that supported large populations while adequately maintaining ecological integrity. Some of

[1] I wish to thank Edward C. T. Chao, Nelson Chou, John Ryther, John Edmond, and Ralph Luken for helpful insights and Richard C. Howard for making otherwise unavailable Chinese materials available to me.

[2] M. Yoshida, "The Chinese Concept of Nature," in Nathan Sivin, ed., *Chinese Science: Exploration of an Ancient Tradition* (Cambridge, Mass.: MIT Press, 1973), pp. 71–89.

[3] B. Boxer, "The Geography of China: A Report on a Seminar," *Social Science Research Council Items,* 28, 3 (Sept. 1974): 44–46.

China's most innovative and comprehensive post-1949 attempts to deal with pollution and environmental management problems are built to some degree on traditional interests and skills.

While Mao was not especially sensitive about environment, his writings showed traces of the influence of Marx and Engels, who did anticipate some elements of modern ecological theory. In his classic essay "On Contradiction," for example, Mao provided a theoretical basis for later ideological endorsement of environmental protection efforts.

In contrast to most other subjects surveyed in this volume, environmental science has no clearly defined disciplinary foundation and is diffuse and uncoordinated. It includes work in many areas of the physical and biological sciences, particularly chemistry, biomedicine, and theoretical and applied branches of engineering. Recognition of the discipline of "environmental science" and emphasis on environmental research are fairly recent phenomena in the PRC. Until 1978, when the first journal on environmental science, *Huanjing kexue* (Environmental Science), began publication in Beijing, pertinent information was sparse and scattered. Consequently, this chapter is oriented more to Chinese environmental policy than to environmental science. The interested reader should consult the chapters on medicine, plant protection, oceanography, and geography in this volume for additional details and insights about environmental efforts under way in these disciplines.

### *Environmental Policy*

China only recently recognized the need for a cohesive national environmental policy. This does not mean, however, that China was not taking measures that benefited the country's environment, but simply that all too often environmental gains were by-products of other goals and policies more directly related to major national priorities. For the first two decades of the PRC, for example, environmental concern focused on improving the living conditions of the Chinese people. There were almost perpetual health campaigns, mobilization of the masses to eliminate "the four pests" (flies, mosquitoes, rats, and sparrows), and educational campaigns regarding sanitation and safe methods of using human excrement as fertilizer. In the late 1960's, a major national campaign calling on the Chinese population to "transform wastes into treasures" and to recycle and recover the "three wastes"—liquids, gases, and solids—was launched. This policy combined Mao's principles of self-reliance and frugality and at the same time made environmental protection and conservation a concrete responsibility of the "worker masses."

Mao's policy of industrial and urban dispersion was primarily inspired by security concerns and economic and social considerations, but it also served environmental protection objectives, since pollution is normally a direct reflection of population densities. By controlling the growth of major cities and building small rural industries, the Chinese could reduce waste loads and transform them more efficiently into "treasures."

However, high-level attempts to coordinate China's environmental policies and research efforts to devise regulatory approaches commensurate with the needs of a developing economy became evident only in the 1970's. By then, air and water pollution had become serious, and concern that environmental mismanagement was polluting and degrading soil and water resources and thereby endangering agricultural production was growing. Because environmental threats to agricultural stability and productivity have always been a strong inducement for remedial measures, the scientific and technological establishment was formally asked to contribute to a national program of environmental protection and reconstruction.

The result was a national conference on environmental protection, personally endorsed by Chairman Mao, held in Beijing in 1973. The delegates considered the results of a national survey that revealed the extent of air and water pollution in industrialized areas, the threat to the natural environment due to improper agronomic practices, the neglect and de-emphasis of environmental factors in factory location and construction, the ignoring of environmental considerations in order to satisfy production goals, and the inadequate technical capability for managing disposal and recovery of solid, liquid, and gaseous wastes. The recommendations of this conference, published as a State Council directive, formed the basis of China's national environmental protection program.

Top priority among the administrative recommendations was given to the establishment of an organization under the State Council to coordinate and implement the national program. Similar lower-level bodies were recommended for provinces, autonomous regions, and provincial-level municipalities (Beijing, Shanghai, and Tianjin), although their functional responsibilities were left vague. The national-level Environmental Protection Office (EPO) was formed in 1974 under the State Council with an initial staff of 20 (up to 50 by 1979). Its relatively low status as a coordinating and planning body was reflected in its designation as an "office" rather than as a more prestigious "commission" or "special agency."

To pursue its coordinating and planning missions, the EPO has a science and technology division to supervise research, monitoring, and assessment in research institutions and a planning division to devise long-range environmental protection plans and to monitor compliance with regulations. The 1973 State Council directive authorizing the EPO made some reference to its standard-setting functions, but did not spell them out. Impending organizational changes should centralize monitoring and standard setting in the EPO. New divisions—water, air, nature protection, and foreign affairs—were to be added in 1979. These changes will probably shift responsibility for air and water quality from the Ministry of Public Health.[4]

In its role as national coordinating agency, the EPO regularly brings together officials from various ministries to consider the potential environ-

[4] R. A. Luken, personal communication.

mental effects and risks of industrial and resource development projects under their jurisdictions and to oversee expenditures of funds for environmental protection. Since production has always had the highest priority, however, protection needs have suffered, and there are few incentives for industrial ministries to invest funds in pollution control. Disputes over priorities are probably referred to the Standing Committee of the State Council for resolution.[5] The EPO also has little influence over environmental research and development in the ministries and limited authority to levy fines. Furthermore, the extent of its influence in setting research priorities in provinces and provincial-level municipalities is unclear.

Provincial-level organization varies from region to region. There is some evidence to suggest that the best administrative integration of environmental research and control exists in major industrial regions. In Jilin province, for example, by-product recovery and multiple-use procedures were apparently used in most large industries by 1974.[6]

China's institutional matrix for environmental protection was augmented in the mid-1970's by the establishment of small (15–20 persons) environmental bureaus (headed by deputy ministers) in each ministry concerned with manufacturing and resource development (including Public Health, Metallurgy, Public Security, Agriculture, Forestry, Light and Heavy Industry, and Transportation and Communications). In the industrial ministries, the environmental bureaus were made responsible for pollution control in government-owned plants.

Other environmental research bodies that were instituted in response to the 1973 directive included 15 institutes affiliated with provincial and municipal governments to study local or regional problems, a few research units in industrial ministries, and some 20 research groups in universities. There are few details of the administrative and policymaking links among these organizations. The 1973 directive also called for the development of environmental science curricula in colleges and universities and a national environmental research institute. As of 1979, however, formal education and training in environmental science continued to be based, for the most part, in individual disciplines. There was no evidence that the recommendations had been implemented. Thanks to the new, formal endorsements of environmental protection by the central government and the Party, however, there were indications in the late 1970's that the EPO was about to assume a more substantive policymaking role.

In a report to the Fifth National People's Congress on February 26, 1978, Chairman Hua Guofeng called for "elimination of pollution and the protection of the environment [as] a major issue involving the people's health, an issue to which we must attach great importance." He also called for "regu-

[5] D. M. Lampton, "Administration of the Pharmaceutical Research, Public Health, and Population Bureaucracies," *China Quarterly*, 74 (June 1978): 398.

[6] "Kirin Province Improves Environmental Hygiene," *Foreign Broadcast Information Service (FBIS)*, Aug. 12, 1974, pp. L4–5.

lations to protect the environment" and to "make sure that related problems are satisfactorily solved."[7] Further commitment was indicated by the inclusion of the following clause in the new constitution adopted on March 5, 1978: "The state protects the environment and natural resources and prevents and eliminates pollution and other hazards to the public."[8]

These general pronouncements were amplified in Vice-Premier Deng Xiaoping's opening address at the March 1978 National Science Conference. Environmental concerns were mentioned first in connection with agriculture, where "national exploitation and utilization of resources" and "protection of the ecological system" were advocated. The multipurpose utilization principle, with particular emphasis on the role of science and technology in helping to realize the benefits of recycling and by-product use, was also reaffirmed. Finally, "environmental protection" was designated as one of 27 "research spheres" to be pursued as part of the 1978–85 outline plan for the development of science and technology. Continued concern about environmental protection is reflected in Hua's June 1979 report on the work of the government to the National People's Congress, in which he mentioned work on an environmental protection law (promulgated September 1979) and the need for "taking environmental protection into account in the construction of new projects." Hua further stated that "pollution problems caused by existing enterprises must be solved step by step."[9]

To a large extent, the success of the new emphasis on environmental protection will depend on Beijing's ability to set and enforce environmental standards—until now a serious weakness of the program. Since the early 1970's, industrial standards for environmental protection have presumably been applied in the design, construction, and operation of factories, but compliance has been erratic, and there have been numerous reports of increased industrial pollution. In Beijing, for example, a number of major polluting industries were identified in response to widely publicized letters of complaint sent to the municipal environmental protection office.[10]

In 1978, the EPO, with the powerful backing of the State Planning and State Economic Commissions, censured 167 industrial enterprises throughout the country and ordered them to take "proper measures" to control pollution by 1982.[11] The order warned that if pollution was not reduced sufficiently during this period, the factories would be closed down and their managers held responsible. Obviously more significant, however, was the authorization of funds and equipment to some of the enterprises to assist in waste management. This appears to be the first substantial financial commitment by the central government to pollution control in individual enterprises.

[7] *Peking Review (PR)*, 1978, 10 (Mar. 10): 26.
[8] *PR*, 1978, 11 (Mar. 17): 8.
[9] *Beijing Review*, 1979, 27 (July 6): 17.
[10] "Peking Residents Appeal for Pollution Control," *PR*, 1978, 23 (June 9): 30.
[11] *Ibid.*

At the end of 1978, environmental authorities in Beijing expressed interest in the American approach of using economic analyses to set standards.[12] However, the methods followed in the United States are likely to be irrelevant to the Chinese situation because ambient and effluent standards in the United States are developed in response to specific legal mandates and are based upon economic data that reflect American conditions.

*Research and Application*

As of the late 1970's, Chinese notions of the purposes and objectives of environmental science and technology reflected (1) a continuing interest in comprehensive approaches to conservation and management of soil, water, mineral, and material resources; (2) concern for protection of natural areas and endangered plant and animal species; (3) acknowledgment of the urgent need for scientific and technical work to ameliorate the harmful effects of industrialization on ecological systems and on human health; and, of course, (4) the application of Marxist ideas of the social and economic value of environmental protection. Environmental research is also seen as contributing to the nation's ability to derive economic rewards from frugality in resource use during a period of rapid economic expansion.

As has already been pointed out, many scientific disciplines touch on problems involving the study of environmental processes or on the nature and effects of environmental changes on plants, animals, or man. It is difficult, however, to find references to common methodological or theoretical approaches that integrate disciplinary work into a coherent, preconceived "environmental" framework. Furthermore, in any problem area, wide gaps separate research aimed at understanding basic processes of environmental change from parallel attempts to apply this knowledge in designing control measures to minimize environmental damage.

Since the early 1970's, references to the mission and objectives of environmental research and descriptions of work in various disciplines have made subtle but, in the Chinese context, important distinctions between "protection of the environment" and "pollution control problems." Calls at various times for modification of existing research strategies suggest that national policymakers recognize the different environmental management objectives associated with the two approaches.

Generally speaking, the disciplines primarily supportive of "protection" objectives include plant and animal ecology, botany and geobotany, environmental geology, geography, hydraulic and civil engineering, and geochemistry. Those disciplines that are identified in the literature with what the Chinese call "control" objectives are hydrobiology, analytical (or pollution) chemistry, chemical engineering, microbiology, and soil science. Although public health improvement is frequently mentioned as an objective of environmental research, the environmental health sciences are not usually

[12] Luken, personal communication.

referred to in the scientific literature as environmental sciences per se. This is an important distinction, especially with regard to the criteria employed in environmental standard setting.

It is important to note, however, that environmental researchers seem to be encouraged to bridge the implicitly acknowledged "protection-control" dichotomy by pursuing research that explores new areas beyond accepted disciplinary boundaries. This has led to innovative work on moisture budgeting, water balance, composting efficiency, pesticide removal from soil and plants, and other problem areas, many of which build on traditional methods and technologies. This is not to say that "interdisciplinary research" in the Western sense is being encouraged. The Chinese approach seems to draw on existing strengths without worrying about overlap and the problems of theoretical communication and methodological inconsistency associated with interdisciplinary work as we know it.

The only Chinese Academy of Sciences (CAS) institute with an environmental designation is the Institute of Environmental Chemistry in Beijing. Other CAS institutes that sponsored research on environmental protection or control in the 1970's were the Institutes of Botany, Chemical Physics (Dalian), Entomology, Geochemistry, Geography, Geology, Meteorology, Microbiology, Plant Physiology, and Zoology. Some institutes, such as Geochemistry and Geography, have bureaus or departments specializing in environmental research. Environment-related work also takes place under the aegis of many institutes in the agricultural, engineering and technical, medical, and life sciences, and in the provincial-level environmental research institutes mentioned earlier. The Shenyang Research Institute of Industrial Hygiene, for example, has been conducting experiments since the mid-1960's to improve air quality in factories.[13]

A 1978 article in the new journal on environmental science implicitly distinguished between environmental protection and pollution control.[14] Environmental protection research assumes that man is only one element in the natural world, subordinate but not necessarily passive. It serves to illuminate the effects of landscape modification (e.g. water conservation, land reclamation, mining, afforestation) on man's health and well-being. Environment, in this research context, is roughly equated with nature. The main emphasis lies in trying to understand, predict, and manage the social, health, and related physical and biological effects of large-scale engineering projects, such as the planned diversions of Chang Jiang (Yangtze) water to the north or the alteration of natural habitats in northeastern and southwestern China in connection with mineral exploration and exploitation and hydropower generation.

Research on pollution control, on the other hand, aims at management of problems arising directly from human activities, such as elimination or neu-

[13] "Air Pollution Reduced in Industrial City of Shenyang," *Ta kung pao* (Hong Kong), Aug. 28, 1975, p. 6.

[14] Y. L. Tang, "Huanjing he huanjing kexue" (The Environment and Environmental Sciences), *Huanjing kexue* (Environmental Science), 1978, 1 (Feb.): 47–48.

tralization of harmful substances introduced into the environment. Studies focus primarily on the physical or chemical properties of a given substance (e.g. mercury, DDT) as it is transformed through environmental transport and modification processes or on the control processes (e.g. wastewater treatment, microbial degradation) that must be introduced in a given agricultural, urban, or industrial setting to meet occupational safety, public health, or multiple-use objectives.

In general, it seems that China's allocation of research support and personnel for environmental science research reflects far greater concern with the short-term effects of localized air and water pollution than with the less obvious but geographically more extensive degradation problems stemming from resource development. So far, analytical chemistry has provided the means of studying the origins, distribution, movement, transformation, and quantitative changes of specific pollutants.

In 1974 Yan identified five pollutant categories that should be studied: (1) industrial wastes such as sulfur dioxide, hydrogen, fluoride, heavy metals, cyanide wastes from electroplating, suspended materials from paper, fiber, and foodstuff manufacturing, and radioactive wastes; (2) oil and petroleum hydrocarbons; (3) toxic substances in pesticides, food additives, and other daily use and industrial items; (4) toxic products of naturally aged or transformed materials like rubber and plastics; and (5) bacterial and viral organisms in domestic sewage.[15] But as of 1979, research focused only on studies of natural cycles of groundwater, surveys of effluent quality, analyses of drinking water and human blood for trace-element contamination, and catalytic reactor development.

*Water Pollution*

The major focus of China's environmental effort in the 1970's was (and undoubtedly will continue to be) the maintenance and improvement of water quality. Most descriptions of water pollution control are industry-specific and illustrate the benefits of cooperation between laboratory researchers, engineers, technicians, and workers in simultaneous achievement of multiple-use, cleanup, and public health goals. Industries in which the greatest attention has been paid to pollution control are iron and steel, oil and petrochemical, metallurgical, chemical fertilizer, plastics, synthetic fiber, and paper and pulp.

One discussion, in 1975, of the development and conceptual basis for applied work in water and wastewater chemistry classified applied research concerns as follows: (1) forms of impurities in water and their conversions; (2) rational indices for water standards; (3) analytical and statistical methodology; (4) fundamental and special chemical and chemical engineering principles in water treatment; and (5) chemical characteristics of sewage

[15] H. Y. Yan, "On Environmental Protection and Chemistry," tr. Joint Publications Research Service (JPRS) from *Huaxue tongbao* (Chemistry Bulletin), 1974, 1 (Jan.): 4–8, *JPRS*, 62,812, Aug. 26, 1974.

and wastewater that enhance integrated control, recycling, and comprehensive utilization measures.[16]

Some of the most effective applications of research along these lines have been carried out since the early 1970's by the Dalian Institute of Chemical Physics in coking and dyestuffs plants. In treating phenolcyanide wastewater from a coking plant, techniques that removed hydrogen cyanide with blast furnace gas, followed by biological treatments, were employed. Similarly, in treating nitrochlorobenzene- and nitrophenol-containing wastewater from dyestuff processing, carbon absorption treatment was enhanced by recycling chlorobenzene.[17] The Haerbin Architectural Engineering College's Water Treatment Research Laboratory has also applied methods for recovery of waste acid and ferrous sulfate, and used electrolysis, evaporation concentration, and ionic exchange in treating wastewater from metal fabrication industries.[18] For new iron- and steel-manufacturing facilities, there is some evidence of control systems being imported along with the plants, since Japanese manufacturers have sold steel mill drainage and disposal systems to China. Until 1976, however, application of indigenous technologies had been the common practice.

The Daqing oil field in Heilongjiang province, China's largest, is a good example of the integrated development of pollution control and material-recycling facilities in a complex of extractive and fabricating industries. Since the opening of the oil field in 1960, conservation and treatment of domestic and industrial wastewater have been coordinated. By the mid-1970's, 16 primary and secondary treatment plants were in operation, and it was reported that in the period 1970–76, 300,000 tons of crude oil were recovered from refinery effluents.[19] Processes for treating sulfur-containing refinery effluents included oxidation, the use of copper chloride as a catalyst for low concentrations of sulfur, and steam stripping at high temperature and pressure, which permits recovery of hydrogen sulfide and ammonia.[20] Active carbon absorption purifiers have also been used to treat refinery effluents for recycling.[21]

Similar examples could be cited for the chemical, paper, and fertilizer industries. The most cost-effective and technically efficient examples of combined control and recycling shared two common features: application in industrial settings close to research institutes or academic research units and application in new plants or plants rebuilt in connection with the develop-

[16] H. H. Tang, "On Water and Wastewater Chemistry," tr. JPRS from *Huaxue tongbao*, 1975, 1 (Jan.): 26–30, *JPRS*, 65,466, Aug. 14, 1975.

[17] B. L. Zhu, "Research Work on Environmental Protection," in B. Boxer and D. Pramer, eds., *Environmental Protection in the People's Republic of China* (New Brunswick, N.J.: Rutgers University, 1978), pp. 12–20.

[18] F. X. Zhang and S. F. Zhou, "Treatment and Utilization of Metal Finishing Wastes Sewage," tr. JPRS from *Huaxue tongbao*, 1975, 5 (Sept.): 25–30, *JPRS*, 66,229, Nov. 26, 1975.

[19] "Taching Oilfield Protects Environment," *FBIS*, Mar. 31, 1976, pp. L1–2.

[20] M. C. Yao, "Treatment of Sulphur-Containing Refinery Sewage," *Huaxue tongbao*, 1974, 2 (Mar.): 21–25.

[21] "New Device for Treating Effluents," *PR*, 1977, 20 (May 20): 32.

ment of integrated urban-industrial-agricultural complexes. H. F. Yang's work at the CAS Institute of Microbiology in applying sophisticated techniques to treating polyacrylonitrile wastewater from synthetic fiber plants is a noteworthy example of fruitful cooperation between laboratories and technologically advanced industry.[22]

Useful and environmentally relevant research has also been carried out on problems of quality maintenance in underground water storage schemes and in the design of water purification plants capable of discharging and treating effluents for agricultural use.[23]

*Air Pollution and Waste Management*

In contrast to water pollution, air pollution received relatively little attention in China during the 1970's. And yet because of the predominant use of coal for domestic and industrial purposes, urban air pollution was severe. Given China's limited capability for widespread installation of modern air pollution control devices, the prospects for reduction of unacceptably high levels of sulfur dioxide, nitrogen dioxide, and particulates are poor.[24]

Although there are reports that some institutions, such as the Jiangsu Institute of Plant Research, have done some important work in this field, there is little evidence of any coordinated scientific and technical research relating to either the monitoring or the control of air pollution. Areas of interest in the late 1970's included instrumentation and procedures for determining trace toxicants, laser scanning to measure particulate concentration, analysis of flue gases from steel-smelting electric arc furnaces, gravimetric analysis, and chemical member filter measurement. On the whole, however, the Chinese appeared to be relying (quite sensibly) on air pollution technology available in the West.

Many scientific accounts of materials recovery research and environmental improvement have emphasized the economic and health rewards of cooperative efforts by scientists, technicians, and workers to conserve and reuse industrial raw materials and to control pollution. In support of this, traditional techniques for composting organic and recycling inorganic materials have been improved. Since 1949, however, in the larger setting of environmental protection, this aspect of environmental sanitation (along with industrial hygiene) most commonly falls under public health councils and not under environmental science institutions.

[22] H. F. Yang, "The Application of Acrylonitrile-Oxidizing Microorganisms for Polyacrylonitrile Wastewater Treatment," in Boxer and Pramer, eds., *Environmental Protection*, pp. 71–84.

[23] "Chinese Scientists Find Way to Store Underground Water," *World Environment Report*, May 8, 1978, p. 7; and "Gravity Flow-Type Purification Plant Designed and Built," tr. JPRS, *People's Republic of China Scientific Abstracts, JPRS*, 61,933, May 8, 1974.

[24] Z. Z. Zhang, "Application of Ring-Oven Technique in Air Pollution Research," *Huaxue tongbao*, 1974, 2 (Mar.): 26–30.

*Integrated Environmental Management*

Most of China's environmental research is done in conjunction with specific, economically important projects. Two excellent examples of integrated environmental research, assessment, and management are associated with Guanting Reservoir and Yaer Lake.

Guanting Reservoir, located in an arid region 100 km northwest of Beijing, generates hydropower and supplies water and fish to the capital. Development of industries—metallurgical, mining, chemical, paper, tanning, and synthetic fiber—in the river valleys above the reservoir led to serious pollution. Under the combined authority of the reservoir administration and industrial environmental protection units, steps were taken to solve this problem. One report described the organization of a cooperative project, under way since 1972 and involving physical geographers, environmental chemists, and biologists, to describe and model the effects of pollutants on biological communities and ecosystems in the reservoir and to study the processes that govern links between pollutant sources and receptors.[25] One aspect of this work was an analysis of the form, content, distribution, and chemical transformation of toxic heavy metals in reservoir sediments and their biogeochemical migration in the water-sediment-aquatic organism system.[26] Overall research and management achievements in the Guanting Reservoir area provide evidence of successful integration of agricultural and industrial production and environmental protection.

Integrated pollution control in a major fowl-breeding, aquaculture, and aquatic products area, Yaer Lake in Hubei province, began in 1976 under a plan drawn up at China's leading center for research on fresh-water pollution, the Institute of Hydrobiology (also referred to as the Institute of Aquabiology) in Wuhan. Here the problem was to restore a major inland fishery that had been severely polluted since the late 1950's by effluents from pesticide and chemical plants built along its shores. Engineering work involved installation of a network of pipes to drain polluted lake waters into four oxidation ponds that were formed by damming off a 200-ha area of the lake. Treatment was then carried out through bacterial degradation and removal by algae. The Institute of Hydrobiology carried out wide-ranging experiments on the use of plant materials for removal of toxic chemicals from industrial effluents and the breakdown of chlorinated hydrocarbon pesticides from agricultural runoff. It also illustrated the complementarity of pollution control and economic productivity by harvesting *Phomidium ambiguum,* a blue-green alga used to remove ammonia nitrogen from acrylonitrile wastewater for use as a fertilizer.[27]

[25] J. H. Wang, "Guanting shuiku shuiyuan baohu de yanjiu" (Preservation of Water Sources at Guanting Reservoir), *Dili zhishi* (Geographical Knowledge), 1978, 4 (Apr.): 13.
[26] *Ibid.*
[27] D. M. Wang, "Experiment on the Removal of Ammonia Nitrogen by Algae," in Boxer and Pramer, eds., *Environmental Protection,* pp. 66–70.

*

The Chinese government's reaffirmation of the importance of environmental protection in 1978–79 and the publication of a new environmental science journal and environmental protection law are hopeful signs that environmental science will receive greater support, in keeping with the national effort to improve science and technology. However, documentary analysis and discussions with scientists and businessmen interested in pollution control lead to less sanguine conclusions for several reasons.

Chinese environmental research is only indirectly responsive to abatement goals and regulatory objectives. Because of the organizational structure and priority-setting mechanism of research units and government bodies, scientific and technical work are unlikely to serve national environmental protection needs in the immediate future. There obviously will be local exceptions to this, and some research areas, such as environmental chemistry, environmental geology, and aquatic biology, could make significant contributions to the development of industry-specific control technologies and in environmental chemical analysis. However, in the late 1970's, China was just beginning to think about environmental standards and regulations. Evidence of bureaucratic lethargy and mismanagement in implementing existing research and enforcement programs abounded. It was also questionable whether China's central government planners wish to include environmental factors in development planning decisions at high management levels.

The problem derives partially from funding priorities. From 1976 through 1978, China committed itself to billions of dollars worth of plant and technology purchases from foreign suppliers. These capital purchases will hasten the growth of the iron and steel industry, mining, telecommunications, oil production, energy development, and transportation. Although pollution control is of great concern to Chinese authorities, expenditures on indigenous research and development and on direct imports of control devices are likely to be far lower than needed. There is little reason to expect that this will change under the new policy for readjusting and restructuring the economy announced in 1979.

The direction followed by China in modifying its domestic and international environmental policies in line with its modernization drive will be closely watched by the global community. We have seen that despite its domestic environmental problems, in bilateral relations and in the style of participation in international organizations, China has done much to enhance its national image. By sharing its experience in disease control, recycling, energy conservation, marginal land management, and small-scale water use, China has gained the admiration of many developing countries and foreign advocates of conservation and soft technologies. Because of its size, population, and physical diversity, China's emerging environmental policies will have a major effect on growing international efforts to find a viable basis for environmental cooperation in the interest of human survival.

# *SOCIAL SCIENCES*

## *INTRODUCTION*

*Harry Harding*

Western social science arrived in China surprisingly early. Translations of Western books on sociology, economics, and politics began appearing in China around the turn of the century, followed by a second wave of translated Marxist works in the 1920's.[1] Anthropology was added to the Qing curriculum of higher education in 1903, and many Western-style universities established departments of economics, political science, and sociology. Chinese scholars began writing textbooks to introduce Western social science to their colleagues and students. These activities merely supplemented, of course, an indigenous tradition of historical scholarship that was centuries old.

Despite these early beginnings, the development of social science in China was uneven. Before 1949 sociology, Chinese history, and economics developed most rapidly. With sociology departments in missionary colleges leading the way, Chinese social scientists developed an interest in social survey techniques and used both statistical and anthropological methods to study village life, labor problems, urban conditions, and demographic trends.

[1] Much of the information in this article was obtained during a nine-day visit to the Chinese Academy of Social Sciences in December 1978. I would like to express my gratitude to John Lewis and Douglas Murray for arranging my visit; to Mr. Tang Kai, then director of the Foreign Affairs Bureau of the Academy, for organizing my stay in Beijing; and to my hosts, Liang Wensen of the Institute of Economics and Yang Aiwen of the Academy, for their hospitality, friendship, and assistance.

After the arrival of Marxist scholarship in China in the 1920's and 1930's, historians began to apply Marxist theories of social development to China and attempted to create a Marxist periodization of Chinese history. Economists produced important studies of agricultural and industrial development in China and histories of capitalism and the Chinese land system.

Other disciplines, in contrast, developed much more slowly. As Albert Feuerwerker points out elsewhere in this section, interest in Chinese history was not paralleled by any significant scholarship on the history of foreign countries. And while a number of universities established political science departments, the study of politics in China did not develop as rapidly as sociology or even economics.[2]

The social sciences underwent sweeping reorganization in the early 1950's as part of the curricular reforms undertaken by the new Communist government. Sociology was criticized for its alleged failure to understand the crisis of the old society, and political science was described as the attempt of Westernized intellectuals to promote "bourgeois" political theory. Both disciplines were effectively abolished by 1952. Anthropology, which had previously emphasized the study of Chinese peasant life, was redefined so as to concentrate on non-Han minority peoples. In place of these discredited fields, the new regime promoted history, philosophy, and economics, and to a lesser extent, law, international relations, and nationalities studies. In each case, Chinese scholarship was heavily influenced not only by Marxist theory, but also by contemporary Soviet research.

These trends were criticized, albeit futilely, during the Hundred Flowers period of 1956–57. Some of China's most prominent sociologists, among them Fei Xiaotong, tried to revive sociology as an academic discipline and proposed that the Chinese Academy of Sciences establish a committee for research on domestic social problems. Qian Duansheng, China's best-known political scientist, advocated the creation of an institute of law modeled after the London School of Economics, to conduct teaching and research on politics, sociology, economics, and jurisprudence. Others called for the resumption of the study of Western social science and criticized the domination of Chinese social science by Marxist theories. When the Hundred Flowers period ended in mid-1957, however, these proposals were rejected, and their proponents denounced as "bourgeois rightists." Some, like Fei Xiaotong, found themselves under a cloud of suspicion that was not removed for 20 years.[3]

[2]For the historical development of the social sciences in China, see Morton H. Fried, "China," in Joseph S. Roucek, ed., *Contemporary Sociology* (New York: Philosophical Library, 1958), pp. 993–1012; Leonard Shih-lien Hsü, "The Sociological Movement in China," *Pacific Affairs,* 4, 4 (Apr. 1931): 283–307; Sun Pen-wen, "Sociology in China," *Social Forces,* 27, 3 (Mar. 1949): 247–51; and Wang Yü-ch'üan, "The Development of Modern Social Science in China," *Pacific Affairs,* 11, 3 (Sept. 1938): 345–62.

[3]For social science in China in the 1950's, see C. J. Ch'en, "Chinese Social Scientists," *Twentieth Century* (London), 163, 976 (June 1958): 511–22; Maurice Freedman, "Sociology in and of China," *British Journal of Sociology,* 13, 2 (June 1962): 106–16; and G. William Skinner, "The New Sociology in China," *Far Eastern Quarterly,* 10, 4 (Aug. 1951): 365–71.

By 1961, therefore, an American survey of Chinese science—the predecessor of this volume—could decide to exclude most of Chinese social science from its coverage on the grounds that there was little of academic merit to describe: "Social science research in an academic sense has become unimportant and been restricted to supporting the propaganda activities of the regime, or to working in a few applied fields, such as developing educational methods. . . . Clearly, social scientists in China cannot be expected to make more than the most meager contributions in theory and application to their disciplines."[4] In retrospect, this harsh description appears to have been overstated, particularly for history and perhaps for linguistics and economics as well. But it had become an understatement by the height of the Cultural Revolution. By 1967 social science came to an abrupt halt in both the universities and the research institutes attached to the Academy of Sciences. Scholars were sent to May 7th cadre schools for "reeducation," were placed under detention, or were arrested. Subscriptions to foreign periodicals were terminated. Research materials were lost, dispersed, or destroyed, and private scholarly collections were confiscated.

When Chinese universities reopened in the early 1970's, some social science research resumed. But the research was to a great extent in the service of political leaders engaged in the struggle for succession to Mao Zedong. Political economists developed theories to prove that the policies advocated by Deng Xiaoping would produce a new bourgeoisie under socialism. Historians produced studies that, in allegorical terms, criticized prevailing policies toward the Soviet Union and the emerging tendency to import advanced technology from the West. They also sought to show that the contemporary struggle between "revolutionaries" and "revisionists" in China was the latest stage in a 2,000-year-old struggle between a "progressive" legalist tendency and a "retrogressive" Confucianist line. Philosophers compiled the *Concise Philosophical Dictionary,* which, it was later charged, overemphasized the role of human will in social and technological change, exaggerated the degree of class struggle in socialist society, and overstated the impact of the political superstructure on the economic base. To be sure, not all scholarship was put to political ends. As Albert Feuerwerker reminds us, historians did produce some new textbooks on modern Chinese history and world history during this period. But most social science in the early 1970's appears to have been an attempt by radical political leaders to create a theoretical justification for their social, economic, and organizational policies.

With the fall of the Gang of Four in October 1976, a renaissance in social science began. An independent Chinese Academy of Social Sciences (CASS) was established in May 1977. An important essay by Guo Mojuo, published in March 1978, called for more research by social scientists on various public policy issues.[5] In November 1978 the *Renmin ribao (People's Daily)* in-

[4] Sidney H. Gould, ed., *Sciences in Communist China* (Washington, D.C.: American Association for the Advancement of Science, 1961), pp. 49–50.

[5] New China News Agency dispatch, Mar. 11, 1978.

troduced the hitherto unorthodox idea that the "laws" of social science, like those of natural sciences, were "independent of the desires or standpoint of any individual or class."[6] Graduate study in the social sciences was resumed, as were limited exchanges with Western social scientists. Academic ranks were reestablished. And Chinese leaders announced plans to train a corps of some 200,000 social scientists by 1985.

These recent developments make it highly appropriate that this survey of science and technology include the social sciences. Moreover, the expansion of academic exchanges between China and the West makes it imperative that Western social scientists understand the activities of their Chinese counterparts. As the contributions to this section indicate, Chinese social science in the late 1970's remained constrained by outdated research methods, ignorance of scholarly developments in the West, and above all the continuing tendency to produce exegeses or justifications of established Party policy. But there were signs that Chinese social scientists were beginning to assume influential roles as policy analysts and to explore certain policy questions more independently than in the past.

The most important center for social science research in China is the CASS, located in Beijing. Although it became an independent organization only in 1977, the Academy has a relatively long history. It was originally the Department of Philosophy and Social Sciences, one of five research departments organized under the Academy of Sciences in the early 1950's. Under the Department some 15 research institutes were gradually established, encompassing the humanities as well as the social sciences. The most important of the social science institutes were the Institute of Economics, the branch Institute of Economics in Nanjing, the three Institutes of History (of ancient, medieval, and modern Chinese history), the Institute of International Relations, and the Institute of World Economics.

During the Cultural Revolution, virtually all research activities in the Department halted. The only exceptions appear to have been the Institute of Archaeology, which continued investigations of finds uncovered during the Cultural Revolution, and the Institutes of History, which continued work on Fan Wenlan's general history of China. Otherwise, the Department's researchers were sent to May 7th cadre schools for "reeducation," and its buildings were reassigned to other uses. The Chinese now charge that the Gang of Four planned to disband the Department of Philosophy and Social Sciences, but that this proposal was vetoed by Premier Zhou Enlai.

As a result of Zhou's intervention, the Department's research staff were recalled from the May 7th cadre schools in 1972, but not, however, to resume their research. Chinese social scientists say that they did little but engage in political study and participate in the various political movements launched between 1972 and the fall of the Gang of Four in October 1976.

[6] "Does Truth Have a Class Character?" *Renmin ribao,* Nov. 28, 1978, p. 3.

In May 1977 the Department of Philosophy and Social Sciences was removed from the jurisdiction of the Academy of Sciences, renamed the Academy of Social Sciences, and placed under the veteran Party historian and propagandist Hu Qiaomu. Research work gradually resumed in 1977, but by the end of 1978 some serious problems still remained. The CASS library was still integrated with that of the Academy of Sciences; the libraries of the various research institutes were still in serious disarray; and not all research materials had been recovered after the chaos of the Cultural Revolution. Some institutes did not have adequate office space or in some cases any offices at all. And exchanges with foreign scholars and universities had just resumed.

But the basic organizational structure of the CASS was established. As of December 1978, its central administration consisted of the Political Office (personnel and ideological training), and the Bureaus of Research Organization (coordination of research activities within the Academy), Research Planning and Liaison (long-term planning and liaison with other research institutions), Foreign Affairs (international exchanges), and Administration (finances and facilities). The CASS also administered the Postgraduate Institute and the Social Science Publishing House, which published some (although not all) of the journals and books written or edited at the Academy's research institutes.

The heart of the CASS is, of course, its research institutes. As of early 1979, the Academy had 20 such organizations. One of these, the Institute of Scientific Information, collected research materials and data and developed bibliographic and information-processing techniques, while the other 19 institutes were rather evenly divided among economics, social sciences, history, and humanities (see Appendix B).

All together, the Academy had a staff of some 1,500 to 2,000 people, some two-thirds of whom were described as researchers.

By the end of 1978, the Academy had developed ambitious plans for the future. These included, first of all, plans for more staff. By retaining virtually all students enrolled in the Postgraduate Institute and presumably by hiring researchers from graduate programs in universities, the Academy planned to increase its staff to 10,000 (not all of whom would be researchers) by 1980. These new researchers were, in turn, to be divided among more research institutes, which would be responsible for contemporary Chinese history (i.e. history since 1919), foreign-area studies, economic management, Marxism-Leninism, Mao Zedong's thought, and domestic social problems. All told, there may be as many as 60 research institutes under the CASS by 1985. The expansion of the Academy will require, of course, more buildings. In 1978, the CASS was constructing a 24-story building near its current headquarters in Beijing, principally to house the research institutes in the humanities. Other institutes may be moved into larger quarters. The Academy also planned to issue more publications, including more academic journals, scholarly dictionaries and glossaries, textbooks on Chinese and world history, and a social science encyclopedia. Finally, there were also plans for library development, including the separation of the CASS library from that of the Academy of Sciences and the expansion of the individual research libraries attached to each institute.

Research in the social sciences is also conducted under a variety of other auspices. First, many universities have established social science research institutes. Most of these appear to specialize in various aspects of economics or history, but there are a number involved in foreign-area studies. These include the Institute of Social Sciences at the University of Heilongjiang (Haerbin), which reportedly specializes in Soviet studies; institutes on Japan and Korea at Jilin University (Changchun); an Institute of Asian and African Studies at Beijing University; and an Institute of American Studies at Nankai University in Tianjin. The Chinese People's University in Beijing probably has more social science research institutes than any other university. After reopening in 1978, it established institutes on Soviet and Eastern European affairs, economic management in foreign countries, population theory, Qing history, the history and development of Marxism-Leninism, and Chinese literature and language reform.

Second, a number of government and Party agencies have established research institutes in the social sciences. These include the Institute of International Problems of the Ministry of Foreign Affairs, the Institute of Economics of the State Planning Commission, and the Institute of Foreign Trade of the Ministry of Foreign Trade. Before the Cultural Revolution, the Party Central Committee had several research offices that conducted studies of various policy problems, and some of these have probably been reestablished. The Higher Party School, in addition to offering training classes for Party officials and managers, conducts research on the history of the Chinese Communist Party.

Finally, under plans drawn up in 1978, each province was eventually to establish its own social science institute or, if more human and material resources were available, a full-fledged social science academy. The research work undertaken at existing provincial institutions is similar to that done at the CASS, but often has a local connection. Research in applied economics, for example, appears to focus on policy problems of particular interest to provincial government and Party officials. And historians at provincial research institutes often engage in studies of local history, such as the work on the peasant movements of the 1920's being undertaken by the Guangdong Provincial Institute of Philosophy and Social Science. Provincial institutions also conduct some joint projects with each other. A prominent example is a dictionary of economic terms that is reportedly being compiled by the Guangdong Institute in cooperation with comparable institutes in several other provinces.[7]

[7]Brantly Womack, "Report on a Visit to the Guangdong Provincial Philosophy and Social Science Research Institute."

These three types of institution—the university research institutes, the research institutes under Party and government agencies, and the provincial social science academies—are formally independent of the CASS. But it appears that their research will be coordinated by a single set of national plans, which the CASS will play an important role in formulating.

Graduate training in the social sciences resumed in 1978. Graduate students are enrolled both in the Postgraduate Institute of the CASS and in the teaching departments and research institutes of major universities.

The Postgraduate Institute, established in October 1978, is divided into 12 academic departments: philosophy, economic theory, practical economics, world economics, Chinese literature, foreign literature, history, law, language, nationalities studies, journalism, and religion. This division closely parallels that of the CASS research institutes. Each academic department is further divided into several specialties, again corresponding roughly to the departments within the Academy's research institutes.

The first class of 405 graduate students was admitted to the Postgraduate Institute through a two-stage examination process. First, there was a national examination for all prospective graduate students, including those who wished to apply to programs other than those offered by the Postgraduate Institute. This examination tested the applicants' basic knowledge of Chinese and foreign history, Chinese language, political theory, and foreign languages. Any applicant who passed this examination, received satisfactory letters of recommendation, and was in adequate physical condition was then given a second examination in his chosen field by the department to which he had applied.

Students admitted to the Postgraduate Institute enroll in a three-year curriculum. During their first year, they are given supplemental basic courses in Marxist philosophy, economics, scientific socialism, and two foreign languages. They may also begin course work in their specialities. During the second year, students enroll in advanced courses offered by the Institute's academic departments. These courses are primarily tutorial in format and are organized around reading reports prepared by each student. Faculty for these courses are drawn mainly from the research staff of the CASS, with occasional lecturers from universities or other research institutions. Finally, during their third year, the students are expected to write dissertations. Although the Postgraduate Institute does not yet award formal degrees, its graduate program appears comparable to a more specialized and abbreviated version of an American Ph.D.

Most of China's major universities have also begun to offer graduate training in the social sciences. At Beijing Normal University, for example, the department of political education planned to enroll about 400 graduate students by the early 1980's, most of whom would become faculty members in other higher-level normal schools. At Chinese People's University, there were some 100 graduate students in the social sciences at the end of 1978, and graduate enrollments were expected to rise to 3,000–4,000 by the early 1980's.

The curriculum and admission procedures at the universities are virtually identical to those of the Postgraduate Institute. Students are screened by the same national examination and then given a supplementary examination by the institution to which they have applied. Graduate training involves a three-year curriculum, culminating in a dissertation. Training includes a combination of advanced tutorials in the students' specialities, basic graduate courses in philosophy and economics, and foreign languages.

As a general rule, graduate students in China are trained to do the same work as their professors. Most graduate students in the Postgraduate Institute will be assigned to the CASS research staff after completing their training. (One exception is the Department of Journalism, most of whose graduates will be assigned to the New China News Agency.) Similarly, those who receive graduate training in the research institutes attached to major universities usually work as academic researchers. Graduate students in normal universities become professors of education, and those trained in the teaching departments of comprehensive universities become university teachers. In this sense, graduate training in China is divided into a number of separate tracks, which mirror the compartmentalization of the academic community as a whole.

Undergraduate training in the social sciences is difficult to summarize, largely because as in the United States, there is such wide variation between universities. In particular, there are important differences between the comprehensive universities, which offer courses in virtually all the humanities and sciences, and the normal universities, which primarily train teachers for Chinese middle schools.

Almost all comprehensive universities offer majors in the social sciences, but their departmental structures vary widely. Most universities have departments of history, economics, and philosophy. Majors in political economy may be trained in economics departments (as in Beijing and Zhongshan [Guangzhou] Universities) or in separate departments of politics (as at Nanjing University) or political economy (as at Fudan University [Shanghai]). Some universities have departments of international relations, which offer courses not only on international affairs, but also on comparative politics and political philosophy. There are few, if any, departments of anthropology or sociology.

The courses in these departments are usually restricted to students majoring in them. But most comprehensive universities also have courses in the social sciences that are required of all students, regardless of major. Usually there are three: philosophy (including dialectical and historical materialism), political economy, and Party history. These required courses are usually offered by special nondepartmental "teaching groups." The Ministry of Education has commissioned standard textbooks for these three subjects, as well as a fourth text on the international Communist movement.

Compared with the comprehensive universities, China's normal universities have a slightly different structure for undergraduate training in the social sciences. Many normal universities have a single large social science department that trains students to become political instructors in high schools. For example, the 350 students in Beijing Normal University's Department of Political Education take a wide range of courses on such subjects as Party history, philosophy, the international Communist movement, political economy, scientific socialism, economic history, formal logic, and the pedagogy of political education. Because students do not major in a single speciality, their training appears to be broader than that received by their counterparts in comprehensive universities.

Finally, at least one major university in China specializes in the social sciences: the Chinese People's University. Because it trains students for careers as ideological theoreticians, managers, Party and state officials, and academicians, the People's University offers a more comprehensive curriculum in the social sciences than other undergraduate institutions. As of late 1978, it had 12 social science departments: philosophy, political economy, industrial economics, agricultural economics, finance economics, trade economics, journalism, law, the history of China, the history of the Chinese Communist Party, scientific socialism, and economic information processing. Each of these was subdivided into a number of specialities. The People's University is formally under the jurisdiction of the Ministry of Education, but there is said to be "close coordination" between it and the CASS.

The quality of undergraduate course work offered in Chinese universities is difficult to gauge. Available information, however, is not reassuring. The sole reading assignment of students taking a course on the theory of the three worlds offered by Beijing University's Department of International Relations in late 1978, for example, was the official exegesis of the theory in the *People's Daily.* Students majoring in political economy at Zhongshan University in 1976 read only Marx and Lenin when they studied capitalism and only Lenin and Mao when they studied socialism. (They did read some Smith, Ricardo, and Keynes when they studied economic theory.) New teaching materials were said to be under preparation, however, and students were spending more time in the classroom (and less time in open-door education outside school) than they did before the purge of the Gang of Four.

Almost all social science research in China—whether conducted at the CASS, at provincial social sciences academies, or at university research institutes—is coordinated by a national planning effort that began in 1978. When completed, the national plan will have several components. First, there will be a general plan for the social sciences, extending to 1985. Second, there will be more specific plans for each of 22 disciplines and subdisciplines. Each of these disciplinary plans will identify general theoretical questions, historical topics, and practical policy problems that require research and assign them to research institutes and individual scholars. Third, the national plan will contain a section on library development. In addition, there will be a plan for what the Chinese call the "science of science": the study of the boundaries separating various disciplines and subdisciplines and the development of research methodologies.

As of the beginning of 1979, the planning and coordination of economic research had proceeded furthest.[8] The CASS sponsored six regional meetings in 1978, one in each of China's economic planning regions, to formulate plans for economic research. These were followed by a central forum to combine the regional efforts into a single national plan. Each meeting was attended by representatives of Party propaganda departments, government planning and administrative agencies, and universities and research institutes. The Party propaganda officials apparently were present to help identify policies for which economists could contribute a sophisticated theoretical justification. Government officials helped select specific policy problems on which academic research could be useful. The representatives of the universities and research institutions described the resources and interests of their institutions and their staffs.

The regional and central meetings reportedly drew up a three-year plan (1978–80) and an eight-year plan (1978–85) for economic research, which contained 480 research projects. The principal projects in general economic theory included studies of the political economy of socialism, imperialism, and social imperialism. In economic history, the projects included studies of the development of capitalism in China, the history of Chinese economic thought, Chinese economic history since 1919, and the history of foreign economic theory. The plans also identified a large number of practical policy problems, including economic management, forestry, agriculture, the motor vehicle industry, urban economics, conservation, tourism, and energy. Although individual researchers may pursue their own scholarly interests outside the plan, such work does not appear to be highly valued or especially common.

Historical studies aside, social science research in China falls into two principal categories: research on general theory and applied research on practical policy problems. Much of the research on general theory amounts to an exegesis of current Party policy. Indeed, many such research projects are initiated by Party propaganda officials seeking sophisticated explanations and justifications of policies that intellectuals, Party cadres, and government officials will find persuasive. Several prominent examples of such research in 1977 and 1978 can be identified. One is the discussion of Mao Zedong's 1956 speech "On the Ten Major Relationships" broadcast over Radio Beijing in early 1977. This commentary, written by the CASS Institute of Economics, laid the foundation for a number of important Party policies that year, particularly the decisions to resume gradual moderniza-

[8]For the planning process in economics, see *Guangming ribao,* May 22, 1978; in *Foreign Broadcast Information Service Daily Report: People's Republic of China (FBIS),* May 31, 1978, pp. E14–15.

tion of the armed forces and to increase industrial wages.[9] A second example is the exegesis of the theory of the three worlds published in the *People's Daily* in November 1977. Although it appeared under the by-line of the newspaper's editorial department, this article on the theoretical foundations of Chinese foreign policy was actually prepared by a CASS group led by Hu Qiaomu.[10] A third example is the study of democracy and class struggle conducted by the Institute of Law in 1978, which helped to justify the political liberalization undertaken at the end of that year. Other research of this type includes articles on "practice as the criterion of truth" prepared at the Institute of Philosophy and work on "distribution according to labor" conducted by the Institute of Economics.

This research tends to be deductive and conceptual in nature. While the articles embodying the research may represent slightly different viewpoints, they consist almost entirely of commentaries on writings of Marx, Lenin, Engels, and Mao. Only rarely is empirical evidence presented to confirm or refute propositions derived deductively from doctrine.

The second category of social science research consists of projects on practical policy problems. This may involve research on the situation in China and recommendations for change and reform. In such cases, researchers may be given access to government data or to particular basic-level units (factories, offices, enterprises, etc.). Occasionally, the research may be conducted jointly by social scientists from a research institute and officials or research personnel attached to the relevant government agency. Examples of this kind of domestic research include the studies of economic law and contracts, the electoral system, and the civil and criminal codes undertaken by the Institute of Law, and the research on bonuses, interest, and profits conducted by the Institute of Economics.

Sometimes the study of practical problems involves research on foreign institutions and procedures. In late 1977 and early 1978 Yugoslavia and Romania received particular attention. Since then, however, Chinese social scientists have also studied the United States, Japan, Western Europe, and even the Soviet Union. Examples of comparative research include the study of development strategies undertaken by the CASS Institute of World Economics and the research on foreign management techniques by the Institute of Industrial Economics.

There is increasing evidence that research on specific policy problems is taken seriously by government and Party officials. According to CASS administrators, some government agencies hold occasional meetings on current policy problems at which they seek opinions from the scholarly community. One example of this is an important conference on economic theory held in late 1978, which involved an active exchange of views between offi-

[9] CASS, Institute of Economics, *Guanghui de wenxian, qiangda de wuqi* (A Brilliant Document, a Powerful Weapon) (Beijing: Renmin Chubanshe, 1977).

[10] "Chairman Mao's Theory of the Differentiation of the Three Worlds Is a Major Contribution to Marxism-Leninism," *Peking Review (PR)*, 1977, 45 (Nov. 4): 10–41.

cials responsible for rural policy and scholars studying agricultural economics. The scholars apparently criticized tendencies toward "egalitarianism" in rural work, pointed out the shortcomings in the Dazhai model of remuneration according to political attitude, and called for greater use of material incentives in agriculture.[11] A second and even more prominent example of scholarly participation in policymaking is a report on the national economy presented to the State Council in July 1978 by Hu Qiaomu, president of the CASS. Hu's report, later published in the *People's Daily,* was a hard-hitting criticism of Party and government officials for letting administrative whim, rather than economic laws, govern economic activities.[12]

No doubt there are significant constraints on the ability of scholars to criticize prevailing government and Party policy. Indeed, many scholars may still be reluctant to express their views unless they can be reasonably sure that their opinions will receive a sympathetic hearing from an important segment of officialdom. But the tendency for researchers to express themselves on important issues and for government and Party officials to seek out academic opinion and suggestions is growing.

The results of social science research are currently disseminated through three channels. One is the scholarly conference. Unlike meetings of professional associations in the United States, scholarly conferences in China tend not to be disciplinary-wide affairs, but rather meetings that concentrate on particular historical, theoretical, or policy problems. In principle, any university, research institute, or professional society has the ability to sponsor a national conference, although authorization from the CASS or other government or Party agency may be required.

Various scholarly conferences were held in 1978. Among the most important were meetings on the principle of distribution according to labor, the periodization of Chinese history, the role of Confucius in Chinese history, the Taiping Rebellion, the philosophical thought of Mao Zedong, the origins of a new "bourgeois" class in contemporary China, practice as the criterion of truth, democracy and the socialist legal system, and general economic theory.

Professional societies and associations constitute a second channel for the communication of scholarly research. Inactive during the decade of the Cultural Revolution, professional societies will reportedly be reestablished for each major discipline, drawing members from university faculties, research staffs, and Party and government officials trained in the social sciences. As of early 1979, at least three such societies had been organized: a Chinese Association of Enterprise Management, a Chinese Society of Sociology

[11] New China News Agency dispatch, Nov. 30, 1978.

[12] For an English translation, see "Observe Economic Laws, Speed Up the Four Modernizations," *PR,* 1978, 45 (Nov. 10): 7–12; 1978, 46 (Nov. 17): 15–23; and 1978, 47 (Nov. 24): 13–21.

(headed by Fei Xiaotong), and a Chinese Futures Research Society.[13] These disciplinary societies will be coordinated at the provincial level by broader organizations, usually called associations of social science societies or associations of philosophical and social science societies. At the national level, some of the professional societies appear to be linked to the Science and Technology Association.

Scholarly journals are another channel for the dissemination of research results. Of these, the most prominent are those edited at the various institutes of the CASS, including:

*Zhexue yanjiu* (Philosophical Studies)
*Jingji yanjiu (Economic Studies)*
*Lishi yanjiu (Historical Studies)*
*Jingji guanli (Economic Management)*
*Shijie jingji* (World Economy)
*Zhongguo yuwen (Chinese Language)*
*Shijie lishi* (World History)
*Wenxue pinglun* (Literary Criticism)
*Shijie wenxue* (World Literature)
*Zhongguoshi yanjiu* (Studies in Chinese History)
*Lishi xue* (The Study of History)
*Minzu yuwen (Minority Languages)*
*Fangyan (Dialects)*
*Faxue yanjiu (Studies in Law)*

Some CASS institutes also publish journals of translations from foreign scholarly work, and the CASS was considering the publication of its own general social science journal, possibly entitled *Zhongguo shehui kexue* (Chinese Social Science).

Other institutions independent of the CASS also publish scholarly journals. Most major universities publish academic journals, and many of these have separate social science editions. Examples are the journals of Beijing University, Beijing Normal University, Jilin Normal University (Changchun), Liaoning University (Shenyang), Zhongshan University, Gansu Normal University (Lanzhou), Nanjing University, and Nankai University. In addition, some of China's provincial social science academies and professional organizations publish journals, such as the quarterly *Shehui kexue zhanxian* (Social Science Front), published by the Jilin Provincial Association of Philosophical and Social Science Societies. Before the Cultural Revolution, professional societies edited some of China's most important journals in the social sciences (including *Zhengfa yanjiu* [Research in Politics and Law]), and it is likely that they will resume this function in the 1980's. The Chinese Futures Research Society, for example, has already announced

[13] *PR*, 1978, 12 (Mar. 24): 7–8; *PR*, 1979, 13 (Mar. 30): 29–30; and *Renmin ribao*, Mar. 13, 1979, p. 3.

plans to begin publication of a journal called *Weilai yu fazhan* (Future and Development).

Not all Chinese scholarly journals are circulated publicly. *Sulian wenti ziliao* (Materials on Soviet Questions), published by the Institute of Soviet and Eastern European Studies of the Chinese People's University, is distributed only to a select group of scholars and officials concerned with foreign affairs.

The state of Chinese social science brings to mind the classic metaphor of the half glass of water: Should it be considered half full or half empty? Should we emphasize the progress that social science has made in China since 1976, or should we stress the constraints on its further development?

Chinese social science certainly remains plagued by serious shortcomings, not all of which are political. Some involve shortages of human and material resources—inadequate office space, poor library facilities, inability to use computers to process research data, and, above all, the absence of an entire generation of younger scholars who were denied academic training between 1966 and 1976. Other constraints are intellectual—the low level of general education even among the best graduate students, the ignorance of developments in Western social science, and inadequate knowledge of relevant statistical techniques and research methodologies. Morale may be a problem, too, for promotions to the higher ranks of research staffs and the professoriate are extremely slow in coming.[14]

There are also institutional and structural constraints. One of the most important is the compartmentalization of Chinese social science and the resultant inbreeding. Because Chinese educational and scientific institutions are still patterned after the Soviet model, the line between academic research and university teaching is drawn much more rigidly in China than in the United States. Few professors do research beyond that required for the preparation of lectures or the compilation of course materials. Similarly, few researchers do much teaching, except to participate in graduate training programs. This barrier between teaching and research even separates teaching departments and research institutes within a university. Teaching and research staffs are given different titles, work in separate offices, and apparently have little regular scholarly interaction.

This compartmentalization further affects the career patterns of Chinese social scientists. Graduate students trained in a university teaching department can expect to become university professors, usually at the same institution at which they did their graduate work. They have little opportunity to move, even for a year or two, to a research institute to conduct scholarly research. Similarly, a graduate student trained as a researcher will probably

[14] These intellectual constraints are emphasized in Chou Fan, "Chung-kung she-hui k'o-hsüeh yen-chiu ti hsien-chuang chi ch'ien-ching" (The Present Situation and the Future Prospects of Social Science Research in Communist China), *Ch'i-shih nien-tai* (Hong Kong), 107 (Dec. 1978): 42–50.

be given little opportunity to teach undergraduates, even on a temporary basis. Chinese university administrators say that they are aware of the problems created by this rigid compartmentalization, but they have thus far announced few reforms that might improve the situation.

This list of human, material, and structural constraints does not mean that political limits on scholarship are no longer a problem. Most research is still linked directly to the need of the Party and government for persuasive exegeses of prevailing policy or for analyses of pressing policy problems. Although some of this work has shown more independence and has even questioned some basic assumptions, for the most part Chinese social scientists have not yet developed the ability to conduct autonomous scholarly research. They have not yet reached the crucial point, identified by one Russian political scientist in discussing research in his own country, where they can "raise questions whose answers are not known beforehand."[15]

On the other hand, there has been progress in Chinese social science during the 1970's. Quantitatively, the numbers of scholarly journals, academic conferences, research institutes, graduate students, and international exchanges have increased significantly. Qualitatively, there have been important improvements. Published research includes more investigative reports and statistical data, indicating a revival of inductive scholarship and empirical research. Scholarly journals contain differing views on various research questions, demonstrating that the existence of such differences is no longer automatically assumed to reflect "class struggles" within the academic community. Social scientists have shown a greater interest in the concepts, theories, and techniques of their counterparts in the West. And, above all, political leaders seem increasingly aware that social scientists might contribute to the "four modernizations" by careful study and analysis of both Chinese and foreign economic, political, and social experiences.

It is still unlikely that Chinese social scientists, given the constraints mentioned above, will make important theoretical contributions to their disciplines in the near future. If recent trends continue, however, Chinese scholars may help to increase the effectiveness of Chinese policymaking and to restructure their country's political and economic systems. Already some social scientists have begun to call for the application of "social engineering" techniques to policymaking and the use of social science theory in drawing up long-range social and economic development plans.[16] If such developments do occur, social science in China will still resemble the policy analysis practiced by the RAND Corporation and the Brookings Institution, rather than the pure research undertaken at Berkeley or Harvard. But that will be a significant and encouraging development in itself.

[15] F. Burlatskii, "Politics and Science," *Pravda*, Jan. 10, 1965, p. 4; quoted in David E. Powell and Paul Shoup, "The Emergence of Political Science in Communist Countries," *American Political Science Review*, 64 (1970): 575–76.

[16] *Jingji guanli (Economic Management)*, 1979, 1 (Jan.); in *FBIS*, Mar. 29, 1979, pp. L10–12.

# *LINGUISTICS*

*Jerry Norman*

China is a multilingual country. The language of its Han majority is fragmented into an almost incredible number of dialects, some of which are almost like different languages. This linguistic diversity presents the government with many problems, of which the chief is the need to promote the common national language (*putonghua*) to facilitate efficient communication among the country's disparate groups. Yet despite all this diversity, linguistics has not reached the status of a fully autonomous discipline in China, as it has in the United States. There is little interest in human language as a topic worthy of study purely for its own sake. Linguistic studies are nearly always justified in practical terms: grammar serves the teaching of correct Chinese in schools; dialect surveys are useful in the campaign to promote *putonghua* in areas where nonstandard forms of Chinese are spoken; the non-Han languages of the national minorities are studied in order to spread literacy among these people.

As far as is known, there are no departments of linguistics as such in Chinese universities. But linguists are generally found in Chinese language and literature departments. The Institute of Linguistics under the Chinese Academy of Social Sciences (CASS) seems to be the only body entirely concerned with the study of language. However, the Institute's name is somewhat misleading, since it is devoted to the study of Chinese and not to general linguistics. This Institute was largely inactive for the 12 years of the Cultural Revolution and its aftermath (1966–78). It is now functioning again under the leadership of China's premier linguist, Lü Shuxiang. The Institute is divided into several departments: Modern Chinese, Classical Chinese, Dialectology, Lexicography, Linguistic Experimentation, and Machine Translation.

The Institute's Publishing Section is responsible for producing the prestigious linguistic journal *Zhongguo yuwen (Chinese Language)*, which resumed publication in 1978 after a 12-year hiatus. It has published articles on a wide range of topics, including the teaching of Chinese, grammar, dialectology, epigraphy, script reform, and problems of standardization. As can be seen from this brief list, its scope is considerably wider than that of comparable linguistic journals in the West. In early 1979 two new linguistic journals, *Fangyan (Dialects)* and *Minzu yuwen (Minority Languages)*, were established. Apparently *Zhongguo yuwen* was to be devoted to articles on *putonghua* and the history of Chinese, leaving other topics to the new journals.

China's native linguistic tradition has been intimately connected to textual studies; language interested the traditional Chinese scholar insofar as it helped to explain difficult ancient texts. Epigraphy was an important field of study whose roots date to about A.D. 100 when the famous character dictionary of Xu Shen, the *Shuowen jiezi*, was completed. A native school of phonology, strongly influenced by Indian Buddhist ideas, developed in the Tang dynasty. But the study of grammar was virtually unknown in China before the late nineteenth century.

With few exceptions, linguistics, in the current understanding of the term, is a Western importation. Modern ideas about language were mostly introduced to China by Chinese scholars who had studied in Europe or the United States. For example, Y. R. Chao, perhaps the most influential and well-known Chinese linguist of the twentieth century, received his higher education in the United States, was active in China in the 1920's and 1930's, and taught Chinese linguistics at the University of California at Berkeley after the Second World War. Wang Li, a linguist of amazingly wide interests and achievement who is still active at Beijing University, received his early linguistic training in France.

After 1949 Soviet ideas became influential in Chinese linguistic circles. The Chinese minority language policy has been largely modeled on that of the Soviet Union. A. A. Dragunov's *Issledovaniia po grammatike sovremennogo kitaiskogo iazyka* (Research on the Grammar of Modern Chinese) has exercised considerable influence on the study of Chinese grammar, and Joseph Stalin's well-known essay "Marxism and the Problems of Linguistics" remains influential despite its long eclipse in the Soviet Union. Thus, the study of linguistics in China has been heavily influenced by the West and has lagged behind developments there.

Another noteworthy characteristic of Chinese linguistics is its sinocentric nature. Linguists in the PRC study only Chinese languages or those foreign languages deemed important to national needs, such as Russian, English, and Japanese. A Chinese scholar would not undertake the study of an American Indian language or comparative Semitic, for example, unless he could show that it had immediate and practical relevance to China's needs.

The study of grammar was not a traditional Chinese concern, but after Western ideas and methods of grammatical analysis were introduced in the late nineteenth century, Chinese grammar became a recognized field of research. By 1949 there was a small but nonetheless impressive body of grammatical writing in existence. During the 1950's and 1960's activity in the field increased greatly. Pedagogical grammars, monographs dedicated to particular problems of morphology and syntax, and numerous journal articles on virtually every facet of Chinese grammar testify to the vitality of grammatical research during this period.

The articles that have appeared in *Zhongguo yuwen* since its reappearance indicate that this tradition is still very much alive; in the first two issues of the journal published in 1978, a major study on the function of the particle *de* in declarative sentences appeared, written by a leading Chinese grammarian, Zhu Dexi. There have also been shorter studies of other facets of Chinese grammar.

To the average American linguist, the model employed in most Chinese grammatical writing will appear quite antiquated. The typical grammar is a mixture of traditional European grammatical notions tempered occasionally by ideas taken from American structuralism. Soviet models have also been influential. The history of Chinese linguistics is yet to be written. Until the completion of such a specialized study, it is difficult to specify the degree to which various foreign ideas have influenced Chinese linguists. However, it is interesting to note that the two American linguistics delegations that visited China in 1974 and 1977 found linguists in Beijing and Shanghai quite well informed about linguistic issues in the United States.

A few linguists have periodically tried to introduce a more advanced structuralist approach to grammatical description, but so far without much impact. More recent developments in American linguistics, although not unknown, have had little noticeable effect on Chinese grammatical research. It is only fair to point out, however, that despite this lack of interest in keeping up with the latest theoretical fashions of the West, Chinese grammarians have succeeded in producing a number of works that display not only an intimate knowledge of Chinese linguistic structure, but also a great deal of analytic sensitivity. Examples of such works are Wang Huan's *Bazi ju he beizi ju* (Pretransitive and Passive Sentences) and Lu Zhiwei's *Hanyu de goucifa* (Chinese Morphology). Needless to say, these and other studies are highly valuable and even indispensable to the American or European linguist concerned with the Chinese language.

A great number of Chinese linguists have good training in articulatory phonetics, and many are able and experienced field-workers. This is evident from the numerous excellent monographs and articles on Chinese dialects and minority languages. Phonetic training is carried out using the International Phonetic Alphabet with some adaptations developed by Chinese phoneticians for the special problems encountered in China. There seems to be little interest or activity in experimental phonetics or theoretical phonology. Neither the 1974 nor the 1977 American linguistics delegation discovered any work in either of these fields at the universities and institutes they visited.

Dialectology has always been a major concern of Chinese linguists. Chinese dialectal diversity is probably unequaled anywhere in the world. In 1956 and 1957 a national survey, ordered by the State Council, was carried out to obtain a preliminary idea of the type and extent of variation. The results of this survey were widely used in devising handbooks for teaching *putonghua*. From a scientific point of view, however, this survey left much to be desired; the results were very uneven due to the various techniques employed and the diverse skills of the people engaged in collecting data. Unfortunately, most of the results were never published or made available to scholars; the field notes are now scattered and apparently in some cases were lost or destroyed during the Cultural Revolution. In 1979 many of the veterans of this earlier survey, along with their students, were planning to undertake new and more extensive local surveys.

Unlike some other fields of linguistic research, dialectology is well represented at several provincial universities. The 1977 American Applied Linguistics Delegation met a surprisingly large number of scholars engaged in dialect research or planning to launch new survey work. The head of the Institute of Linguistics, Lü Shuxiang, was at work on a description of tonal sandhi in his own native dialect of Danyang, which belongs to the important Wu group of dialects; in 1978 his colleague Li Rong published an extremely interesting description of tone change in his native dialect of Wenling, also part of the Wu group.[1] Other equally interesting and significant projects were described to the Delegation by scholars in Xian, Nanjing, Shanghai, Hangzhou, and Guangzhou. Because such large areas of China are still virtually unknown to dialectologists, the results of these new projects and surveys should be very exciting.

About 6 percent of China's population is non-Han. Most of these peoples have their own languages, several of which possess long literary traditions. Government policy dictates that where possible, minorities should be allowed to have their own written languages. In many cases these languages have very unsatisfactory writing systems or have never been written down at all. In the 1950's Chinese linguists began to study the minority languages systematically. They devised practical orthographies for several languages that had not been written previously and compiled textbooks, dictionaries, and grammars to facilitate the study and teaching of languages like Uighur, Tibetan, Mongolian, and Korean that already had writing systems. They also conducted dialect surveys of several important minority languages to help standardize them. On the basis of such research, the Baarin dialect was chosen as the standard form of pronunciation for Mongolian in the Inner Mongolian Autonomous Region.

Study on the minorities in general is carried out in the CASS Institute of Nationalities and in eight Institutes of Nationalities, of which the most important is the Central Institute of Nationalities in Beijing. Here research is carried out on a wide range of minority languages from all over China; several well-known specialists are on the Institute's staff. There are other institutes in the southwest, Tibet, Qinghai, and the northwest.

Chinese linguists have probably studied virtually all of the country's 50 or so minority languages, and several important dictionaries, grammars, and descriptive sketches have been published. Others are in preparation. The 1977 American Applied Linguistics Delegation heard about work on an impressive number of dictionaries, and an article in the national newspaper *Guangming ribao (Guangming Daily)* described a large-scale project to compile a comprehensive dictionary of classical Tibetan in Chengdu.[2] It is

[1] *Zhongguo yuwen,* 1978, 2: 96–103.
[2] *Guangming ribao,* July 7, 1978.

hoped that future publications on China's minority languages will be made accessible to Western scholars.

Language planning and reform occupy a prominent place in China's national policy. The Committee on Script Reform is directly under the State Council, the government's highest administrative organ. Since the 1950's there has been a continuous effort to simplify the complicated Chinese script. This has involved reducing the number of strokes in complex characters, eliminating superfluous characters, and rationalizing character structure. By 1964 more than 2,000 characters had been simplified. In December 1977 a further list of 200 simplified characters was promulgated for trial use. A number of widely read publications, including the *Renmin ribao (People's Daily)* and the ideological journal *Hongqi (Red Flag),* began to employ them. However, there was obviously a great deal of dissatisfaction with the new reforms. Several articles in *Zhongguo yuwen* were devoted to criticizing the simplified characters.[3] The major criticisms were that there had not been sufficient consultation with the masses, especially with linguists and others intimately concerned with problems of language; that the reform had been launched too quickly, without sufficient preparation; and that many commonsense principles had been ignored in preparing the new list. Perhaps because of these criticisms or for other reasons, the use of the new list of simplified characters was suddenly withdrawn from use in the autumn of 1978.

The Committee on Script Reform has also made a great effort to promote the use of the standard language, *putonghua,* which is based on the dialect of the capital, Beijing. Several prominent linguists have settled on definitive norms for *putonghua,* particularly in pronunciation and lexicon. This standard language is now taught and used extensively in every part of China. However, this has not meant the elimination of local dialects; in cities like Shanghai and Guangzhou they still exhibit great vitality despite the widespread use of *putonghua* in government and education.

Since 1957 China has had an official spelling system using the Latin alphabet. This system, commonly referred to as *pinyin,* is widely used in textbooks, dictionaries, and reference works to indicate proper pronunciation. Official policy states that *pinyin* (or a system like it) will eventually replace the present writing system, but no timetable has been given. There is little indication at present that the switchover will take place in the foreseeable future. It will be interesting to see whether greater use of computers in China will accelerate the universal adoption of *pinyin* or some other form of romanized Chinese. In 1979 *pinyin* was made official in China's dealings with foreign countries, replacing in the process the nineteenth-century Wade-Giles system traditionally used by sinologists in the English-speaking world. From 1979 all foreign-language publications originating in China, including

[3]See especially Yu Xialong, "Guanyu dierci hanzi jianhua gongzuo de yixie yijian" (Some Views on the Second Simplification of Chinese Characters), *Zhongguo yuwen,* 1978, 2: 127–29.

dispatches from the New China News Agency, employed *pinyin*. This change created a certain amount of consternation among Western journalists, librarians, and scholars concerned with China, but they could do little but accept the new system with good grace.

Lexicography has been surprisingly slow in developing in China. During the 1950's and 1960's no large, comprehensive dictionaries—either monolingual or bilingual—were published, despite the fact that dictionaries then in use were seriously out of date. An excellent but small dictionary of single characters, the *Xinhua zidian,* appeared in 1954 and went through many editions. Several of China's outstanding linguists participated in the compilation and subsequent revision of this dictionary. It is an extremely important work, which now serves as the standard authority on pronunciation. It is still in print and is widely used both in China and in the West by students and scholars of modern Chinese.

A comprehensive dictionary of single characters and multicharacter words, the *Xiandai hanyu cidian,* was produced in the early 1960's. A trial edition of 1,000 copies was issued in 1965, but publication of the dictionary for general use was suppressed during the Cultural Revolution. In 1973 more than 200,000 copies were printed, but distribution was halted due to the alleged interference of Yao Wenyuan, one of the Gang of Four. Fortunately, this fine lexicon is now available outside China in Japanese and Hong Kong reprint editions. As of 1979, it was still not widely available in China, but there were signs that it would soon be made available to the general Chinese public. (The reasons for the delay probably involved the unsettled nature of script reform in the late 1970's and the need to review the political acceptability of the definitions of certain ideologically loaded terms.)

Now that lexicographers are free from the intense ideological and political restraints that they suffered during the Cultural Revolution, a veritable lexicographic renaissance appears to be under way. The 1977 Applied Linguistics Delegation was told of more than 100 dictionary projects in progress.

A new English-Chinese dictionary appeared in Shanghai in 1975. The study of English has recently acquired a tremendous importance in China. It is a crucial part of China's campaign to modernize its science and technology, and it is being emphasized at all levels of the educational system. This new orientation toward the English-speaking world will require a major effort to improve both the human and written resources for English teaching. Russian, which was the major foreign language studied in the 1950's, has virtually disappeared from the schools, forcing in some cases the retooling of Russian teachers into English teachers.

China's linguists clearly have very different concerns from those of most of their Western counterparts. Almost all are engaged in practical activities, such as language planning or teaching Chinese and foreign languages. Few have the leisure (or perhaps even the inclination) to engage in the highly theoretical pursuits that are currently so popular among linguists in the

United States. Communist ideology undoubtedly also places certain restraints on what can be viewed as legitimate topics of research. It is well to keep in mind, however, that great changes are taking place in China's academic world. The field of linguistics may well present a very different aspect in the future. The developments will be fascinating to follow.[4]

## *ARCHAEOLOGY*

### *Kwang-chih Chang*

Archaeology has a long history in China and occupies an important place in Chinese society and scholarship, but until the opening of China to Western visitors in the 1970's, most information about the reasons for archaeology's prominence, the priorities placed on various kinds of archaeological activities, the training and institutional settings of archaeologists, and theoretical and methodological approaches had to be derived from painstaking library research.[1]

This chapter, which attempts to describe the current state of Chinese archaeology, is based on archaeological publications, reports of delegations that visited China to observe archaeological workings firsthand,[2] and my own visits to China in May–June 1975, June–August 1977, and October–November 1978.[3]

Archaeology is taught in university history departments, and the Institute of Archaeology is, along with the Institute of Chinese History, included in the Chinese Academy of Social Sciences. Contemporary Chinese historiography depends on archaeology for much of its data, and it may be correct to say that archaeological exhibits in historical museums throughout the country supply a substantial portion of the historical knowledge of the Chinese populace.

A newly published archaeological manual distinguishes between history, which is based on documents, and archaeology, which is history based on artifacts.[4] It is my impression that in the last 30 years most of the major

[4]For further information about Chinese linguistics, see Winfred P. Lehmann, ed., *Language and Linguistics in the People's Republic of China* (Austin: University of Texas Press, 1975).

[1]Cheng Te-k'un, "Archaeology in Communist China," *China Quarterly*, 23 (1965): 67–77.

[2]F. P. Lisowski, "Report on a Visit to Medical and Palaeontological Institutions in China, 1973," unpublished report; Jeannette Hope, ed., *Australian Quaternary Newsletter*, 7 (1976), Special issue "devoted entirely to a report on a visit to the People's Republic of China last November by six Quaternary scientists from the ANU"; and William W. Howells and Patricia Tsuchitani, eds., *Paleoanthropology in the People's Republic of China* (Washington, D.C.: Committee on Scholarly Communication with the People's Republic of China, 1977).

[3]Kwang-chih Chang, "A Palaeoanthropological Tour in the People's Republic of China," *Discovery*, 12 (1977): 42–51.

[4]Jilin University, Department of History, and Hebei, Cultural Relics Administration, *Gong nong kaogu jichu zhishi* (Basic Archaeological Knowledge for Workers and Peasants) (Beijing: Wenwu, 1978).

achievements in Han history—which are enormous and important—have either been in archaeology or been made possible through it. The contributions of archaeology to history in China are perhaps unparalleled and are probably due to the extraordinary importance given archaeology by Chinese society.

Since 1949 archaeology, perhaps alone among all the humanistic and social sciences, has never ceased to grow, although there were periods—notably from 1966 to about 1970—when its activities were curtailed. The position of archaeology in Chinese social sciences—which is also indicated by the government's unsparing financial support—is largely explained by the fact that cultural relics continue to turn up because of extensive construction of roads, dams, reservoirs, irrigation canals, and houses throughout China. But archaeology's strength can also be explained by such factors as the traditional Chinese reverence for antiquities, the personal interest of influential national leaders in China's historical heritage,[5] and the national regulations for preserving antiquities.

The "Provisional Regulations for the Protection and Management of Cultural Relics," which were proclaimed by the State Council in 1961, are very important; their observance has resulted in the generation and preservation of countless cultural relics. These regulations state that all cultural relics having historical, artistic, and scientific value belong to the state and should not be dismantled or transported abroad without permission. Categories of cultural relics protected by the state include architectural sites of historical importance; ancient cultural sites, such as tombs and stone cave temples; valuable art objects, books, and documents; and representative objects that reflect the social life of different eras. The State Council also specified 180 archaeological sites designated for state protection. These comprised 33 revolutionary sites and buildings, 14 stone cave temples, 77 ancient buildings, 11 stone carvings and other monuments, 26 ancient sites, and 19 ancient tombs. A plaque is posted at each site, marking its "national protection" status. In most cases there is an office at the site to administer protective measures, undertake research, and manage exhibits.

According to the regulations, all units of local government must establish special institutes for the protection and management of cultural relics and are responsible for conducting excavations. Before any large-scale engineering work, the government unit must investigate the planned work area for buried cultural relics and include archaeological requirements in the engineering plans.[6]

[5] Xia Nai, "Jingai de Zhou zongli dui kaogu wenwu gongzuo de guanhuai" (Premier Zhou Enlai's Interest in and Concern for Archaeological and Cultural Relics Work), *Kaogu (Archaeology)*, 1977, 1: 5–8.

[6] "Wenwu baohu guanli zhanxing tiaoli" (Provisional Regulations for the Protection and Management of Cultural Relics), *Wenwu*, 1961, 4/5: 7–9; "Di yi pi quanguo zhongdian wenwu baohu danwei mingdan" (List of the First Batch of Cultural Relics Designated for National Protection), *Wenwu*, 1961, 4/5: 10–16; and Louis Dupree, "The Antiquities Regulations of the People's Republic of China," *Archaeology*, forthcoming.

New relics are constantly being discovered in the course of agricultural or industrial work.[7] Many important archaeological discoveries were made during the construction of dams and reservoirs along the Yellow River valley. For example, at Liujiaxia in Yongjing county in eastern Gansu, where a large reservoir and a hydroelectric power station were built between 1964 and 1974, archaeologists began work along the riverbanks as soon as engineers began surveying; more than a year was spent on investigation, another year on excavations. Some sites were sheltered from the rising water level; others were excavated before their submergence. The best-known site that had to be protected with a dam was the Bingling Temple Grottoes, with their stone Buddhist sculptures of the Northern Wei. Important prehistoric sites that had to be cleared included Jijiachuan and Qinweijia. Manpower for the work came from the Institute of Archaeology and the Gansu Provincial Museum. Funding for the project came largely from the reservoir budget. This funding pattern is typical; the budgets of construction projects automatically include funds for archaeological work. Cultural relics agencies bear the financial burden only when archaeological work is done outside a project area or in the countryside.

Salvage archaeology—because of its urgency—often preempts a significant amount of archaeological manpower at the local and provincial level. Archaeologists are also on call to investigate reports of accidental discoveries made by farmers, workers, or just about anybody who thinks he has stumbled on something of importance. Archaeological investigations resulting from research programs formulated on purely scholarly considerations are relatively rare and are undertaken mostly by the research institutes of the Chinese Academy of Sciences and the Chinese Academy of Social Sciences.

Although there are no data on the number of professional, full-time archaeologists in China, there are at least several thousand. Most of them work in museums, national or regional research institutes, and universities.

Museums and other governmental archaeological agencies have been established at the national, provincial, and subprovincial levels. The National Bureau for the Administration of Cultural Relics Institutions in Beijing sits at the top of the museum system. Led by Wang Yeqiu, it administers the Museum of Chinese History, the Palace Museum, the Beijing Library, the Institute of Cultural Relics Conservation Science and Technology, the Wenwu Press, and the Work Team for Overseas Exhibitions of Cultural Relics. None of these Beijing establishments engage in archaeological fieldwork, but the two national museums have large numbers of research personnel, and the Cultural Relics Conservation Institution is becoming an important conservation center and has a radiocarbon laboratory. The Bureau also sponsors and organizes national conferences of museum workers and interdisciplinary research workshops. In addition, it supplies considera-

[7] Yang Jianfang, "Wenwu kaogu gongzuo yu shengchan jianshe" (Cultural Relics and Archaeological Work and Productive Constructions), *Wenwu*, 1976, 10: 79–83.

ble funding to support archaeological work and the construction of exhibition halls at the provincial and subprovincial levels.

Each province and major city has a bureau of culture and within the bureau of culture there is generally a cultural relics department responsible for protecting antiquities. The bureaus also manage provincial and municipal museums, which are either administratively parallel or subordinate to the cultural relics departments. Another governmental agency, the archaeological team (sometimes called a cultural relics work team), is mainly responsible for archaeological fieldwork, sometimes as part of a research project, but more often as part of a conservation or salvage operation. Another related organization, the commission for the administration of cultural relics, once a part of each provincial and major municipal government, has in most cases been absorbed into the museum system at the bureau, department, or team level. The organization of museum work at the provincial and subprovincial levels is not uniform throughout the country, but all archaeologists who work within this system are government workers.

Many senior archaeologists working within the museum system are graduates of archaeological workers training institutes. But many middle-level and junior staff members (and an occasional senior administrator) are university graduates who majored in archaeology.

Two research institutes do archaeological work at the national level: the Institute of Archaeology of the Chinese Academy of Social Sciences (CASS) and the Institute of Vertebrate Paleontology and Paleoanthropology of the Chinese Academy of Sciences (CAS). Both are located in Beijing.

The Institute of Archaeology, established in 1950, is commonly regarded as the most important and authoritative research organization in Chinese archaeology. Its director, Xia Nai, received a doctorate in archaeology from London University. Within the Institute are three research sections: Primitive Society, Slave Society, and Feudal Society. In addition, the Institute has a library, a laboratory (which began radiocarbon dating in 1965), and an editorial office. The Institute publishes monographs and two periodicals, *Kaogu xuebao (Archaeologica Sinica),* a quarterly, with Xia Nai as chief editor; and *Kaogu (Archaeology),* a bimonthly, edited by An Zhimin. The Institute also maintains three permanent field stations at Anyang, Luoyang, and Xian and sends archaeological teams throughout the country, but particularly to North China, to undertake reconnaissance and excavations.

The Institute of Vertebrate Paleontology and Paleoanthropology has a research section that deals with man. Its employees are paleolithic archaeologists, mostly graduates of Beijing University (where they majored in archaeology) or of Fudan University in Shanghai (where they majored in anthropology). The Peking Man localities at Zhoukoudian, in the southwestern part of Beijing, are administered by the Institute. New large-scale excavations began in October 1978 under the leadership of Wu Rukang. The Institute publishes monographs, collections of scientific essays, the quarterly *Vertebrata Palasiatica,* and a popular magazine, *Huashi* (Fossils).

The Institute of Vertebrate Paleontology and Paleoanthropology was established in 1957—a direct descendant of the Laboratory for Cenozoic Research. Its director, Yang Zhongjian (who died in January 1979), and its two senior staff members, Pei Wenzhong and Jia Lanpo, were on the staff of the Laboratory. Because of this tradition, Early Man research is conducted in an interdisciplinary setting in which paleontologists, paleolithic archaeologists, paleoanthropologists, and Cenozoic geologists work together, both in the field and in the laboratory. Paleolithic archaeology is thus separate from postpaleolithic archaeology, which is practiced under the Institute of Archaeology.

In 1949, when the PRC was founded, only a handful of trained archaeologists could be found in China. The more notable included Guo Baojun, Liang Siyong, Xia Nai, Su Bingqi, Pei Wenzhong, Li Wenxin, and Tong Zhuchen. These scholars confronted the gigantic archaeological tasks facing them in three ways. First, they personally conducted fieldwork at such important locations as Anyang and Huixian, training field archaeologists on the spot. Second, beginning in 1952, they established an archaeology program at Beijing University to give systematic, professional training to undergraduates. Third, in 1953 they set up a training institute to give three-month intensive courses to cultural relics workers. This institute suspended operation after graduating four classes, but its graduates now include many of China's leading museum workers.

By the late 1950's Beijing University had graduated a sufficient number of archaeology students to fill many of the country's archaeological posts. A significant percentage of cultural relics cadres and archaeologists still consists of Beijing University archaeology graduates. Ten other universities subsequently started archaeology training programs: Northwest (Xian), Nanjing, Shandong (Jinan), Jilin (Changchun), Xiamen, Sichuan (Chengdu), Zhongshan (Guangzhou), Wuhan, Zhengzhou, and Liaoning (Shenyang). The programs at some of these universities are still new and experimental, and some only admit students every other year. A twelfth archaeological program is being planned, probably at Shanxi University in Datong.

In addition to the required university courses such as modern history, political economy, foreign languages, and physical education, each university offers a rather broad range of general and specialized courses on archaeology. Many of the specialized courses naturally reflect the specialities of the faculty. But the training archaeology students receive does have some common features—in particular a virtually exclusive preoccupation with Chinese prehistory and history, emphasis on fieldwork and field experience, and close attention to Marxist historical materialism. University-trained archaeologists are highly competent in handling Chinese materials and issues, but are sometimes unfamiliar with archaeological issues of general interest, such as comparative problems. Many university programs also contain geological and paleontological components, and most students display a strong interest in the natural environment that ancient man lived in.

Each year there are more than one hundred archaeology graduates; these now constitute the mainstay of the country's professional archaeology manpower. Graduates tend to be employed in areas near their universities, apparently because they are more familiar with the region's archaeology. On the other hand, students enrolling in field archaeology courses are often sent to distant provinces to obtain experience in the archaeology of different areas. Thus, Nanjing University students often participate in excavations near Xian under the supervision of Northwest University staff members, and students from Xiamen University are sometimes seen near Chengdu doing fieldwork with Sichuan University faculty members.

In many provinces graduates of universities and the short-term training institutes of the mid-1950's are assisted by graduates of the worker-peasant-soldier archaeology training programs of the 1970's. These programs, organized by the various museums and universities, give elementary archaeological training to people who have other careers, but are interested in learning about archaeology. After taking the course, they return to their own jobs, but perform some preservation tasks, serve as permanent scouts for accidental finds, and are available as trained and experienced excavators when important local excavations take place.

Since archaeology in China is a branch of history, the basic theory and methodology governing archaeological analysis and interpretation are based on some of the tenets of Marxist historical materialism. There is no significant deviation from these tenets insofar as archaeologists' philosophical orientations are concerned, although the degree to which their writings conform to these tenets varies. At a conference in the summer of 1978, in Changchun in Jilin province, historians and archaeologists engaged in intensive debate about the line dividing slave society from feudal society in Chinese history; no fewer than six different views were aired that placed the line at (1) the beginning of the Western Zhou period (ca. 1100 B.C.); (2) the beginning of the Spring-Autumn period (ca. 700 B.C.); (3) the transition from the Spring-Autumn to the Warring-States period (ca. 600–450 B.C.); (4) the beginning of the Qin dynasty (221 B.C.); (5) the beginning of the Eastern Han dynasty (first century); and (6) the beginning of the Jin dynasty (third century). The employment of the Marxist evolutionary scheme that recognizes these two stages of societal development was, however, not questioned. With increasing contact with the West, there is no doubt that Chinese historians will frequently encounter non-Marxist historiographic paradigms, but whether these will affect archaeological thinking and writing is difficult to say.

Chinese archaeological writings usually consist of descriptions of artifacts and sites, generalizations about ancient society and ancient life derived from such Marxist classics as Lewis Henry Morgan's *Ancient Society* (1877) and Friedrich Engels' *The Origin of the Family, Private Property, and the State* (1884), or hypotheses about the classification and interrelations of prehistoric and ancient cultures. There is little evidence of the sophisticated use of classificatory or quantitative methods, although the use of computers in archaeological research has begun.[8] But because Chinese archaeologists are seldom preoccupied with methodology or proving preconceived hypotheses, the data they present in archaeological publications are exemplary for their precision and detail of description. The data therefore provide useful raw material for study by other archaeologists, whatever their theoretical and methodological orientation. Despite their Marxist persuasion, Chinese archaeological publications are remarkably free of theoretically derived subjective distortion of data.

At the level of data recovery techniques, Chinese archaeologists at their best are comparable with the best in the West, but at their worst they do not always meet professional standards. They are not much different from archaeologists elsewhere, but sometimes lag behind in areas where modern science and technology are involved. Here is a brief and impressionistic assessment of Chinese archaeological techniques:

1. Reconnaissance and prospecting. Chinese archaeologists rely on the good old-fashioned method of looking around on foot or on horseback (or camelback or whatever). In the loess lands of North China, the old tomb-plunderer's tool, the "Luoyang spade," has proved a handy and efficient aid. It is in fact a portable core-borer, and an experienced archaeologist can insert it deep into loessial ground, pull it out with a sample cross-section, and determine if the site is sterile or rich in residential or burial remains. By using the Luoyang spade at Yinxu in Anyang, archaeologists have been able to reconstruct the grid plan of substantial portions of the 1928–1937 Xiaotun and Xibeigang excavations without reexcavating.

2. Excavation. Excavation techniques for carpet-style horizontal lifting of natural layers and special features, such as house floors, tombs, and clay impressions of chariots, appear to be highly sophisticated. Artifacts are usually recorded according to grids and layers, but the use of precise plans to plot artifacts with reference to house floors or locations appears to have begun more recently. Techniques for the recovery of very small pieces of bone, seeds, and other macrofossils of the biota do not seem to be highly developed; flotation is not done, and refuse soil is not always sifted.

3. Chronometry. The first Chinese radiocarbon laboratory was established in 1965 at the Institute of Archaeology by a husband-wife physicist team, Chou Shihua and Cai Lianzhen. Their laboratory remains the most experienced in radiocarbon work. But samples are also processed at four other radiocarbon laboratories, located at the Institute of Geology (Beijing), the Department of History at Beijing University, the Institute of Cultural Relics Conservation Science and Technology (Beijing), and the Guiyang Institute of Geochemistry (Guizhou).[9] Some of these laboratories have begun

[8] Tong Enzheng, Zhang Shengkai, and Chen Jingchun, "Guanyu shiyong dianzi jisuanji zhuihe Shang dai pu jia suipian de chubu baogao" (A Preliminary Report on the Use of Electronic Computers for Piecing Together Oracle Shell Fragments of the Shang Dynasty), *Kaogu,* 1977, 3: 205–9.

[9] Xia Nai, "Tan shisi ceding niandai he Zhongguo shiqian kaoguxue" (Carbon-14 Dating and the Prehistoric Archaeology of China), *Kaogu,* 1977, 4: 217–32.

experiments on thermoluminescence as well as other isotopic dating techniques.[10] Paleomagnetic dating results have been reported from laboratories at the Institutes of Geology, Geodynamics, and Geochemistry.[11]

Radiocarbon dates are reported in the following journals: *Kaogu* (Institute of Archaeology), *Wenwu* (Bureau of Culture), *Scientia Geologica Sinica* (Institute of Geology), *Acta Geologica Sinica* (Geological Society of China), *Acta Geophysica Sinica* (Geophysical Society of China), and *Geochimica* (Guiyang Institute of Geochemistry).

These chronometric operations started in China during the period when self-reliance in science and technology was the prevailing national goal; no foreign equipment or technical personnel are known to have been involved. The recent policy changes, however, may allow foreign technology to contribute to Chinese advances in archaeological chronology.

4. Science and archaeology. Chinese archaeologists have traditionally worked with geological scientists, such as Pleistocene stratigraphers, structural and glacial geologists, geomorphologists, and paleontologists. In the Institute of Vertebrate Paleontology and Paleanthropology, paleolithic archaeologists, paleoanthropologists, and Quaternary paleontologists and geologists all work within the same section, and most field research projects are interdisciplinary. Scientists from no fewer than 16 research institutes are participating in the new Zhoukoudian excavations that were begun in late 1978.

In the biological sciences, archaeologists have used palynology to reconstruct regional vegetational histories and the environmental history of archaeological sites.[12] Most archaeologists receive training in zooarchaeology and human osteology, but wisely confine their encounters with animal remains to identifying jawbones of dogs, cattle, deer, and other common beasts and leave the rest to mammalogists. Zoologists have recently compiled massive catalogues of animals and fishes on regional and system (river or lake) bases, but archaeologists have just begun to use these resources.

The use of physical sciences focuses on such tasks as the chemical analysis of ceramics and bronze, copper, and iron artifacts, although sophisticated metallurgy has been employed in studying the history of iron metallurgy in ancient China.[13] Chemical methods for conserving bronzes and bamboo slips are particularly well developed in China.[14]

[10] Chou Shihua and Cai Lianzhen, "Taoqi de reshiguang ceding niandai jieshao" (An Introduction to the Thermoluminescence Dating of Pottery), *Kaogu*, 1978, 5: 344–51.

[11] Ma Xinghua and Qian Fang, "Gu di ci yu jiu shiqi shidai kaogu" (Paleomagnetic Dating and Paleolithic Archaeology), *Kaogu*, 1978, 5: 352–57.

[12] Zhou Kunshu et al., "Huafen fenshi fa ji qi zai kaoguxue zhong de yunyong" (Pollen Analysis and Its Application in Archaeology), *Kaogu*, 1975, 1: 65–70.

[13] Li Zhong, "Guanyu Gaocheng Shang dai tong yue tie ren de fenxi" (The Analysis of the Iron Blade of the Shang Dynasty Bronze Axe from Gaocheng), *Kaogu xuebao (Archaeologica Sinica)*, 1976, 2: 17–34.

[14] "Baohu gu qingtongqi de yizhong xin fangfa" (A New Method for Preserving Ancient Bronzes), *Kaogu*, 1975, 3: 195–96; and "Gudai zhu jian de tuoshui chuli" (Dehydrating Ancient Bamboo Tablets), *Kaogu*, 1976, 4: 279–80.

With few exceptions, notably Zhoukoudian, Anyang, Juyan, and Dunhuang, the most important archaeological sites in China have been found in the last 30 years. Archaeology has changed the face of Chinese historiography. Some finds command the attention of foreign archaeologists because of their unique importance. Some of the richest and most impressive of these finds are the following:

1. Fossils of probable *Ramapithecus* in Yunnan.[15]
2. New fossils of *Homo erectus* in Yuanmou (Yunnan), Lantian (Shaanxi), and Zhoukoudian (Beijing).[16]
3. Neolithic sites throughout China dating from 5000–10,000 B.C., where the decisive transitional step from food gathering to food producing was occurring.[17]
4. Yangshao Culture sites at Banpo and Jiangzhai near Xian, where whole village sites were excavated.[18]
5. Hemudu site near Yuyao, northern Zhejiang, where advanced rice domestication occurred before 5000 B.C.[19]
6. The discovery of a Neolithic culture in eastern North China, called the Dawenkou Culture, displaying credible evidence of significant social stratification and incipient statehood.[20]
7. The discovery of the Erlitou Culture in western Henan, possibly the first true state-type culture in Chinese archaeology, perhaps that of the Xia dynasty of traditional Chinese history.[21]
8. New finds at Anyang, Henan, site of the last Shang capital, including the tomb of Lady Hao, consort of Shang king Wu Ding, which contained several hundred bronze vessels and several hundred jade carvings.[22]
9. Inscribed oracle bones from Zhou Yuan, in central Shaanxi, the tra-

[15] Xiang Jiang, "Yunnan Lufeng faxian guyuan xiahegu huashi" (Mandible fossil of *Ramapithecus* Discovered in Lufeng, Yunnan), *Vertebrata Palasiatica*, 15, 2 (1977): inside back cover; and Xu Qinghua, "Lufeng lama guyuan" (The *Ramapithecus* of Lufeng), *Huashi*, 1978, 2: 1–3.

[16] Kwang-chih Chang, "Chinese Palaeoanthropology," *Annual Review of Anthropology*, 6 (1977): 137–59.

[17] Kwang-chih Chang, "Kung-yüan-ch'ien wu-ch'ien nien tao i-wan nien ch'ien Chung-kuo yüan-ku wen-hua tzu-liao" (Ancient cultures in China from 5000 to 10,000 B.C.: The Current Evidence), *Bulletin of the Institute of Ethnology, Academia Sinica*, 6 (1978): 97–104.

[18] "Shanxi Lintong Jiangzhai yizhi di er san ci fajue de zhuyao shouhuo" (Major Results of the Second and Third Excavations at the Jiangzhai Site in Lintong, Shaanxi), *Kaogu*, 1975, 5: 208–84.

[19] "Hemudu faxian yuanshi shehui zhongyao yizhi" (Important Primitive Society Site Discovered at Hemudu), *Wenwu*, 1976, 8: 6–14; and "Hemudu yizhi di yi qi fajue baogao" (Report on the First-Phase Excavations of the Hemudu Site), *Kaogu xuebao*, 1978, 1: 39–94.

[20] "Tantan Dawenkou wenhua" (On the Dawenkou Culture), *Wenwu*, 1978, 4: 58–66.

[21] Wu Ruzuo, "Guanyu Xia wenhua ji qi laiyuan de chubu tansuo" (A Preliminary Investigation into Xia Culture and Its Origins), *Wenwu*, 1978, 9: 70–73.

[22] Kwang-chih Chang, "Yin-hsü fa-chüeh wu-shih nien" (Fifty Years of Excavations at Yin-hsü), in *Chung-yang yen-chiu-yüan ch'eng-li wu-shih chou-nien chi-nien lun-wen chi* (Papers Presented in Celebration of the Fiftieth Anniversary of the Founding of the Academia Sinica) (Taipei: Academia Sinica, 1978), pp. 291–311; and "Anyang Yinxu wu hao mu de fajue" (The Excavation of Tomb No. 5 at Yinxu in Anyang), *Kaogu xuebao*, 1977, 2: 57–98.

ditional early Western Chou capital, bringing to light for the first time historical documents of the Chou people before their conquest of the Shang in the twelfth century B.C.[23]

10. The underground warriors' vaults east of the mausoleum of Qin Shi Huangdi in Lintong, Shaanxi, in which more than 6,000 life-size terra-cotta figures of warriors, officers, and horses were found.[24]

11. A Qin dynasty tomb at Shuihudi, in Yunmeng, Hubei, in which legal codes and regulations transcribed on bamboo slips were found.[25]

12. Three early Han dynasty tombs at Mawangdui in Changsha, Hunan, which contained a large number of Han books (written on silk), maps, paintings, and lacquerwares. The books included many that had long been lost, including a number that validated the authenticity of previously suspect extant works.[26]

These and other similarly important findings are also noteworthy for the light they shed on Chinese archaeological practices. Archaeology as practiced in China is justified by Chairman Mao's saying: "The past should be made to serve the present." This, however, is not as revolutionary as it may appear. As early as the first century B.C., the great Chinese historian Sima Qian (Ssu-ma Ch'ien) had already stated that "events in the past, when not forgotten, can teach us about the future." Archaeology is particularly suited to this traditional role given to history because it appeals to the senses as well as to the mind and is therefore an effective tool in political education and propaganda. This is no cause for intellectual concern as long as archaeological facts are not distorted to serve transient political interests. Even during the years 1975–76, when the ancient Confucian-Legalist struggle was debated during political struggles, archaeology was only marginally involved.

Archaeology in China has strong state backing, and because the state is usually more powerful and efficient than the academic world, Chinese archaeology is well run in the organizational sense. The plundering of the nation's buried treasures for personal profit, long a way of life, has been eliminated. The citizenry has been educated to be archaeologically aware, and

[23] New China News Agency dispatch, in *Hua-ch'iao jih-pao* (New York), Nov. 1, 1977.

[24] "Lintong xian Qin yong keng shi jue di yi hao jian bao" (Preliminary Report No. 1 on the Test Excavation of the Qin Figures Pit in Lintong County), *Wenwu*, 1975, 11: 1–18; and "Qinshihuang ling dongce er hao bing ma yong keng zuan tan shi jue jian bao" (Preliminary Report on the Borings and Test Excavations of the Second Warriors and Horses Pit East of the Mausoleum of Qin Shi Huangdi), *Wenwu*, 1978, 5: 1–19.

[25] Ji Xun, "Yunmeng Shuihudi Qin jian gai shu" (A General Description of the Qin Dynasty Written Slips from Shuihudi, Yunmeng), *Wenwu*, 1976, 5: 1–6; and "Yunmeng Qin jian shi wen" (Transcription of the Qin Dynasty Written Tablets from Yunmeng), *Wenwu*, 1976, 6: 11–14; 7: 1–11; 8: 27–37.

[26] Hunan Provincial Museum, and Chinese Academy of Sciences, Institute of Archaeology, *Changsha Mawangdui yihao Han mu* (Han Tomb No. 1 at Mawangdui, Changsha) (Beijing: Wenwu, 1973); and "Changsha Mawangdui er san hao Han mu fajue jian bao" (Preliminary Report on the Excavations of Han Tombs Nos. 2 and 3 at Mawangdui, Changsha), *Wenwu*, 1974, 7: 39–48.

when anything of apparent historical value is discovered, it is immediately reported to the authorities. Most of the finds enumerated above were first discovered in this fashion. Important discoveries are then handled by the museums or the research institutes. When interagency collaboration is desirable or necessary, teams can be organized at the state's expense to do assigned jobs, such as the Mawangdui silk books study teams organized by the Bureau of Cultural Relics. Sites of particular visual value are often protected to preserve their in-situ appearance; the Banpo Neolithic village site and the Han tomb no. 3 at Mawangdui are both covered by large domes and permanent shelters. The Qin warriors' vault was also being covered in the late 1970's, with a permanent structure 230 m long, 72 m wide, and 22 m tall designed to shelter the entire pit. The job was expected to be completed in 1979, when the Chinese planned to resume work on the excavation and restoration of the figures inside the structure. Eventually the entire pit is to become a vast exhibition hall.

Because of the enormous number of historic and prehistoric relics, the professional manpower and financial resources allocated to archaeology must be employed in accordance with established priorities. I received the strong impression that within the museum system top priority is assigned to the protection or excavation of long-known relics and to new, accidental discoveries threatened by construction projects. Research on theoretical problems or even on recovered relics has a much lower priority. Within the national research institutes, there is more room for active research projects—as opposed to passive response to salvage or rescue needs—but even here archaeologists' movements are more often dictated by circumstances than by deliberate research designs. With recovered material, top priority goes to descriptive publications; archaeologists often complain that they are too busy to do research or to synthesize already published materials. A synthesis of Chinese prehistory and historic archaeology published in 1962 (*Archaeological Results of the New China*) has long been out of date, but a new book was planned by the Institute of Archaeology for publication in 1979 to commemorate the thirtieth anniversary of the founding of the PRC. For these reasons, there were no plans as of 1979 to excavate the imperial mausoleums of Qin Shi Huangdi, Tang Gaozong, and Wu Zetian, even though spectacular finds were expected. These were considered major national treasures, safely protected by the state, and limited manpower and financial resources were allocated to more urgent needs. One can only sympathize with such rational use of the nation's limited archaeological assets.

Rational planning of archaeological activities of this magnitude may inhibit individual initiatives that lead to important scholarly advances, but on balance it eliminates waste and enables the nation's archaeologists to concentrate on accomplishing urgent tasks. The result is that more cultural relics get protected, preserved, and researched, and fewer get destroyed than under any other system imaginable for China. On the other hand, one can hope that with the certain increase of archaeological manpower in the next few years, more archaeologists will be spared to concentrate on issues of general concern.

Chinese archaeologists are well equipped for their immediate needs. They are, in fact, universally admired for the gigantic achievements of the last 30 years. But over the long run, they may wish to import methods of data recovery and analysis and to acquaint themselves with the accomplishments and activities of foreign archaeologists. Both steps would further enhance the quality of archaeological work in China.

## *HISTORY*

*Albert Feuerwerker*

During the 1950's, history shared the general expansion of higher education and scholarly research in the PRC. A substantial number of historical publications—2,032 titles in all, many of them multivolume works—were listed in the *Quanguo zong shumu* (National Bibliography of China) between 1950 and 1958. Many of these works were popular pamphlets or books for elementary readers, but there were also important new collections of source materials, hundreds of scholarly monographs, and some reprints of older studies. World history, as opposed to Chinese history, was sparsely represented and consisted largely of translated works—mainly Russian—or textbooks based on Soviet models.

Specialized history periodicals included *Lishi yanjiu* (Historical Research), 1954– , published by the Chinese Academy of Sciences (CAS); *Lishi jiaoxue (Teaching of History)*, 1951– , a publication intended for middle-school teachers; *Jindai shi ziliao (Source Materials on Modern [Chinese] History)*, 1954– , which made available previously unpublished documents; and *Wen shi zhe (Literature, History, and Philosophy)*, 1951– . The bi-weekly history supplement, *Shixue (Study of History)*, to the Beijing daily newspaper *Guangming ribao (Guangming Daily)* included articles on a wide range of historical subjects, as did the scholarly journals published by several universities.

Most universities and colleges had departments of history; in some cases there were two departments, one for Chinese history and one for foreign history. The editor of *Lishi yanjiu* estimated in 1957 that there were 1,400 "teachers of contemporary history and history of the Revolution in institutions of higher education in all parts of the country."

Historical research was conducted by the CAS Institute of Historical Research, by the universities, by ad hoc extra-academic bodies, to some extent by the Communist Party itself, and by national and local government archival offices. The Institute of History had three offices. The First Office (ancient history) was headed by Guo Mojuo (Kuo Mo-jo), president of the CAS and concurrently director of the CAS Department of Philosophy and Social Science. The head of the Second Office (medieval history) was Chen Yuan, a specialist on the Tang period (A.D. 618–907) and the history of Buddhism. Fan Wenlan, a scholar who began his career as a student of the Chinese classics, headed the Third Office (modern history). Research on economic history was also carried on at the CAS Institute of Economics in Beijing. There were two major repositories of historical documents: the Ming-Qing archives at the Palace Museum in Beijing and the 1911–49 archives of the Beijing and Nanjing governments located in Nanjing.[1]

For over two millennia, the writing of history in China has never been totally divorced from contemporary political concerns. For example, each dynasty composed the "official history" of its predecessor; this helped to justify the transfer of the "Mandate of Heaven" to the new rulers. In addition to these *Twenty-four Standard Histories,* the educated elite and to a lesser extent the general populace were made aware of the "lessons of history" by exposure to a vast corpus of official and private historical works and by the presence of historical themes in written and oral literature. "Good" and "bad" emperors and ministers were stock characters, for example, in the Beijing opera. Such historical novels as *The Romance of the Three Kingdoms* were widely read and endlessly explicated.

Thus, when PRC historians set about rewriting Chinese history in the 1950's, they were following a long-standing historical tradition. In their attempts to create a new and popular Marxist historical tradition, they emphasized several major themes: the interpretation of peasant rebellions as the real motive force of historical development in China's feudal society; the early development of "incipient capitalism" in Ming (1368–1644) and Qing (1644–1911) China, which was diverted from its normative Marxist course by a combination of foreign aggression and domestic reaction; the problem of the formation of the "Han nation"; the onus borne by foreign imperialism for the shameful record of political weakness, economic chaos, and cultural discord in China in the century ending in 1949; and an effort to establish an acceptable periodization for China's history.

The problem of finding meaning in this reinterpreted past was not easily solved. Peasant rebellion as the main content of Chinese history proved to be a chimera. The emphasis on China's independent development toward

[1]Perhaps 15 percent of the materials in the prewar Ming-Qing archives were shipped to Taiwan in 1949, including much of the palace memorial collection and the Grand Council record books; the bulk of the remainder was left in considerable disorder, and it is not known to outsiders how accessible these documents have been. The modern history archive at Nanjing, formerly called the Historical Materials Compilation Department of the Third Office of the Institute of History, was established in 1951. During the 1950's it was engaged primarily in assembling and processing its collection. American historians who visited China in recent years were denied access to both archives. It now appears that the Ming-Qing collection will be opened to foreign scholars, at least in part.

capitalism had unwanted political implications; it raised doubts about the historical necessity of a proletarian revolution led by the Communist Party to overthrow feudal society. Discussions of the Han nation's formation led either to the postulation of China's uniqueness, which orthodox Marxism could not allow, or to metaphysical nonsense. If imperialism was the key to modern Chinese history, history was in danger of losing its autonomy and meaning. And a generally acceptable periodization was impossible, given the lack of careful scholarly research on China's premodern history.[2]

On the eve of the Great Leap Forward in 1958, Chinese historians were engaged in extensive self-criticism over their failure to reinterpret the past. Their political rulers were also unhappy. From early 1958 the slogan "emphasize the present and de-emphasize the past" was adopted and broadcast in all important historical and political periodicals as a means of remedying this unsatisfactory state of affairs. But this slogan had a double edge. On the one hand, it was meant to be used as a guide for allocating available historical manpower and facilities. But it also indicated a tightening of the political screws following the Hundred Flowers episode of 1957. In the Party's view, historians had done a poor job because they were ideologically backward. This was attributed to their isolation from the great struggles of the masses to build a socialist society. Historians had failed to solve the problems of the past because they did not adequately comprehend the present. The only solution was to politicize history and involve historians in current struggles.

After 1958, in response to the call to emphasize the present and de-emphasize the past, PRC historians increasingly wrote factory and commune histories and semipopular propaganda pieces about current events. History departments throughout China, like factories and farms, enthusiastically reported a "great leap forward" in their production of lecture materials, study guides, articles, and books. Historians were trying to redeem themselves. As the History Department at Nankai University in Tianjin put it:

> Our faculty and students have all come deeply to realize that the open field for historical science is not in heaps of old documents or in ivory towers, but in the factories and villages, in the midst of the sharp struggles of the masses. Historical science must be in the service of the political struggles and in the service of production. Historical scientific workers must heed the word of the Party and combine with the workers and peasants. . . . Only then can they demonstrate their utility in the socialist revolution and in the construction of socialist society.[3]

After 1960 the number of academic and popular historical works published in China decreased markedly. This was partly the result of a severe

[2]The continued paucity of solid, modern research on ancient China was one reason the Gang of Four was able to perpetrate the fallacy that the Confucian-Legalist struggle (allegedly a conflict between "slave" and "feudal" society) was the main drama of Chinese history from the sixth century B.C. to the fall of the later Han in the third century A.D. (See *Lishi yanjiu*, 1978, 3 [Mar.]: 43.)

[3]*Lishi yanjiu*, 1958, 10 (Oct.): 73–75.

paper shortage caused by the economic dislocations of the Great Leap Forward. But it also reflected the increasingly low priority assigned the social sciences and humanities in the task of building socialism and the suspicion with which "bourgeois" intellectuals were regarded. A survey of *Quanguo xin shumu* (National Catalogue of New Publications) for the years 1962–66 indicates that little more than 50 new historical titles were published annually during this period, far less than the 200–300 per year of the 1950's. Some works on foreign history appeared, but Chinese history, mostly modern, was predominant.

The discussions of the early 1960's centered first on the problem of whether historical personages should be evaluated in terms of the problems and values of their own times or by the standards of present-day socialist China. Later discussions broadened to encompass the question of the "critical inheritance" of China's entire "cultural legacy": that is, how should Chinese relate to 2,000 years of a society that, despite its remarkable political and cultural achievements, was "feudal" and thus oppressive to the majority of the population? And finally the discussions became a philosophical and political debate between proponents of "historicism"—what we might call historical relativism—and those who called for the strict application of a present-oriented "class viewpoint" to all historical judgments. Although these historical debates suffered from more than occasional heaviness, they were serious and had substantive intellectual content.

This historical dynamism ended with the Cultural Revolution. Ironically, its start was signaled by an attack on a Ming historian, Wu Han, who was chairman of the Beijing Historical Society and deputy mayor of Beijing. In November 1965 Wu was attacked for writing a historical drama, *The Dismissal of Hai Rui,* in which he covertly criticized Chairman Mao's economic policies and implicitly identified the hero, a mid-sixteenth-century official of almost legendary popularity, with Marshal Peng Dehuai, who had been ousted by Mao in 1959 for opposing the Great Leap Forward and for advocating better relations with the Soviet Union. The campaign against Wu Han spread to other historians as well, including Deng Tuo, an official of the Beijing Municipal Committee of the Communist Party and editor of its theoretical magazine, *Qianxian (Frontline).*

*Lishi yanjiu,* China's leading historical journal, ceased publication after April 1966. In a front-page editorial on June 3, 1966, the *Renmin ribao (People's Daily)* attacked "bourgeois 'authorities' in the field of historical studies" for opposing the "scientific theses" of Mao Zedong, for denying the class struggle, and for suppressing truly revolutionary historians through their control of the leading academic positions. Prominent historians, among them Li Shu, editor-in-chief of *Lishi yanjiu,* were similarly attacked in the same newspaper on October 23, 1966.

The three offices of the Institute of History were either closed or had their activities drastically curtailed between 1966 and 1970. This epoch of confused struggle between the supporters of Mao and their opponents was hardly conducive to academic scholarship. Historians, like other intellectuals, were reeducated either in May 7th cadre schools in the countryside or at their places of work. For some, the Cultural Revolution meant more than a forced hiatus in their academic work; the torture and death in 1969 of the eminent 71-year-old historian Jian Bozan, a major intellectual and political leader in the 1950's, was described in gruesome detail by wallposters at Beijing University in April 1978 and subsequently confirmed in official publications. Needless to say, no scholarly publications appeared during the Cultural Revolution.

What were the consequences for history of the post–Cultural Revolution political conflicts associated with Lin Biao and the Gang of Four? In the current version, the Gang's power was based largely on its control of the media—interpreted very broadly to include historical publications. And indeed, it does seem that historical interpretations now identified with the Gang's ideology and policies dominated Chinese historiography—such as it was—in the years 1973–76. *Lishi yanjiu*, which resumed publication in 1974, carried an editorial in late 1976 entitled "The Misfortune of *Lishi yanjiu* and the Plot of the Gang of Four to Utilize History to Oppose the Party" that instructed its readers to disregard everything published in the 12 preceding numbers.

History, like other academic fields, recovered slowly from the effects of the Cultural Revolution. As the universities and colleges gradually reopened, the first task was the preparation of new teaching materials and textbooks. The texts used in the 1950's, many written under the influence of Soviet historiography, could no longer be used. Works by Chinese authors included many by older historians who were still suspected for their "bourgeois" outlook.

Few serious historical works appeared between 1971 and 1977. For example, Shanghai Normal University and Beijing University brought out new world history texts in 1973 and 1974 respectively. The first is an 800-page work treating Europe, North and South America, Asia (except China), and Africa since the seventeenth century in conventional Marxist terms. The second is a much shorter collection of popular lectures. Publication of the excellent punctuated edition of the *Twenty-four Standard Histories,* which had been halted in 1966, resumed in 1972, and by the end of 1977 all the dynastic histories were in print. A much larger number of popular historical pamphlets aimed at the masses started to appear in bookstores in early 1971. A typical first edition of a 50- to 100-page "small" history runs from 200,000 to 500,000 copies; a second printing frequently brings the total to nearly a million. Pamphlet series included *Zhongguo jindai shi zongshu* (History of Modern China Series) prepared by the history departments of Fudan University and Shanghai Normal University;[4] *Lishi zhishi duwu* (Readings for Historical Knowledge)—22

[4] Available with "some editorial changes" in English translations published by the Foreign Languages Press, Beijing.

titles on Chinese history and 30 titles on foreign history were in print in 1977, some of these being revised versions of pamphlets that had appeared before 1966; and *Xue dian lishi zongshu* (Learn a Little History Series)—26 titles on Chinese and foreign history in print by 1977. The pamphlets on Chinese history mostly reflect the version of "China's history in Marxian dress" elaborated in the 1950's.[5] But those published between 1973 and 1976 echo some of the historical interpretations ascribed to the Gang of Four. Those on foreign history introduce Chinese readers to the Western revolutionary tradition through brief biographies of, for example, Luxemburg, Bakunin, Lassalle, and Zeitlin, or simplified treatments of the Paris Commune, the First and Second Internationales, the revolutions of 1848, the American War of Independence, the independence struggles of Latin America, and slave revolts in ancient Rome.

However, since 1977 there are hopeful signs that academic research and scholarship are being revalued for their potentially enormous contribution to achieving the "four modernizations," which are the explicit goals of the Hua-Deng leadership. In these new circumstances, it is reasonable to predict that the serious study and writing of history will flourish—at least to the extent that they did in the 1950's. Pre–Cultural Revolution journals have reappeared, and new ones have been started—for example, *Shijie lishi* (World History) and *Lishixue* (Historiography). Some scholarly monographs are now being published, and plans for extensive documentary publications are being considered.

It is significant that Chinese historians have acknowledged the worth of foreign scholarship on China and said that they desire contact with foreigners. For example, China's most eminent senior historian, Gu Jiegang, in an article in the *Guangming ribao* of March 11, 1978, conceded that "many foreigners have now surpassed us in the study of Chinese history—including historical geography, nationalities, culture, and all aspects of the arts. Now is the time to greatly activate our international academic exchanges. But only if we catch up from behind can we mutually advance and progress together." In the same issue of that newspaper, Cai Meibiao, head of the General History Department at the Institute of Modern History wrote: "In recent years foreign scholars have published many works, including reference books and research tools, on many areas of the social sciences and especially on the history and culture of China. They are worthy of being studied and used by us. In line with the spirit of 'let a hundred schools of thought contend,' we should create the conditions to strengthen international academic and cultural exchanges. This is fully necessary in order to develop China's social sciences and historical science."

[5]For history in the PRC before the Cultural Revolution, see Albert Feuerwerker, "China's History in Marxian Dress," *American Historical Review*, 66, 2 (Jan. 1961): 323–53; Feuerwerker and S. Cheng, *Chinese Communist Studies of Modern Chinese History* (Cambridge, Mass.: Harvard University Press, 1961); Feuerwerker, ed., *History in Communist China* (Cambridge, Mass.: MIT Press, 1968); and Frederic Wakeman, Jr., "Historiography in China After 'Smashing the Gang of Four,' " *China Quarterly*, 76 (1978): 891–911.

It is also significant that institutional structures for historical research and teaching are expanding. Almost nothing is known about the Institute of History during the Cultural Revolution and early 1970's.[6] But a new Chinese Academy of Social Sciences (CASS) was established in 1977. Four of the Academy's institutes are directly concerned with historical research: the Institute of History, the Institute of Modern History, the Institute of World History, and the economic history department of the Institute of Economics. They enrolled research students for advanced training in the fall of 1978 for the first time since the Cultural Revolution.

In early 1978, the CASS history institutes, sometimes in conjunction with universities, began to convene meetings of historians to develop eight-year plans for research. The Institute of Modern History, for example, has outlined a major project to compile a "dynastic history" of the Republic of China (1912–49). Symposia have also been held on such subjects as the Taiping Rebellion, the 1911 Revolution, the reasons for China's economic and technological backwardness after the Ming dynasty, and the historical role of middle and small landlords in premodern China. The older historians, who were major contributors to the historical output of the 1950's and were once reviled as "bourgeois" and "capitalist roaders," were very much in evidence at these historical meetings. Little is known about younger scholars in their thirties, forties, and fifties. They have had scant opportunity since 1958 for research and publication. The history of Chinese historiography in the post-Mao era still lies ahead.

## *ECONOMICS*

### *Dwight H. Perkins*

Economics is one of the few social sciences whose right to exist has not been questioned in China. *Das Kapital,* after all, is a book about economic development, and China's contemporary economy has been constructed on principles that are said to derive from Marx or from Marx, Lenin, and Mao. It follows that some group had to be given the task of interpreting what Marxist or Marxist-Leninist-Maoist economic principles were in the Chinese context. The issues involved were too fundamental to be left to professional economists, and at times everyone from members of the Politburo to workers and peasants has participated in discussions of the correct line in economics. Even though professional economists have not had the field to themselves in China, they have often played an important role in debates about economic organization.

[6]Pichon P. Y. Loh, "Report from China: The Institute of Modern History, Peita and the Central Institute of Nationalities," *China Quarterly,* 70 (1977): 383–89.

In China it is not easy to specify just who is and who is not a professional economist. Academic economists in universities and research institutes clearly belong in this category, but what about the far larger number of people involved in planning and managing the economy? A planner who draws up the material balances (a method for relating inputs and outputs) is doing the work of an economist even if he is an engineer. One important issue is whether a distinction should be made between economics and management. In the United States, university departments of economics are usually separate from management or business schools, but the subject matter covered increasingly overlaps. The Chinese currently make a distinction between economics and industrial economics that is similar to the American economics–business school categorization.

The following discussion analyzes to some extent the economics profession in China in the broadest sense of the term, but the principal emphasis is on the activities of the major economics research institutes and university departments of economics and on industrial economics. Any attempt to deal with economics in the broader sense would become a general discussion on economic planning and organization.

There were quite a few well-trained economists in academic and government jobs in China before 1949. Although some of these economists emigrated to the United States or Taiwan in the 1940's, others stayed in China and sometimes held important posts. Ma Yinchu, the president of Beijing University in the 1950's and at the time an active participant in economics debates, held a Ph.D. in economics from Columbia University. Even in the 1970's, the economics faculty of Beijing University contained two men who had received Ph.D.'s in economics from Harvard, and there was at least one Yale graduate teaching at Nankai University.[1] Similarly, in the Chinese Academy of Sciences (CAS), a deputy director of the Institute of Economics in the 1960's, Wu Baosan, held a Ph.D. from Harvard.[2]

Although some economists trained in the West continued to hold important posts in China after 1949, their role was in no way comparable to that of Western-trained natural scientists. Whereas the natural sciences have a reasonable claim to universality regardless of the economic or social system, economics is more bound to institutions. Marxist political economy and economics as it is practiced in the West may overlap in some areas, but they are decidedly not the same. Similarly, the principles governing the behavior of state-owned enterprises directed by a central plan differ from those of private enterprises operating in response to market forces. Many, perhaps most, of the economists staffing Chinese economic research institutes and university economics departments either acquired their

[1] See John Kenneth Galbraith, *A China Passage* (Boston: Houghton Mifflin, 1973), p. 31. Galbraith visited the Beijing University economics faculty in 1972, and Bruce Reynolds and others met the Nankai University faculty in 1973.

[2] Chu-yuan Cheng, *Scientific and Engineering Manpower in Communist China, 1949–1963* (Washington, D.C.: National Science Foundation, 1965), p. 554.

knowledge through direct experience in economic affairs, such as land reform and planning, or were trained in economic subjects in Chinese universities after 1949.[3]

Before the Chinese Academy of Social Sciences (CASS) was established in 1977, the premier center of economic research was the CAS, which had Institutes of Economics and World Economy. In addition, the Chinese Academy of Agricultural and Forestry Sciences had an Institute of Agricultural Economics. With the formation of the CASS, existing institutes of economics were consolidated and new ones created. As of 1978, there were Institutes of Economics, World Economics, Industrial Economics, Agricultural Economics, and Trade and Commerce. As of 1973, most universities only had departments of political economy, although at least two also had departments of industrial economics.[4] Since 1973 departments specializing in economics have probably proliferated, but details about them will have to await further research and visitors' reports.

The division of labor among the various economics institutes is reasonably clear in some cases and somewhat murky in others. The Institute of Economics is concerned with the history and content of both Marxist and capitalist economic theory and with Chinese economic development before and after 1949. The Institute of World Economics is as its name implies. The Institute of Industrial Economics covers such topics as enterprise management, industrial statistics, and industrial accounting. Graduates in industrial economics, like American business school graduates, are directed toward jobs in industry, while political economy graduates are more apt to become teachers.[5] Presumably the Institutes of Agricultural Economics and Trade and Commerce have a practical orientation.

In appraising the activities of these various institutes and departments of economics, it is well to remember that the social sciences failed to progress during the Cultural Revolution (1966–69) and its aftermath.[6] For example, the head of the Institute of Economics, Sun Yefang, was severely criticized during the Cultural Revolution and did not reappear in public until 1978. The Institute's major journal, *Jingji yanjiu (Economic Studies)*, ceased publication at the beginning of the Cultural Revolution, only resuming in 1978. Thus, any discussion of economic research in China is about research in the 1950's and early 1960's and a research agenda for the 1970's and 1980's.

In 1978 the Institute of Economics announced a plan for the study of eco-

[3]This statement is based on a limited sample of economists and the fact that 49,100 students in economics and finance were graduated in China in the 1950's (Leo Orleans, *Professional Manpower and Education in Communist China* [Washington, D.C.: National Science Foundation, 1961], p. 128).

[4]Bruce Reynolds, "Notes on a Study Trip to China" (1973).

[5]*Ibid.*

[6]See Mu Shih, "Research Work in Philosophy and Social Sciences Unshackled," *Peking Review (PR)*, 1978, 19 (May 22): 17.

nomics during the next three to eight years.[7] Apparently, over 100 of the 500 topics listed on the agenda are "of immediate concern for China's socialist construction."[8] These include national economic planning, price theory, accounting theories, and statistics. If the past is a guide, much of the work under these headings will be highly practical in nature, dealing with such questions as what kinds of statistics should be collected or what accounting categories should be used by industry. In the 1950's this kind of work was mostly published in such journals as *Tongji gongzuo (Statistical Work)* or *Jihua jingji (Economic Planning)*.

Economists' contributions to socialist construction, however, are not confined to these applied topics. A major task of economic theorists in the late 1970's, for example, was "to refute the distorted Marxist political economy spread by the Gang of Four."[9]

Probably the greatest theoretical controversy throughout the PRC's history has concerned the role of profits in a socialist economy. Chinese planners began experimenting with various means of enhancing the role of profits as a success indicator for managers of state-run enterprises in the mid-1950's. After a brief eclipse during the Great Leap Forward (1958–59), profits were once again a major topic for discussion in theoretical journals. During the Cultural Revolution, Sun Yefang was attacked for his theory of "putting profits in command" only to be restored to an influential position when it was pointed out that he was actually showing how profits under socialism differed from those under capitalism. The issue was not whether profits were an effective means of eliciting good performance from managers, but whether certain kinds of profits were appropriate in an economy operated on the principles of Marx, Lenin, and Mao.

A second major category of economic analysis in China involves summarizing China's economic development. The great variety of studies falling under this category ranges from broad interpretative studies to narrower, more focused efforts. For example, in 1962 Xu Dixin published a broad study of China's economy during the transitional period (1949–57), and Li Chengrui published a book on agricultural taxation.[10] Both books are written from a Marxist perspective, but are different in style. Xu's study is essentially a nonquantitative description of the changes in economic institutions in the 1950's. Li's book, in contrast, is highly quantitative (in the sense of using many statistics) and goes well beyond simply describing the tax system to analyzing such topics as the declining burden of agricultural tax.

[7] *PR*, 1978, 22 (June 2): 4.

[8] New China News Agency dispatch, Beijing, May 29, 1978; in *Foreign Broadcast Information Service (FBIS)*, May 31, 1978, p. E16.

[9] *Ibid.*

[10] Xu Dixin, *Zhongguo guodu shiqi guomin jingji de fenxi* (Analysis of China's Transition Period Economy) (Beijing: Renmin Chubanshe, 1962); and Li Chengrui, *Zhonghua renmin gongheguo nongye shi shigao* (A Draft History of the Agricultural Tax in the People's Republic of China) (Beijing: Caizheng Jingji Chubanshe, 1962). There is a Japanese translation published by the Institute of Developing Economics (Tokyo, 1968).

Chinese economists are also engaged in studying China's pre-1949 economic history. The large compilations of data from the nineteenth and twentieth centuries are particularly valuable. The three-volume *Zhongguo jindai nongye shi ziliao* (Materials on the History of Modern China's Agriculture), published in 1957, for example, contains several thousand pages of qualitative and quantitative data.[11] The organization of these volumes is in accordance with Marxist categories, but the text consists mainly of straightforward statistical tables on everything from food prices to tenancy levels and of excerpts from primary sources. There are similar collections of data for modern industry and handicrafts. Work on Chinese economic history, however, is not confined to compilations, statistical or otherwise. There are also interpretative essays, usually designed to show that China's economy has passed through stages similar to those described by Marx for the West, or that imperialism was responsible for China's failure to develop before 1949.[12]

Works on foreign capitalist economies have concentrated on negative aspects. Inflation and unemployment, for example, have received a good deal of attention. Works on non-Marxist economic theory have also been critical in nature. Titles of articles from the 1950's include "A Critique on Keynes's Economic Thought" and "A Critique on the National Income Studies of China by Bourgeois Economists."[13] In less public settings, however, the Chinese have expressed considerable interest in some aspects of Western economic theory. Wassily Leontief, for example, has lectured in China on input-output analysis, and CASS officials have indicated a desire to learn more about systems analysis. As of 1978, there was still considerable emphasis on criticizing bourgeois economic theories and some stress on the not inconsistent goal of learning something about them.[14]

This discussion of some of the major categories of economic research in China is in no sense exhaustive, but does indicate the wide range of topics that are of interest to economists in the People's Republic. The topics on the research agenda of the late 1970's were not dramatically different from those of the 1950's and early 1960's. Furthermore, all analysis continued to be based on Marxist political economy. Whether China's economists in the 1980's will make positive use of at least some aspects of Western economics, however, remains an open question.

[11] Chinese Academy of Sciences, Institute of Economics (Li Wenzhi comp.), *Zhongguo jindai nongye shi ziliao* (Beijing: Sanlian, 1957).

[12] See Albert Feuerwerker and S. Cheng, *Chinese Communist Studies of Modern Chinese History* (Cambridge, Mass.: Harvard University Press, 1961), pp. 168–207.

[13] *Jingji yanjiu*, 1957, 6 (Dec.): 39–59; and *Jingji yanjiu*, 1958, 2 (Feb.): 51–69. The later article is by Wang Jingyu, Ma Liyuan, Huang Fangzhang, and Zhang Zhuoyuan.

[14] See notes 6 and 8.

# *POLITICAL SCIENCE*

*Harry Harding*

Political science does not, as yet, have a separate identity as an academic discipline in China.[1] As of early 1979, there was no institute of politics at the Chinese Academy of Social Sciences (CASS), no journal of political science, and no professional association for political scientists. Relatively few universities had departments of politics. Although many normal universities had departments of "political education," these tended to encompass all the social sciences and bore little resemblance to political science departments in the West. Even the Chinese translation of the Western term "political science" (*zhengzhi kexue*) sounded unfamiliar to most Chinese social scientists.

Nevertheless, a chapter on political science in China still has a place in this volume. First, Chinese social scientists do study problems, although under different headings, that Western scholars consider to be facets of political science. Chinese legal scholars who study the Chinese electoral system, foreign-area specialists who study the structure and operation of the Soviet Union's Communist Party, and economists who analyze the role of developing countries in world politics have much in common with political scientists in Western Europe, Japan, and the United States.

Second, it seems highly likely that political science will gradually develop a distinct disciplinary identity. At the Fifth National People's Congress in February 1978, Premier Hua Guofeng called for a "national development plan for philosophy and the social sciences" that would encompass research into the "past as well as the present state of Chinese and world politics."[2] In early 1979 the CASS established an Institute of World Politics to conduct research on international and comparative politics and foreign political philosophy. The Academy was also exploring the possibility of creating other institutes for such subjects as Marxism-Leninism-Mao Zedong Thought and scientific socialism—both of which would have significant parallels with Western political science.

Third, and most pragmatically, a chapter such as this can facilitate the growing exchanges between Chinese and American social scientists. The Chinese have already begun inviting American political scientists to lecture

[1] Much of the information in this article was obtained during a nine-day visit to the CASS in December 1978. I would like to express my gratitude to John Lewis and Douglas Murray for arranging my visit; to Mr. Tang Kai, then director of the Academy's Foreign Affairs Bureau, for organizing my stay in Beijing; and to my hosts, Liang Wensen of the Institute of Economics and Yang Aiwen of the Academy, for their hospitality, assistance, and friendship. Ann Fenwick of Stanford University provided invaluable help in surveying the major Chinese social science journals during the 1950's, 1960's, and 1970's. Her assistance, in turn, was made possible by a grant from Stanford's Center for East Asian Studies.

[2] Hua Guofeng, "Unite and Strive to Build a Modern Powerful Socialist Country: Report on the Work of the Government Delivered at the First Session of the Fifth National People's Congress [Feb. 26, 1978]," in *Peking Review (PR)*, 1978, 10 (Mar. 10): 28.

at the CASS. Many American political scientists, for their part, would like to engage in scholarly exchanges with their Chinese counterparts. But to make this possible, it is clearly necessary to understand the disciplinary headings under which Chinese social scientists conduct research on political issues and the major institutions in which such work is undertaken.

Political science has never been a highly developed academic discipline in China. There was, to be sure, some interest in Western political science before 1949. Western works on politics were translated into Chinese. A few Chinese scholars, such as Qian Duansheng (Ch'ien Tuan-sheng), wrote original works on Chinese politics, comparative government, and international relations. Quite a few universities established departments of political science and offered courses on public administration, international relations, comparative politics, and political theory. But political science did not develop as rapidly as other branches of social science, such as sociology and economics. In large part, this was because those who taught politics also sought either to practice it or to influence it. In times of political upheaval, this was conducive to the publication of a wide range of writings on China's political and social problems, but not to the development of political science as an academic discipline.

With the establishment of the PRC in 1949, even this weak base was almost completely destroyed. Political science was described as little more than an attempt, fostered by Westerners, to propagate bourgeois political theory in China. Most university departments of political science were abolished, and the disciplinary subject matter was divided, largely along Soviet lines, among other branches of social science. Thus, general political analysis, particularly the study of capitalist and socialist political systems, was assigned to political economy. The study of the Chinese political and administrative structure was incorporated into jurisprudence. Normative political theory became the province of philosophy. And the study of international politics and law was taken over by university law faculties and occasionally by separate departments of international relations.

During the Hundred Flowers period, several political scientists, including Qian Duansheng, proposed a partial restoration of political science as an academic discipline. Qian advocated the creation of a law institute similar to the London School of Economics, containing a department of political science, as well as departments of economics, sociology, and law. Others called for the resumption of the study of comparative politics and Western political science, a revamping of the field of international politics, and the reestablishment of political science departments in major Chinese universities. These proposals were rejected and described as a backhanded effort to restore the "reactionary" form of political science taught before 1949.[3]

[3] Wu Enyou, "Expose the Reactionary Nature of the Politics Departments in the Old Universities," *Zhengfa yanjiu*, 1957, 6 (Dec.): 47–49. See also H. Yuan Tien, "Is Sociology Dead in Communist China?" *American Sociological Review*, 27, 3 (June 1962): 413–14; and C. J. Ch'en, "Chinese Social Scientists," *Twentieth Century* (London), 163, 916 (June 1968): 511–22.

Divided among a number of new disciplinary homes, the academic analysis of politics languished in the 1950's. *Jingji yanjiu (Economic Studies),* the most important scholarly journal in economics, published relatively little even on Marxist political economy. A few textbooks and lecture series appeared, but Chinese universities relied principally on translations of standard Soviet texts, including A. Leontiev's *Political Economy.* The leading journal of international affairs, *Shijie zhishi* (World Knowledge), served mainly as a chronicle of current events and published virtually nothing of an academic character. Even the journal of the Chinese Society of Politics and Law, *Zhengfa yanjiu* (Research in Politics and Law), dealt primarily with judicial matters, such as the role of lawyers, rules of evidence, trial procedures, and the operation of state control agencies. Only a minority of its articles were concerned with political structure and process.

The early 1960's saw a slight change of emphasis, as Chinese political economists sought to develop a theory of socialist society that would be clearly distinct from the "revisionist" variants of political economy and scientific communism produced in the Soviet Union. A major stimulus to this endeavor was Mao Zedong's own criticism of a Soviet textbook on political economy. In a commentary written in the early 1960's, Mao charged that Soviet political economists had systematically ignored the role of the superstructure in promoting economic development, and the corresponding need for continuing revolution in the superstructure and in production relations in order to maintain progress toward communism.[4] Perhaps as a result of Mao's intervention, Chinese scholarly journals, particularly *Zhengfa yanjiu,* did begin to publish more articles on political economy in the early 1960's, addressing such topics as the continuation of class struggle under socialism, the management of contradictions in socialist society, and the role of mass movements in economic construction.

The onset of the Cultural Revolution in 1966–67 brought this work—along with work in virtually every other scholarly discipline—to an abrupt halt. With the reopening of the universities in the early 1970's, however, theoretical work in political economy was resumed and given perhaps more attention than at any time since 1949. Several university-level textbooks were prepared, one of which was published in Shanghai in 1975. Others were circulated widely in draft form.[5] The radical Shanghai journal *Xuexi yu pipan* (Study and Criticism) also ran a number of articles on various as-

[4] Mao Zedong, "Reading Notes on the Soviet Union's *Political Economics,*" in *Miscellany of Mao Tse-tung Thought (1949–1968), Part II,* tr. Joint Publications Research Service, JPRS-61269–2 (Arlington, Va.: JPRS, Feb. 10, 1974), pp. 247–313.

[5] Draft textbooks were prepared by a writing group under the Shanghai Party Committee and by the Institute of Economics of Nankai University. See *Hongqi,* 1978, 4 (Apr.): 75–81; and *Tianjin ribao,* Apr. 9, 1978, in *Foreign Broadcast Information Service Daily Report: People's Republic of China,* Apr. 12, 1978, pp. K5–7. The published text is *Zhengzhi jingjixue jichu zhishi,* 2 vols. (Shanghai: Shanghai Renmin Chubanshe, 1975). For an English translation, see George C. Wang, ed., *Fundamentals of Political Economy* (White Plains, N.Y.: M. E. Sharpe, 1977).

pects of political economy, including some written by political economists at Fudan University. The basic purpose of these materials was to offer a theory of the emergence of a new bourgeois class in socialist society. Most traced it to the maintenance of "bourgeois rights" under socialism (such as distribution according to labor) and called for steady progress in reducing all social and economic differences between leaders and led, city and countryside, and mental and manual labor. Retrospectively, these books and articles have been criticized as the theoretical foundation for the Gang of Four's efforts to defend the "newborn things" of the Cultural Revolution and to identify veteran cadres as a new bourgeois class within the Party.[6]

Since the fall of the Gang of Four in October 1976, research in China on political questions has taken a new set of directions. Before discussing them, however, we must first identify the disciplines that roughly correspond to political science in the West and the major institutions in which research and teaching are conducted.

Research into political questions in China falls under no fewer than seven academic disciplines:

1. Political economy. Recent Chinese textbooks define political economy as the analysis of the patterns of ownership of the means of production, the relations among people engaged in production, and the patterns of distribution of production output. Much of political economy, then, corresponds more to Western economics than to Western political science. But the study of the relations of production requires, the Chinese say, the analysis of two further aspects of society that are more political in nature. These are the forms of class struggle in various types of societies and the relationship of the political, legal, and ideological superstructure to a society's economic base. In this sense, Chinese political economy does roughly correspond to the study of political development and comparative politics in the West, and particularly to Western attempts to create general models of the political process.

Virtually every university in China offers courses in political economy, either through political economy departments or as specialties in departments of economics or political education. Research in political economy is conducted in the institutes of economics attached to universities and the CASS. The Academy, for example, has a Department of Political Economy under its Institute of Economics. (Some of the research projects of that department are discussed briefly below.)

2. International relations. A broader discipline than its counterpart in the United States, international relations in China often includes the study of Western political philosophy and the political systems of foreign countries, as well as relations among nations. Departments of international relations,

[6]For an authoritative set of essays criticizing the political economy of the Gang of Four, see CASS, Institute of Economics, *"Sirenbang" dui Makesizhuyi zhengzhi jingjixue de cuangai* (The Tampering with Marxist Political Economy by the Gang of Four) (Taiyuan: Shanxi Renmin Chubanshe, 1978). Other critical articles appeared in *Hongqi* and *Jingji yanjiu* in 1978.

such as those at Beijing and Fudan Universities, may therefore offer courses on Chinese foreign policy, capitalist and socialist forms of imperialism, foreign political thought, international law and organization, comparative politics, national liberation movements, and the history of the international communist movement.

As for research, the CASS has two institutes concerned with international relations. The Institute of World Politics, established in early 1979, plans to undertake research on foreign political systems, international politics, and foreign political thought, with particular attention to international security issues and Soviet affairs. And although most of the research conducted at the Academy's Institute of World Economics emphasizes economic developments, some projects, to be discussed below, address important political issues. Studies of international relations reportedly are also undertaken at the Institute of International Problems attached to the Ministry of Foreign Affairs.

3. Law. Discussed in much greater detail in Jerome Cohen's contribution to this volume, the study of law in China appears to parallel Western political science at several points. Legal institutes address the more structural aspects of Chinese politics, such as the system of elections to people's congresses, the electoral process within the Party, and the relationship between the Party and the state. Within the CASS Institute of Law, such questions appear to be the province of the Departments of Constitutional Law and Theory of the State and of Law. Instructional programs in law, designed principally to train officials for government service, have been greatly expanded since 1977. Undergraduate courses in law are offered by specialized institutes of political science and law and by law departments in a number of comprehensive universities. Graduate programs are offered by university law departments and legal research institutes.[7]

4. Philosophy. Philosophers in China often consider some of the more normative aspects of political structure and process in socialist systems. These include such issues as the appropriate scope of the dictatorship of the proletariat, the proper extent of socialist democracy, and the correct relationship between the Party and the state. Moreover, the discipline of philosophy shares an interest in the relationship between the superstructure and the economic base with political economy and a concern with foreign political thought with the discipline of international relations. Philosophy departments also address the epistemological questions common to all the social sciences, particularly the degree to which the laws of social science are independent of class standpoint or possess a "class character."

In the CASS Institute of Philosophy, these issues are addressed principally by three research departments: Dialectical and Historical Materialism, Marxist-Leninist Philosophy and History, and Mao Zedong Thought. Because of the Institute's unwieldy size, it is likely that some of these depart-

[7] *PR*, 1979, 23 (June 8): 6–7.

ments, particularly those dealing with Marxism-Leninism and Maoism, may eventually become independent research institutes. Instruction in philosophy is usually offered by philosophy departments in comprehensive universities and by departments of political education in normal schools.

5. Foreign-area studies. Foreign-area studies is a growing interdisciplinary field in China, with research institutes at a number of universities devoted to the study of particular countries and regions. These include the Institute of Soviet and Eastern European Studies at Chinese People's University, an institute of American studies at Nankai University, and institutes on Japan and Korea at Jilin University. Beijing University reportedly has an Institute of Asian and African Studies and operates an Institute of South Asian Studies jointly with the CASS. The Academy, for its part, is likely to establish additional area studies institutes in the relatively near future. All these institutes, of course, deal with the government and politics of the countries and regions in which they specialize.

6. Contemporary Chinese history. For Western political scientists interested in China, scholars studying contemporary Chinese history may be natural counterparts. In China, contemporary history is defined as beginning with the May Fourth Movement of 1919 and is further divided into the history of the Chinese Communist Party, the Chinese Revolution, and the PRC. All universities offer courses in these areas, either under a department of history or, in the case of normal universities, under a department of political education. (The Chinese People's University has a separate department of Chinese Communist Party history.) At the national level, research on contemporary Chinese history is divided between the Higher Party School, which conducts work on Party history, and the CASS Institute of Modern History, which engages in research on the history of the Revolution and the PRC. Textbooks on all three topics are in preparation.

7. Scientific socialism. Of great potential interest is the emergence of a new social science discipline in China called scientific socialism, which roughly corresponds to the discipline of scientific communism in the Soviet Union. In essence, scientific socialism is a branch of political economy dealing primarily with the dynamics of socialist society and, to a lesser degree, with the process of proletarian revolution in capitalist societies and the transformation of socialism into communism.

As of late 1978, there was a debate among Chinese social scientists over the boundaries of this new discipline. One view was that scientific socialism should be defined rather narrowly as the study of the international communist movement, proletarian revolution, and class relations and class struggle under socialism. Another view held that scientific socialism should be considered the vehicle for a comprehensive analysis of the economic, political, and sociological aspects of socialist society, regardless of their relation to "class struggle." At that time, the first view appeared to be prevailing, largely because a narrower definition of the scope of scientific socialism would prevent it from competing with such established disciplines

*Correspondences Between Chinese Social Sciences and Western Political Science*

| Chinese discipline | Equivalent branches of Western political science | Place in university curriculum: Comprehensive universities | Place in university curriculum: Normal universities | Place in Chinese Academy of Social Sciences (CASS) |
|---|---|---|---|---|
| Political economy | Corresponds to political economy and to general theories and models of comparative politics | Separate department or part of economics department | Part of department of political education | Department of Political Economy under the Institute of Economics |
| International relations | Corresponds to international politics; in some Chinese universities contains foreign political philosophy and comparative politics | Often offered as a separate department | Apparently rarely offered | Institutes of World Economics and World Politics |
| Law | Includes public law and some aspects of comparative politics and political philosophy | Occasionally exists as a separate department | None known | Departments of Constitutional Law and Theory of the State and Law under the Institute of Law |
| Philosophy | Includes political philosophy, epistemology, and some aspects of comparative politics | Usually exists as a separate department | Part of department of political education | Departments of Dialectical and Historical Materialism, Marxist-Leninist Philosophy and History, and Mao Zedong Thought under the Institute of Philosophy |
| Foreign-area studies | Includes comparative politics | Some research institutes attached to universities | None known | Institute of South Asian Studies; more area studies institutes are highly likely |
| Contemporary Chinese history | Corresponds to contemporary Chinese politics | Usually exists as part of history department | Part of department of political education and department of history | Department of Contemporary History under the Institute of Modern History; may become an independent institute |
| Scientific socialism | Corresponds to political sociology, comparative politics (especially comparative communism), and some aspects of international relations | Offered relatively rarely | Part of department of political education | None |

as political economy and philosophy. But there were clearly some social scientists who believed that China needed a stronger disciplinary base for research on contemporary political and social problems and hoped that scientific socialism might eventually serve that function.

All normal universities have received instructions from the Ministry of Education to introduce courses on scientific socialism in their departments of political education. In the comprehensive universities, however, scientific socialism has a lesser role to play. With the exception of the Chinese People's University, which established a department of scientific socialism when it reopened in 1978, few other universities appear to offer courses in this new discipline. Nor does an institute of scientific socialism exist in the CASS, although it is possible that one might be created in the early 1980's.

The accompanying table summarizes this discussion by listing the seven social science disciplines discussed above and noting the ways in which they correspond to aspects of Western political science. It also indicates the place of each discipline in the curriculum of Chinese comprehensive and normal universities and in the CASS.

Since the purge of the Gang of Four in October 1976 and especially since the establishment of the CASS in May 1977, new research trends have been evident in dealing with various international and domestic political questions. This section seeks to summarize these. Rather than proceeding discipline by discipline, however, it groups research trends into five categories more familiar to Western political scientists.

1. International relations. Some of the most interesting research currently being conducted in China involves problems of international relations. Much of this work is done by the CASS Institute of World Economics, but some is conducted at various foreign-area studies centers. The establishment of the CASS Institute of World Politics should be a further stimulus to work in this area.

International relations research focuses on three principal topics. The first is the analysis of the general global system within the framework of the "theory of the three worlds" set forth by Mao Zedong and Deng Xiaoping in 1974. The important exegesis of the theory that appeared in *Renmin ribao (People's Daily)* in November 1977 was drafted, I was told, by a CASS writing group headed by the Academy's president, Hu Qiaomu.[8] Since then, scholars have attempted to find clearer economic and political criteria for distinguishing the superpowers, the developed capitalist and socialist countries, and the Third World developing nations.

Second, scholars of international relations have been studying questions of international political economy, particularly Third World demands for a new international economic order. They have been trying to understand the

[8] "Chairman Mao's Theory of the Differentiation of the Three Worlds Is a Major Contribution to Marxism-Leninism," *PR*, 1977, 45 (Nov. 4): 10–41.

origin and content of these demands and to forecast if they will be successful. Scholars at the Institute of World Economics have been especially interested in the future of OPEC (Organization of Petroleum Exporting Countries) and the possibility of creating comparable producer cartels in other export commodities.

Third, international relations specialists at a number of institutions have attempted to define social imperialism, identify the major characteristics of social-imperialist states, and explain why and how the Soviet Union became such a power.[9] As of December 1978, the latter question had aroused substantial controversy. Most scholars argued that the Soviet Union became a social-imperialist country only with the rise of Brezhnev in 1964, and that this transformation was first reflected in the Soviet invasion of Czechoslovakia in 1968. Others, in contrast, pointed out that the origins of Soviet expansion can be found in the Khrushchev and even Stalin periods.

2. Western politics. The study of Western capitalist systems has been concentrated in the Institute of World Economics, with additional work conducted at the Institute of World History and in some university institutes of foreign-area studies. Of particular interest to Chinese social scientists has been the changing relationship among the state, the monopoly capitalists, and the proletariat in Western capitalist countries. Two specific issues have apparently caused some debate. The first is whether the relationship between the state and monopoly capital has reached a new stage in recent years. Some argue that it has, and that growing governmental intervention in the economy has intensified the contradictions between capitalists and the state. Others reply that no fundamental change has occurred, and that capitalists have found mechanisms for evading or manipulating governmental controls. The second issue concerns the status of the proletariat. Some scholars assert that the condition of the proletariat in most Western countries is getting progressively worse, while others believe that the proletariat's living standard may be falling relative to those of other classes, but is rising in absolute terms. The first group of scholars conclude that the capitalist system is "decadent and dying," while the others insist that it is "not totally decadent," and that it is "dying, but not yet dead."

3. Third World politics. In studies of Third World countries, Chinese social scientists have shown increasing interest in creating typologies of political and economic systems. Increasingly they seem to be aware that the developmental stages posited by Marxist-Leninist political economy do not apply to the Third World and that some new categories are needed. The main questions for research, then, concern the principal roads of political and economic development taken by Third World countries since independence and the factors that have shaped the path a particular nation has followed. Researchers at the Institute of World Economics have differed over the like-

[9]See, for example, CASS, Institute of World Economics, "Soviet Social-Imperialism: Most Dangerous Source of World War," *PR*, 1977, 29 (July 15): 4–10, 21.

lihood of a Third World nation ever evolving into a typical capitalist country. Some argue that this will happen, while others reply that capitalism is a phenomenon restricted to existing developed capitalist states in the West.

4. Soviet politics. Descriptive work continues, as it has for some years, on the political system of the Soviet Union. This involves analysis of class relations and "national contradictions" in the Soviet Union, as well as descriptions of Soviet law, political theory, state organization, Party development, and mass administration. Although the Soviet Union is still regarded as revisionist, there has been some debate over the time at which the Soviet Union became a revisionist society. The most common date given is 1956, when Khrushchev gave his secret speech denouncing Stalin to the Twentieth Party Congress. But some scholars have begun to trace the roots of revisionism to earlier dates, pointing to the emergence of excessive wage differentials during the Stalinist period. Even more interesting is an emerging reevaluation of the Soviet political and economic system. Compared with the period from 1963 to 1976, when Chinese writers described all aspects of Soviet life as revisionist, Chinese scholars have more recently begun to ask whether certain features of Soviet politics and administration remain socialist, rather than revisionist in nature. No conclusions have yet been reached, but Chinese political scientists appear to be aware that such a question demands a more sophisticated and discriminating description of Soviet society than they have yet developed.

5. Socialist politics and socialist political theory. In addition to engaging in criticism of the theories of political economy allegedly developed under the influence of the Gang of Four, Chinese scholars have also conducted research on the class structure of Chinese society and the characteristics of class struggle under socialism. As mentioned above, radical political economists argued in the mid-1970's that the principle of distribution according to labor, along with other "bourgeois rights," was an important source of new bourgeois elements in socialist society. To refute this view and thus to lay the theoretical foundation for a restoration of material incentives, four national conferences were held in 1977 on the question of distribution according to labor, another national conference was held in Changsha on the new bourgeoisie in China, and a research project was conducted in the Wenzhou area on the origins of new bourgeois elements. After much discussion, the conferences reportedly concluded that new bourgeois elements would not emerge from a graduated wage system if wage differentials were not excessive. The Wenzhou investigations found that new bourgeois elements emerged not from the application of bourgeois rights, but rather from the breakdown of law and discipline under the influence of the Gang of Four, which permitted officials to misuse their positions for personal or factional gain.[10]

On the related question of class struggle in China, social scientists have

[10] On the Wenzhou investigations, see *PR*, 1979, 5 (Feb. 2): 9.

been discussing the proper scope of the dictatorship of the proletariat. The Gang of Four, particularly Zhang Chunqiao, have been accused of advocating an all-round dictatorship and thereby exaggerating the scope of class struggle. In reaction to this, some philosophers and political economists have argued that in a relatively mature socialist system, such as China's, few social and political tensions reflect class struggle. In the same vein, historians have begun to reinterpret the record of intra-Party disputes and have suggested that relatively few actually represented instances of class struggle. Both conclusions stem directly from the new Party policy, announced at the Third Plenum of the Central Committee in December 1978, that the "large-scale turbulent class struggles of a mass character have in the main come to an end."[11]

The redefinition of the extent of class struggle in China raises the related question of the scope for socialist democracy. One topic under consideration by the Institute of Law has been the possibility of reforming the Party and state electoral systems. Scholars at the Institute have considered such changes as allowing more candidates for legislative offices than the number of seats available; holding direct elections of delegates to provincial-level people's congresses and even to the National People's Congress; and providing opportunities for write-in candidates. Direct election of such state officials as county magistrates and provincial governors is reportedly under preliminary consideration.

The expansion of political research in China since the purge of the Gang of Four can be traced to a number of factors. China's increasingly active foreign policy has created a demand for more scholars knowledgeable about foreign affairs and international issues. The need to refute the concepts and doctrines of the Gang of Four has increased the need for skilled propagandists. And the renewed interest in political liberalization and administrative reform has stimulated the development of policy analysis.

Despite these encouraging developments, political science in China still operates under a double set of constraints. As is true of virtually all the social sciences, the study of politics is largely devoted to developing sophisticated explanations and justifications of Party policies and investigating practical policy problems of interest to Party and state officials. In this sense, China's equivalent of political scientists spend most of their research time as publicists, propagandists, and policy analysts, rather than as academics. Moreover, political science also suffers from a second constraint: its lack of identity as a separate academic discipline. The study of political questions is, as we have seen, divided among a number of disciplines, with little apparent coordination.

There are preliminary signs, however, that both these constraints have

[11] "Communique of the Third Plenary Session of the 11th Central Committee of the Communist Party of China [Dec. 22, 1978]," in *PR*, 1978, 52 (Dec. 29): 6–16.

been loosened in recent years. Increasing discussions of scientific socialism and even of politics as legitimate foci of academic inquiry suggest that the study of political questions has been placed to an unprecedented degree on the Chinese academic agenda. The creation of an Institute of World Politics by the CASS will do much to reinforce the notion that political science can be a worthy academic discipline.

In addition, discussion of both theoretical and practical policy questions has become noticeably freer since the purge of the Gang of Four. Debates over the scope of scientific socialism, the date when the Soviet Union became a social-imperialist country, the nature of the "crisis of capitalism," the degree to which Soviet society still contains socialist elements, and the possibility of reform of the state electoral process suggest a heartening increase in freedom of discussion and inquiry. Political science in China still has a long road to travel, but its progress along that road may be of increasing interest to Western scholars.

## LAW

### *Jerome Alan Cohen*

At the time of the Russian Revolution of 1917 many Bolsheviks hoped that their new society would produce a truly revolutionary legal system—one that did not have to rely on complicated laws, formal institutions, and personnel trained in the law. Yet within five years that hope proved to be an illusion; the New Economic Policy that followed the initial chaotic era of War Communism gave rise to legal codes, to European-style courts, prosecutors' offices, arbitration organs, notaries, and other institutions, and to law schools that educated the personnel to staff those agencies.

One of the critical factors in this change was the need to enlist foreign cooperation in the new regime's economic development. Lenin made it clear that a formal legal system was necessary to attract British and French capitalists. Without legal codes, institutions, and specialists to rely on, foreign businessmen would lack confidence in the new government and find it difficult to conduct trade, extend loans, and make investments.

In the late 1970's China's leaders made a similar volte-face. The need to create a legal environment that would inspire confidence in foreign businessmen was an important factor in precipitating this development, and the *Renmin ribao (People's Daily)*, the voice of the Central Committee of the Chinese Communist Party, frequently quoted Lenin to justify a change of legal policy that must have been as bewildering to many Chinese as it was welcome to others.

China's legal history since the founding of the PRC in 1949 has been extremely varied. Soviet-style legislation, institutions, and legal education prevailed during the 1950's. Codes of civil and criminal law as well as procedure were drafted, although no comprehensive codes of law were actually promulgated. The National People's Congress (NPC) enacted organizational laws for the courts, the procuracy, and local government, and an elaborate court system was developed, including offices that provided legal advice and representation for people living in large cities. Lawyers in all their varied roles—law reformers, prosecutors, judges, notaries, law teachers, scholars—began to play a prominent role, and scholarly books and law reviews gradually appeared.

This early trend toward a formal legal system for the PRC was reversed by the "antirightist" campaign launched in mid-1957 and by the Great Leap Forward of 1958–59. The Cultural Revolution of 1966–69 administered the coup de grace to the system, when the *People's Daily* published such editorials as "In Praise of Lawlessness."

After Chairman Mao Zedong's death and the arrest of the Gang of Four in the fall of 1976, however, Chinese leaders gradually reasserted the importance of creating a formal legal system. Their principal reason was their determination to modernize China as rapidly as possible. They realized, as Max Weber and other social scientists have long pointed out, that law plays a key role in facilitating economic development. It provides guidelines for action, instruments for channeling transactions, and means for settling disputes. Predictability and security of expectations are crucial to development. For example, buyers must know that sellers will deliver commodities of the desired quantity and quality and at the prescribed time, place, and price. Sellers have similar needs for assurance. Unless foreign traders, bankers, and investors can feel confident that a country will act in a stable, predictable fashion, they will prove reluctant to conclude capital export contracts and to make long-term loans and investments.

In addition, law can protect the basic rights of a country's citizens. China's leaders recognize that unless talented people feel secure against arbitrary detention, suppression of thought, and confiscation of property, they will be afraid and reluctant to contribute the initiatives, innovations, and criticisms that spur progress.

The PRC began to respond to these needs in 1978. A new constitution was issued in March, and the legislative roles of the NPC and its Standing Committee were renewed after more than a decade of inactivity. A Committee for the Legal System composed of both politicians and academics was set up under the NPC Standing Committee and charged with responsibility for submitting draft legislation to the parent body. The Ministry of Justice, abolished a decade earlier, was also restored. The law departments of Beijing University and the Chinese People's University (Beijing) reopened, and an Institute of Law was created within the Chinese Academy of Social Sciences (CASS). By late 1979 two law journals were being published—*Faxue yanjiu (Studies in Law)* and *Minzhu yu fazhi* (*Democracy and Legality*)—the first since 1966. In Beijing lawyers returned to public life, and the city's offices for legal advice and its lawyers' association were reestablished, along with public notaries' offices, which play a useful role in a variety of civil transactions.

By late 1979 the Chinese press was carrying many essays on legal problems, scholars and propagandists were preparing books and pamphlets to educate the public and the bureaucracy, and traditional operas and television programs were being used to popularize the law. All these activities were guided by the Political and Legal Commission of the Party's Central Committee.

In early 1979 the PRC announced that some 30 pieces of legislation were in need of either revision or immediate drafting. In February the NPC Standing Committee revised the 1954 regulations for the detention and arrest of suspected offenders, which had long been blatantly ignored. The new legislation provides that only the police can arrest persons and imposes new limits on the length of time that the police can detain a person prior to arrest and trial. It also prescribes strict standards for arrest. This was followed by the first major piece of economic legislation of the post-Mao era, a lengthy provisional law for the allocation of forestry resources.

In July the NPC enacted revised laws concerning local government, nationwide elections, the courts, and the procuracy (the prosecutorial agency that also is responsible for checking on the legality of official conduct). In addition, it promulgated a substantial criminal law code that not only provided many badly needed general principles, but also sought to give more precise meaning to a number of vaguely defined offenses, such as "smuggling" and "speculation." The code also made punishable various other economic offenses, including fraud, tax evasion, and violation of registered trademarks. To reduce foreign exchange outlays and allow direct foreign investment in China, a law regulating joint ventures between Chinese and foreigners was also passed.

Revisions of early 1950's legislation relating to the family, trade unions, and land policies and the enactment of many new laws are expected in the 1980's. PRC leaders are especially eager to enact further economic legislation, such as corporation, patent, copyright, tax, banking, commercial, and factory laws, and an environmental protection law to cope with industrial pollution has already been promulgated. Legislation establishing export-processing zones in certain areas of China and allowing foreign firms special privileges is emerging from the pipeline. Chinese tax authorities are also planning how to deal with various types of joint ventures and compensation trade arrangements and with the complexities of royalties, construction contracts, production sharing in oil and gas and other extractive industries, bank loans, withholding taxes, and other phenomena new to a nation long isolated from economic cooperation with other countries.

As of 1979, the Chinese were confident that by 1981 their country would have a complete set of laws. In marked contrast to the law reform era of the 1950's, when the U.S.S.R. was the principal model, the Chinese in the late

1970's were consulting a wide variety of foreign models in both the industrialized and the developing world. For example, in late 1978 a delegation of experts went to Japan's Patent Office in Tokyo to study its laws. Tax specialists have been to Romania, Yugoslavia, and Japan. Experts on banking law have consulted with the Bank of Tokyo's legal documentation office.

Although the activities of the newly revived "people's lawyers" will be largely confined to domestic affairs, they may play a role in international negotiations and transactions, in addition to officials trained to represent Chinese ministries and companies. The PRC is likely to take an interest in international law and to adhere to various multilateral and bilateral arrangements protecting patents and copyrights, avoiding double taxation, and arbitrating trade and investment disputes. China may also join the International Monetary Fund and the World Bank, if the question of Taiwan's representation can be settled to Beijing's satisfaction.

Of course, China has carried on trade successfully for many years despite the lack of economic legislation. Although the difficulties of adjusting conventional international business arrangements to the PRC's unique circumstances should not be minimized, much can be accomplished even during this transitional period. The Chinese are wise to proceed with deliberate speed. They know that in East as well as West haste makes waste.

# *APPENDIXES*

# *Appendix A*

## *MAJOR SPEECHES AT THE NATIONAL SCIENCE CONFERENCE, March 18–24, 1978*

In March 1978 China held its first National Science Conference, a gigantic affair attended by some 6,000 scientists, technicians, and administrators. It was called by the Central Committee of the Communist Party to signal a major shift in science policy and to stress the important role that Chinese scientific and technical personnel would be expected to play in overcoming a decade of political disruption and achieving the government's ambitious modernization goals.

The three major speeches at this conference were made by Vice-Premier Deng Xiaoping, who, as the leading force behind China's modernization program, made the keynote address; by Fang Yi, vice-premier and minister in charge of the State Scientific and Technological Commission, who spelled out some of the national plan's scientific targets for developing science and technology; and by Party Chairman–Premier Hua Guofeng, who concluded the conference with a speech that was directed as much at the Party cadres—the managers of science—as at the scientists themselves. The importance of these speeches—which are frequently referred to by the authors of this volume and are reproduced here—has not diminished, despite some modifications in the modernization goals. The speeches reveal the thrust and priorities of China's new science policies, and barring another major political digression, they will guide China's development in science and technology for years to come.

The speeches by Deng Xiaoping and Hua Guofeng, as well as the "abridgment" of Fang Yi's report, were reprinted in the March 24, March 31, and April 7, 1978, issues of the *Peking Review.* Since the magazine did not become *Beijing Review* and convert from Wade-Giles to pinyin until January 1979, the transliteration of Chinese names was not changed.

### *Deng Xiaoping's speech at the opening ceremony of the National Science Conference, March 18, 1978*

Comrades!

The successful convocation of the national science conference is a matter of great joy for us and for the people throughout the country. The very fact that today we are holding this grand gathering, unparalleled in the history of science in China, clearly indicates that the days are gone for ever when the gang of Wang Hung-wen, Chang Chun-chiao, Chiang Ching, and Yao Wen-yuan could willfully sabotage the cause of science and persecute the intellectuals. Never before has work in science and technology received such attention and concern from the whole Party and the whole people. Vast numbers of scientists and technicians, the workers, the peasants, and the armymen are actively participating in the movement for scientific experiment. Enthusiasm for science and its study is becoming popular among the young people. The entire nation is embarking with tremendous enthusiasm on the march toward the modernization of science and technology. Splendid prospects lie before us.

Among those attending the present conference are outstanding scientists and technicians from various fronts, first-rate technical innovators, model laborers who excel in scientific farming, and cadres devoted to the Party's scientific undertakings. You have worked diligently for the progress of science and technology in our socialist motherland and made outstanding contributions. On behalf of the Central Committee of the Communist Party of China, I thank you and pay you tribute.

Comrades!

Our people face the great historic mission of comprehensively modernizing agriculture, industry, national defense, and science and technology within this century, making our country a modern, powerful socialist state. We have waged a sharp and bitter struggle against the "gang of four" on whether or not to accomplish the four modernizations. The "gang of four" made the absurd claim that "if the four modernizations are carried through, capitalist restoration will happen on the same day." Their wild sabotage brought our national economy for a time to the brink of collapse and was increasingly widening our distance from advanced world scientific and technological standards. Were they really opposed to the restoration of capitalism? Not at all. On the contrary, wherever their influence was most rampant, signs of capitalist restoration were most widespread. What they did serves as a negative example, making us appreciate more deeply that under conditions of proletarian dictatorship, if we do not modernize our country, raise our scientific and technological level, develop the social productive forces, strengthen our country, and improve the material and cultural life of the people, our socialist political and economic system cannot be fully consolidated and there will be no sure guarantee of our country's security. By adhering to the Party's basic line formulated by Chairman Mao, the more up-to-date our agriculture, industry, national defense, and science and technology, the greater our strength in the struggle against capitalism and all forces of restoration, and the more our people will support the socialist system. Only by making our country a modern, powerful socialist state can we more effectively prevent capitalist restoration, cope with aggression and subversion by social-imperialism and imperialism, and be more certain of gradually creating the material conditions for the advance to the great ideal of communism.

The crux of the four modernizations is the mastery of modern science and technology. Without modern science and technology it is impossible to build modern agriculture, modern industry, or modern national defense. Without high-speed development of science and technology, it is impossible to develop the national economy at high speed. On the proposal of Chairman Hua, the Central Committee of the Party has decided to call this national science conference to bring home to the whole Party and the whole country the importance of science, map out a program, commend the advanced units and individuals, and discuss measures for speeding up the development of science and technology. Today, I am going to give some opinions on pertinent questions.

The first question—the question of understanding that science is part of the productive forces. On this point, the "gang of four" raised a hue and cry, confounding right and wrong and causing much confusion. Marxism has consistently held that science and technology are part of the productive forces. More than a century ago, Marx said: "Wider use of machines in production calls for a conscious application of natural science." He also pointed out: "Science too [is] among these productive forces." The development of modern science and technology has bound science and production ever more tightly together. Science and technology as productive forces are manifesting their tremendous role ever more obviously.

Modern science and technology are undergoing a great revolution. The last three decades have not just seen advances in some aspects of scientific theory and production techniques, nor has this period been merely the general run of progress and reform. No, there have been profound changes and new leaps in almost all areas of science and technology. A whole series of new, rising sciences and technologies have emerged and are still doing so. Modern science has opened the way for the progress of production techniques and determined the direction of their development. Many new instruments of production and technological processes have come into being first in the laboratory. A series of newborn industries, including high polymer synthesis, atomic energy, electronic computers, semiconductors, astronautics, and lasers, have been founded on the basis of newly emerged science and technology. Of course there are now and there will be many theoretical research topics with no practical application in plain sight for the time being. But a host of historical facts have proved that once a major breakthrough is scored in theoretical research, it means tremendous progress for production and technology sooner or later. Contemporary natural science is being applied to production on an unprecedented scale and at a higher speed than ever before. This has given all fields of material production an entirely new look. In particular, the development of electronic computers, cybernetics, and automation technology is rapidly raising the level of automation in production. With the same amount of manpower and in the same number of work hours, people can turn out scores or hundreds of times more products than before. How have the social productive forces made such tremendous advances and how has labor productivity increased by such a big margin? Mainly through the power of science, and the power of technology.

Everyone knows that the basic factors in the productive forces are the means of production and manpower. What is the relationship of science and technology to the means of production and to manpower? Throughout history, the means of production have always been linked with science and technology of one kind or another, and likewise, manpower has always meant manpower armed with a certain knowledge of science and technology. We often say that man is the most active factor among the productive forces. "Man" here refers to people who possess a certain scientific knowledge, experience in production, and skills in the use of tools to produce material wealth. There were great differences in the instruments of production man used, his mastery of scientific knowledge, and his production experience and skills in the stone, bronze, and iron ages and in the seventeenth, eighteenth, and nineteenth centuries. Today, the rapid progress of modern science and technology is accelerating the renewal of production equipment and changes in technological processes.

Many products are superceded by a new generation of products in a matter of a few years. Only by acquiring a higher level of scientific and general knowledge, rich experience in production, and advanced skills can a worker play a bigger role in modern production. In our society, the laborers, who have a high degree of political awareness, study conscientiously and assiduously to raise their scientific and general level and thus will surely be able to achieve a higher labor productivity than that attained under capitalism.

The recognition that science and technology are productive forces brings the following question in its train: How should we regard the mental labor involved in scientific pursuits? Since science is becoming an increasingly important part of the productive forces, are people engaged in scientific and technological work to be considered workers or not?

There are various kinds of brain workers in societies under the rule of exploiting classes. Some are entirely in the service of the reactionary ruling classes and have thus set themselves against workers engaged in manual labor. But even in those cases, as Lenin said, there are many intellectuals engaged in scientific and technical work who themselves are not capitalists but scholars, although they are permeated with bourgeois prejudices. The fruits of their work are used by the exploiters, but, generally speaking, this is determined by the social system, and not by their own free choice. They are entirely different from politicians who rack their brains to advise the reactionary ruling classes directly. Marx pointed out that ordinary engineers and technicians join in the creation of surplus value. That is to say, they, too, are exploited by the capitalists.

In a socialist society, brain workers trained by the proletariat itself differ from intellectuals in any exploiting society in history. In the course of socialist transformation in China, Chairman Mao pointed out that intellectuals from the old society faced the question of what kind of "skin" they attached themselves to. Class contradictions and class struggle exist throughout the historical period of socialism, and the intellectuals face throughout the need to solve the question of what kind of "skin" to attach to and whether to keep to the proletarian stand. But, generally speaking, the overwhelming majority of them are part of the proletariat. The difference between them and the manual workers lies only in a different role in the social division of labor. Those who labor, whether by hand or by brain, are all working people in a socialist society. With the advancement of modern science and technology and progress toward the four modernizations, a great deal of heavy manual work will gradually be replaced by machines. Manual labor will steadily decrease for workers directly engaged in production, and mental work will continuously increase. Moreover, there will be an increasing demand for more people in scientific research and for a larger force of scientists and technicians. The "gang of four" distorted the division of labor between mental and manual work in our socialist society today, calling it class antagonism. Their aim was to attack and persecute the intellectuals; undermine the alliance of workers, peasants, and intellectuals; disrupt the social productive forces; and sabotage our socialist revolution and construction.

Correctly understanding that science and technology belong to the productive forces and that brain workers who serve socialism are a part of the working people has a close bearing on the rapid development of our scientific undertakings. Since we accept these two premises, we must naturally put great effort into developing scientific research and science education and give full play to the revolutionary initiative of the scientific and technical workers and the educational workers in order to accomplish the four modernizations in the short space of 20-odd years and bring about a tremendous growth of our productive forces.

Our science and technology have progressed enormously since the founding of New China and played an important role in economic construction and national defense construction. In old China, this would have been unthinkable. There is no way for anyone to deny this great achievement. But we must see, with a clear head, that there is still a very big gap between our science and technology and advanced world levels and that our scientific and technical forces are still very weak, far from meeting the needs of modernization. We have lost a lot of time, in particular, as a result of sabotage by Lin Piao and the "gang of four."

How do things stand with the technical level of our production? Several hundred million people are busy producing food. We still have not really solved the grain problem. Average annual output of grain per farm worker is about 1,000 kilograms in China, whereas in the United States the figure is over 50,000 kilograms, a disparity of several dozen times. Labor productivity in our iron and steel industry, too, is only a small percentage of advanced levels abroad. The gap in the newly emerged industries is still wider. A lag in this field of only eight to ten years, or even three to five years, makes a big gap, let alone a lag of ten to twenty years.

Chairman Mao often reminded us: "China ought to make a greater contribution to humanity." In ancient times, China had brilliant achievements in science and technology; its four great inventions played a significant role in the advance of world culture. But our ancestors' achievements can serve only to confirm our confidence in catching up with and surpassing advanced world levels and not to console us on our backwardness today. Our contributions in science and technology at present are highly incommensurate with the position of a socialist country like ours.

Will factually pointing out this backwardness make people lose heart? There might be such people. They do not have half a whiff of Marxism about them. As for us proletarian revolutionaries, by stating the facts and making a serious analysis of the historical and the present causes of this situation, we can accurately draft our strategic plan, deploy our forces, and strive for a rapid change in the situation. Only in this way, moreover, can we activate people to study modestly and speedily master the world's latest science and technology.

Backwardness must be perceived before it can be changed. A person must learn from the advanced before he can catch up and surpass them. Of course, to raise China's scientific and technological level we must rely on our own efforts, develop our own inventions, and adhere to the policy of independence and self-reliance. But independence does not mean shutting the door on the world, nor does self-reliance mean blind opposition to everything foreign. Science and technology are a kind of wealth created in common by all mankind. Any nation or country must learn from the strong points of other nations and countries, from their advanced science and technology. It is not just today, when we are scientifically and technically backward, that we need to learn from other countries; after we catch up with the advanced world levels in science and technology, we will still have to learn from the strong points of others.

China's revolution has attracted all the world's revolutionary people, who live and breathe with it. Our socialist modernization has won their interest and support and will do so on a widening scale. We must actively develop international academic exchanges and step up our friendly contacts with scientific circles of other countries. We express heartfelt thanks to all our friends abroad who have helped us in science and technology.

That is the first question on which I want to speak.

The second question concerns the building of a mammoth force of scientific and technical personnel who are both red and expert.

For the modernization of science and technology, we must have a mighty scientific and technical force of the working class which is both red and expert, and a large number of scientists and experts in engineering and technology who are first rate by world standards. We have a heavy task before us to build such a force.

An important question here is that we must have a correct understanding of being both red and expert, and set reasonable standards for it.

The "gang of four" made the absurd statement, "The more knowledgeable, the more reactionary." They said they "preferred laborers with no culture" and they boosted as a "model of being red and expert" an ignorant counterrevolutionary clown who handed in a blank examination paper. On the other hand, they vilified as being "white and expert" good comrades who studied diligently and contributed to the motherland's cause of science and technology. This reversal of right and wrong and of ourselves and the enemy seriously muddled people's minds for a time.

Chairman Mao advocated intellectuals becoming both red and expert, encouraging everyone to remold the bourgeois world outlook and acquire the proletarian world outlook. The basic question about world outlook is whom to serve. If a person loves our socialist motherland and is serving socialism and the workers, peasants, and soldiers of his own free will and accord, it should be said that he has initially acquired a proletarian world outlook and, in terms of political standards, cannot be considered white but should be called red. Our scientific undertakings are an important part of our socialist cause. To devote oneself to our socialist science and contribute to it is an important manifestation of being red, the integration of being red with being expert.

Imbued with Mao Tsetung Thought, our scientists and technicians have made truly rapid progress in the last 28 years. The overwhelming majority of them love the Party and love socialism, strive to integrate themselves with the workers, peasants, and soldiers, work wholeheartedly and fruitfully at their posts. Their faith in the Party and in socialism never wavered, no matter how Lin Piao and the "gang of four" persecuted and tormented intellectuals; they kept working on science and technology under extremely difficult conditions. Many showed a high level of political awareness in the eleventh struggle between the two lines. The smashing of the gang unleashed in them great revolutionary enthusiasm. They wholeheartedly support the Party Central Committee headed by Chairman Hua and work still harder for the four modernizations. How invaluable are these scientists and technicians! They are worthy of the title "red and expert," fit to be called our working class's own scientific and technical force. Chairman Hua once stressed with great satisfaction that such a force is an important factor in our confidence that we will catch up and surpass advanced world standards. This is a realistic, scientific appraisal.

This appraisal naturally does not mean that these scientists and technicians all have a very high level of political and ideological consciousness or that there are no shortcomings and mistakes of one kind or another in their ideology, their work style, or their specific work; it means that judged by the basic criterion of political stand, the overwhelming majority of them take the stand of the working class, and these revolutionary intellectuals constitute a force our Party can rely on. They should not be complacent or come to a halt, but should continue the effort, constantly seeking new progress both politically and in their specific fields. Their shortcomings and mistakes are a matter for education and assistance, something to be overcome through criticism and self-criticism. No one is free from shortcomings and mistakes. Take people like us, our cadres doing political work and our veteran cadres who have been in the Party for decades; do we not also have shortcomings or errors of this kind or that? Why be especially exacting toward vocational cadres and technical experts? As for scientists and technicians with undesirable family backgrounds or who committed mistakes in the past or whose families and social contacts present problems, we should judge them mainly by their own basic political attitude, by the way they acquit themselves, and by their contributions to socialist revolution and construction.

There is a section of scientists and technicians whose bourgeois world outlook has not fundamentally changed, or who are rather deeply influenced by bourgeois ideology. They often waver in the midst of sharp, fierce, and complicated class struggle. As long as they are not against the Party and against socialism, we should, in line with the Party's policy of uniting with, educating, and remolding the intellectuals, bring out their specialized abilities, respect their labor, and take an interest in their progress, giving them a warm helping hand. Chairman Mao consistently held that the more people in our revolutionary ranks the better, that we should respect those who have knowledge and specialized skills or have made contributions, and that our attitude toward any person who has made mistakes should be first to observe and second to give help and not to look down on him. We must earnestly implement these teachings of Chairman Mao's.

In our socialist society, everyone should remold himself. Not only those who have not changed their basic stand should remold, but everybody should study and constantly remold himself, study new problems, absorb what is new and consciously guard against corrosion by bourgeois ideology, so as better to shoulder the glorious and arduous task of building a modern, powerful socialist country.

To catch up and surpass advanced world levels within the century means that we should cover the distance in the next 22 years that took others 40 or 50 years or more. Scientists and technicians should concentrate their energy on scientific and technical work. When we say that at least five-sixths of their work time should be left free for their scientific and technical work, this is meant to be the minimum demand. It is still better if even more time is available for this purpose. If some persons work seven days and seven evenings on end to meet the needs of science or production, that shows their lofty spirit of selfless devotion to the cause of socialism. We should learn from them, commend them, and encourage them. Innumerable facts prove that only he can mount the pinnacles of science who devotes himself heart and soul, constantly strives for perfection, fears neither hardship nor disappointment. We cannot demand that scientists and technicians, or at any rate, the overwhelming majority of them, study a lot of political and theoretical books, participate in numerous social activities, and attend many meetings not related to their work. Lin Piao and the "gang of four" frequently attacked scientists and technicians, accusing them of "being divorced from politics" and labeling people "white and expert" when they studied diligently to improve their knowledge and skills. "White" is a political concept. Only political reactionaries who are against the Party and against socialism can be called "white." How can you label as "white" a man who studies hard to improve his knowledge and skills? Scientists and technicians who have flaws of one kind or another in their ideology or their style of work should not be called "white," if they are not against the Party and socialism. How can our scientists and technicians be accused of being divorced from politics when they work diligently for socialist science? The cause of socialism calls for a division of labor. On condition that they keep to the socialist political stand, comrades of different trades and professions are not divorced from politics when they do their best at their posts; on the contrary, this is a concrete manifestation of their service to proletarian politics and of their socialist consciousness. A few years ago, Lin Piao and the "gang of four" made it quite difficult for the workers to do their jobs, for the peasants to till the land, for the armymen to do their military training, and for the students to study or scientists and technicians to do research in their work. What heavy losses this meant for our socialist cause! Was that not a profound lesson?

While striving to raise the level of our present scientific and technical force, and making full use of their abilities, we must also exert ourselves to train new personnel. Owing to sabotage by Lin Piao and the "gang of four," there is an age-gap in this force which makes the training of a younger generation of scientific and technical personnel all the more pressing.

We have a vast supply and a great potential in matters of selecting and training talented personnel. With the recent reform of the college enrollment system, we have discovered fine young people who are diligent, hardworking, and talented. We are pleased to see their outstanding accomplishments. Though the "gang of four" ran wild for a time, they failed to dampen the enthusiasm of the youngsters for study, nor could they stifle the revolutionary zeal of the teachers to educate the next generation assiduously for the Party and the people. Today the Central Committee of the Party headed by Chairman Hua is paying close attention to science and education and laying strong emphasis on training and selecting talented people. We can foresee the dawn of a new era, with a multitude of outstanding people like the stars in the sky. The future of science lies with the youth. The growth of the younger generation is the hope of our flourishing cause.

Education is basic for training scientific and technical personnel. We must comprehensively and correctly carry out the Party's policy on education, determine its orientation, and make a good job of the educational revolution, to ensure a tremendous expansion and improvement. Education concerns not only the educational departments; Party committees at all levels must attend to it earnestly as a major issue. People of all trades and professions must support it and put great effort into running schools and colleges. People's teachers are gardeners tending the successors to the revolution. Their creative labor should be held in respect by the Party and the people. Their teaching time must be guaranteed and care and attention must be given to their political life, working conditions, and professional studies. Teachers with outstanding contributions in pedagogy should be commended and awarded.

On the question of talented people, we must particularly stress the need to break with convention in the discovery, selection, and training of those with outstanding talent. This was one of the big issues muddled by the "gang of four." They vilified scientists, professors, and engineers distinguished for their contributions as bourgeois academic authorities, and all outstanding young and middle-aged scientists and technicians trained by our Party and state, as revisionist sprouts. We must thoroughly eliminate the pernicious influence of the gang and take up the important task on the scientific and educational fronts of training in the shortest possible time a group of experts in science and technology who are first-rate by world standards. In the early period of the War of Resistance Against Japan, Chairman Mao said that our Party's fighting capacity would be much greater and our task of defeating Japanese imperialism would be more quickly accomplished if there were one or two hundred comrades with a grasp of Marxism-Leninism which was systematic and not fragmentary, genuine and not hollow. The revolutionary cause needs outstanding revolutionaries, and so does the scientific cause need outstanding scientists. Working-class persons with outstanding talent come from the people and serve the people. Only an extensive mass base can provide a continuous flow of talent, and outstanding talents will, in turn, help raise China's scientific and cultural standards as a whole.

The discovery or training of talented people by our scientists and teachers is in itself an achievement and a contribution to the state. The history of science shows what great results can be produced in the field of science from the discovery of a genuinely talented person! Some of the world's scientists have looked upon their discovery and training of new talent as the greatest achievement of a lifetime. There is much to be said for this view. A number of outstanding mathematicians in China today were discovered in their youth by older-generation mathematicians who helped them mature. Some of the newcomers may have surpassed their teachers in scientific achievement, but the teachers' contributions are indelible, nonetheless.

The third question I want to discuss is how to make the system of division of responsibilities under the leadership of Party committees work in scientific and technical departments.

Rapid development of science and technology hinges on good Party leadership in these fields.

Our country has entered a new period of development in socialist revolution and construction. According to the Constitution adopted by the Fifth National People's Congress and Chairman Hua's Report on the Work of the Government to the Congress, the general task in this new period is to steadfastly continue the revolution under the dictatorship of the proletariat; deepen the three great revolutionary movements of class struggle, the struggle for production, and scientific experiment; and transform China into a great and powerful socialist country with modern agriculture, industry, national defense, and science and technology by the end of the century. To accomplish this general task we must wage a great political and economic revolution and a great scientific and technical revolution. This is the new content for continuing the revolution under the dictatorship of the proletariat in the new period of development.

To meet the requirements of the new situation and the new task, there must be corresponding changes in the center of gravity for Party work and in the Party's work style. During the unprecedented Great Proletarian Cultural Revolution, our Party concentrated maximum efforts on the political revolution. Today, after victory in the struggle to expose and criticize the "gang of four," while continuing to eliminate their pernicious influence and deepen the socialist revolution on the ideological and political fronts, the whole Party must take firm hold of the work of modernization and carry out the great political and economic revolution and the great scientific and technical revolution, tasks which history has conferred on us.

The Party committees at various levels should learn from Taching and Tachai and make an earnest effort to grasp simultaneously the three great revolutionary movements of class struggle, the struggle for production, and scientific experiment. Following the examples of Taching and Tachai, they should unfold mass movements for scientific experiment, with new technical progress and new production records every year. There are several hundred thousand enterprises and several hundred thousand production brigades in our country. Extensive application of advanced science and technology to industry and agriculture and a greater, faster, better, and more economical growth of production can come about only through large-scale technical transformation and scientific experiments in every enterprise and every production brigade. At the same time, we must work energetically for the success of specialized scientific research institutes. Professional scientists and technicians form the mainstay of the revolutionary movement for scientific experiment. Without a strong contingent of professional scientific researchers of high caliber, we could hardly scale the heights of modern science and technology and it would be difficult for the scientific experiment movement of the masses to advance wave upon wave in a sustained way. We must get the specialists integrated with the masses.

The Central Committee has stipulated that a system of individual responsibility for technical work be established in scientific research institutes, and that the system of division of responsibilities among institute directors under the leadership of Party committees be set up. These are important organizational measures which help strengthen the leading role of the Party committees while bringing into full play the role of the specialists.

The basic task of scientific research institutes is to produce scientific results and train competent people. They must show more scientific and technical achievements of high quality and train scientific and technical personnel who are both red and expert. The main criterion for judging the work of the Party committee of a scientific research institute should be the successful fulfillment of this basic task. Only when this is well done has the Party committee really done its duty to consolidate the dictatorship of the proletariat and build socialism. Otherwise, putting politics in command will remain mere empty talk.

A lot of work has to be done to fulfill this basic task. It is impossible for Party committees to handle and solve all these matters. We must honestly admit that in scientific and technical work, there are many things we do not know. Even should we know them, it would still be impossible for Party committees to do everything. There must be a division of responsibilities and a system of individual responsibility at each post from top to bottom. This is the only way to make our work orderly and efficient and bring about high-speed development; and this is the only way to define the duties incumbent on each post and to mete out the proper awards and penalties, at the same time obviating procrastination or evasions of responsibility and avoiding getting in each other's way.

The leadership given by Party committees is primarily political leadership, that is, to ensure the correct political orientation and the implementation of the Party's line, principles, and policies and to bring out the initiative of all concerned. At the same time, leadership is exercised through the plan. Good plans must be drawn up for scientific research, personnel must be carefully appraised and placed where they can do the best work, and all forces must be well organized. In order to follow out the plans and push forward our scientific research, it is also necessary to guarantee the supporting services and supplies and to provide the necessary working conditions for scientific and technical personnel. This is also part of the work of the Party committees. I am willing to be the director of the logistics department at your service and to do this work well, together with the leading comrades of Party committees at various levels.

We should give the director and the deputy directors of research institutes a free hand in the work of science and technology according to their division of labor. Party committees should back up the work of all Party and non-Party experts in administrative positions and try to bring out all their capacities so that they really have powers and responsibilities commensurate with their positions. These experts are also cadres of the Party and the state. We must never look askance at them. Party committees should get acquainted with their work and examine it but should not attempt to supplant them.

We must give full scope to democracy and follow the mass line, heeding opinions from scientific and technical personnel in such things as evaluating scientific papers, examining the competency of scientific and technical personnel, working out plans for scientific research, and appraising research findings. As to divergent views on academic questions, we must follow the principle of letting a hundred schools of thought contend and encourage free discussion. We must listen closely to experts' opinions and enable them to play their full role so that we can do better at scientific and technical work and reduce our errors as much as possible. This is an important aspect of the mass line for Party committees of scientific research institutes.

Do we mean to lighten the load of our political work or to lower its standards when we stress that scientific and technical personnel must concentrate on their specific work? No, we do not. This means a demand to raise the level of our political work, improve the method, do away with everything that smacks of formalism, eliminate the poisonous influence of the "gang of four," and conscientiously learn the fine traditions of Liberation Army political work. We must support whatever is conducive to the development of socialist science, and criticize and educate those who seek personal gain, hide their findings, refuse to work in coordination, or even resort to monopoly and plagiarism, and those who display other erroneous ideas and styles of work which are detrimental to the development of socialist science. As we are engaged in socialist modernization and are advancing toward the mastery of modern science and technology, the important task for our political work today is to make every scientist and technician understand how his work relates to the grand goal of the four modernizations, to encourage and mobilize them to work together with one heart, and to coordinate their efforts in the spirit of revolution, so as to storm the citadels of science.

Although our Party has accumulated some experience in leading scientific and technological work over the past 20-odd years, we must admit that we confront a very large realm of necessity, an area we still do not know, with regard to how to effectively organize, manage, and lead socialist science and technology. Until there is a change in this state of affairs, we can hardly have major achievements and the initiative will not be in our hands. Chairman Mao taught us time and again that persons in the dark cannot light the way for others. Leading Party cadres at various levels must not be content to remain laymen. They must study their work and gradually learn the ropes. We must apply ourselves to the study of Marxism and raise our political level, and we must also strive to acquire scientific knowledge, sum up experience, both positive and negative, study and grasp the objective laws governing scientific and technological work, and implement the Party's principles and policies correctly and comprehensively. Our Party was able to lead the people to the overthrow of the system of exploitation and to the transformation of society, and it will certainly be able to grasp the laws governing scientific and technological work and lead our people to the heights of world science.

The rights and wrongs in regard to political line have been basically clarified; we have mapped out a program with the measures for its execution; the masses are already on the move. The task now confronting our Party organizations at all levels is to inspire real drive in the masses, to find down-to-earth solutions to problems, and to do good, solid work. In a word, we must put everything on a solid footing. We must stop all the manifestations of formalism, which go in for ostentation but disregard practical results, real efficiency, actual speed, quality, or cost. Bad habits like empty talk, boasting, and lying must be stamped out.

Comrades,

The Eleventh Party Congress, the Fifth National People's Congress, and the Fifth Chinese People's Political Consultative Conference, coming one after the other, fully demonstrated the great unity of our whole Party and the great unity of the people throughout the country. This national science conference is likewise a gathering of unity. The unity of the Party and the unity of the people—these are the basic guarantees for the sure triumph of our cause. Let us hold high the great red banner of Mao Tsetung Thought and, under the leadership of the Party Central Committee headed by Chairman Hua, march forward, unswerving and victorious, moving valiantly toward the grand goal of a modern, powerful socialist country!

May science in China flourish and grow! I wish the conference complete success!

## *Excerpts from Fang Yi's speech at the National Science Conference, March 18, 1978: "Outline National Plan for the Development of Science and Technology (1978–85); Relevant Policies and Measures"*

Chairman Mao and Premier Chou mapped out a gigantic plan for us to make China a modern, powerful socialist country. By the end of this century, all departments and localities in China that can use machines must be fully mechanized, electrification must be realized in both urban and rural areas, the production processes in major industrial departments automated, advanced techniques extensively applied, labor productivity raised by big margins, and a radical change brought about in industrial and agricultural production so that our national economy can take its place in the front ranks of the world. We must equip our armed forces with the advanced achievements in science and technology and greatly enhance our national defense capabilities. We shall build a vast army of working-class scientists and technicians who are both red and expert, and we must have our own experts in science and technology who are first rate by world standards. It is also necessary to acquire the most sophisticated equipment for scientific experimentation so that we can approach advanced world levels in most branches of science and technology, catch up with them in some branches, and take the lead in certain others. With the accomplishment of these tasks, we can say that we have realized by and large our objective of modernizing agriculture, industry, national defense, and science and technology. China will radiate even more brilliantly throughout the world.

The eight years from now through 1985 are crucial for this long-term plan. We must foster lofty ideals and set high goals, work out a strategic plan, fully mobilize all positive factors, and organize all our forces well.

Our plan should serve the needs of realizing the four modernizations, which hinge on modernizing science and technology. The plan on science and technology must dovetail with the plan on production and construction, and the two must be organically combined. Research in applied sciences and in basic theories and the immediate and long-term tasks must be properly arranged to avoid over-emphasizing one to the neglect of the other.

Our plan should be aimed at high-speed development. Compared with advanced world levels in science and technology, our country is now lagging 15 to 20 years behind in many branches and more still in some others. Modern science and technology are developing rapidly. Only by developing at a higher speed can we catch up with or surpass the capitalist countries. We fulfilled five years ahead of schedule the major targets specified in the 12-year plan for the development of science and technology mapped out in 1956. In the mid-1960's we approached advanced world levels at the time in some scientific and technical spheres, which helped the popularization of some new techniques and the building of some new, rising industries. Now that we have much better conditions and a more solid foundation than in those days, a much higher speed is entirely possible.

Our plan should be one with the present-day advanced levels as its starting point. We must be good at learning from the advanced. We should conscientiously assimilate the experience and lessons of our predecessors so as to avoid the detours they went through. In carrying out the first plan for the development of science and technology, we took semiconductor technology, which was an advanced branch of science at the time, as our starting point for studying and developing electronic computers. As a result, we soon passed the stage of the electron tube and gained time. In the years to come, we must work hard to raise all our scientific research work to advanced levels as quickly as possible. Scientific experiments by the masses should also be steadily improved on the basis of popularization.

Since last June, departments under the State Council and various localities and units have, through repeated discussions and revisions, mapped out a draft outline National Plan for the Development of Science and Technology, 1978–85.

The draft outline plan sets forth the following goals to be attained in the next eight years:

(1) Approach or reach the advanced world levels of the 1970's in a number of important branches of science and technology.

(2) Increase the number of professional scientific researchers to 800,000.

(3) Build a number of up-to-date centers for scientific experiment.

(4) Complete a nationwide system of scientific and technological research.

The eight-year outline plan (draft) makes all-around dispositions for the tasks of research in 27 spheres, including natural resources, agriculture, industry, national defense, transport and communication, oceanography, environmental protection, medicine, finance and trade, culture, and education, in addition to the two major departments of basic and technical sciences. Of these, 108 items have been chosen as key projects in the nationwide endeavor for scientific and technological research. When this plan is fulfilled, our country will approach or reach the advanced world levels of the 1970's in a number of important branches of science and technology, thus narrowing the gap to about ten years and laying a solid foundation for catching up with or surpassing advanced world levels in all branches in the following fifteen years.

The eight-year outline plan (draft) gives prominence to the eight comprehensive scientific and technical spheres, important new techniques, and pace-setting disciplines that have a bearing on the overall situation. It calls for concentrating all forces and achieving remarkable successes so as to promote the high-speed development of science and technology as a whole and of the entire national economy.

*Agriculture.* In accordance with the principle of "taking grain as the key link and ensuring an all-around development," we will in the next three to five years actively carry out comprehensive surveys of our resources in agriculture, forestry, animal husbandry, sideline production, and fisheries, study the rational exploitation and utilization of the resources and the protection of the ecological system, and study the rational arrangement of these undertakings.

We should implement in its entirety the Eight-Point Charter for Agriculture (soil, fertilizer, water conservancy, seeds, close planting, plant protection, field management, and improved farm tools) and raise our level of scientific farming. We should study and evolve a farming system and cultivating techniques that will carry forward our tradition of intensive farming and at the same time suit mechanization; and study and manufacture farm machines and tools of high quality and efficiency. We will study science and technology for improving soil, controlling water, and making our farmland give stable and high yields. In order to improve as quickly as possible the low-yielding farmland that accounts for about one-third or more of the country's total, we must make major progress in improving alkaline, lateritic, clay, and other kinds of poor soil, in preventing soil erosion, and in combating sandstorms and drought. We will study projects for diverting water from the south to the north and the relevant scientific and technical problems; study and develop new compound fertilizers and biological nitrogen fixation, methods of applying fertilizer scientifically, and techniques for drainage and irrigation; breed new seed strains, work out new techniques in seed breeding, and improve the fine crop varieties in an all-around way so that they will give still higher yields, produce seeds of better quality, and can better resist natural adversities. We should quickly develop new insecticides that are highly effective and are harmless to the environment, and devise comprehensive techniques for preventing and treating different kinds of plant diseases and pests.

We need to step up scientific and technological research in forestry, animal husbandry, sideline production, and fisheries. We should provide new tree varieties and techniques that will make the woods grow fast and yield more and better timber; develop multipurpose utilization of forest resources, and study techniques and measures for preventing and extinguishing forest fires; step up research on building pasturelands, improving breeds of animals and poultry, mechanizing the processes of animal husbandry, increasing the output of aquatic products, breeding aquatic products, and marine fishing and processing.

We will set up up-to-date centers for scientific experiments in agriculture, forestry, animal husbandry, and fisheries.

We must lay great emphasis on research in the basic theories of agricultural science, and step up our studies of the applications of agricultural biology, agricultural engineering, and other new techniques to agriculture.

*Energy.* We must make big efforts to accelerate the development of energy science and technology so as to carry out full and rational exploitation and utilization of our energy resources.

We have our own inventions in the science and technology of the oil industry, and in some fields we have caught up with or surpassed advanced levels in other countries. We must continue our efforts to catch up with and surpass advanced world levels in an all-around way. We should study the laws and characteristics of the genesis and distribution of oil and gas in the principal sedimentary regions, develop theories of petroleum geology, and extend oil and gas exploration to wider areas; study new processes, techniques, and equipment for exploration and exploitation and raise the standards of well drilling and the rate of oil and gas recovery; and actively develop crude oil processing techniques, use the resources rationally and contribute to the building of some ten more oilfields, each as big as Taching.

China has extremely rich resources of coal, which will remain our chief source of energy for a fairly long time to come. In the next eight years, we should mechanize the key coal mines, achieve complex mechanization in some of them, and proceed to automation. The small- and medium-sized coal mines should also raise their level of mechanization. Scientific and technical work in the coal industry should center around this task, with active research in basic theory, mining technology, technical equipment, and safety measures. At the same time research should be carried out in the gasification, liquefaction, and multipurpose utilization of coal, and new ways explored for the exploitation, transportation, and utilization of different kinds of coal.

To improve the power industry is another pressing task. We should take as our chief research subjects the key technical problems in building large hydroelectric power stations and thermal power stations at pit mouths, large power grids, and super-high-voltage power transmission lines. We must concentrate our efforts on comprehensive research in the techniques involved in building huge dams and giant power-generating units and in geology, hydrology, meteorology, reservoir-induced earthquakes, and engineering protection, which are closely linked with large-scale key hydroelectric power projects.

New sources of energy should be explored. We should accelerate our research in atomic power generation and speed up the building of atomic power plants. We should also step up research in solar energy, geothermal energy, wind power, tide energy, and controlled thermonuclear fusion, pay close attention to low-calorie fuels, such as bone coal, gangue, and oil shale, and marsh gas resources in the rural areas, and making full use of them where possible.

Attention should be paid to the rational utilization and saving of energy, such as making full use of surplus heat, studying and manufacturing fine and efficient equipment for this purpose, and lowering energy consumption by every means and particularly coke consumption in iron smelting, coal consumption in power generation, energy consumption in the chemical and metallurgical industries.

*Materials.* Steel must be taken as the key link in industry. It is imperative to make a breakthrough in the technology of intensified mining and solve the scientific and technological problems of beneficiating hematite. We should speed up research work on the paragenetic deposits at Panchihhua, Paotow, and Chinchuan, where many closely associated metals have been formed; solve the major technical problems in multipurpose utilization; intensify research on the exploitation of copper and aluminum resources; make China one of the biggest producers of titanium and vanadium in the world; and approach or reach advanced world levels in the techniques of refining copper, aluminum, nickel, cobalt, and rare-earth metals. We should master new modern metallurgical technology quickly, increase varieties, and improve quality; study and grasp the laws governing the formation of high-grade iron ore deposits and the methods of locating them; establish a system of ferrous and non-ferrous materials and extend it in the light of the characteristics of our resources.

We should make full use of our rich natural resources and industrial dregs and increase at high speed the production of cement and new types of building materials which are light and of high strength and serve a variety of purposes; step up research in the technology of mining and dressing non-metal ores and in the processing techniques; lay stress on research in the techniques of organic synthesis with petroleum, natural gas, and coal as the chief raw materials; step up our studies of catalysts and develop the technology of direct synthesis; renovate the techniques of making plastics, synthetic rubber, and synthetic fiber; and raise the level of equipment and automation in the petrochemical industry. We must solve the key scientific and technical problems in producing special materials necessary for our national defense industry and new technology, and evolve new materials characteristic of China's resources.

We should devote great efforts to basic research on the science of materials, develop new experimental techniques and testing methods, and gradually be able to design new materials with specified properties.

*Electronic computers.* China must make a big new advance in computer science and technology. We should lose no time in solving the scientific and technical problems in the industrial production of large-scale integrated circuits, and make a breakthrough in the technology of ultra-large-scale integrated circuits. We should study and turn out giant computers, put a whole range of computers into serial production, step up study on peripheral equipment and software of computers and on applied mathematics, and energetically extend the application of computers. We aim to acquire by 1985 a comparatively advanced force in research in computer science and build a fair-sized modern computer industry. Microcomputers will be popularized, and giant ultra-high-speed computers put into operation. We will also establish a number of computer networks and data bases. A number of key enterprises will use computers to control the major processes of production and management.

*Lasers.* We will study and develop laser physics, laser spectroscopy, and nonlinear optics in the next three years. We should solve a series of scientific and technical problems in optical communications, raise the level of the routine laser quickly, and intensify our studies of detectors. We expect to make discoveries and creations in the next eight years in exploring new types of laser devices, developing new laser wavelengths, and studying new mechanisms of laser generation, making contributions in the application of the laser to studying the structure of matter. We plan to build experimental lines of optical communications and achieve big progress in studying such important laser applications as separation of isotopes and laser-induced nuclear fusion. Laser technology should be popularized in all departments of the national economy and national defense.

*Space.* We should attach importance to the study of space science, remote sensing techniques, and the application of satellites; build modern centers for space research and systems for the application of satellites; step up the development of the vehicle series, and study, manufacture, and launch a variety of scientific and applied satellites; actively carry out research in the launching of skylabs and space probes; and conduct extensive research in the basic theory of space science and the application of space technology.

*High energy physics.* We expect to build a modern high energy physics experimental base in ten years, completing a proton accelerator with a capacity of 30,000 million to 50,000 million electron volts in the first five years and a giant one with a still larger capacity in the second five years.

We should from now on set about the task in real earnest and make full preparations for experiments in high energy physics, with particular stress on studying and manufacturing detectors and training laboratory workers. We should step up research in the theory of high energy physics and cosmic rays; consciously promote the interpenetration of high energy physics and the neighboring disciplines; actively carry out research in the application of accelerator technology to industry, agriculture, medicine, and other spheres; and pay attention to the exploration of subjects which promise important prospects of application.

*Genetic engineering.* We must in the next three years step up the tempo of building and improving the related laboratories and conduct basic studies in genetic engineering. In the next eight years, we should combine them with the studies in molecular biology, molecular genetics, and cell biology and achieve fairly big progress. We should study the use of the new technology of genetic engineering in the pharmaceutical industry and explore new feasible ways to treat and prevent certain difficult and baffling diseases and evolve new high-yield crop varieties capable of fixing nitrogen.

We must grasp firmly and effectively the above eight important spheres. But this in no way means that we can neglect work in other spheres. All branches of science and technology have their specific positions and roles in our socialist construction, and none can be dispensed with or replaced. We should grasp the key spheres well on the one hand and make overall planning and give all-around consideration on the other. Make all-around arrangements while laying emphasis on the key points—this is our policy.

Here I would like to mention in particular the question of multipurpose utilization. Chairman Hua has given this instruction: "We must attach importance to multipurpose utilization which makes full use of natural resources and alleviates pollution of the environment. The three industrial wastes (liquids, gases, and dregs) will bring harm if they are discarded but will become treasures if they are turned to good account." People working in all professions and trades should go in for multipurpose utilization, and the departments concerned should be organized to concentrate their forces on tackling major scientific and technical problems.

The march toward the modernization of science and technology means in essence a comprehensive and fundamental technical transformation of all fields of material production in our country. This is a great technical revolution that history has entrusted to us. Accomplishment of this revolution depends on leadership by the Party and on the people of the whole country. Our Party organizations at all levels, first of all the leading Party groups of the ministries and commissions under the State Council and the Party committees of the provinces, municipalities, and autonomous regions, must earnestly implement the instructions of Chairman Hua, simultaneously grasp the three great revolutionary movements of class struggle, the struggle for production, and scientific experiment, and do the following work well in a down-to-earth manner.

### *(1) Consolidate the Scientific Research Institutions and Build Up a Scientific and Technological Research System*

In the next eight years, we must create a nationwide scientific and technological research system which covers all branches of study—which should complement each other and be rationally distributed and developed in coordination—and which integrates professionals with the masses and military research efforts with those undertaken in the civilian sector.

The Chinese Academy of Sciences, the various departments under the State Council, and the key universities and colleges must restore, strengthen, and build a number of key scientific research institutions. They must pay particular attention to strengthening research in those disciplines where the work has been weak, and to building and expanding a number of research institutions in the basic sciences and new branches of science and technology where there is an urgent need.

The provinces, municipalities, and autonomous regions must establish and strengthen research institutions. Special attention should be paid to the establishment and development of scientific research institutions in the interior and in areas inhabited by the minority nationalities.

All large industrial and mining enterprises should take active steps to establish and strengthen research institutions. Small- and medium-sized factories and mines may establish research institutions according to the situation. Scientific research groups, technical innovation groups, or teams for tackling difficult scientific and technical problems should be set up wherever possible.

Big efforts should be made to strengthen research institutions in agronomy and farm machinery and tools at the county level and, with these as the nucleus, consolidate and expand the network of agro-science institutions and agrotechnical stations at county, commune, production brigade, and production team levels.

No time should be lost in consolidating the existing scientific research institutions, particularly the key ones. We must first of all consolidate their leading bodies. It is imperative that Party committee secretaries are selected from among comrades who have a good understanding of Party policies and are eager to promote science, that experts or near-experts are appointed to leading professional posts, and that conscientious and hard-working comrades are given charge of supply services.

The system of institute directors assuming responsibility under the leadership of the Party committees must be applied in scientific research institutions. The orientation, tasks, and personnel of the research institutes and their subdivisions should be determined after extensive discussions by the masses.

*(2) Open Broad Avenues to Able People and Recruit Them Without Overstressing Qualifications*

Resolute measures must be taken to train able people in greater numbers and at a faster rate.

Conscientious efforts must be made to run universities and colleges, middle schools, and primary schools well, and the key colleges and schools must be run successfully. It is necessary to modernize the means of education step by step, develop television and radio courses, and increase and improve the equipment of school laboratories. We should take active steps to set up college television courses, correspondence universities, and evening schools. Diverse forms must be employed to expand student enrollment.

Great efforts must be made to strengthen postgraduate training. The Chinese Academy of Sciences and institutions of higher learning should take active steps to increase postgraduate enrollment, and all production departments and local research institutions that have the required facilities should do the same.

We should discover and select talented people from among young people who participate in scientific contests, readers of scientific journals, people recommended by various departments, and inventors and innovators on the industrial and agricultural fronts.

Outstanding students can graduate from school ahead of time. Key colleges and universities can break with conventional practices and enroll exceptionally outstanding young people at any time.

We must take resolute steps to transfer to scientific and technical posts those competent and well-trained scientific and technical personnel whose specialties are not being put to use.

*(3) Institute Regulations for Training, Appraising, Promoting, and Rewarding Scientific and Technical Personnel*

We must encourage scientific and technical personnel to make a diligent study of Marxism-Leninism-Mao Tsetung Thought, constantly raise their political consciousness, and remold their world outlook in the three great revolutionary movements, so as to make steady progress along the road of becoming both red and expert.

It is highly important to train a core force of scientific workers and top-notch scientists. Scientific and technical personnel who have attained the level of assistant research fellow, lecturer, and engineer or higher should be given a certain period of time for advanced study every two or three years. Plans should be worked out to select and send scientific and technical personnel abroad for advanced study or for short-term work. Various measures should be taken to help scientific and technical workers in general study basic theory and acquire specialized knowledge.

We must firmly establish the moral code under which it is an honor to produce good results in scientific research for the socialist motherland. Technical titles should be restored, the system of individual responsibility established for all technical work, and the appraising and promotion of scientific and technical personnel undertaken at regular intervals.

Scientific and technical personnel who have made important contributions to the country should be rewarded.

*(4) Uphold the Policy of Letting a Hundred Schools of Thought Contend*

"Let a hundred schools of thought contend" is the correct policy Chairman Mao formulated for developing China's socialist cause of science. Free contention among different schools should be encouraged and fostered in science. With regard to academic problems, we should have both freedom for criticism and freedom for refuting criticism; we should foster the attitude of upholding truth and correcting mistakes, and strictly prohibit the practice of affixing political labels indiscriminately. Scientific papers and reports must not be withheld from publication, so long as they do not divulge state secrets or involve charlatanry. Those scientific and technical personnel who have aired erroneous views on academic questions should not be discriminated against, but should be helped so that they will work better.

The National Scientific and Technical Association and the societies of natural sciences should broaden their academic activities. Scientific research institutions, universities and colleges, and all scientific and technical organizations should make it a regular practice to carry out academic activities.

We actively advocate the study of Marxist philosophy by scientific and technical workers, and we should encourage and help them to do so. It is necessary to hold different kinds of forums regularly, begin publishing journals on the dialectics of nature, carry out research on the history of natural science, and encourage scientific and technical personnel to guide their scientific research with Marxist philosophical concepts.

### *(5) Learn Advanced Science and Technology from Other Countries and Increase International Academic Exchanges*

An important way to develop science and technology at high speed is to utilize fully the latest achievements in the world in science and technology and absorb their quintessence. We should introduce selected advanced techniques that play a key and pace-setting role.

We must strengthen scientific and technical cooperation and academic exchanges with other countries and keep abreast of the results, trends, policies, and measures of their scientific and technological research as well as their experience in organization and management. We should actively and systematically enlarge the scope of sending scientific and technical personnel, returned students, and postgraduates abroad to study, receive advanced training, make study tours, and take part in international academic conferences and other academic activities. Meanwhile, we will also invite foreign scientists, engineering, and technical experts to China to give lectures, serve as advisers, or join us in scientific research.

### *(6) Ensure Adequate Work Hours for Scientific Research*

We must make sure that scientific research workers can devote at least five-sixths of their work hours each week to professional work.

Scientific research workers should be encouraged to study and improve their professional competence in their spare time.

Core research workers should be provided with assistants and their administrative duties cut down.

We should carry out lively political work and strive to achieve practical results in all our work.

Supply services in scientific research institutions must serve scientific research and improve the living as well as working conditions for scientific and technical workers.

### *(7) Strive to Modernize Laboratory Facilities and Information and Library Work*

In the next eight years, we should build a number of modern experimental installations and centers. We should give a high priority to refitting the existing laboratories so as to modernize them as quickly as possible.

Emergency measures must be taken to push forward the design and production of instruments and equipment. Efforts must be made to expand, renovate, and build a number of factories specializing in scientific instruments and chemical reagents. Great attention should be paid to the study of new principles, new techniques, and new instruments and equipment.

It is essential to strengthen the management of the designing, production, distribution, and use of scientific instruments and bring them under an overall national plan.

We should improve and strengthen our scientific and technical information institutions, and collect foreign scientific and technical information and data extensively through diverse channels. Exchanges of domestic research results between various disciplines should be well organized, and the analysis and study of scientific and technical information improved.

It is essential to equip informational institutions with modern facilities in the shortest possible time. In eight years we will set up a number of documentation retrieval centers and data bases, and build a preliminary nationwide computer network of scientific and technical information and documentation retrieval centers. We should also strengthen the publication of scientific and technological material.

### *(8) Close Cooperation with an Appropriate Division of Labor*

The Chinese Academy of Sciences is the overall center for research in natural science throughout the country. Its main task is to study and develop new theories and techniques and to solve major scientific and technical problems involving many fields in our economic construction, in cooperation with the departments concerned. It should lay stress on basic theoretical research and aim at raising standards.

The institutions of higher learning serve as both educational and research centers; they are an important force in scientific research, covering both the basic and the applied sciences.

Research institutions of the various departments and localities should devote themselves mainly to the applied sciences, but they should also undertake appropriate research in basic science.

The above institutions and the nonprofessionals engaged in scientific experiments should work in close cooperation with an appropriate division of labor.

We must give play to the spirit of subordinating ourselves to the national interest. It is imperative to give scope to the initiative of both the central and local authorities. Scientific research must be integrated with production and utilization. All departments, regions, and organizations should display a communist spirit of cooperation.

### *(9) Speed Up Popularization and Application of Scientific and Technical Achievements and New Techniques*

Work on the exchange and popularization of scientific and technical achievements needs improving.

Close attention should be paid to the intermediate links between scientific research and industrial and agricultural production, and essential pilot factories and workshops to trial-produce new products should be built or improved.

We should study and formulate appropriate technical and economic policies and encourage the popularization and application of scientific and technical achievements.

### *(10) Make Big Efforts to Popularize Science*

We must arm our cadres and the masses with modern scientific and technical knowledge.

We should organize popular science groups which combine the efforts of both professionals and the masses; expand the publication of popular science readers and the production of science and education films; run successfully halls of science and technology, museums, exhibition centers, technique-exchanging teams, and various kinds of scientific and technical clubs. The press, radio, and television should devote more space and time to the dissemination of science and technology. Public establishments should gradually extend their programs to include popular science activities.

Special efforts should be made to interest our cadres at various levels in science and technology. Arrangements should be made for scientific and technical workers and teachers to acquaint cadres with the latest trends, basic knowledge, and current research results in China and the rest of the world.

In popularizing science, we must give full scope to the active role of educated young people settling in the countryside.

All sectors must pool their efforts to foster among the cadres, the masses, and the young people the habit of loving, studying, and applying science.

## *Hua Guofeng's speech at the National Science Conference, March 24, 1978: "Raise the Scientific and Cultural Level of the Entire Nation"*

Comrades!

This National Science Conference convened by the Central Committee of our Party has been in session for seven days. At the opening session, Comrade Teng Hsiao-ping made a speech and Comrade Fang Yi delivered a report, both being very important and very good. Comrades attending the conference have discussed them in earnest and have all expressed hearty support. Everyone is greatly encouraged, in high spirits, and free from anxiety. A national plan for the development of science and technology will be worked out, and advanced collectives and individuals on the scientific and technical front commended at this conference. We are all fully confident that the conference will be a great success and will have a tremendous and far-reaching impact on the development of our science and culture, the growth of our national economy, and the building of a modern, powerful socialist country.

The Central Committee attaches great importance to this conference. Shortly after the smashing of the anti-Party "gang of four" of Wang Hung-wen, Chang Chun-chiao, Chiang Ching, and Yao Wen-yuan, it convened the Second National Conference on Learning from Tachai in Agriculture and later the National Conference on Learning from Taching in Industry. Even then, the Central Committee was considering the convocation of a science conference after those on agriculture and industry. We officially announced this decision in the political report to the Eleventh Party Congress. Later the Central Committee issued a circular to the whole Party and the whole country on holding the conference. It is the first time in the history of our Party and our People's Republic that the Party Central Committee has convened a conference of such a gigantic scale and broad representation in order to mobilize the whole Party, the whole army, and the people of all nationalities throughout the country to march toward the modernization of science and technology. This is an important measure adopted by our Party to carry out this general objective for the new period in our country's socialist revolution and socialist construction. Ours is a conference of tremendous immediate and historical significance.

Our country has basically eliminated the chaos created by the "gang of four" and is moving toward great order across the land. This has come about through a great, deep-going political revolution aimed at exposing and criticizing the "gang of four" over the last year or so, through the collective efforts of the Party and the people on the political, economic, military, and cultural fronts, and through the several historic conferences for carrying forward the revolutionary cause pioneered by our predecessors and forging ahead into the future—the Eleventh Party Congress, the Fifth National People's Congress, and the meeting of the Fifth National Committee of the Chinese People's Political Consultative Conference. Now, the line and the general task for the new period have been clearly formulated. The new Constitution has been promulgated. The fundamental tasks and policies for various fields of work, the three-year and eight-year plans, and a 23-year outline for the development of the national economy have been mapped out. Although some specific regulations, policies, and plans are yet to be worked out or improved, our major political guideline, namely, grasping the key link of class struggle and bringing about great order across the land, has in the main been set. What is required at present is to follow the line, policies, and plans already laid down and to work hard and in a down-to-earth manner; to sweep away interference, surmount difficulties, and fulfill the tasks before us step by step.

The general task laid down for the whole Party and the whole people in the new period by the Eleventh Party Congress and the Fifth National People's Congress has been written into the fundamental law of the state. The task is: To persevere in continuing the revolution under the dictatorship of the proletariat; carry forward the three great revolutionary movements of class struggle, the struggle for production and scientific experiment; and make China a great and powerful socialist country with modern agriculture, industry, national defense, and science and technology by the end of the century. It shows that we must follow the road of socialism unswervingly, grasp the three great revolutionary movements simultaneously, and accomplish the splendid goal of the four modernizations. The world has witnessed different roads to modernization. There is capitalist or imperialist modernization and revisionist or social-imperialist modernization. What we want is socialist modernization, to be attained by steadfastly continuing the revolution under the dictatorship of the proletariat. Only this kind of modernization conforms to the common aspirations and the fundamental interests of the people of all our nationalities. Only this can bring genuine happiness to our people and gladden the people the world over.

Socialism is the only way out for China. This was proved long ago by hard realities. In old semi-colonial, semi-feudal China, many persons with lofty ideals sought to develop science in China and make the country independent, strong, and prosperous, but they all failed. Not until the Chinese Communist Party led the people of the whole country in winning complete victory in the new-democratic revolution and in establishing the socialist system did China build up a fairly modern industrial base, which provides the conditions for going on to the four modernizations. The "gang of four" were sworn enemies of socialism. They opposed the four modernizations in a criminal attempt to subvert the dictatorship of the proletariat and restore capitalism. If China were to follow the counterrevolutionary revisionist line of the gang, it could only be reduced to a colony or semi-colony of social-imperialism and imperialism. To achieve the four modernizations, China, an independent and socialist country, must adhere to Chairman Mao's thought and persist in continuing the revolution under the dictatorship of the proletariat, which means sticking to the socialist road. For us, socialism and the four modernizations are inseparable from each other. Only by persevering in socialist revolution and continuing to transform that part of the superstructure and the relations of production not in correspondence with the growth of the productive forces can we constantly promote the four modernizations. Only by building up a modern agriculture, industry, national defense, and science and technology can we provide our socialist system with a powerful material base, steadily consolidate and develop this system, effectively defeat capitalism at home, and find ourselves in a stronger position to resist aggression by social-imperialism and imperialism from abroad.

The general task for the new period calls for hard work in every field by the whole

Party, the whole army, and the people of all our nationalities. Here I would like to go into one question in particular, that is, the necessity for greatly raising the scientific and cultural level of the entire Chinese nation.

The people of all nationalities in our country are industrious, brave, and rich in creative talent. Under the guidance of Marxism-Leninism-Mao Tsetung Thought, our people have acquired many highly valuable capabilities in revolution and construction, performed great deeds, and made tremendous progress in the course of their long and arduous struggle. Now we must start a new and sustained study movement in order to extend our battle with nature, march toward the four modernizations, and fulfill the general task history has assigned us in the new period. It is necessary to raise the study of Marxism-Leninism-Mao Tsetung Thought to a new level and, at the same time, strive to improve our general educational standard, acquire modern scientific knowledge, and master the skills and the methods of management, which are indispensable to modern production. Raising the scientific and cultural level of the entire Chinese nation is a colossal task facing all our people. It is a task of strategic importance. Unless it is accomplished, our general task for the new period cannot be fulfilled.

The "gang of four" willfully undermined our socialist undertakings in science and culture, and they even babbled that "the more knowledgeable, the more reactionary" and that "it is preferable to have laborers with no culture." What they practiced was a fascist cultural autocracy, sinister and rotten to the core. Owing to their disruption, the enthusiasm of the scientific and cultural workers and the masses was dampened, our scientific and cultural undertakings fell far short of the needs of our socialist revolution and construction, and the gap between the level of science and technology in China and advanced world levels, which had narrowed at one time, widened again. This teaches us by negative example that raising the scientific and cultural level of the people is not a matter solely of imparting knowledge but is a great class struggle. We must carry through to the end the struggle to expose and criticize the "gang of four," eliminate the pernicious influence of their counterrevolutionary revisionist line, and clear the way for raising the scientific and cultural level of the entire Chinese nation.

It is in the vital interest of hundreds of millions of people to raise the scientific and cultural level of the entire Chinese nation. This can be achieved only by drawing in and relying on vast numbers of people, and only by effectively organizing all the people on every front on a country-wide scale. What we need is thousands upon thousands of skilled workers, skilled peasants, and other skilled working people with both socialist consciousness and the ability to master modern production techniques; enormous numbers of revolutionary intellectuals in different trades and professions; and revolutionary cadres capable of managing a modern economy and modern science and technology. We need mighty contingents for industry, agriculture, science and technology, culture, and national defense—people who are both red and expert and who are particularly good at fighting. It won't do to have only a small number or a section of the people; hundreds of millions of people, the entire Chinese nation, must reach a much higher level.

Obviously, if our workers lack scientific and cultural knowledge and fail to learn new production skills, they can hardly master modern industrial production processes. If our rural people's commune members lack scientific and cultural knowledge, do not know how to use electricity, machinery, chemical fertilizer, insecticides, etc., or have no knowledge of scientific farming, they cannot keep themselves abreast of the needs in modernizing agriculture. If our People's Liberation Army commanders and fighters lack knowledge of modern military science and techniques, they cannot use modern arms and equipment and cannot organize and direct modern warfare well. On the other hand, if hundreds of millions of people grasp such knowledge and skills, they will become competent workers, peasants, and armymen, and large numbers of technical specialists, innovators, inventors, and scientists will emerge from among them. We should therefore pay great attention to raising the scientific and cultural level of the whole nation. The modernization of science and technology should not be regarded as a matter only for scientific and technological organizations, nor should it be left to a few people in research institutions or universities. The most powerful base and inexhaustible source of strength for the modernization of science and technology in our country are the masses of the people in their hundreds of millions who, fired with enthusiasm, are determined to do away with blind faith, emancipate their minds, rid themselves of inferiority complexes, call up the courage to break new ground and to think, speak, and act, and exert themselves in study and work.

While we stress the need to rely upon the masses in their hundreds of millions, we must also make vigorous efforts to expand our ranks of professional scientists and technicians. We already have a working-class contingent of scientists and technicians who are both red and expert. We should unite all revolutionary and patriotic scientific and technical workers. Effective measures should be taken to train new forces and expand the professional contingent quickly. It is necessary to raise the level of the professionals and train large numbers of scientists and technicians who are top-notch by world standards. We hope our scientists and technicians will keep raising their political consciousness, serve socialism wholeheartedly, and integrate with the workers, peasants, and soldiers while at the same time devoting themselves to their professional work, constantly improving their capabilities, combining personal effort with collective wisdom, and striving to reach the summits of science and technology. We also hope that more and more people will have a better understanding of Marxism and firmly establish a proletarian, communist world outlook through studying Marxist theory and through class struggle and practical work. In that case, we will be speaking the same language, not only the common language of patriotism and the socialist system but increasingly that of the communist world outlook. The professional contingent is the vanguard and the backbone in raising the scientific and cultural level of the entire Chinese nation. It has the duty of taking the strongholds of science and technology and popularizing scientific and technological knowledge among the people. Our people's armed forces have always had a system under which there is a "three-in-one" combination of the field armies, the regional forces, and the militia, forming an impregnable bastion in people's war. This should be applied to the scientific and technical field as well. The hundreds of millions of people who are studying diligently to master science and culture can be likened to a vast militia force on the front of science and technology, while the ranks of professionals are like the field armies and the regional forces. A general rise in the scientific and cultural level of the masses will provide the base and conditions for the professionals to make advances, and the professionals, for their part, will guide the mass forces, crystallize their experience and wisdom, and raise their standards. This should be the system under which our country's scientific and technical forces operate; it is the road to victory through people's war on our scientific and technological front. Advancing our science and culture is a people's cause. By spreading scientific and cultural knowledge to raise the level of the entire nation, combining popularization with the raising of standards, and integrating professional with mass forces, we will form a mighty army for science and culture and greatly speed up our socialist modernization.

The education of young people is another very important aspect that merits special attention in raising the scientific and cultural level of the entire Chinese nation. The young people are our successors in the proletarian revolutionary cause. Starting from an early age they should develop themselves physically, foster communist values and work style, and show heroism in the interests of the collective. They should also cultivate, from an early age, the good habit of loving, studying, and using science. Our Party and our state must show particular concern for the healthy growth of the young people; make a good job of running primary and middle schools, universities, and other types of schools at various levels; open all kinds of channels for study; create the conditions for bringing up the young people as laborers who have both socialist consciousness and culture; and constantly train from among them scientific and technical personnel who are both red and expert. As talented young people keep coming to the fore in large numbers, our science and technology will flourish.

The task of raising the scientific and cultural level of the entire Chinese nation involves higher demands on our cadres, first of all on leading cadres at all levels. Chairman Mao taught us: "Conditions are changing all the time, and to adapt one's thinking to the new conditions, one must study. Even those who have a better grasp of Marxism and are comparatively firm in their proletarian stand have to go on studying, have to absorb what is new and study new problems." Leading organs and cadres at all levels should be good at adapting themselves to the requirements of our advances in socialist modernization and must improve their methods of leadership and of work. Far from being weakened, political and ideological work should be strengthened in the new period of development in our socialist revolution and construction. Our Party has fine traditions in political and ideological work; we should carry them forward and eliminate the pernicious influence of Lin Piao and the "gang of four." We should do our political and ideological work in a more meticulous and deep-going way so as to constantly prevail over the ideological influence of the bourgeoisie and other exploiting classes, overcome the force of habit characteristic of the petty producer, and make our political and ideological work an important guarantee of bringing about socialist modernization. Politics is the commander, the soul in everything, and it won't do not to grasp political and ideological work; but neither will it do if we concern ourselves solely with politics and remain laymen, without any knowledge of technical and professional work. Chairman Mao taught us in all earnestness in 1958: "We must exert ourselves, we must study and carry through to the end this great technological revolution which history has assigned us. This question should be brought up for discussion among the cadres, and a cadre conference should be called to discuss what else we have in the way of capabilities. In the past we had the capabilities of waging war and carrying out land reform, but these capabilities alone are not enough now. We must acquire new ones and achieve a real understanding of professional work, of science and technology, or we cannot possibly exercise effective leadership." Following Chairman Mao's instructions, quite a number of our comrades have pitched in and obtained very good results in their study. But there are some comrades who have failed to understand the profound significance of Chairman Mao's instructions. When Lin Piao and the "gang of four" were engaged in disruption and sabotage, dishing up all sorts of fallacies and creating much confusion, they suppressed or attacked all those who paid attention to professional work or production. At that time, it was out of the question for people seriously to tackle modern science and technology. Now many cadres have emancipated their minds and are diligently studying politics, economics, military affairs, professional work, and technology, and the situation is most encouraging. Cadres at all levels in various professions and trades should do the same. Our cadres holding leading positions or doing political and administrative work on the scientific and technological front in particular should devote greater energy to study and, in the light of the characteristics of scientific and technical work, should do a good job in political and ideological work, organization and management, and rear service work. We should respect the labor of the intellectuals and show concern for them politically and for their work and life as well. We should get close to them, understand them, be familiar with them, and forge close friendships. We should create favorable conditions for their work and give full scope to their initiative and creativity. Our comrades must do well in all these respects and ensure the comprehensive and correct implementation of the Party's line, principles, and policies so as to make new contributions to our socialist revolution and socialist construction.

In order to raise the scientific and cultural level of our nation, it is necessary to reiterate Chairman Mao's slogan of learning from foreign countries. Our principle is to learn the strong points of all nations and countries, to learn from them all that is truly good in politics, economics, military affairs, science, technology, literature, and art. While upholding independence and self-reliance, we should learn from other countries analytically and critically. We have always opposed the slavish comprador philosophy which holds that anything foreign is good, while nothing Chinese is any good, fancying that even the moon looks better over foreign lands, and that China can only move along at a snail's pace behind other countries. The "gang of four," out of malicious intent, slandered our effort to learn from foreign countries as "slavish comprador philosophy." This was nothing but turning matters upside down and confusing right and wrong. Their purpose was to create counterrevolutionary opinion so that they could usurp Party and state power and overthrow the central leading comrades who correctly followed Chairman Mao's principle of learning from foreign countries. If we indiscriminately refused to learn from foreign countries, China would remain backward for ever. What socialist modernization could one speak of then? It is obvious that all nations and countries in the world have strong points and weaknesses. They should learn from one another, drawing on the strong points of others to make up for their own weak points, so as to make steady progress. Can we refuse to study Marxism because its birthplace was in the West? Can we refuse to learn from the Great October Socialist Revolution because it took place in Russia? As for natural science and technology, we are behind advanced world levels. We admit our backwardness but we refuse to lag behind; we must catch up. This requires us to be good at absorbing whatever is good in things foreign, take them over and turn them to our account, and combine our learning from foreign countries with our own inventiveness so that we can catch up with and surpass advanced world levels as soon as possible. We should learn from foreign countries now, but should we do so when we overcome our backwardness and become advanced? Yes, because even then other countries will still have points worth learning, and we should still learn from them. What is wrong with that? After 10,000 years, we must still learn from others!

The first eight years are the key to accomplishing the four modernizations in 23 years, that is, by the year 2000. This is true also for raising the scientific and cultural level of the entire Chinese nation. We should work out plans for the next three and the next eight years and an outline for 23 years. From now on, we should encourage diligent efforts throughout our society to study politics, raise the level of education, and learn science and technology. Science means honest, solid knowledge and allows no hypocrisy and complacency. Only with honesty, modesty, and perseverance can one really learn something. It is imperative for all our people to foster and develop the habit of studying hard. It is an honor to love to study, and it is a shame to refuse to study. It is an honor to be red and expert, and it is a shame to refuse to make progress. It is an honor to work hard and contribute more to socialism, and it is a shame to indulge in ease and comfort, dislike labor, and live off socialism. Our entire country should be turned into a great school.

Tremendously raising the scientific and cultural level of the entire Chinese nation is not only a prerequisite for the four modernizations. We should look at its significance in a still deeper, broader, and longer perspective.

As the scientific and cultural level of the entire nation rises, we shall be able to use Marxism-Leninism-Mao Tsetung Thought still better to arm the cadres and the masses. In natural science, neither theory nor experiment can be cut off from materialism and dialectics. We should urge all research workers in natural science to make conscious use of the Marxist world outlook to guide their work, and at the same time to spread materialism and dialectics far and wide among the masses through study of science and technology and participation in scientific experiments. Marxism has its source in the entire reservoir of human knowledge. It was by drawing critically on all the knowledge provided by previous science that Marx confirmed his revolutionary conclusions. That is why raising the scientific and cultural level is very important in studying Marxism, in gaining a deeper understanding of it, and in applying it in a still better way.

Raising the scientific and cultural level of the whole nation will help arouse the masses to participate in managing the economic, cultural, and educational undertakings as well as affairs of the state. It will also help develop socialist democracy in the political life of the country. Lenin put it this way: We are perfectly aware of what Russia's cultural underdevelopment is doing to Soviet power—which in principle has provided an immensely higher proletarian democracy—how this lack of culture is reducing the significance of Soviet power and reviving bureaucracy. The Soviet apparatus is accessible to all the working people in word, but actually it is far from being accessible to all of them. And not because the laws prevent it from being so, as was the case under the bourgeoisie; on the contrary, our laws assist in this respect. But in this matter laws alone are not enough. A vast amount of educational, organizational, and cultural work is required; this demands a vast amount of work over a long period. How profound are these words of Lenin's! The task we set today of raising the scientific and cultural level of the entire nation is closely related to a full development of socialist democracy.

In socialist society, we must create the conditions for gradually narrowing the differences between town and country, between industry and agriculture, and between physical and mental labor. From a long-term point of view, tremendously raising the scientific and cultural level of the entire nation means training in the whole nation hundreds of millions of working people who have both socialist consciousness and culture. They are the kind of working people who are politically minded and are educated, who can combine mental labor with physical labor, who are both red and expert with an all-around development, and who are at once worker-intellectuals and intellectual-workers. This is the direction for our advance.

Comrades! On the eve of the founding of New China, our great leader Chairman Mao said: "We can learn what we did not know. We are not only good at destroying the old world, we are also good at building the new." What we are now engaged in is this great cause of building a new world.

Our country has a long history of thousands of years. Our nation once created a splendid science and culture. In the last few hundred years, owing to the corruption of the feudal system and aggression by colonialists and imperialists, science and culture fell behind in our country. Since the founding of the People's Republic of China, with its advanced socialist system and under the leadership of the Communist Party, there has been rapid progress in our science and culture. The economic and technical blockade enforced by imperialism failed to strangle us; the tearing up of contracts and withdrawal of specialists by Soviet revisionists failed to subdue us. We have developed our science and technology independently through our own efforts. Have we not made our own atom bombs, hydrogen bombs, and man-made satellites? Have we not trained a contingent of outstanding scientific and technical workers who are both red and expert? Our people have deep respect for the many scientists who have made important contributions to science and technology in China, including the late Comrades Li Szu-kuang and Chu Ko-chen. Facts past and present show that we Chinese too have a head and two hands and are no stupider than others. The key lies in a correct line. Delays and setbacks in the development of our science and technology resulted from interference and sabotage by the counterrevolutionary revisionist line of Liu Shao-chi, Lin Piao, and above all the "gang of four." Now that the "gang of four" has been smashed following the shattering of the two bourgeois headquarters of Liu Shao-chi and Lin Piao, Chairman Mao's proletarian revolutionary line can be implemented correctly and in an all-around way. Class struggle, however, is protracted, and we will still have to remove obstacles from our path. But the greatest hindrance to our advance has now been cleared away. Several hundred million people are now marching toward the modernization of science and technology and thousands of contingents of professional scientific and technical workers are sweeping forward without hindrance. We are fully determined to accomplish the important tasks facing the scientific and cultural fronts in the new period and we are entirely confident of success. We will emerge as a nation with a high standard of culture in the world.

Comrades! By comprehensively and correctly implementing Chairman Mao's revolutionary line, we can give full scope to the superiority of the socialist system, fire the enthusiasm of the masses of people for studying new things and building a new world, unite all forces in society that can be united with and get them organized, and march forward under a unified plan to the common goal. This is the basic guarantee for the sure triumph of our cause. As we advance, we must study many things we do not know and overcome many difficulties. We can learn anything, provided we rely on the initiative of the masses. No difficulty can deter us so long as we rely on the united strength of the masses. Our slogan is: Study, study, and once more study; unite, unite, and once more unite. Let the whole Party, the whole army, and the people of all our nationalities hold high the great banner of Chairman Mao, rally closely around the Party Central Committee, make concerted efforts to raise tremendously the scientific and cultural level of the entire Chinese nation, and successfully fulfill the great historic mission of building a modern and powerful socialist state.

# Appendix B

## RESEARCH INSTITUTES UNDER THE CHINESE ACADEMIES OF SCIENCES

The most important scientific work in China has always been conducted in the national-level research institutes of the Chinese academies of sciences. It is here that the overwhelming majority of China's most competent scientists and engineers hold positions—albeit often with concurrent assignments at universities or industrial ministries. Universities have never been important in the Chinese research scheme, although the creation of "key" colleges and universities and the new emphasis on graduate education should lead to better research facilities in the future. Some important research in engineering and technology is supported by industrial ministries; it is mostly conducted in large enterprises under their jurisdiction.

The number of research institutes under the academies is still growing, and the list that follows is valid as of late 1979. It excludes the institutes under the provincial branches of the academies.

### THE CHINESE ACADEMY OF SCIENCES

#### Astronomy

Astronomical Observatory, Beijing
Shahe (Solar) Station, Beijing
Xinglong (Stellar) Station, Hubei
Miyun (Radio-Astronomy) Station, Beijing
Yunnan Astronomical Observatory, Kunming
Purple Mountain Astronomy Observatory, Nanjing
Shaanxi Astronomical Observatory
Shanghai Astronomical Observatory
Astronomical Instruments Factory, Nanjing
(See also the various institutes of optics and precision instruments under "Physics.")

#### Chemistry

Chemistry Institute, Beijing
Chemistry Institute, Guangzhou
Chemistry Institute, Xinjiang
Organic Chemistry Institute, Shanghai
Organic Chemistry Institute, Chengdu
Chemical Physics Institute, Lanzhou
Chemical Physics Institute, Lüda
Biochemistry Institute, Shanghai
Applied Chemistry Institute, Changchun
Chemical Engineering and Metallurgy Institute, Beijing
Photochemistry Institute, Beijing
Environmental Chemistry Institute, Beijing
Geochemistry Institute, Guiyang
Silicate Chemistry and Technology Institute, Shanghai
Carbon Chemistry Institute, Shaanxi

#### Earth and Environmental Sciences

Geology Institute, Beijing
Geology Institute, Lanzhou

Geology Institute, Wuhan
Geophysics Institute, Beijing
Survey and Geophysics Institute, Wuhan
Earth Structure Institute, Changsha
Water and Soil Conservation Institute, Shaanxi
Rock Soil Mechanics Institute, Wuhan
Pedology (Soils) Institute, Nanjing
Salt Lakes Institute, Xining, Qinghai
Glaciology, Permafrost, and Deserts Institute, Lanzhou
Geology and Paleontology Institute, Nanjing
Oceanography Institute, Qingdao
South China Sea Oceanographic Institute, Guangzhou
Atmospheric Physics Institute, Beijing
Geography Institute, Beijing
Geography Institute, Ürümqi, Xinjiang
Geography Institute, Changchun
Geography Institute, Chengdu

*Engineering*

Computer Center, Beijing
Computer Technology Institute, Beijing
Computer Technology Institute, Shenyang
Automation Institute, Beijing
Automation Institute, Shenyang
Electronics Institute, Beijing
Electronic Technology Institute, Guangzhou
Electrical Engineering Institute, Beijing
Electron Optics Institute, Shanghai
Semiconductors Institute, Beijing
Metallurgy Institute, Shanghai
Metals Institute, Shenyang
Structure of Matter Institute, Fuzhou
Mechanics Institute, Beijing
Engineering Mechanics Institute, Haerbin
Applied Chemistry Institute, Changchun
Chemical Engineering and Metallurgy Institute, Beijing
Photochemistry Institute, Beijing
Space Physics Institute, Xian
Energy Institute, Guangzhou
Optics and Precision Instruments Institute, Hefei
Precision Instruments Institute, Haerbin
Optics and Precision Instruments Institute, Shanghai
Optics and Precision Instruments Institute, Changchun
Optics and Precision Instruments Institute, Xian
Photoelectricity Institute, Chengdu
Acoustics Institute, Beijing

*Life Sciences*

Genetics Institute, Beijing
Biophysics Institute, Beijing
Microbiology Institute, Beijing
Cell Biology Institute, Shanghai
Hydrobiology Institute, Wuhan
Virology Institute, Wuhan
Biology Institute, Chengdu
Physiology Institute, Shanghai
Psychology Institute, Beijing
Materia Medica Institute, Shanghai
Zoology Institute, Beijing
Zoology Institute, Kunming
Entomology Institute, Shanghai
Botany Institute, Beijing
Botany Institute, Guangdong
Botany Institute, Kunming
Botany Institute, Wuhan
Plant Physiology Institute, Shanghai
Tropical Plants Institute, Yunnan
Forestry and Pedology Institute, Shenyang
Biology, Pedology, and Desert Institute, Ürümqi
Vertebrate Paleontology and Paleoanthropology Institute, Beijing

*Mathematics and Computer Sciences*

Mathematics Institute, Beijing
Computer Center, Beijing
Computer Technology Institute, Beijing
Computer Technology Institute, Shenyang
Automation Institute, Beijing
Automation Institute, Shenyang

*Physics*

Physics Institute, Beijing
Physics Institute, Changchun
Technical Physics Institute, Shanghai
Physics Institute, Wuhan
Physics Institute, Xinjiang
Theoretical Physics Institute, Beijing
Atomic Energy Institute, Beijing (jointly administered by the Second Ministry of Machine Building)
Nuclear Physics Institute, Shanghai
Modern Physics Institute, Lanzhou

High Energy Physics Institute, Beijing
Plasma Physics Institute, Hefei
Cosmic Ray Institute, Kunming
Semiconductors Institute, Beijing
Mechanics Institute, Beijing
Space Physics Institute, Xian
Chemical Physics Institute, Lanzhou
Chemical Physics Institute, Lüda
Atmospheric Physics Institute, Beijing
Acoustics Institute, Beijing
Photoelectricity Institute, Chengdu
Optics and Precision Instruments Institute, Hefei
Precision Instruments Institute, Haerbin
Optics and Precision Instruments Institute, Shanghai
Optics and Precision Instruments Institute, Changchun
Optics and Precision Instruments Institute, Xian

## *THE CHINESE ACADEMY OF AGRICULTURAL SCIENCES*

The list of institutes under the Chinese Academy of Agricultural Sciences was provided by that Academy in November 1979. The few locations of individual institutes included in the original list were supplemented when possible. There are now academies of agricultural sciences in each of the provinces, autonomous regions, and municipalities, which, in turn, have their own specialized research institutes.

Agricultural Economy Research Institute
Agricultural Mechanization Research Institute
Agricultural Meteorology Research Laboratory
Animal Husbandry Research Institute
Animal Husbandry Research Institute, Lanzhou
Animal Schistosomiasis Research Laboratory
Apiculture Research Institute
Biological Control Research Laboratory
Citrus Research Institute, Sichuan
Cotton Research Institute, Henan
Crop Research Institute, Beijing
Fiber (cotton excluded) Research Institute
Field Irrigation and Drainage Research Institute
Fruit Trees Research Institute, Xingcheng
Fruit Trees Research Institute, Zhengzhou
Germplasm Research Institute
Grass Land Research Institute
Oil Bearing Crops Research Institute
Plant Protection Research Institute, Beijing
Research Institute for the Division into Districts in Agriculture (Institute of Agro-region Classification)
Research Institute for the Utilization of Atomic Energy
Research Section for the History of Chinese Traditional Cultural Practices
Sericulture Research Institute, Zhenjiang
Soil and Fertilizer Research Institute, Dezhou
Sugar-beet Research Institute
Tea Research Institute, Hangzhou
Tobacco Research Institute, Jinan
Traditional Chinese Veterinary Science Research Institute
Vegetable Research Institute
Veterinary Sciences Research Institute, Haerbin
Veterinary Sciences Research Institute, Lanzhou

## *THE CHINESE ACADEMY OF FORESTRY SCIENCES*

In January 1979 the Ministry of Agriculture and Forestry, which controlled the Chinese Academy of Agriculture and Forestry Sciences, was split up into the Ministry of Agriculture and the Ministry of Forestry. A similar split occurred at the Academy, but while the research institutes under the new Academy of Agricultural Sciences have become available, no list of research institutes under the Academy of Forestry has so far been published.

## *THE CHINESE ACADEMY OF MEDICAL SCIENCES*

Basic Medical Theory Institute, Sichuan
Blood Transformation Institute, Sichuan
Cancer Research Institute and Ritan Cancer Hospital, Beijing

Capital Hospital (includes a Clinical Medicine Institute), Beijing
Epidemiology Institute, Beijing
Fuwai Cardiovascular Institute and Hospital, Beijing
Hematology Institute, Sichuan
Labor Health Institute (environmental health, nutrition, hygiene), Beijing
Materia Medica Institute, Beijing
Medical Biology Institute, Shanghai
Medical Information Institute, Beijing
Medical Instruments Institute, Sichuan
Medical Radiology Institute, Sichuan
Parasitology Institute, Shanghai
Pediatrics Institute, Beijing
Virology Institute, Beijing

## *THE CHINESE ACADEMY OF TRADITIONAL MEDICINE*

Acupuncture and Moxibustion Institute, Beijing
Traditional Medicine Institute, Beijing
Traditional Medicine Institute, Nanjing
Traditional Medicine Institute, Shanghai

## *THE CHINESE ACADEMY OF SOCIAL SCIENCES (all in Beijing)*

Institute of Information

### *Economics*

Institute of Economics
Institute of Industrial Economics
Institute of Agricultural Economics
Institute of Finance, Trade, and Materials Economics
Institute of World Economics

### *History*

Institute of Archaeology
Institute of History (ancient Chinese history)
Institute of Modern History (Chinese history since 1840)
Institute of World History

### *Humanities*

Institute of Linguistics
Institute of Nationalities
Institute of World Religions
Institute of Literature (Chinese literature)
Institute of Foreign Literature

### *Social Sciences*

Institute of Philosophy
Institute of Law
Institute of Journalism
Institute of World Politics
Institute of South Asian Studies (administered in conjunction with Beijing University)

# *Appendix C*

## *SCIENCE AND ENGINEERING SOCIETIES*

The role, and consequently the importance, of professional societies in the People's Republic of China has fluctuated over the years. In certain periods, these associations have been little more than political arms of the Party; in others, they have played an effective role in facilitating interaction between Chinese intellectuals and providing a focal point for a variety of activities within a given discipline. Slow to regain their footing immediately after the Cultural Revolution, professional societies were enjoying a strong resurgence by the late 1970's. They were also taking increasing initiative in establishing international contacts and scholarly exchanges.

The number of professional societies in China is growing. The following list of Chinese science and engineering societies at the national level is fairly complete as of the fall of 1979, but it does not include local societies, which are most common in such fields as agriculture, medicine, and technical sciences. The multiple listings of some societies represent common alternative translations. The list is based on Robert Boorstin's compilation (*China's Professional and Industrial Societies*; National Council for U.S.–China Trade, Washington, D.C., 1979).

Chinese Medical Association/Chinese Society of Medical Science/Chinese Medical Society
Chinese Society of Agricultural Economics
Chinese Society of Agricultural Machinery
Chinese Society of Agronomy/Chinese Association of Agriculture
Chinese Society of Anatomy
Chinese Society of Animal Husbandry and Veterinary Science
Chinese Society for the Application of Atomic Energy in Agriculture
Chinese Society of Architecture
Chinese Society of Astronomy
Chinese Society of Automation
Chinese Society of Aviation/Chinese Society of Aeronautics and Aeronautical Engineering
Chinese Society of Botany
Chinese Society of Chemical Engineering
Chinese Society of Chemical Industries
Chinese Society of Chemistry/Chinese Society of Chemicals
Chinese Society of Civil Engineering/Chinese Society of Railways
Chinese Society of Coal/Chinese Society of Coal Mining
Chinese Society of Electrical Engineering
Chinese Society of Electronics
Chinese Society of Entomology
Chinese Society of Environmental Science/Chinese Environmental Society
Chinese Society of Fisheries/Chinese Society of Marine Products/Chinese Society of Aquatic Products
Chinese Society of Forestry
Chinese Society of Futural Research
Chinese Society of Genetics
Chinese Society of Geodetics and Cartography/Chinese Society of Surveying and Cartography
Chinese Society of Geography

Chinese Society of Geology
Chinese Society of Geophysics
Chinese Society of Gynecology and Obstetrics
Chinese Society of Horticulture
Chinese Society of Hydraulic Engineering (Hydrology)/Chinese Society of Water Conservation
Chinese Society of Internal Medicine
Chinese Society of Light Industry
Chinese Society of Mathematics
Chinese Society of Mechanical Engineering
Chinese Society of Mechanics/Chinese Dynamics Society
Chinese Society of Metals/Chinese Society of Metallurgy
Chinese Society of Meteorology
Chinese Society of Metrology and Instrumentation/Chinese Society of Standardization
Chinese Society of Microbiology
Chinese Society of Modern Management Research
Chinese Society of Navigation/Chinese Society of Maritime Navigation
Chinese Society of Neurology and Psychiatry
Chinese Society of Nursing
Chinese Society of Oceanography/Chinese Society of Oceanography and Limnology
Chinese Society of Oncology
Chinese Society of Ophthalmology
Chinese Society of Otorhinolaryngology
Chinese Society of Paleontology (and Paleoanthropology)
Chinese Society of Paper-Making
Chinese Society of Pathology
Chinese Society of Pediatrics
Chinese Society of Pedology/Chinese Society of Soils and Fertilizers
Chinese Society of Petroleum
Chinese Society of Pharmacology
Chinese Society of Photography/Chinese Society of Light-Sensitive Materials
Chinese Society of Physics/Chinese Physical Society
Chinese Society of Physiology/Chinese Society of Physiological Sciences
Chinese Society of Plant Physiology
Chinese Society of Plant Protection/Society of Plant Pathology
Chinese Society of Plants/Chinese Society of Crop Research
Chinese Society of Psychology
Chinese Society of Public Health
Chinese Society of Radiology
Chinese Society for Research in Dialectics of Nature
Chinese Society of Sericulture/Chinese Society of Silkworm Research
Chinese Society of Shipbuilding Engineering/Chinese Society of Naval Architecture and Marine Engineering
Chinese Society of Silicates/Chinese Society of Silicate Research
Chinese Society of Space Flight/Chinese Society of Astronautics
Chinese Society of Stomatology
Chinese Society of Surgery
Chinese Society of Tea (Research)
Chinese Society of Technical Economy Research
Chinese Society of Textile Engineering
Chinese Society of Traditional Chinese Medicine
Chinese Society of Tropical Crops
Chinese Society of Tuberculosis Prevention/Chinese Anti-Tuberculosis Society
Chinese Society of Zoology

# *Appendix D*

## *KEY INSTITUTIONS OF HIGHER EDUCATION*

After prolonged debates about the educational system, China's practical needs apparently triumphed over ideology and in March 1978 Beijing announced that 88 colleges and universities were being designated as "key" institutions. The subsequent reopening of Chinese People's University in Beijing raised the number to 89. These schools, selected from more than 600 institutions of higher education, were to be given the necessary resources to make them the most advanced teaching and research universities in China with the best facilities, the most up-to-date equipment available, and the most competent teaching staffs. Perhaps even more importantly, they were to enroll only academically promising students—those with the highest grades in the national college entrance examinations. As of 1979, there were considerable qualitative differences between these select institutions. They were training the scientific and technical manpower needed for China's modernization plans and some of them already enjoyed close scholarly contacts and exchanges with academic institutions in the West.

The 89 key colleges and universities are listed below. The list indicates the city in which the institution is located if it is not part of the name.

### *COMPREHENSIVE UNIVERSITIES*

Beijing University (and its branches)
Chinese People's University, Beijing
Fudan University, Shanghai
Jilin University, Changchun
Lanzhou University
Nanjing University
Nankai University, Tianjin
Inner Mongolia University, Huhehot
Northwest University, Xian
Shandong University, Jinan
Sichuan University, Chengdu
Wuhan University
Xiamen University
Xiangtan University
Xinjiang University, Ürümqi
Yunnan University, Kunming
Zhongshan University, Guangzhou

### *SCIENCE UNIVERSITIES*

Chinese University of Science and Technology, Hefei
Zhejiang University, Hangzhou

## POLYTECHNIC UNIVERSITIES

Beijing Engineering College
Central China Engineering College, Wuhan
Changsha Engineering College
Chongjing Construction Engineering College
Chongjing University
Dalian Engineering College
East China Engineering College, Shanghai
Haerbin Polytechnic University
Nanjing Engineering College
Northwest Industrial University, Xian
Qinghua University, Beijing (and its branches)
South China Engineering College, Guangzhou
Tianjin University
Tongji University, Shanghai
Xian Jiaotong University

## POLYTECHNIC COLLEGES

### *Aeronautics*

Beijing Aeronautical College
Nanjing Aeronautical Engineering College

### *Agricultural and Forestry*

Dazhai Agricultural College
Jiangsi Communist Labor University
North China Agricultural University, Zhuoxian, Hebei
Yunnan Forestry College, Kunming

### *Chemical and Petroleum Engineering*

Beijing Chemical Engineering College
Daqing Petroleum College
East China Petroleum College, Shanghai
Guangdong Chemical Engineering College, Guangzhou
Shanghai Chemical Engineering College

### *Electric Power*

East China Water Conservancy College, Shanghai
Hebei Electric Power College, Shijiazhuang
Wuhan Hydroelectric Power College

### *Electronics and Telecommunications*

Beijing Post and Telecommunications College
Chengdu Telecommunications Engineering College
Northwest Telecommunications Engineering College, Xian

### *Geology*

Changchun Geology College
Wuhan Geology College

### *Light Industry*

Hubei Construction Industry College, Wuhan
Northwest Light Industry College, Xian
Shanghai Textile Engineering College

### *Machine-building*

Hefei Industrial University
Hunan University, Changsha
Jilin Industrial University, Changchun
North China Agricultural Mechanization College, Beijing
Northeast Heavy Machinery College, Shenyang
Zhenjiang Agricultural Machinery College

### *Meteorology*

Nanjing Meteorological College

### *Mining and Metallurgy*

Beijing Iron and Steel College
Central South Mining and Metallurgical College, Changsha
Fuxin Coal Mining College, Liaoning province
Northeast Engineering College, Shenyang
Sichuan Mining College, Chengdu

### *Oceanography*

Shandong College of Oceanography, Qingdao

### *Shipbuilding*

Haerbin Shipbuilding College
Shanghai Jiaotong University

*Transportation*

Dalien Maritime College
Northern Jiaotong University, Beijing
Southwest Jiaotong University, Chengdu

## MEDICAL COLLEGES

Beijing College of Chinese Medicine
Beijing Medical College
Shanghai No. 1 Medical College
Sichuan Medical College, Chengdu
Zhongshan Medical College, Guangzhou

## NORMAL COLLEGES

Beijing Normal College
Shanghai Normal College

## MISCELLANEOUS

Beijing Foreign Languages Institute
Beijing Foreign Trade College
Beijing Physical Culture College
Central Music College, Beijing
Central Nationalities College, Beijing
Shanghai Foreign Languages Institute
Southwest Political and Law College, Chongqing
Wuhan Surveying and Cartography College

*Appendix E*

# SCIENTIFIC AND TECHNICAL JOURNALS

The complete cessation of all scholarly publication during the Cultural Revolution is the clearest evidence of the anti-intellectual spirit of that period. Only a few national newspapers and journals continued to publish. The rest of China's printing facilities were occupied in turning out millions of copies of the writings of Chairman Mao, including the ubiquitous "little red book." The resumption of publication was slow in the early 1970's, accelerated in the mid-1970's, and exploded in the late 1970's. Xinhua news agency reported that by mid-1979, 940 journals with a circulation of 112 million had resumed or started publication, and new titles were being added monthly. Nearly 700 of the journals were said to "belong to the category of natural science and social science." However, this broadly defined category included scores of provincial publications for schools (from science magazines for lower grades to magazines used in middle school science courses) and a variety of local periodicals with limited circulation.

The following list of journals was compiled by Han-chu Huang of the Library of Congress and is based on the Library's holdings as of October 1979. Although it is quite comprehensive in terms of journals pertinent to this volume, it is in no sense complete. There are many journals known to be published in the PRC that the Library of Congress does not receive. Some will surely become available in due course; others are intended only for domestic distribution. Whenever possible, the entry includes the name of the journal in pinyin and in translation, its periodicity, and either the year in which the publication was resumed (R precedes the date) or the year in which publication began. Unless otherwise noted, the place of publication is Beijing.

The letters that precede each of the entries attempt to reflect the level of the journal and the audience for which it is intended: A stands for Academic, I for Intermediate, and P for Popular. These are obviously very approximate categories and rather subjective designations, but as someone who daily handles these publications, Mr. Huang is in the best position to make the judgments. The code should help readers to identify the more serious journals published in China.

*Agriculture and Forestry*

A *Dongbei caoben zhiwuzhi* (Records of northeastern vegetation). Annual, R 1976

I *Guizhou nongye keji* (Guizhou agricultural technology). Bimonthly, Guiyang, 1979

I *Liaoning nongye kexue* (Liaoning agricultural science). Bimonthly, Shenyang, Nov. 1978
A *Linye kexue (Scientia silvai sinicae).* Quarterly, R 1979
P *Nongcun kexue shiyan* (Rural scientific experimentation). Monthly, 1978
P *Nongye jixie* (Agricultural machinery). Monthly, R 1978
P *Nongye kexue shiyan* (Agricultural scientific experimentation). Monthly
I *Yunnan nongye keji* (Yunnan agricultural technology). Bimonthly, Kunming
A *Zhongguo linye kexue* (Journal of Chinese forestry science). Quarterly, R 1976
A *Zuowu xuebao* (Journal of crops). Quarterly, 1973

*Archaeology*

A *Gemin wenwu* (Revolutionary relics). Bimonthly, 1976
A *Gugong bowuyuan yuankan* (Journal of the Palace Museum). R 1979
I *Huashi* (Fossils). Quarterly, 1974
A *Kaogu* (Archaeology). Bimonthly, R 1972
A *Kaogu xuebao (Archaeologica sinica).* Quarterly, R 1972
A *Wenwu* (Relics). Monthly, R 1972

*Architecture*

A *Jianzhu xuebao* (Journal of architecture). Bimonthly, R 1973

*Astronomy*

P *Tianwen aihaozhe* (Astronomical amateur). Monthly, R 1978
A *Tianwen jikan (Studia astronomica sinica).* Annual, 1978
A *Tianwen xuebao (Acta astronomica sinica).* Quarterly, R 1974

*Biology*

A *Shiyan shengwu xuebao (Acta biologiae experimentalis sinica).* Quarterly, R 1978
A *Shuisheng shengwuxue jikan (Acta hydrobiologica sinica).* Quarterly, R 1975

*Botany*

A *Yuanyi xuebao* (Journal of horticulture). Quarterly
A *Zhiwu baohu xuebao* (Journal of plant protection). Quarterly
A *Zhiwu bingli xuebao (Acta phytopathologica sinica).* Semi-annual
A *Zhiwu fenlei xuebao (Acta phytotaxonomica sinica).* Quarterly
A *Zhiwu shengli xuebao (Acta phytophysiologica sinica).* Quarterly
A *Zhiwu xuebao (Acta botanica sinica).* Quarterly, R 1978
P *Zhiwu zazhi* (Journal of botany). Quarterly, R 1977

*Chemistry*

A *Diqiu huaxue (Geochimica).* Quarterly, 1973
A *Fenxi huaxue* (Journal of analytical chemistry). Quarterly, Changchun, 1973

A *Gaofenzi tongxun* (Polymer communications). Bimonthly, R Aug. 1978
A *Guisuanyan xuebao* (Journal of silicates). Quarterly
A *Huaxue shiji* (Chemical reagents). Bimonthly, Tianjin, 1979
I *Huaxue tongbao* (Chemistry bulletin). Quarterly, R 1973
A *Huaxue xuebao (Acta chimica sinica).* Quarterly, 1975
A *Shengwu huaxue yu shengwu wuli jinzhan* (Advances in biochemistry and biophysics). Bimonthly, 1974
A *Shengwu huaxue yu shengwu wuli xuebao (Acta biochimica et biophysica sinica).* Quarterly, R 1975

*Economics and Finance*

A *Jihua jingji* (Economic planning). Monthly, 1979
A *Jingji guanli* (Economic management). Monthly, 1979
A *Jingji yanjiu* (Economic studies). Monthly, R 1978
A *Shijie jingji* (World economy). Monthly, R 1979
I *Zhongguo duiwai maoyi* (China's foreign trade). Bimonthly, R 1974
I *Zhongguo jinrong* (Chinese finance). Monthly, R 1979

*Education*

A *Lishi jiaoxue* (Teaching history). Monthly, Tianjin, R 1979
I *Renmin jiaoyu* (People's education). Monthly, R 1977
I *Shanghai jiaoyu* (Education in Shanghai). Monthly, Shanghai, R 1978
I *Shanxi jiaoyu* (Education in Shanxi). Monthly, Taiyuan, R 1979
A *Waiyu jiaoxue yu yanjiu* (Teaching of and research in foreign languages). Quarterly, R 1978
I *Xinjiaoyu* (New education). Monthly, Guangzhou, 1975

*Electronics*

I *Dianzi jishu* (Electronic technology). Monthly, Shanghai, R 1979
I *Dianzi kexue jishu* (Electronic science technology). Monthly
I *Dianzi shijie* (Electronic world). Monthly, Oct. 1979
A *Dianzi xuebao (Acta electronica sinica).* Quarterly, R 1979
I *Dianzixue tongxun* (Communications on electronics). Quarterly, Mar. 1979
A *Jisuanji xuebao* (Chinese computer journal). Quarterly, 1978
I *Wuxiandian* (Radio). Monthly, R 1974

*Engineering*

I *Chuanbo gongcheng* (Ship engineering). Bimonthly, Shanghai, 1979
A *Dalien gongxueyuan xuebao* (Journal of the Dalien Institute of Technology). Quarterly, Dalien
A *Huazhong gongxueyuan xuebao* (Journal of the Central China Engineering Institute). Quarterly, Wuhan, 1974

A *Nanjing gongxueyuan xuebao* (Journal of the Nanjing Engineering Institute). Quarterly, Nanjing

### Entomology

A *Kunchong xuebao (Acta entomologica sinica).* Quarterly, R 1973
I *Kunchong zhishi* (Entomological knowledge). Bimonthly, R 1974

### Environment

I *Huanjing baohu* (Environmental protection). Bimonthly
I *Huanjing baohu zhishi* (Knowledge of environmental protection). Monthly, Guangzhou, 1978
I *Huanjing kexue* (Environmental science). Bimonthly, 1978

### General Science

P *Keji shijie* (World science and technology). Monthly, Guangzhou, Apr. 1979
A *Kejishi wenji* (Collected papers on the history of Chinese science and technology). Irregular, Shanghai, Nov. 1978
P *Kexue puji* (Popular science). Monthly, Shanghai, 1974
P *Kexue shijie* (Scientific world). Monthly, Guangzhou, Mar. 1979
I *Kexue shiyan* (Scientific experimentation). Monthly, R 1971
A *Kexue tongbao* (Science bulletin). Monthly, R 1973
P *Kexue zhichun* (The beginning of science and technology). Monthly, Guangzhou, 1979
P *Shaonian kexue* (Young people's science). Monthly, Shanghai
P *Shaonian kexue huabao* (Science pictorial for youth). Monthly, Jan. 1979
P *Xiandaihua* (Modernization). Monthly, 1979
P *Zhongxue keji* (Science and technology for high schools). Bimonthly, Shanghai
A *Zhongguo kexue (Scientia sinica).* Monthly, R 1972

### Genetics and Heredity

I *Yichuan (Hereditas).* Bimonthly, Jan. 1979
A *Yichuan xuebao (Acta genetica sinica).* Quarterly, 1974
A *Yichuan yu yuzhong* (Genetics and breeding). Bimonthly, 1976
I *Yichuanxue tongxun* (Bulletin of genetics). Quarterly, 1972

### Geography

A *Dili xuebao (Acta geographica sinica).* Quarterly, R 1978
I *Dili zhishi* (Geographical knowledge). Monthly, R 1973

### Geology

A *Changchun dizhi xueyuan xuebao* (Journal of the Changchun Geology Institute). Quarterly, Shenyang, 1976

A *Dicengxue zazhi (Acta stratigraphica sinica).* Quarterly, 1977
A *Dizhen yanjiu* (Research on earthquakes). Quarterly, Kunming, 1978
A *Dizhi kexue (Scientia geologica sinica).* Quarterly, R 1973
A *Dizhi lixue luncong* (Collected papers and notes on geomechanics). Annual, R 1977
A *Dizhi xuebao (Acta geologica sinica).* Quarterly, R 1977
A *Zhongguo kexueyuan Nanjing dizhi gushengwu yanjiusuo jikan* (Memoirs of the Nanjing Institute of Geology and Paleontology, Academia Sinica). Annual, 1974

*History*

A *Jindaishi ziliao* (Materials for modern history). Monthly, R 1978
A *Lishi yanjiu* (Historical studies). Monthly, 1974
A *Shijie lishi* (World history). Bimonthly, 1979
A *Wen shi zhe* (Arts, history, and philosophy). Quarterly, Jinan, R 1974

*Law*

A *Faxue yanjiu* (Studies in law). Bimonthly, 1979
A *Minzhu yu fazhi* (Democracy and legality). 1979

*Linguistics*

I *Fangyan* (Dialects). Quarterly, 1979
A *Yuwen xuexi congkan* (Language learning series). Annual, 1978
I *Yuwen zazhi* (Journal of language). Quarterly, 1979
A *Zhongguo yuwen* (Chinese language). Bimonthly, R 1978

*Mathematics*

A *Jisuan shuxue (Mathematicae numericae sinica).* Quarterly, Feb. 1979
A *Shuxue de shijian yu renshi* (Practice and understanding of mathematics). Quarterly, 1975
I *Shuxue jiaoxue* (Teaching mathematics). Monthly, 1979
A *Shuxue tongbao* (Mathematics bulletin). Monthly
A *Shuxue xuebao (Acta mathematica sinica).* Bimonthly, 1974
A *Yingyong shuxue xuebao (Acta mathematicae applanatae sinica).* Quarterly, Feb. 1978

*Medicine*

I *Anyi xuebao* (Journal of Anhui medicine). Bimonthly, Hefei
A *Baiqiuen yike daxue xuebao* (Journal of the Bethune Medical College). Quarterly, Changchun
P *Chijiao yisheng zazhi* (Barefoot doctor journal). Monthly, Dec. 1972
P *Dazhong yixue* (Popular medicine). Monthly, Shanghai, R 1978

I *Fujian yiyao zazhi* (Journal of Fujian medicines). Bimonthly, Fuzhou
I *Heilongjiang yixue cankao* (Heilongjiang medical reference bulletin). Monthly, Haerbin, 1978
I *Huli zazhi* (Journal of nursing). Bimonthly, place of publication unknown
A *Jiefangqun yixue zazhi* (Journal of Chinese Liberation Army medicine). Bimonthly
A *Jilin yike daxue xuebao* (Journal of Jilin Medical College). Bimonthly, Zhangchun, 1978
I *Liaoning zhongji yikan* (Liaoning journal of intermediate medicine). Monthly, Shenyang
P *Qunzhong yixue* (People's medicine). Monthly, Shanghai, 1974
I *Shandong yiyao* (Shandong medicine). Monthly, Jinan, R date unknown
P *Shanghai chijiao yisheng zazhi* (Shanghai barefoot doctor journal). Bimonthly, Shanghai, 1975
I *Shanghai yixue* (Shanghai medicine). Monthly, Shanghai, 1978
I *Shanghai zhongyiyao zazhi* (Shanghai journal of traditional Chinese medicine). Bimonthly, Shanghai, R 1979
I *Shanxi yiyao zazhi* (Medical journal of Shanxi). Bimonthly, Taiyuan
I *Tianjin yiyao* (Tianjin medicine). Monthly, Tianjin, 1973
I *Tianjin yiyao: guke fukan* (Tianjin medicine: osteology). Quarterly, Tianjin
I *Tianjin yiyao: zhongliu fukan* (Tianjin medicine: tumors). Quarterly, Tianjin
P *Xinyixue* (New medical science). Monthly, Guangzhou, 1970
I *Xinyiyaoxue zazhi* (Journal of new medicopharmaceutical science). Monthly, 1972
P *Xinzhongyi* (New traditional Chinese medicine). Bimonthly, Guangzhou, 1971
A *Xumushouyi xuebao* (Journal of animal veterinary medicine). Quarterly, place of publication unknown
A *Yaoxue xuebao* (Journal of pharmacology). Monthly
A *Zhongguo shouyi zazhi* (Chinese veterinary journal). Monthly, R 1979
A *Zhonghua erbiyanhouke zazhi* (Chinese journal of otorhinopharyngolaryngology). Quarterly
A *Zhonghua erke zazhi* (Chinese journal of pediatrics). Quarterly
A *Zhonghua fangshexue zazhi* (Chinese journal of radiology). Quarterly
A *Zhonghua fuchanke zazhi* (Chinese journal of gynecology and obstetrics). Quarterly
A *Zhonghua jiehe he huxixi jibing zazhi* (Chinese journal of tuberculosis and respiratory diseases). Quarterly
A *Zhonghua kouqiangke zazhi* (Chinese journal of stomatology). Quarterly
A *Zhonghua neike zazhi* (Chinese journal of internal medicine). Bimonthly
A *Zhonghua shenjing jingshenke zazhi* (Chinese journal of neuropsychosis). Quarterly
A *Zhonghua waike zazhi* (Chinese journal of surgery). Bimonthly
A *Zhonghua xinxueguanbing zazhi* (Chinese journal of cardiovascular disease). Quarterly
A *Zhonghua yanke zazhi* (Chinese journal of ophthalmology). Quarterly
A *Zhonghua yixue jianyan zazhi* (Chinese journal of medical laboratory technology). Quarterly

A *Zhonghua yixue zazhi* (Chinese medical journal). Monthly, Shanghai, R 1973

A *Zhonghua yufang yixue zazhi* (Chinese journal of preventive medicine). Quarterly

A *Zhonghua zhongliu zazhi* (Chinese journal of oncology). Quarterly

P *Zhongji yikan* (Intermediate level medicine). Monthly, 1979

A *Zhongliu fanzhi yanjiu* (Research and cancer prevention). Bimonthly, Shanghai, 1979

I *Zhongyi zazhi* (Journal of traditional Chinese medicine). Monthly, place of publication unknown

*Microbiology*

A *Weishengwu xuebao (Acta microbiologica sinica).* Quarterly, R 1973

A *Weishengwuxue tongbao* (Microbiology bulletin). Bimonthly, R 1977

*Natural Science*

A *Beijing daxue xuebao zirankexue (Acta scientiarum naturalium, Universitatis Pekinensis).* Quarterly, R 1977

A *Fudan xuebao* (Journal of Fudan University, natural sciences edition). Shanghai, R 1973

A *Jilin daxue zirankexue xuebao (Acta scientiarum naturalium, Universitatis Jilinensis).* Bimonthly, Changchun, R 1974

A *Lanzhou daxue xuebao, zirankexue* (Journal of Lanzhou University, natural sciences edition). Quarterly, Lanzhou, 1978

A *Nanjing daxue xuebao, zirankexue* (Journal of Nanjing University, natural sciences edition). Quarterly, Nanjing, 1978

A *Qinghua daxue xuebao, zirankexue* (Journal of Qinghua University, natural sciences edition). Quarterly

A *Shanghai Jiaotong daxue xuebao* (Journal of Jiaotong University, natural sciences edition). Quarterly, Shanghai

A *Xiamen daxue xuebao, zirankexue* (Journal of Xiamen University, natural sciences edition). Quarterly, Fujian

A *Xian Jiaotong daxue xuebao* (Journal of Xian Jiaotong University, natural sciences edition). Quarterly, Xian, Mar. 1978

A *Zhongshan daxue xuebao, zirankexue* (Journal of Zhongshan University, natural sciences edition). Quarterly, Guangzhou, R 1973

I *Ziran zazhi* (Nature). Monthly, Shanghai, 1978

I *Ziran ziyuan* (Natural resources). Quarterly, place of publication unknown

*Oceanography*

I *Haiyang* (The ocean). Monthly, Jan. 1979

A *Haiyang kexue jikan (Studia marina sinica).* Annual, R 1974

A *Haiyang yu huzhao (Oceanologia et limnologia sinica).* Quarterly, R 1978

I *Haiyang zhanxian* (Ocean front). Monthly, 1977

*Paleontology*

A *Diceng gushengwu lunwenji* (Professional papers on stratigraphy and paleontology). Annual

A *Gujizui dongwu yu gurenlei (Vertebrata palasiatica).* Quarterly, R 1973

A *Gushengwu xuebao (Acta palaeontologica sinica).* Bimonthly, R 1976

*Philosophy and Social Science*

I *Anhui shida xuebao: zhexue shehui kexueban* (Journal of Anhui Normal College: philosophy and social science edition). Quarterly, Wuhu, 1976

A *Beijing daxue xuebao: zhexue shehui kexueban* (Journal of Beijing University: philosophy and social science edition). Quarterly, 1974

A *Fudan xuebao: shehui kexueban* (Journal of Fudan University: social science edition). Monthly, Shanghai, 1978

I *Gansu shida xuebao: shehui kexueban* (Journal of Gansu Normal College: social science edition). Quarterly, Lanzhou, 1976

A *Jilin daxue xuebao: zhexue shehui kexueban* (Journal of Jilin University: philosophy and social science edition). Quarterly, Changchun, 1974

I *Jilin shida xuebao: shehui kexue* (Journal of Jilin Normal College: social science edition). Bimonthly, Changchun, R 1979

I *Jilin shida xuebao: zhexue shehui kexueban* (Journal of Jilin Normal College: philosophy and social science edition). Quarterly, Changchun, 1975

A *Liaoning daxue xuebao: zhexue shehui kexueban* (Journal of Liaoning University: philosophy and social science edition). Monthly, Shenyang, 1974

A *Lilun xuexi* (Theoretical learning). Bimonthly, Changchun, 1975

A *Lilun yu shijian* (Theory and practice). Monthly, Shenyang, 1979

A *Nankai daxue xuebao: zhexue shehui kexueban* (Journal of Nankai University: philosophy and social science edition). Bimonthly, Tianjin, 1975

A *Qinghua daxue xuebao* (Journal of Qinghua University). Quarterly, R 1974

I *Tianjin shiyuan xuebao* (Journal of Tianjin Normal College). Bimonthly, Tianjin, 1975

A *Wuhan daxue xuebao: zhexue shehui kexueban* (Journal of Wuhan University: philosophy and social science edition). Bimonthly, Wuchang, 1974

A *Xiamen daxue xuebao: shehui kexueban* (Journal of Xiamen University: social science edition). Quarterly, Xiamen, 1979

A *Zhexue yanjiu* (Philosophical studies). Monthly, 1978

A *Zhongshan daxue xuebao: zhexue shehui kexueban* (Journal of Zhongshan University: philosophy and social science edition). Bimonthly, Guangzhou, 1973

*Physics*

A *Daqi kexue (Scientia atmospherica sinica).* Quarterly, 1977

A *Diqiu wuli xuebao (Acta geophysica sinica).* Quarterly, R 1973

A *Diwen wuli* (Journal of low temperature physics). Quarterly, Mar. 1979

A *Gaoneng wuli* (High energy physics). Quarterly, 1977

A *Gaoneng wuli yu hewuli (Physica energiae fortis et physica nuclearis).* Bimonthly, 1977

I *Jiguang* (Lasers). Monthly, Shanghai, 1974
A *Lixue xuebao* (Journal of mechanics). Quarterly, R 1978
A *Lixue yu shijian* (Mechanics and practice). Quarterly, Mar. 1979
A *Shengxue xuebao* (Journal of acoustics). Quarterly, Mar. 1979
I *Wuli* (Physics). Bimonthly, 1972
A *Wuli xuebao (Acta physica sinica).* Bimonthly, 1974
A *Zidonghua xuebao* (Journal of automation). Quarterly

*Physiology*

A *Shengli xuebao (Acta physiologica sinica).* Quarterly, R 1978
I *Shenglikexue jinzhan* (Progress in physiological science). Quarterly, R 1978

*Psychology*

A *Xinli xuebao* (Journal of psychology). Quarterly

*Seismology*

A *Dizhen dizhi* (Seismology and geology). Quarterly, Mar. 1979
A *Dizhen xuebao* (Journal of seismography). Quarterly
I *Dizhen zhanxian* (Seismographical front). Bimonthly, 1974
A *Xibei dizhen xuebao* (Northwestern seismology journal). Quarterly

*Social Science (see Philosophy)*

*Soil*

I *Turang* (Soil). Bimonthly, Nanjing, 1977
A *Turang xuebao (Acta pedologica sinica).* Quarterly, R 1978

*Zoology*

A *Dongwu fenlei xuebao (Acta zootaxonomica sinica).* Quarterly, R 1979
A *Dongwu xuebao (Acta zoologica sinica).* Quarterly, R 1973
I *Dongwuxue zazhi* (Journal of zoology). Quarterly, R 1974

*INDEXES*

# INDEX OF PERSONAL NAMES

This index lists all twentieth-century mainland Chinese mentioned in the text and notes. Names are given in the pinyin romanization, with the Wade-Giles equivalent in parentheses. Where a name is known only from English-language sources and the correct pinyin form is unascertainable, or where initials are used and the surname is identical in pinyin and Wade-Giles, the name is given as it appears in the text.

# INDEX OF INSTITUTIONS

Research institutes, science and engineering societies, and key institutions of higher learning are also listed in Appendixes B, C, and D, pp. 564–72.

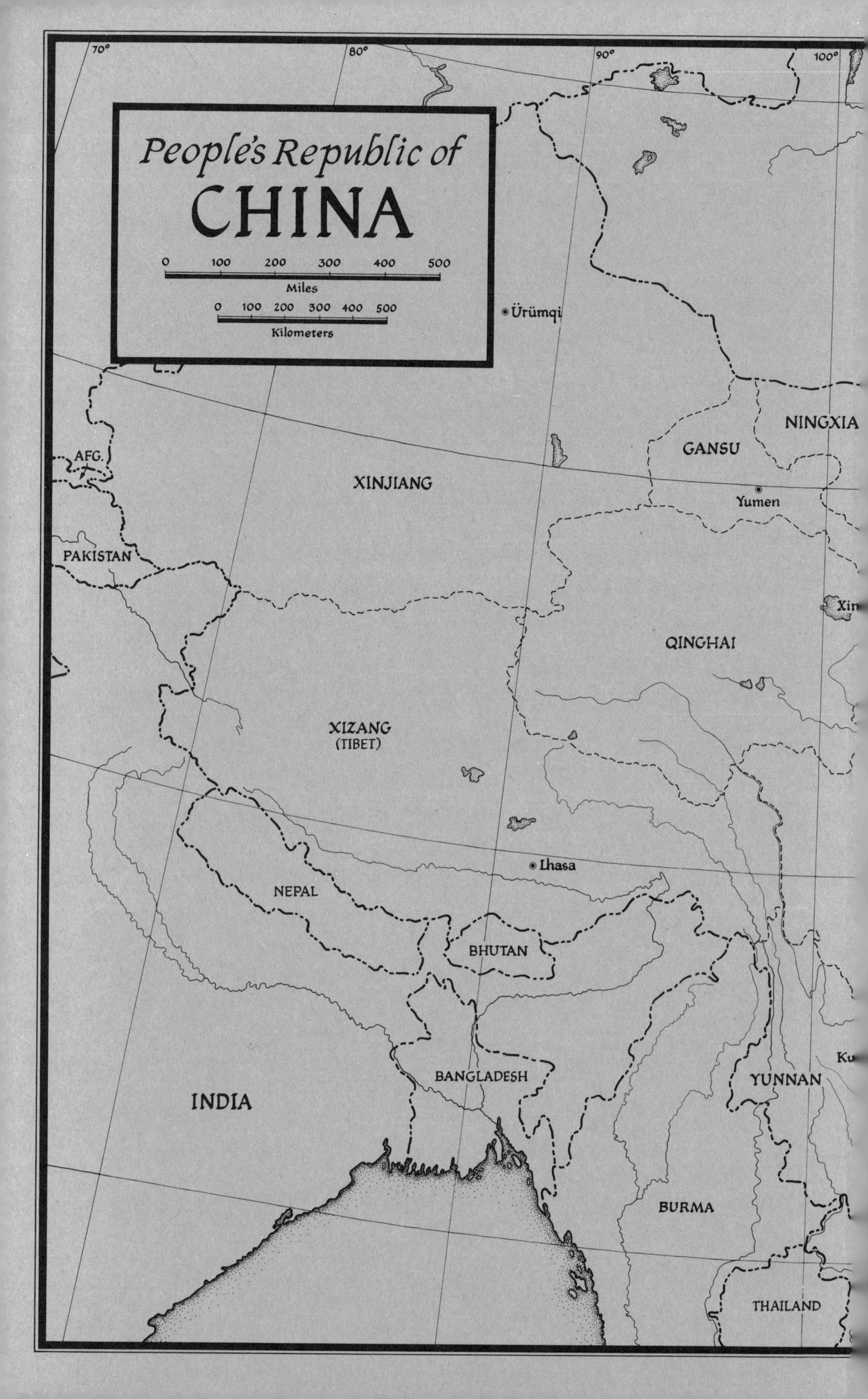

People's Republic of
CHINA
0
100
200
300
400
500
Miles
0
100
200
300
400
500
Kilometers
70°
80°
90°
100°
Ürümqi
NINGXIA
GANSU
AFG.
XINJIANG
Yumen
PAKISTAN
QINGHAI
XIZANG
(TIBET)
Lhasa
NEPAL
BHUTAN
BANGLADESH
YUNNAN
INDIA
BURMA
THAILAND